SUSTAINABLE AGRICULTURE AND THE INTERNATIONAL RICE-WHEAT SYSTEM

Soil Biochemistry, Volume 1, edited by A. D. McLaren and G. H. Peterson
Soil Biochemistry, Volume 2, edited by A. D. McLaren and J. Skujiņš
Soil Biochemistry, Volume 3, edited by E. A. Paul and A. D. McLaren
Soil Biochemistry, Volume 4, edited by E. A. Paul and A. D. McLaren
Soil Biochemistry, Volume 5, edited by E. A. Paul and J. N. Ladd
Soil Biochemistry, Volume 6, edited by Jean-Marc Bollag and G. Stotzky
Soil Biochemistry, Volume 7, edited by G. Stotzky and Jean-Marc Bollag
Soil Biochemistry, Volume 8, edited by Jean-Marc Bollag and G. Stotzky
Soil Biochemistry, Volume 9, edited by G. Stotzky and Jean-Marc Bollag
Soil Biochemistry, Volume 10, edited by Jean-Marc Bollag and G. Stotzky

Organic Chemicals in the Soil Environment, Volumes 1 and 2, edited by C. A. I. Goring and J. W. Hamaker
Humic Substances in the Environment, M. Schnitzer and S. U. Khan
Microbial Life in the Soil: An Introduction, T. Hattori
Principles of Soil Chemistry, Kim H. Tan
Soil Analysis: Instrumental Techniques and Related Procedures, edited by Keith A. Smith
Soil Reclamation Processes: Microbiological Analyses and Applications, edited by Robert L. Tate III and Donald A. Klein
Symbiotic Nitrogen Fixation Technology, edited by Gerald H. Elkan

Plant Toxicology: Fourth Edition, Revised and Expanded, edited by Bertold Hock and Erich F. Elstner

Additional Volumes in Preparation

SUSTAINABLE
AGRICULTURE
AND THE
INTERNATIONAL
RICE-WHEAT
SYSTEM

edited by
RATTAN LAL
The Ohio State University
Columbus, Ohio, U.S.A.

PETER R. HOBBS
Cornell University
Ithaca, New York, U.S.A.

NORMAN UPHOFF
Cornell University
Ithaca, New York, U.S.A.

DAVID O. HANSEN
The Ohio State University
Columbus, Ohio, U.S.A.

CRC Press
Taylor & Francis Group
Boca Raton London New York

CRC Press is an imprint of the
Taylor & Francis Group, an **informa** business

First published 2004 by Marcel Dekker, Inc.

CRC Press
Taylor & Francis Group
6000 Broken Sound Parkway NW, Suite 300
Boca Raton, FL 33487-2742

First issued in paperback 2020

© 2004 by Taylor & Francis Group, LLC
CRC Press is an imprint of Taylor & Francis Group, an Informa business

No claim to original U.S. Government works

ISBN 13: 978-0-367-57834-3 (pbk)
ISBN 13: 978-0-8247-5491-4 (hbk)

Visit the Taylor & Francis Web site at
http://www.taylorandfrancis.com

and the CRC Press Web site at
http://www.crcpress.com

Foreword

Rattan Lal and his coeditors have brought together a distinguished international panel of authors to write on a topic of major importance to the maintenance of world food supplies. Rice and wheat are certainly the two most important cereal crops, providing the staple diet for more than half the world's population.

The rice–wheat system lies at the boundaries of the climatic conditions where the crops can be most readily grown. It is also at the boundary between the areas where the two oldest and largest of the international agricultural research centers operate.

Among the chapter authors are major contributors to the intercenter program of the CGIAR, including IRRI and CIMMYT. Borlaug and Dowswell contribute a chapter setting the rice–wheat system into the context of future world food prospects, and Dr. Lal himself reviews the soil and water resources of south Asia in the light of future needs, and he has also contributed an important chapter on the history of no-till farming. This sets the scene for a more detailed discussion of the problems and challenges of no-till farming for the rice–wheat system.

Because most rice produced is grown under irrigation, it places heavy demands on water resources. To minimise water losses soils are traditionally puddled, by cultivating them when wet. This destroys most of the larger pores in the soil. Wheat is produced in dry areas, and although producing greater yields under irrigation, most is grown using water stored in the soil. It is therefore important that all the water falling on the soil enter the soil and not be lost as runoff. This requires that the larger pores in the topsoil be preserved. Hence if a wheat crop is to be grown successfully after rice a suitable structure has to be created, after it has been previously destroyed by puddling. This is a challenge for both farmers and scientists, and it is surprising that it has received less attention from scientists than it deserves.

This book brings together both the practical experience of many farmers from different parts of the world and the insights of scientists as to how best to manage the contrasting soil requirements. Among the management options that have received the most recent attention from scientists are no-till systems. The book covers experiences in South Asia, South America, and the United States in machinery design, as well as the social and economic issues related to adoption of the no-till system. Not least of the advantages is the effect on soil and water conservation.

Dr. Lal and his colleagues are to be warmly congratulated on bringing together the wide range of information in this volume. It should be invaluable to many agricultural advisers, as well as farmers' leaders and scientists concerned with the optimization of no-till systems.

Dennis J. Greenland, Ph.D.
Tropical Agriculture Consultant
Reading, and
Former Director
Scientific Sciences
CAB International
Wallingford, England

Preface

No-till farming* was first developed on a large scale in the U.S. corn belt during the 1960s to reduce risks of soil erosion from sloping lands planted to row crops. It has been widely adopted in North and South America, Canada, Australia, South Asia, and other regions. It is now practiced on over 72 million hectares (Mha) of cropland around the world, and the area is gradually increasing. In addition to soil and water conservation, conversion from plow-till to no-till with crop residue mulch enhances soil biodiversity in terms of both macro- and microfauna, increases the soil organic matter pool, and improves the environment. Soil carbon sequestration through no-till farming reduces the rate of increase of atmospheric carbon dioxide and mitigates global climate change. Other important advantages of no-till farming are the saving in time and resources required for seedbed preparation and a longer growing season.

---

*The term "no-till" farming is used in this book to describe an agricultural system with minimal soil disturbance coupled with surface residue mulch. Rotations, cover crops, and green manures also figure in no-till farming where appropriate to control biotic factors and produce the needed residues if previous crop residues are not sufficient. FAO uses the term "conservation agriculture" instead of no-till farming. "Conservation" tillage is another term that combines minimal soil disturbance with residue mulch, but doesn't include rotations, cover crops or green manures and is usually restricted to a single crop rather than a system. For minimal soil disturbance, "zero-tillage" and "no-tillage" are often used in the literature. "Reduced tillage" is used when soil is disturbed less than in conventional tilled soils but more than in no-till or zero-till.

The rice–wheat system (RWS) is vital to food security for South Asia, especially for densely populated regions of Afghanistan, Pakistan, India, Nepal, and Bangladesh. It is used by millions of farmers in South Asia (13.5 Mha) and China (10 Mha). This system increased in importance with the advent of the Green Revolution: shorter-duration varieties of rice and wheat that enabled both crops to be grown in one calendar year in the Indo-Gangetic Plains of South Asia and central and southern areas of China. No-Till farming has a specific advantage and place in this RWS research in both South Asia and China. And the no-till experiences in South Asia figure prominently in this book, with some mention of technology utilized from Rice–Wheat Consertium (RWC) interactions with China. The 13.5 Mha of RWS in South Asia accounts for 25% of the rice and 33% of the wheat produced in the region. Approximately 32% of the total rice area and 42% of the total wheat area in South Asia are cultivated using this system.

Populations in the region are expected to continue to grow rapidly in the foreseeable future. From 1995 to 2025, projections are that India's population will grow from 967 to 1392 million, that of Pakistan from 132 to 285 million, that of Bangladesh from 125 to 196 million, and that of Nepal from 23 to 41 million. Thus, total population in the region is expected to increase by 53%, from 1247 to 1914 million during this period. Annual cereal consumption in the region is expected to grow at a rate of 2% to 2.5% until population growth stabilizes toward the middle of the 21st century.

However, arable land area in the region is limited, and per capita land devoted to food grain production is progressively decreasing. The total amount of cultivated land area in 1998 was 169 Mha in India; 21.3 Mha in Pakistan, 9.6 Mha in Bangladesh, and 2.4 Mha in Nepal.

The RWS of the Indo-Gangetic and Brahmaputra flood plains and of the foothills of the Himalayas was a key contributor to the Green Revolution during the 1970s. Rapid yield increases in rice and wheat averted the starvation common in this region in the 1950s and early 1960s. Rice production in South Asia grew from 67 million tons in the early 1960s to 178 million tons in 2000. Wheat production rose from 15 to 90 million tons during the same period.

Despite these impressive results, grain yields in the region remain low when compared with their potential and with yields obtained elsewhere in the world. The average yield of wheat is 2.6 Mg/ha in India, 2.2 Mg/ha in Pakistan and Bangladesh, and only 1.6 Mg/ha in Nepal. The average rice yield is 2.9 Mg/ha in India; 2.8 Mg/ha in Pakistan and Bangladesh, and only 2.4 Mg/ha in Nepal.

Changes in management practices indicate that a potential exists to at least double the yields of both crops. However, even current yield levels may not be sustainable using existing technology, mainly because the negative effects of resource degradation offset the positive effects of productivity-enhancing technical change. However, data presented in this book show that changed management practices have already demonstrated the potential for such an increase through more efficient use of natural resources.

Soil degradation and water quality and quantity are severe problems in South Asia. The total land area affected by water erosion is 11.2 Mha in Afghanistan, 1.5 Mha in Bangladesh, 32.8 Mha in India, 1.6 Mha in Nepal, and 7.2 Mha in Pakistan. Land area affected by wind erosion is estimated at 2.1 Mha in Afghanistan, 10.8 Mha in India, and 10.7 Mha in Pakistan. A serious problem of groundwater depletion also exists in the Punjab regions of India and Pakistan, and salinity is a major problem that affects 4.1 Mha in India and 3.8 Mha in Pakistan. Yields of wheat and rice per unit input of fertilizer are declining because of the decline in soil organic matter content and soil structure quality. The response of wheat yield to fertilizer use in Punjab, Pakistan, declined from 60 Kg of grain per Kg of N input in the 1970s to 20 Kg of grain per Kg of N input in the 1990s. Similar trends are reported in India.

In addition to the decline in soil quality, drought and climatic factors are causes of declines in output per unit input. Wheat yields in the region are affected by the length of the growing season. Wheat is sown soon after the rice harvest, and its sowing prior to mid-November is optimal in order to obtain high yields of good quality. If sown late, wheat is subject to higher temperatures in March and April. Wheat shows a linear decline from 1% to 1.5% per day from sowing late in November. Late sowing also reduces the efficiency of expensive inputs such as chemicals and water. Crop yields are also adversely affected by a progressive decline in soil structure, nutrient and elemental imbalance, and perturbations in the hydrologic cycle. Soil puddling—the cultivation of soil while inundated or near saturation—is a deliberate attempt to destroy soil structure in order to decrease water infiltration rates for the rice crop. However, the coarse texture of alluvial soils still results in high percolation losses and low water-use efficiency. The rice intensification system is relevant to making further improvements. Puddling may have some beneficial effects on the yield of semi-aquatic rice, but the poor soil structure it generates adversely affects the growth and yields of subsequent wheat. Adverse effects of poor soil structure also exacerbate ever-declining levels of soil organic matter content. Puddling also delays transplanting rice, and eventually sowing wheat. These are reasons why new systems of rice intensification and management are relevant to making further improvements in RWS as discussed in various chapters in the book.

The RWC for the Indo-Gangetic Plains is a partnership among (a) national agricultural research systems of Bangladesh, India, Nepal, and Pakistan; (b) several major international agricultural research centers (CIMMYT, IRRI, ICRISAT, CIP, and IWMI); and (c) several universities, including Cornell and other U.S. universities, IAC Wageningen, IACR Rothamsted, CABI-UK, and Melbourne University. The RWC's goal is to improve the sustainable productivity of rice–wheat systems in South Asia, thereby conserving natural resources, improving livelihoods, and reducing poverty.

The RWC's most evident success has been the development, testing, and deployment of resource-conserving technologies (RCTs) in the Indo-Gangetic Plains.* Adoption of a no-till system of seedbed preparation facilitates early sowing of wheat, conserves soil water, and improves soil structure. RWC partners, in collaboration with local manufacturers, have developed appropriate seeding equipment that cuts through the stubble to allow no-till seeding. In areas harvested by combines, loose rice straw creates a problem. In traditional systems, it is either removed or burned. Wheat residue is also removed for animal feed or burned prior to puddling for transplanted rice. Burning and removal of crop residues contribute to a rapid decline in soil organic matter content, in addition to having adverse impacts on air quality. Thus, the direct sowing of rice and elimination of puddling in the RWS also need serious consideration.

This book is a state-of-the-knowledge compendium on no-till farming in South Asia in particular and the world in general. The book identifies knowledge gaps, defines priorities, and formulates recommendations about how to improve the rice–wheat farming system. Specifically, the following topics are addressed in this volume:

Food security and natural resources in relation to food demand, per capita soil and water resources, soil degradation and global warming

Problems and opportunities in no-till farming in relation to soil structure and its dynamics, soil temperature and moisture regimes, root growth and proliferation, and elemental cycling and greenhouse effect

Practical aspects of no-till farming in site-specific situations, with a focus on seeding equipment and residue management, weed control, water- and nutrient-use efficiency, and integrated pest management

Impact of no-till farming on profitability, social and gender equality, and sustainable use of natural resources

Research and development priorities, networking, and modus operandi to establish long-term experiments that evaluate changes in soil quality, water use and environment and crop yield

Future collaboration between RWC partners, U.S. universities, and international organizations

---

*"Resource-conserving technologies" is a term the RWC adopted for various tillage and other techniques that result in more efficient use of resources. No-till or zero-till, reduced till, bed planting, laser leveling, and fertilizer efficiency methods are some examples. The RWC is now promoting no-till farming for the entire system, which will include minimal soil disturbance, residue retention, and rotations.

The book is thematically divided into seven sections. Part I includes three overview chapters. Chapter 1 by Borlaug and Dowswell addresses issues of food security and the role of agricultural technology in enhancing food production. Chapter 2 describes the state of the soil and water resources of South Asia. Chapter 3 addresses environmental concerns to achieving food security. Part II comprises four chapters dealing with challenges and opportunities of no-till farming, including evolutionary developments, problems and challenges, and opportunities and constraints. Part III includes six chapters dealing with specific application of no-till farming to the RWS in South Asia with case studies from India, Nepal, Pakistan, and Bangladesh. Part IV contains eight chapters addressing adaptation of no-till farming to other soil-specific situations (e.g., equipment, mechanization, raised-bed systems, pest management) with case studies from the United States, Brazil, Australia, and South America. Seven chapters in Part V deal with economic and social issues relating to no-till farming. Part VI includes two chapters on networking and international cooperation, and research and development priorities are outlined in the last chapter in the concluding Part VII.

This book is based on a workshop held at The Ohio State University on February 20, 2002. The one-day workshop was jointly organized by The Ohio State University, the RWC, CIMMYT, Texas A&M, and Cornell University. It concluded with a public lecture by Dr. Norman Borlaug, recipient of the 1970 Nobel Peace Prize. In addition to the papers presented at the workshop, several authors were invited to write manuscripts on specific topics.

The workshop was made possible by strong financial support from all the cooperating institutions and the U.S. Agency for International Development (USAID). Special thanks are due to Dr. Charlie Sloger for arranging funding support from USAID missions in South Asia, and to the RWC in providing the logistic support.

We thank Marcel Dekker, Inc., for their efforts to publish this volume in a timely fashion, thereby making the information contained herein quickly available to the world community. We also recognize the invaluable contributions by Mr. Jeremy Alder, Ms. Carol Camm, and other OSU staff and graduate students. Help received from staff of OARDC in organizing the field visit to Wooster is much appreciated. We also give special thanks to Ms. Brenda Swank for her support in organizing the workshop and for her help in the editorial process and preparation of this volume for publication.

Rattan Lal
Peter R. Hobbs
Norman Uphoff
David O. Hansen

Contents

V. Social and Economic Issues

VI. Networking and International Cooperation

VII. Conclusions

Contributors

Maqsood Ahmed   On-Farm Water Management, Lahore, Punjab, Pakistan

Edgar Amezquita   Centro Internacional de Agricultura Tropical (CIAT), Cali, Colombia

Merle M. Anders   University of Arkansas Rice Research and Extension Center, Stuttgart, Arkansas, U.S.A.

Edmundo Barrios   Centro Internacional de Agricultura Tropical (CIAT), Cali, Colombia

José R. Benites   Land and Water Development Division, Food and Agriculture Organization of the United Nations (FAO), Rome, Italy

N. I. Bhuiyan   Soil Science Division, Bangladesh Rice Research Institute, Gazipur, Bangladesh

Norman E. Borlaug   Crop and Soil Sciences Department, Texas A&M University, College Station, Texas, U.S.A.

K. R. Dahal   Institute of Agriculture and Animal Science, Rampur, Nepal

M. H. Devare   Department of Crop and Soil Sciences, Cornell University, Ithaca, New York, U.S.A.

Christopher R. Dowswell   CIMMYT, Mexico City, Mexico

J. M. Duxbury   Department of Crop and Soil Sciences, Cornell University, Ithaca, New York, U.S.A.

D. Lynn Forster   Department of Agricultural, Environmental, and Development Economics, The Ohio State University, Columbus, Ohio, U.S.A.

Pedro Luiz de Freitas   Embrapa Soils, Rio de Janeiro, Brazil

D. L. George   School of Agronomy and Horticulture, University of Queensland, Gatton, Queensland, Australia

Mushtaq A. Gill   Director General of Agriculture (Water Management), On-Farm Water Management, Lahore, Punjab, Pakistan

G. S. Giri   Regional Agriculture Research Station, Bhairahawa, Nepal

Jason Grantham   University of Arkansas Rice Research and Extension Center, Stuttgart, Arkansas, U.S.A.

Raj Gupta   CIMMYT, New Delhi, India

David O. Hansen   The Ohio State University, Columbus, Ohio, U.S.A.

N. I. Hashmi   National Agricultural Research Centre (NARC), Islamabad, Pakistan

Peter Hazell   International Food Policy Research Institute, Washington, D.C., U.S.A.

Paul Heisey   Resource Economics Division, USDA Economic Research Service, Washington, D.C., U.S.A.

Peter E. Hildebrand   University of Florida, Gainesville, Florida, U.S.A.

Peter R. Hobbs   Cornell University, Ithaca, New York, U.S.A.

Jared Holzhauer   University of Arkansas Rice Research and Extension Center, Stuttgart, Arkansas, U.S.A.

M. I. Hossain   Bangladesh Agricultural Research Station, Rajshahi, Bangladesh

Muhammad Azeem Khan   Social Sciences Institute, National Agricultural Research Centre (NARC), Islamabad, Pakistan

J. K. Ladha   International Rice Research Institute, Manila, Philippines

Rattan Lal   Carbon Management and Sequestration Center, School of Natural Resources, The Ohio State University, Columbus, Ohio, U.S.A.

J. G. Lauren   Department of Crop and Soil Sciences, Cornell University, Ithaca, New York, U.S.A.

Pedro L. O. A. Machado   Embrapa Soils, Rio de Janeiro, Brazil

Ram Kanwar Malik   Department of Agronomy, Haryana Agricultural University, Hisar, India

G. N. McCauley   Agricultural Research & Extension Center, Texas Agricultural Experiment Station, Texas A&M University, Beaumont, Texas, U.S.A.

C. A. Meisner   CIMMYT, Dhaka, Bangladesh

John E. Morrison, Jr.   Agricultural Research Service, U.S. Department of Agriculture, Temple, Texas, U.S.A.

J. R. Murray   School of Agronomy and Horticulture, University of Queensland, Gatton, Queensland, Australia

M. S. Nunez   Agricultural Research & Extension Center, Texas Agricultural Experiment Station, Texas A&M University, Beaumont, Texas, U.S.A.

D. K. Painuli   PC (AICRP Soil Physical Constraints and their Amelioration for Sustainable Crop Production), IISS, Bhopal, India

D. S. Pathic   Nepal Agricultural Research Council (NARC), Kathmandu, Nepal

David Pimentel   College of Agriculture and Life Sciences, Cornell University, Ithaca, New York, U.S.A.

Idupulapati M. Rao   Centro Internacional de Agricultura Tropical (CIAT), Cali, Colombia

Paul Robbins   Department of Geography, The Ohio State University, Columbus, Ohio, U.S.A.

Mark Rosegrant   International Food Policy Research Institute, Washington, D.C., U.S.A.

M. A. Saleque   Soil Science Division, Bangladesh Rice Research Institute, Gazipur, Bangladesh

J. S. Samra   ICAR, New Delhi, India

P. K. Sardana   Department of Agronomy, Haryana Agricultural University, Hisar, India

Kenneth D. Sayer   CIMMYT, Houston, Texas, U.S.A.

G. Edward Schuh   HHH Institute of Public Affairs, Minneapolis, Minnesota, U.S.A.

A. Shaheed   Wheat Research Center, Nashipur, Dinajur, Bangladesh

R. K. Shrestha   Department of Soil Science, University of Wisconsin–Madison, Madison, Wisconsin, U.S.A.

Bal Ram Singh   Department of Plant and Environmental Sciences, Agricultural University of Norway, Aas, Norway

S. Singh   Department of Agronomy, Haryana Agricultural University, Hisar, India

Yadvinder-Singh   Department of Soils, Punjab Agricultural University, Ludhiana, India

B. A. Stewart   West Texas A&M University, Canyon, Texas, U.S.A.

M. A. Sufian   Wheat Research Center, Nashipur, Dinajur, Bangladesh

A. S. M. H. M. Talukder   Wheat Research Center, Nashipur, Dinajur, Bangladesh

C. Tripathi Regional Agriculture Research Station, Bhairahawa, Nepal

J. N. Tullberg School of Agronomy and Horticulture, University of Queensland, Gatton, Queensland, Australia

Norman Uphoff Cornell International Institute for Food, Agriculture and Development, Cornell University, Ithaca, New York, U.S.A.

Frank M. Vanclay Tasmanian Institute of Agricultural Research, University of Tasmania, Hobart, Tasmania, Australia

R. G. Wallace Agricultural Research & Extension Center, Texas Agricultural Experiment Station, Texas A&M University, Beaumont, Texas, U.S.A.

M. O. Way Agricultural Research & Extension Center, Texas Agricultural Experiment Station, Texas A&M University, Beaumont, Texas, U.S.A.

Keith D. Wiebe Resource Economics Division, USDA Economic Research Service, Washington, D.C., U.S.A.

Tony E. Windham University of Arkansas Cooperative Extension Service, Little Rock, Arkansas, U.S.A.

A. Yadav Department of Agronomy, Haryana Agricultural University, Hisar, India

1

Prospects for World Agriculture in the Twenty-First Century

Norman E. Borlaug
Texas A&M University
College Station, Texas, U.S.A.

Christopher R. Dowswell
CIMMYT
Mexico City, Mexico

I. INTRODUCTION

The rice–wheat system of South Asia is an extremely important cropping system. It covers nearly 14 million hectares and accounts for one fourth of the rice and one third of the wheat produced in the region. Despite the importance of this system to the regional agricultural economy, its productivity has been limited by soil and water and crop management problems associated with the turnaround phases between the rice and wheat crops. Thus, identification of improved crop management practices that enhance the productivity of the rice–wheat farming system in environmentally sustainable ways is of paramount importance to the region and the world.

II. PERFORMANCE OF THE GLOBAL FOOD SYSTEM

Thanks to a continuing stream of high-yielding varieties that have been combined with improved crop management practices, food production has more than kept pace with global population growth during the past 40 years. Contemporary per capita world food supplies are 23% higher and real prices are 65% lower than they were in 1961 (IFPRI, 2001). Despite these achievements, there is no room for complacency regarding food production and poverty alleviation.

1

Fertilizer use, and especially the application of low-cost nitrogen derived from synthetic ammonia, only became an indispensable component of modern agricultural production after World War II. Nearly 80 million nutrient tons of nitrogen are now consumed annually (FAOSTAT, 2002). It is estimated that 40% of the 6 billion people who inhabit the earth today are alive thanks to the Haber–Bosch process of synthesizing ammonia (Smil, 1999). It would be impossible to replace this amount of nitrogen using only organic sources.

Agricultural intensification has made major contributions to the protection of environmental resources. This is true despite problems such as salinization, caused by poorly engineered and managed irrigation systems, and the local pollution of some ground- and surface water resources, caused in part by excessive use of fertilizers and crop protection chemicals. By increasing yields on the lands best suited to agriculture, world farmers have been able to preserve vast areas of land for other purposes. For example, if cereal yield levels of 1950 had failed to increase to the year 2000, we would have had to use nearly 1.8 billion hectares of land to produce the current global harvest, rather than the 600 million hectares that were used for production. Much of this land would not have been available, especially in highly populated Asia. Moreover, had more environmentally fragile land been brought into agricultural production, the impact on soil erosion, loss of forests, grasslands, and biodiversity, and extinction of wildlife species would have been enormous.

Humans have been able to expand food production to the current level of about 5 billion gross tons per year over the past 10,000 to 12,000 years. By 2020, we will need to expand the current annual harvest by 40%–50%. This will be impossible unless farmers across the world have access to existing high-yielding crop production methods as well as new biotechnological breakthroughs. They offer great promise for improving the yield potential, yield dependability, and nutritional quality of our food crops, as well as for improving human health in general.

III. GREEN REVOLUTION IN ASIA

The breakthrough in wheat and rice production in Asia in the mid-1960s, which came to be known as the Green Revolution, symbolized the process of employing agricultural science to develop modern applications of technology for the Third World. It began in Mexico with the "quiet" wheat revolution in the late 1950s. During the 1960s and 1970s, India, Pakistan, and the Philippines received world attention for their agricultural progress, as reflected by data in Table 1.

South Asian countries have been able to achieve great increases in cereal production, especially in wheat. Today the region is a net exporter of wheat and rice. India's current grain stocks include nearly 30 million tons of wheat and 20

Table 1 Growth in Rice and Wheat Production, South Asia and Developing Regions of Asia

	1961	1970	1980 (million tons)	1990	2000	% Increase 1961–2000
South Asia[a]						
Rice, milled	49	58	74	100	121	+147
Wheat	15	28	44	66	100	+567
Developing Asia[b]						
Rice, milled	122	183	233	31	357	+193
Wheat	44	51	128	202	232	+427

[a] South Asia region includes Bangladesh, Bhutan, India, Nepal, Pakistan, and Sri Lanka.
[b] Developing Asia region includes all developing countries, from Turkey in West Asia to China in East Asia.
Source: FAOSTAT (2001).

million tons of rice. Pakistan has considerably expanded its exports of wheat and rice. They range from 1 to 2 million tons annually. Bangladesh and Sri Lanka are net importers of wheat. Each imports around 1 million tons annually. Bangladesh also imports around 0.5 million tons of rice (FAOSTAT, 2001).

China has been Asia's greatest food success story since 1980. It has achieved enormous productivity-led increases in total food supply, and it has been able to distribute this food more equitably to poor segments of the population. China's agricultural and rural development successes have greatly facilitated rapid industrial and urban development. Indeed, agriculture in the developing Asia region as a whole has made remarkable strides forward since the food crisis of the 1960s. However, despite the great improvements in quality of life for most rural Asians, hundreds of millions still live in poverty. They continue to experience lower levels of health and education than their urban counterparts.

FAO and other data in Table 2 indicate changes in the factors of production that have occurred in developing Asia (developing countries stretching from Turkey in the west to China in the east) over the past four decades. Irrigated area has more than doubled—to 176 million hectares; fertilizer consumption has increased more than 30-fold; and tractor use has increased from 200,000 to 4.6 million units during this same period of time.

The battle to ensure food security for millions of poor people, who live in misery, is far from won, especially in South Asia. This is true despite the important contributions of smallholder Asian farmers to the 300% increase in grain production through the application of Green Revolution technologies, which took place over the past 30 years.

Table 2   Changes in Factors of Production in Developing Asia

	Modern varieties		Fertilizer nutrient		
	Wheat (million ha% area)	Rice	Irrigation (million ha)	Consumption (million tons)	Tractors (millions)
1961	0/0%	0/0%	87	2	0.2
1970	14/20%	15/20%	106	10	0.5
1980	39/49%	55/43%	129	29	2.0
1990	60/70%	85/65%	158	54	3.4
1998	70/84%	100/74%	176	70	4.6

Sources: FAOSTAT (2001); CIMMYT and IRRI impact data; authors' estimates.

Data from China illustrate that increased food production, although a necessary condition for food security, is not a sufficient condition for it. As shown in Table 3, huge stocks of grain have accumulated in India and China. However, hundreds of millions of people in India suffer from hunger, because they lack the purchasing power to buy food, while China continues to make great strides forward in this regard. Nobel Economics Laureate, Professor Amartya Sen, has argued that this difference is due to differences in the relative emphasis being given to broad-based economic growth and poverty reduction. He argues that China's greater success is a result of the higher priority its government has given to investments in rural education and healthcare services. Nearly 85% of the Chinese population is literate compared to only 55% of the Indian population. Forty-five percent of India's population is below the poverty line, as compared

Table 3   Socioeconomic Development Indicators in China and India

Development Indicators	China	India
Population in 1961 (millions)	669	452
Population in 2000 (millions)	1,290	1,016
GDP per capita income in 1999 (US$)	780	440
Annual population growth, 1990–1999	1.1%	1.8%
Population in agriculture in 1995	47.0%	64.0%
Population with income less than $ 1/Day (1984–1999)	4.6%	35.0%
Malnutrition: children underweight (1993–1999)	9.0%	45.0%
Population over 15 illiterate in 1995	16.0%	45.0%

Sources: World Bank (2001); FAOSTAT (2001).

to China's population. And only 9% of the children in China are malnourished as compared to 35% of the children in India. Healthier and better-educated rural populations in China have made it easier for that nation's economy to grow about twice as fast as that of India during the past two decades. Today, China's per capita income is 77% higher than that of India.

High levels of illiteracy among populations over 15 years of age exist in all countries of South Asia, except Sri Lanka. Data in Table 4 also indicate that infant mortality rates in Sri Lanka are much lower than in other nations, which may be related to its lower illiteracy levels. Other South Asian nations also have higher levels of child malnutrition and absolute poverty, suggesting that they need to implement universal, free primary education for boys and girls. They also need to ensure that secular curricula are taught, to allow students to attain minimum levels of proficiency in reading, writing, arithmetic, basic science, and history. The crisis in primary education in South Asia will need to be solved, especially in rural areas, in order to reduce poverty and malnutrition. This is true despite the size of their grain surpluses.

IV. AFRICAN AGRICULTURE STILL IN CRISIS

Sub-Saharan Africa food production is in crisis in comparison with other regions of the world. No Green Revolution has occurred in Africa. Serious food deficits and deteriorating nutrition, especially among the rural poor, are related to high rates of population growth and lack of application of improved agricultural technologies. Recovery is still very fragile despite recent signs that smallholder agriculture is beginning to improve.

Agricultural development in this region has been affected by many factors, including extreme poverty, poor soils, uncertain rainfall, increasing population

Table 4 Quality of Life Indicators in South Asia

Nation	Infant mortality per 1000	Child malnutrition under 5	Adult illiteracy M	F	Population below 1 $/day 1984–1999
Bangladesh	61	56%	49%	71%	36%
India	71	45%	33%	57%	35%
Nepal	75	47%	43%	78%	42%
Pakistan	90	38%	42%	71%	34%
Sri Lanka	15	33%	6%	12%	25%

Source: World Bank (2001).

pressures, changing ownership patterns for land and cattle, political and social turmoil, shortages of trained agriculturalists, inadequate research and technology delivery systems, and HIV/AIDS. Despite these formidable challenges, the agricultural development formula that worked in Latin America and Asia, including use of fertilizers and improved seeds, and improved agronomic practices, will also work there.

To a large degree, Africa's food crisis has been a result of political leaders' long-term neglect of agriculture. Although this sector is the source of livelihood for 70%–85% of the population in most African nations, agriculture and rural development have received low priority. Investments in distribution and marketing systems and in agricultural research and education have been woefully inadequate. Many governments have chosen to pursue a policy of providing cheap food for politically volatile urban populations at the expense of providing production incentives for farmers.

Many African environments, especially the forest and transition areas, are fragile ecological systems. They have deeply weathered, acidic soils that lose fertility rapidly under repeated cultivation. Slash-and-burn shifting cultivation and complex cropping patterns permitted low-yield, but relatively stable, food production. However, expanding populations, increased food requirements, and increased numbers of farms have resulted in shortened bush/fallow periods and have pushed farmers onto marginal lands. More continuous cropping is depleting organic material, nitrogen, marginal phosphorus, and other nutrient reserves. Disastrous environmental consequences are resulting, such as serious erosion and weed invasion, that lead to impoverished fire climax vegetations.

As shown in Table 5, only about 10 kg of fertilizer nutrients are used per hectare of arable land in sub-Saharan Africa. This compares with rates that are

Table 5 Total Fertilizer Use per Hectare of Arable Land, 1999

World region	Total fertilizer nutrient consumption (1000 tons)	Per capita arable land (1000 hectares)	Per capita fertilizer use (kg/hectare)
Sub-Saharan Africa	1,320	138,799	10
Developing Asia	74,079	448,972	165
European Union	17,340	74,740	233
NAFTA nations[a]	24,265	247,310	99
Latin America[b]	10,405	120,396	86

[a] North American Free Trade Association (Canada, United States, Mexico).
[b] Latin American Integration Association (15 nations).
Source: FAOSTAT (2001).

9 to 23 times higher in other regions. Because of the inordinately low fertilizer-use levels, massive nutrient depletion has occurred in most countries. Over the past 30 years, sub-Saharan African soils are estimated to have experienced a net nutrient loss of 700 kg of nitrogen, 100 kg of phosphorus, and 450 kg of potassium per hectare (Sanchez et al., 1996). Resulting deterioration of agricultural productivity will seriously undermine the foundations of economic growth in Africa. Without more chemical fertilizer, sub-Saharan African agriculture is destined to failure.

Transportation is another major constraint to agricultural development in sub-Saharan Africa. High transportation costs often result in fertilizer costs that are generally double and frequently triple those paid in industrialized nations. Similarly, farm gate produce values are often only 50% of those obtained in urban centers. Inadequate infrastructure in sub-Saharan Africa—especially roads, potable water, and electricity—is a major obstacle to rural and economic development. A comparison of road infrastructure in Africa and other regions are found in Table 6. It highlights the vast gulf in infrastructure between most of Africa and the developed world. Most agricultural production in Africa is generated along a vast network of footpaths, tracks, and community roads where the most common mode of transport is "the legs, heads, and leg of women."

Efficient transport facilitates production and enables farmers to bring their products to markets. Intensive agriculture is highly dependent on vehicle access. Improvements in transport systems would reduce rural isolation, thus helping to break down tribal animosities and to facilitate establishment of rural schools and clinics in areas where teachers and health practitioners have been unwilling to settle. Finding ways to provide effective and efficient infrastructure in sub-

Table 6   Kilometers of Paved Roads per Million People in Selected Nations

Country	Kilometers of paved road	Country	Kilometers of paved road
United States	20,987	Guinea	637
France	12,673	Ghana	494
Japan	6,584	Nigeria	230
Zimbabwe	1,586	Mozambique	141
South Africa	1,402	Tanzania	114
Brazil	1,064	Uganda	94
India	1,004	Ethiopia	66
China	803		

Source: *Encyclopedia Britannica* (2001).

Saharan Africa underpins all other efforts to reduce poverty, improve health and education, and secure peace and prosperity.

V. AGRICULTURAL CHALLENGES IN THE 21ST CENTURY

Most experts believe that the world population will increase from the current 6 billion people to over 7.5 billion people by the year 2020. The demand for cereals, which account for 70% of the world's food supply, will increase by 40%–50%. Increases in animal feed use are driven by rising meat and milk demand, and they will account for 35% of the increased demand for cereals between 1997 and 2020 (Pinstrup-Andersen and Pandya-Lorch, 2000). Demand for maize as an animal feed will be greater than that of all other cereal crops in the developing world.

Population growth, urbanization, and rising incomes will result in a massive increase in the demand for animal products (Pinstrup-Andersen and Pandya-Lorch, 2001). Data in Table 7 suggest that developing countries are likely to consume 100 million more tons of meat and 223 million more tons of milk than they did in 1993. The greatest increase in demand will be for poultry products. By 2020, China will become the world's largest meat producer and India will remain the world's largest milk producer.

Table 7   Actual and Projected Meat Consumption by Region

Region	Total meat consumption (1,000,000 tons)		
	1983	1993	2020
China	16	38	85
Other East Asia	1	3	8
India	3	4	8
Other South Asia	1	2	5
Southeast Asia	4	7	16
Latin America	15	21	39
West Asia/North Africa	5	6	15
Sub-Saharan Africa	4	5	12
Developing world	50	87	188
Industrialized world	88	97	115
World	139	184	303

Source: IFPRI (2001).

As the above data suggest, the livestock subsector will become increasingly important. However, supply increases of livestock products are coming primarily from industrial production, because of the undeveloped state of traditional smallholder livestock systems. Appropriate policies that encourage improvements in animal health and nutrition have the potential to benefit smallholder producers.

China is the second-largest maize producer in the world. Nearly 80% of it is fed to livestock. Recognizing that quality protein maize (QPM) can significantly improve feed conversion rates, especially for swine, China embarked on QPM research in the mid-1970s. This research has been continued to date by a new generation of breeders at CAAS and nine provincial research centers. A half-dozen QPM hybrids were released during the 1990s and are now grown on about 350,000–400,000 ha. Some of this production is part of a rural poverty-reduction program promoted by the local extension services. Smallholder farmers feed their swine with QPM grain with rice middlings and green "chop," leading to higher weight gains and resulting additional income.

A. Dwindling Land Resources

Global arable land area and the potential for further expansion is limited in most regions, including densely populated Asia and Europe. Large unexploited tracts exist only in sub-Saharan Africa and South America; and only *some* of this land should eventually come into agricultural production. Very little uncultivated land exists in Asia. Indeed, some land should be taken *out* of cultivation because of high susceptibility to soil erosion.

Thus, most increases in the global food supply, including more than 85% of cereal production growth, must come from increasing yields of agricultural lands already in production (Pinstrup-Andersen and Pandya-Lorch, 2000). These productivity increases will require the introduction of varieties with higher genetic yield potential and greater tolerance of drought, insects, and diseases. Advances in both conventional and biotechnology research will be needed, as are improvements in soil and water conservation, tillage, fertilization, weed and pest control, and postharvest handling.

Bringing potentially arable lands into agricultural production poses great challenges. The Brazilian *Cerrado*, or savannah, is a good example. It spans a geographic area from latitude 24° to 4° S and varies in elevation from 500 m to 1,800 m, with unimodal precipitation that varies from 900 to 1,800 mm annually. A new generation of crop varieties (forage grasses, rice, soybean, maize, and wheat) was developed possessing tolerance to aluminum toxicity, but had low grain yield potential and other defects, especially susceptibility to various diseases.

The creation in 1973 of the Empresa Brasileira de Pesquisa Agropecuaria (EMBRAPA)—the national Brazilian Agricultural Research Corporation—pro-

Table 8 Cereal and Meat Production in the *Cerrado* in 1990

Land use	Area (million ha)	Productivity (t/ha/year)	Production (million ton)
Crops (rainfed)	10.0	2.00	20.0
Crops (irrigated)	0.3	3.00	0.9
Meat (pasture)	35.0	0.05	1.7
Total	45.3		22.6

Source: Macedo (1995).

vided a major impetus to research aimed at the *Cerrado*. During the 1980s, EM-BRAPA and several international agricultural research centers, especially CIM-MYT and CIAT, began more intensive collaboration to develop a third generation of crop varieties combining tolerance to aluminum toxicity with high yield, better resistance to major diseases, and better agronomic type. This new generation of improved crop varieties is now moving on to farmers' fields.

As shown in Table 8, about 10 million ha of rainfed crops were grown with an average yield of 2 t/ha and a total production of 20 million tons in 1990. The irrigated area is only 300,000 ha. The average production yield for it was 3 t/ha, and total production was 900,000 tons. The *Cerrado* also contained 35 million ha of improved pasture, with an annual meat production of 1.7 million tons.

Macedo (1995) notes that farmers had the potential to attain 3.2 t/ha average yield and 64 million tons of production if available technologies were used on the 20 million ha of arable rainfed land in the *Cerrado* in 1995. As shown in Table 9, irrigated area could have been increased to 5 million ha, with an expected average yield 6 t/ha. Meat production could also have been increased fourfold with improved pastures. In total, food production could have been increased from the 22.6 million to 98 million tons.

Table 9 Potential Food Production if Technology Available in 1995 Is Adopted on *Cerrado* Area Already in Production

Land use	Area (million ha)	Productivity (t/ha/year)	Production (million t)
Crops (rainfed)	20.0	3.2	64
Crops (irrigated)	5.0	6.0	30
Meat (pasture)	20.0	0.2	4
Total	45.0		98

Source: Macedo (1995).

The *Cerrado* can help ensure adequate world food supplies for the next two decades if wise policies are used to stimulate production. Advances made in that region have the potential to be expanded into the *llanos* in Colombia and Venezuela and into central and southern African countries where they have similar soil problems.

B. Increasing Water Scarcity

Water covers about 70% of the earth's surface, but only about 2.5% is fresh water. And most is found in the ice caps of Antarctica and Greenland, in soil moisture, and in deep aquifers not readily accessible for human use. Less than 1% of the world's fresh water is readily available for direct human use (World Meteorological Organization, 1997). Irrigated agriculture accounts for 70% of global water withdrawals. It covers about 17% of cultivated land—about 275 million ha—and accounts for nearly 40% of world food production.

The rapid expansion in world irrigation and in urban and industrial water use has led to growing shortages. "About one third of the world's population lives in countries that are experiencing moderate-to-high water stress, resulting from increasing demands from a growing population and human activity" (WMO, 1997). By the year 2025, as much as two thirds of the world's population could be under stress conditions (WMO, 1997). Appropriate initial investments were not made in many irrigation schemes, especially in developing Asia. Serious salinization has resulted on many irrigated soils, especially in drier areas, and serious water logging in more humid areas. Asian irrigation schemes account for nearly two thirds of all irrigated land, and many are seriously affected by both problems. Thus, most of the funds going into irrigation end up being used for stopgap maintenance expenditures for poorly designed systems rather than for new irrigation projects. Adding such costs to new projects will often result in a poor return on investment, so nations will have to decide how much they are willing to subsidize new irrigation development.

Many technologies exist to improve water-use efficiency. Wastewater can be treated and used for irrigation. It is an important source of water for peri-urban agriculture, which is growing rapidly around many of the world's megacities. New crops and improved varieties that require less water, together with more efficient crop sequencing and timely planting, can achieve significant savings in water use. Proven technologies, such as drip irrigation, save water and reduce soil salinity. Various new precision irrigation systems, which will supply water to plants only when they need it, are also on the horizon. Improved small-scale and supplemental irrigation systems can increase the productivity of rainfed areas and offer much promise for smallholder farmers.

Humanity will need to bring about a "Blue Revolution" in the 21st century to complement the so-called Green Revolution of the 20th century in order to feed a growing world population within the parameters of likely water availability. The Blue Revolution will require that water-use productivity be wedded to land-use productivity.

VI. NEW CROP IMPROVEMENT TECHNOLOGY

Agricultural researchers and farmers worldwide will be challenged during the next 20 years to develop and apply technology to increase global cereal yields by 50%–75%. They must do so in ways that are economically and environmentally sustainable. Most yield gains will come from applying underutilized "shelf technology."

Continued genetic improvement of food crops—using both conventional as well as biotechnology research tools—is needed to shift the yield frontier higher. Biotechnology research tools offer much promise, as do conventional plant-breeding methods. The latter continue to make significant contributions to improved food production and enhanced nutrition. In rice and wheat, interrelated changes in plant architecture, hybridization, and wider genetic resource utilization are being pursued to increase genetic maximum yield potential (Rajaram and Borlaug, 1997). Widespread impact on farmers' fields is still probably 10 to 12 years away. New "super rices," in association with direct seeding, can increase rice yield potentials by 20%–25% (Khush, 1995).

New wheat plants with architecture similar to the "super rices"—larger heads, more grains, and fewer tillers—could lead to an increase in yield potential of 10%–15% (Rajaram and Borlaug, 1997). The introduction of wild species genes into cultivated wheat can result in important sources of resistance for several biotic and abiotic stresses, and in higher yield potential, especially if the transgenic wheats are used as parent material in the production of hybrid wheats (Mujeeb-Kazi and Hettel, 1995).

The success of hybrid rice in China (now covering more than 50% of the irrigated area) has led to a renewed interest in hybrid wheat costs. With better heterosis and increased grain filling, the yield frontier of the new wheat genotypes could be 25%–30% above the current germplasm base. Hybrid triticale offers the promise of higher yield potential than wheat for some areas and uses.

Maize production has really begun to take off in many Asian countries, especially China. It now has the highest average yield of all the cereals in Asia, with much of the genetic yield potential yet to be exploited. Moreover, recent developments in high-yielding quality protein maize (QPM) varieties and hybrids using conventional plant-breeding methods stand to improve the nutritional quality of the grain without sacrificing yields. This research achievement offers important nutritional benefits for livestock and humans. With biotechnology tools, further nutritional "quality" enhancements in the cereals are likely in years to come.

VII. IMPROVED CROP MANAGEMENT TECHNOLOGY

Crop productivity depends on both variety yield potential and crop management to enhance input and output efficiency. Productivity gains are possible through improved tillage, water use, fertilization, weed and pest control, and harvesting.

An outstanding example of new Green/Blue Revolution technology in irrigated wheat production is the "bed planting system," which has multiple advantages over conventional planting systems. The system reduces plant height and lodging and results in 5%–10% increases in yields and better grain quality. It reduces water use by 20%–25%, and efficiencies in fertilizer and herbicide use reduce their use by 30%.

Conservation tillage is spreading rapidly in many parts of the world. Monsanto has estimated that farmers used conservation tillage practices on 95 million ha in 2000 (1997 Annual Report). Turnaround time on lands that are double- and triple-cropped can be significantly reduced through its use, leading to higher yields and lower production costs. Conservation tillage also controls weed populations and greatly reduces the time that small-scale farm families must devote to this backbreaking work.

Mulch left on the ground reduces soil erosion, increases moisture conservation, and builds up the organic matter in the soil—all very important factors in natural resource conservation. It does, however, require modification in crop rotations to avoid the buildup of diseases and insects that find a favorable environment in the crop residues for survival and multiplication.

VIII. BIOTECHNOLOGY POTENTIAL

Biotechnology based on recombinant DNA has developed invaluable new scientific methodologies and products in food and agriculture during the last 20 years. The new biotechnology permits hybridization across taxonomically distinct genera, families, orders, or kingdoms. Recombinant DNA methods have enabled breeders to select and transfer single genes. This has reduced the time needed in conventional breeding to eliminate undesirable genes and has allowed breeders to access useful genes from other taxonomic groups. These gene alterations have conferred producer-oriented benefits, such as resistance to pests, diseases, and herbicides. Biotechnology and plant breeding will result in varieties with greater tolerance to drought, water logging, heat, and cold—important traits given current predictions of climate change. Many consumer-oriented benefits, such as improved nutritional and other health-related characteristics, are also likely to be realized over the next 10 to 20 years.

Despite the formidable opposition in certain circles to transgenic crops, commercial adoption of the new varieties reflects one of the most rapid processes of technology diffusion in the history of agriculture. As shown in Table 10, the

Table 10　Transgenic Crop Coverage, 2001

Region	Area (million ha)	Crops	Area (million ha)
United States	35.7	Soybeans	33.3
Argentina	11.8	Maize	9.8
Canada	3.2	Cotton	6.8
China	1.5	Canola	2.7
Others	0.4		
Total	52.6		

Source: James (2002).

area planted commercially to transgenic crops has increased 30-fold between 1996 and 2001. Transgenic crops were planted on 52.6 million ha in 13 countries by 5.5 million farmers in 2001, compared to only 1.7 million ha in 1996 (James, 2002). During this period, herbicide tolerance has been the dominant trait, accounting for 77% of the area. One quarter of the global transgenic crop area is now found in developing countries. The highest year-on-year percentage growth occurred in China between 2000 and 2001. Its Bt cotton area tripled from 0.5 to 1.5 million ha.

Most existing transgenic crops reduce production costs per unit of output. In theory, they are especially appropriate to the developing world, where over half the population is still engaged in agriculture, and where cost-reducing, yield-increasing technologies are the key to poverty reduction. Because biotechnology is packed into the seed, transgenic crops can help to simplify input delivery, often a major bottleneck in reaching smallholder farmers.

Several genetic engineering breakthroughs could bring enormous benefits to the poor producer and the consumer. Rice is unique in its immunity to rusts (*Puccinia spp.*). All the other cereals are attacked by two to three species of rusts, often resulting in disastrous epidemics and crop failures. Imagine the benefits to humankind if the genes for rust immunity in rice could be transferred into wheat, barley, oats, maize, millet, and sorghum. The world could free itself of the scourge of rusts, which have led to so many famines over human history.

Bread wheat has superior dough for making leavened bread and other bakery products due to the presence of two proteins—gliadin and glutenin. No other cereals have this combination. Genes for these proteins have the potential to be identified and transferred to other cereals, especially rice and maize, so that they could be used to make good-quality, unleavened bread. This would help many countries, especially developing countries in the tropics, where bread wheat flour is often the single largest food import.

A growing potential exists for science to improve the nutritional quality of our food supply. Biotechnology can help achieve further nutritional "quality" enhancements in cereals and other foods at a much faster rate. Gene transfers to increase the quantity of Vitamin A, iron, and other micronutrients contained in rice can potentially bring significant benefits for millions of people with deficiencies of Vitamin A and iron. Vitamin A deficiencies are a major cause of blindness, and iron deficiencies are a major cause of anemia.

Plants also have the potential to be used to vaccinate people against diseases, such as hepatitis B and the Norwalk disease, simply by growing and eating them. This offers tremendous possibilities in poor countries (ACSH, 2000). This line of research and development should be pursued aggressively, and probably through private–public partnerships, since traditional vaccination programs are costly and difficult to execute.

To date, no reliable scientific information exists to substantiate that transgenic crops are inherently hazardous. Recombinant DNA has been used for 25 years in pharmaceuticals, with no documented cases of harm attributed to the genetic modification process. This has also been the case in genetically modified foods. Seed industries have ensured that their transgenic crop varieties are safe to plant and that food from them is safe to eat.

Most agricultural scientists anticipate great benefits from biotechnology in the coming decades. However, new forms of public–private collaboration are likely to be needed to ensure that all farmers and consumers have the opportunity to benefit from this new genetic revolution. In particular, public biotechnology research will be needed to balance and complement private sector research investments. This is true for industrialized countries as well as those in the developing world.

An urgent need exists for developing nations to put regulatory frameworks in place to guide the development, testing, and use of transgenic crops in order to protect people and the environment. Intellectual property rights of private companies should be safeguarded by this process to ensure fair returns to past investments and to encourage greater future investments. The regulatory frameworks should neither be overly bureaucratic, nor have unreasonable risk-aversion expectations. The seed industry should have primary responsibility for ensuring the safety of its products.

Private sector investment is the primary driver of agricultural research and development in industrialized countries. Some argue that the fastest way to get a new technology to poor people is to "speed up the product cycle," so that the technology can spread quickly, first among rich people and later among the poor. This interpretation of the diffusion process may have validity. However, private life science companies would need to establish concessionary pricing in the low-income countries so that poor farmers can also benefit from new transgenic products. Large transnational companies will also need to share their expertise with public research institutions and scientists concerned with smallholder agriculture. They will also need to form partnerships to work on crops and agricultural problems not currently of priority interest in the main transnational markets. Indeed, private biotechnology companies are showing considerable willingness to form such partnerships. Monsanto has led establishment of developing country initiatives in agricultural product and technology cooperation; Syngenta is doing likewise, building partnerships with national and international agricultural research centers to address production problems in Africa and elsewhere. The Donald Plant Science Center in St. Louis, Missouri, was co-founded in 1998 by Monsanto, a consortium of universities, public research institutes, and private foundations. It is an especially exciting development, given its strong Third World orientation in its research agenda and in its training programs.

IX. CONCLUSIONS

The current backlash against agricultural science and technology evident in some industrialized countries is hard to comprehend. It shows how quickly humans become detached from soil and agricultural production! Less than 4% of the population in the industrialized countries is directly engaged in agriculture. Low-cost food supplies and urban bias help to explain why consumers fail to understand the complexities of reproducing the annual world food supply and expanding it further to feed the nearly 80 million new mouths that are born each year. This "educational gap" can be treated in urban nations by making it compulsory for students to take biology, science, and technology policy courses in secondary schools and universities.

The fear of science has grown as the pace of technological change has accelerated during the past 50 years. The development of nuclear power and the prospects of a nuclear holocaust added to people's fear and drove a bigger wedge between the scientist and the layman. The book *Silent Spring* (Carson, 1962) reported that poisons were everywhere and struck a very sensitive public nerve. This perception was not totally unfounded. By the mid-20th century, air and water quality had been seriously damaged through wasteful industrial production systems that pushed effluents literally into "our own backyards."

A debt of gratitude is owed to environmental movement in the industrialized nations. It resulted in legislation over the past 30 years to improve air and water quality, to protect wildlife, to control the disposal of toxic wastes, to protect soils, and to reduce the loss of biodiversity. However, as Gregg Easterbrook (1996) argued, "In the Western world the Age of Pollution is nearly over. ... Aside from weapons, technology is not growing more dangerous and wasteful, but cleaner and more resource-efficient. Clean technology will be the successor to high technology." Easterbrook (1996) also warned that, "As positive as trends are in the First World, they are negative in the Third World. One reason why the West must shake off its instant-doomsday thinking about the United States and Western Europe is so that resources can be diverted to ecological protection in the developing world.'' This protection includes clean water and sanitation systems for human settlements, and soil and water conservation.

More recently, Lomborg (2001) has provided a powerful critique of the way many extremist environmental organization distort scientific evidence. He concludes on the basis of his research that more reasons exist for environmental optimism than pessimism. He emphasizes the need for clear-headed prioritization of resources to tackle real, not imagined, problems in the future.

Paarlberg sounded the alarm nearly a decade ago about the deadlock between agriculturalists and environmentalists over what constitutes ''sustainable agriculture'' in the Third World (Paarlberg, 1994). This debate has confused—if not paralyzed—many in the international community. Afraid of antagonizing powerful environmental lobbying groups, they have turned away from supporting science-based agricultural modernization projects, which are still needed in much of smallholder Asia, sub-Saharan Africa, and Latin America.

The agriculturalist–environmentalist deadlock must be broken in order to focus on the enormous job before us to feed future generations. Ninety percent of newborns will begin life in a developing country, and many in poverty. Only through dynamic agricultural development will there be any hope to alleviate poverty, improve human health and productivity, and reduce political instability.

The world can count on available technology and technology that is well advanced in the research pipeline to sustainably feed the 10 billion people projected to inhabit the planet earth by the end of the 21st century. Pertinent questions are (1) whether farmers and ranchers will be permitted access to the continuing stream of new technologies needed to meet agricultural, food, and nutrition challenges that lie ahead and (2) whether the scourge of poverty can continue to be abated, so that the ever-growing proportion of the world's people are assured the minimum nutrition needed for health and human development.

REFERENCES

ACSH (American Council on Science and Health). 2000. Biotechnology and Food, Second Edition. New York.

Carson, R. 1962. Silent Spring. Fawcett Books, by arrangement with Houghton Mifflin, New York.

Easterbrook, G. 1996. A Moment on the Earth. London: Penguin Books.

Food and Agricultural Organization. 2001. FAOSTAT Online, FAO Statistical Databases. Rome, Italy: FAO. http://apps.fao.org.1 FAOSTAT.

IFPRI (International Food Policy Research Institute). 2001.

James, C. 2002. Global Review of Commercialized Transgenic Crops: 2001. International Service for the Acquisition of Agri-Biotech Applications (ISAAA), Brief No.24. Los Banos, The Philippines: ISAAA Southeast Asia Center.

Khush, G.S. 1995. Modern Varieties—Their Real Contribution to Food Supply and Equity. *Geojournal* 35(3):275–284.

Lomborg, B. 2001. The Skeptical Environmentalist. Cambridge, UK: Cambridge University Press.

Macedo, J. 1995. Prospects for the Rational Use of the Brazilian *Cerrado* for Food Production. Goiania, Brazil: EMBRAPA, CPAC.

Monsanto, J. 1997. Annual Report, St. Louis, Missouri.

Mujeeb-Kazi, A., Hettel, G.P. 1995. Realizing Wild Grass Biodiversity in Wheat Improvement—15 Years of Research in Mexico for Global Wheat Improvement (edited). Mexico, DF: CIMMYT, Wheat Special Report No. 29.

Paarlberg, R. 1994. Sustainable farming: A political geography. Policy Brief No. 4. Washington, DC: International Food Policy Research Institute.

Pinstrup-Andersen, R., Pandya-Lorch, R. 2000. Unfinished Agenda: Perspectives on Overcoming, Hunger, Poverty, and Environmental Degradation (edited). Washington, DC: IFPRI.

Rajaram, S., Borlaug, N.E. 1997. Approaches to breeding for wide adaptation, yield potential, rust resistance and drought tolerance, paper presented at *Primer Simposio Internacional de Trigo*, Cd. Obregon, Mexico, April 7–9.

Sanchez, Pedro A., Izac, A-M., Valencia, I., Pieri, C. 1996. Soil fertility replenishment in Africa: A concept note. In: Breth S.A., Ed. Achieving Greater Impact from Research Investments in Africa. Mexico City: Sasakawa Africa Association.

Sen, A. 1999. Knopf, Borzoi Books: Random House, Inc. New York.

Smil, V. 1999. Long-range perspectives on inorganic fertilizers in global agriculture. Travis P: Hignett Memorial Lecture, International Fertilizer Development Center. Muscle Shoals, Alabama.

World Bank. 2001. World Bank Atlas International Bank for Reconstruction and Development/World Bank, Washington, DC.

World Meteorological Organization. 1997. Comprehensive Assessment of the Freshwater Resources of the World. Geneva, Switzerland.

2

Soil and Water Resources of South Asia in an Uncertain Climate

Rattan Lal
The Ohio State University
Columbus, Ohio, U.S.A.

I. INTRODUCTION

The South Asia region consists of seven countries, namely Afghanistan, Bangladesh, Bhutan, India, Nepal, Pakistan, and Sri Lanka. The region's land area of 514 million hectares (Mha), or 3.8% of the world area (Fig. 1), is home to more than 22% of the world's population. Total arable land area of the region is 203 Mha, or 14.7% of the world's total, and the irrigated land area is 81.7 Mha, representing 30.1% of the world's irrigated land area (Table 1).

Several main sustainable development issues in South Asia are (1) a rapidly increasing population, (2) a decreasing per capita arable land area, (3) increasing risks of soil degradation, (4) stagnating crop yields, (5) a decreasing per capita renewable freshwater supply, (6) increasing standards of living and food demands, and (7) increasing per capita emissions of greenhouse gases from fossil fuel combustion and agricultural activities. These issues can partially be addressed by improving agricultural productivity, strengthening agro-based industries, restoring soils and degraded ecosystems, improving quality of natural waters, reducing emissions of greenhouse gases from agricultural activities, and sequestering carbon (C) in terrestrial ecosystems. The challenge to the scientific community lies in developing long-term regional programs to address these issues. This chapter contains a description of the soils and environments of the South Asia region, a historical assessment of crop production trends, an outline of the cause–effect relationship of soil degradation and crop production, and an ecological approach to sustainable management of natural resources for enhancing productivity and improving environment quality.

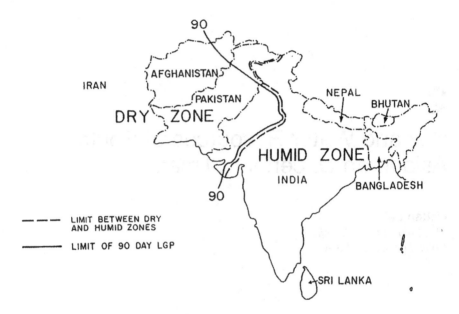

Figure 1 South Asia region. Ninety-day LGP = 90-day length of growing period. (From FAO, 1982.)

Table 1 Population and Arable Land in South Asia in 1998

Country	Population (millions)	Total area (Mha)	Arable land (Mha)	Irrigated land (Mha)
Afghanistan	21.4	65.2	7.9	2.8
Bangladesh	124.8	14.4	7.9	4.2
Bhutan	2.0	4.7	0.14	0.04
India	982.2	328.8	162.0	54.8
Nepal	22.8	14.7	2.9	1.1
Pakistan	148.2	79.6	21.0	18.1
Sri Lanka	18.5	6.6	0.9	0.7
Total	1319.9	514.0	202.7	81.7
World total	5901.1	13387.0	1379.1	271.7
% of the world total	22.4	3.8	14.7	30.1

Source: FAO (1998), Kaosa-Ard and Rerkasem (2000).

II. LAND RESOURCES OF SOUTH ASIA

Per capita arable land area is getting smaller in the South Asia region. It is a relevant land scarcity index (Table 2). Between 1960 and 2025, the arable land area per capita will decline by a factor of 3.9 for Afghanistan, 3.4 for Bangladesh, 3.0 for India, 2.7 for Nepal, 4.9 for Pakistan, and 2.0 for Sri Lanka.

Irrigated cropland area has been one of the bases of the Green Revolution in South Asia. The irrigated land area increased drastically during the last quarter of the past century. Similar to the land scarcity index, the irrigated land area per capita is also declining. It is merely 0.03 ha in Bangladesh, 0.04 ha in Sri Lanka, 0.05 ha in Nepal, and 0.06 ha in India (Table 3). Future expansion of irrigated land area is difficult and expensive. The biggest problem is the scarcity of freshwater resources available for irrigation (Table 4). Similar to the land scarcity index, the annual renewable fresh water per capita will decline between 1950 and 2025 by a factor of 5.1 for Afghanistan, 3.7 for India, and 6.8 for Pakistan. Freshwater availability will indeed continue to be a major challenge for the region during the 21st century and beyond.

Over and above the increase in population, the scarcity of soil and water resources is exacerbated by degradation of soil and pollution/contamination of water. Environmental degradation remains to be a serious problem in South Asia. The land area affected by different degradative processes includes 55.4 Mha by water erosion, 23.6 Mha by wind erosion, 11.0 Mha by fertility decline, 8.7 Mha by waterlogging, and 9.3 Mha by secondary salinization (Table 5). Some of the statistics (especially for fertility decline and groundwater depletion) may be underestimates because of the lack of available information at the country level. Excessive and lack of proper drainage leads to waterlogging and secondary salinization (Table 6). It is caused by excessive input of water into the soil, which has limited natural drainage capacity. The source of excess water may be flood irriga-

Table 2 Arable Land Scarcity Index as Measures by per Capita Arable Land Area (ha/capita)

Country	1960	1990	2025
Afghanistan	0.71	0.54	0.18
Bangladesh	0.17	0.09	0.05
India	0.36	0.20	0.12
Nepal	0.19	0.14	0.07
Pakistan	0.34	0.17	0.07
Sri Lanka	0.16	0.11	0.08

Sources: Engelman and LeRoy (1995); Kaosa-Ard and Rerkasem (2000).

Table 3 Irrigated Area in South Asia

Country	Irrigated area (Mha)				Per capita irrigated land area in 1998 (ha/person)
	1975	1985	1995	1998	
Afghanistan	2.4	2.6	2.8	2.8	0.13
Bangladesh	1.4	2.1	3.2	4.2	0.03
India	33.7	41.8	50.1	54.8	0.06
Nepal	0.2	0.8	0.9	1.1	0.05
Pakistan	13.6	15.8	17.2	18.1	0.12
Sri Lanka	0.5	0.6	0.6	0.7	0.04

Sources: Statistics on irrigation from Kaosa-ard and Rerkasem, (2000); FAO (2000).

tion and at rates more than what is needed, seepage from canal, rainfall, etc. This is a serious problem, especially in areas irrigated with canal water. The water table in Faridkd district, Punjab, increased from 36.5 m below surface in 1895 to 2.4 m in 1983 (personal communication, Dr. P. Lubano, PAU, India). In contrast, however, excessive use of pump irrigation leads to rapid depletion of the water table. The rate of depletion of the water table in some regions of Punjab, India, is more than 0.5 m/y. Similarly, extensive exploitation of groundwater is also occurring in Gujrat, Rajsthan, and Tamil Nadu. In Ahmedabad, Gjurat (India), the water-table level is estimated to have fallen by 2 to 2.5 m annually during the 1980s (Pachauri and Sri Dharan, 1998).

III. CROP YIELDS AND THE GREEN REVOLUTION

Growth in production of rice and wheat has been impressive in South Asia. The production of rice for the region (without Afghanistan) increased 3.7 times from

Table 4 Annual Renewable per Capita Freshwater Availability (m^3) in South Asia

Country	1950	1995	2025	2050
Afghanistan	5,582	2,543	1,105	815
Bangladesh	56,411	19,936	13,096	10,803
Bhutan	129,428	53,672	26,056	18,326
India	5,831	2,244	1,567	1,360
Nepal	7,862	7,923	4,192	3,170
Pakistan	11,844	3,435	1,740	1,310
Sri Lanka	7,678	2,410	1,805	1,600

Source: Gardner-Outlaw and Engelman (1997).

Table 5 Estimates of Soil Degradation in South Asia (Mha)

| Country | Land area affected by | | | | | |
	Water erosion	Wind erosion	Fertility decline	Waterlogging	Salinization	Ground water depletion
Afghanistan	11.2	2.1	—	0	1.3	0
Bangladesh	1.5	0	6.4	0	0	0
Bhutan	0.04	0	0	0	0	0
India	32.8	10.8	3.2	3.1	4.1	—
Nepal	1.6	0	—	0.6	0	0
Pakistan	7.2	10.7	—	1.0	3.8	0.01
Sri Lanka	1.1	0	1.4	0	0.05	0
Total	55.4	23.6	11.0	8.7	9.3	0.01

Source: FAO (1994).

47.0 million tons (Mg) in 1950 to 175.2 million tons in 2000 (Table 7). The magnitude of growth in rice production was sixfold for the same period in Pakistan. The increase in rice production was mainly due to increased grain yields. As shown in Table 8, the increase in rice grain yield between 1977 and 2001 was 1.9 to 3.3 Mg/ha @ 58 kg/ha/y for Bangladesh on to 3.0 Mg/ha @ 25 kg/ha/y for Pakistan (Table 8). The rate of increase in rice yield during the 1990s was lower than those during the 1970s and 1980s and was only 34 kg/ha/y for India, 7 kg/ha/y for Nepal, and 28 kg/ha/y for Sri Lanka, and compared to 54 kg/ha/y for Asia. The higher average yield of rice for Asia on the whole suggests a potential for further increase in yield (Table 9). Increases in wheat yields have been equally impressive (Table 10). Increases between 1977 and 2001 were from 1.6 to 2.4 Mg/ha @ 33 kg/ha/y for Bangladesh, 1.4 to 2.7 Mg/ha @ 54 kg/ha/y

Table 6 Rise of Water Table due to Excessive Irrigation

Irrigation project	Original water table depth (m)	Annual rise (m/y)
SCARP-I[a] Pakistan	40–50	0.4
SCARP-VI Pakistan	10–15	0.2–0.4
Khaipur, Pakistan	4–10	0.1–0.3
Bathinda, India	15	0.6

[a] SCARP: Salinity Control and Reclamation Project
Source: FAO (1990).

Table 7 Trends in Rice Production in South Asia from 1950 to 2000 (10^6 Mg)

Country	1950	1960	1970	1980	1990	1995	1998	2000
Bangladesh	10.9	14.1	15.6	20.9	27.1	26.3	28.2	35.8
India	32.4	49.2	62.3	77.0	109.4	120.5	123.1	128.2
Nepal	2.5	2.3	2.2	2.3	3.1	3.3	3.7	4.0
Pakistan	1.2	1.6	3.4	5.0	4.8	5.7	6.5	7.2
Total	47.0	67.2	83.5	105.2	144.4	155.8	161.5	175.2

Sources: Pingali and Heisey (1999); FAO (2000).

for India, 1.1 to 1.8 Mg/ha @ 29 kg/ha/y for Nepal, 1.4 to 2.3 Mg/ha @ 38 kg/ha/y for Pakistan, and 1.6 to 2.5 Mg/ha @ 38 kg/ha/y for Asia on the whole (Table 10). As for rice, the average yield of wheat in Asia is also higher than that in South Asia and is also an indication of a potential for further yield increases. The average increase in yield of wheat during the 1990s was impressive for Bangladesh (Table 11).

Rapid yield increases of cereal grains in South Asia during the second half of the 20th century have resulted from the use of improved varieties in combination with expanded use of chemical fertilizers and irrigation (Hazel and Ramasamy, 1991). The use of Green Revolution technology quadrupled food grain production in India. The average yield of wheat in India increased from 900 kg/ha in 1964 to 2,300 kg/ha in 1999 (Swaminathan, 2000).

Currently, rates of increase in production and crop yield are slowing down. Data in Table 12 show that the rate of increase in wheat production declined from 3.5%/y in the 1980s to 2.1%/y in 1990. Similarly, the rate of increase in rice production was 10%/y in the 1960s to only 5.0%/y during the 1980s and

Table 8 Increase in Rice Yields Between 1977 and 2000 in South Asia (Mg/ha)

Country	1977	1997	1999	2000	2001
Afghanistan	2.0	1.8	2.0	—	—
Bangladesh	1.9	2.6	3.2	3.3	3.3
India	1.9	2.9	3.0	2.9	3.0
Nepal	1.9	2.3	2.5	2.6	2.7
Pakistan	2.4	2.8	3.0	3.0	3.0
Sri Lanka	2.1	3.4	3.3	3.2	3.2
Asia	2.6	3.8	3.9	3.9	3.9

Sources: Kaosa- ard and Rerkasem (2000); FAO (2001).

Table 9   Land Area and Yield of Rice in 1998

Country	Area (Mha)	Yield (Mg/ha)	Yield increase in 1990s (kg/ha/y)
Afghanistan	0.2	2.5	74
Bangladesh	10.2	2.8	20
India	42.3	2.9	34
Nepal	1.5	2.4	7
Pakistan	2.3	2.8	65
Sri Lanka	0.8	3.2	28

World average yield = 3.7 Mg/ha.
Source: FAO (2000).

Table 10   Increases in Wheat Yield Between 1977 and 2001 (Mg/ha)

Country	1977	1997	1999	2000	2001
Afghanistan	1.2	1.1	1.2	—	—
Bangladesh	1.6	2.0	2.2	2.2	2.4
India	1.4	2.5	2.6	2.7	2.7
Nepal	1.1	1.6	1.7	1.8	1.8
Pakistan	1.4	2.1	2.2	2.5	2.3
Asia	1.6	3.0	2.7	2.5	2.5

Sources: Kaosa-ard and Rerkasem (2000); FAO (2001).

Table 11   Land Area and Yield of Wheat in 1998

Country	Area (Mha)	Yield (Mg/ha)	Yield increase in 1990s (kg/ha/y)
Afghanistan	2.2	1.3	29
Bangladesh	0.8	2.2	72
India	25.6	2.6	45
Nepal	0.6	1.6	29
Pakistan	8.4	2.2	49

World average yield = 2.6 Mg/ha; Ireland = 8 Mg/ha.
Source; FAO (2000).

Table 12 Growth in Cereal Production in South Asia[a]

Year	Rice (milled)		Wheat	
	Production (10^6 Mg)	% increase/y	Production (10^6 Mg)	% increase/y
1961	49	—	15	—
1970	58	2.0	28	9.6
1980	74	2.8	44	5.7
1990	100	3.5	66	5.0
2000	121	2.1	100	5.1

[a] South Asia includes Bangladesh, Bhutan, India, Nepal, Pakistan, and Sri Lanka.
Source: Recalculated from Borlaug (2002).

1990s. Similar observations on declining yield trends were reported by Conway and Toenniessen (1999) for the developing countries generally. Tyford (1994) reports that wheat yields in Pakistan increased by 6.0 kg/kg of fertilizer use during the 1970s, to 2.0 kg/kg of fertilizer use during the 1990s. Declining yield trends are observed despite increasing use of fertilizer in the region (Table 13). Total fertilizer use in the region was merely 0.5 million nutrient tons in 1961–1962, but increased gradually to 21.6 million tons in 2000–2001. Rates of fertilizer use increased in all countries. Total fertilizer use between 1975 and 1995 increased by a factor of 2.5 in Bangladesh, 2.7 in India, 2.0 in Nepal, 3.6 in Pakistan, and 2.4 in Sri Lanka (Table 14).

The declining yield response to fertilizer is attributed to several factors. A major factor is nutrient imbalance. Data in Table 13 show a preferential use of

Table 13 Trends in Fertilizer Consumption in South Asia (10^6 nutrient tons)

Year	Nitrogen	Phosphate	Potash	Total
1961–1962	0.35	0.08	0.06	0.49
1970–1971	1.91	0.64	0.28	2.83
1980–1981	5.00	1.65	0.71	7.36
1990–1991	9.89	3.81	1.50	15.20
2000–2001	14.55	5.24	1.76	21.55
World total (2000–2001)	81.63	32.66	22.16	136.45
% of world total use in South Asia	17.8	16.0	7.9	15.8

Source: Bumb and Berry (2002).

Table 14   Total Fertilizer Consumption in South Asia (kg/ha)

Country	Consumption		
	1975	1985	1995
Afghanistan	15.0	28.2	17.9
Bangladesh	149.5	260.8	372.6
India	103.6	203.6	277.0
Nepal	53.3	57.1	105.9
Pakistan	40.6	95.9	145.8
Sri Lanka	150.8	335.3	363.4

Sources: Kaosa-ard and Rerkasem (2000); FAO (2000).

nitrogen in comparison to phosphate and potash, creating large deficits in the latter. Government subsidies have kept the price of N low relative to P and K. The recommended ratio of nutrient use for N:P:K is 4:2:1. The actual ratio of use for N:K in India is 10:1 (Stauffer and Sulewski, 2000; Roy, 2003). The balanced use of fertilizer nutrients can greatly enhance rice and wheat production in South Asia. In addition to low yield, net nutrient balances are also causing the depletion of micronutrients and soil organic matter (SOM) content.

Other important causes for the decline in crop yields include numerous soil degradative processes including salinization, waterlogging, crusting, compaction, and pollution of soil by agricultural and industrial pollutants containing heavy metals (e.g., As, Cd, Cu, Hg, Ni, Pb, Zn). In addition to undermining the productive capacity of an ecosystem, soil degradation also adversely impacts the environment, especially with regards to the quality of surface and groundwaters and emission of greenhouse gases (GHGs) into the atmosphere.

IV. ENVIRONMENTAL CONSEQUENCES OF SOIL DEGRADATION

Depletion of SOM content is a serious issue in South Asia. It is exacerbated by removal and/or burning of rice straw and wheat stubble especially in the Indian Punjab region where as much as 80% of the rice straw may be burned. Excessive plowing—2 to 4 times for each of the major crops grown—accentuates mineralization/oxidation of SOM. Land-use and soil/crop practices strongly impact the SOM pool (Swarup et al., 2000). Soil organic matter is also influenced by clay content, nature of clay minerals, and temperature and moisture regimes (Velayutham et al., 2000). Depletion of SOM leads to emission of GHGs into the atmosphere, especially CO_2 and N_2O.

Pachauri and Sridharan (1998) estimate that 265 Mha of India's total land area of 329 Mha is used for agriculture, forestry, pasture, and other biomass production. Areas most affected by degradative processes are croplands and grazing lands estimated at 162 and 11 Mha, respectively (FAO, 2000). With an average loss of 30 to 40 Mg C/ha due to historic land use, soil degradation and fertility mining practices, a crude estimate of the C emitted from soil to the atmosphere is 5 to 7 Pg.[*] Similar magnitude of losses may have occurred from soils of other countries of the region. Therefore, conversion of degraded soils and ecosystems to any restorative land use and adoption of recommended management practices could lead to soil carbon sequestration. It is important to recognize, however, that soil C sequestration is a developmental and policy challenge, and especially challenging for soils of dry and warm climates of the South Asian region.

V. STRATEGIES OF SOIL RESTORATION AND SUSTAINABLE MANAGEMENT OPTIONS IN SOUTH ASIA

Soils of the warm, semi-arid, and arid regions of South Asia are coarse-textured and contain predominantly low-activity clays. Sustainable use of soil and water resources of the region, especially of the rice–wheat zone, is a major issue that must be addressed. Widespread use of monospecific crop systems (rice followed by wheat) under intensive irrigation and fertilizer/chemical use is being re-assessed in terms of long-term sustainability. Alternative management systems need to be developed. Some possible technological options follow.

A. No-Till Farming

In the past, the principal benefits of plowing included temporary weed control and turning under crop residue so that it did not clog or accumulate in front of the seed drill. However, the availability of general-purpose and crop-specific herbicides has minimized the importance of plowing as an option for weed control. Similarly, the availability of appropriate seed drills, specifically equipped with devices to cut through a thick layer of crop residue and seed in a rough seedbed, has provided numerous options for nonconventional systems of seedbed preparation. Converting plow-till to no-till with crop residue mulch is such a promising option. No-till is conservation tillage because the soil is left completely undisturbed from harvesting the previous crop to planting the next one, except for placing the seed and injecting the fertilizer. The previous crop residue is left on

---

[*] Pg = petagram = 10^{15} g = 1 billion tons.

the soil surface as mulch, and weed control is achieved by a judicious combination of herbicides and other crop management practices such as ground cover, allelopathy (García-Barrios, 2003). In fact, conservation tillage is defined as any tillage system that provides at least 30% of the ground cover with crop residue mulch. There are several types of conservation tillage depending on the type of machinery used and the degree of soil disturbance. Ridge till involves preparation of the seedbed in the form of a semipermanent ridge-furrow system, and crop residue mulch is maintained on the surface between ridges. Crops (e.g., wheat) are sown on ridges in a single- or double- row pattern. Zonal tillage or strip tillage involves preparation of a narrow strip as a seed/row zone for creating favourable environments for seed germination and seedling establishment.

The interrow zone, also called the soil and water conservation zone, is kept covered by crop residue mulch. The plow plant and chisel plant systems involve seeding after one plowing and chiseling without additional seedbed preparation. Such systems based on low frequency and intensity of mechanical tillage are also called "reduced-tillage" systems. Despite some soil disturbance in most reduced-tillage systems, an important criterion is to achieve at least 30% ground cover by crop residue mulch. Such systems are also called mulch tillage systems. Most conservation tillage systems are designed to control erosion and conserve water in root zones.

Conservation tillage systems evolved in the midwestern United States during the 1960s (Triplett et al., 1968; Van Doren and Triplett, 1969). Since then, conservation tillage systems have been widely adopted in the United States Cropland area under conservation tillage in the United States was only 2.4 Mha in 1968 and increased to 4.1 Mha in1970, 15.8 Mha in 1980, and 29.7 Mha in 1990 (CTIC, 1997). Data in Table 15 show adoption of conservation tillage in the United States during the 1990s when a new definition/criterion was adopted. Cropland area in the United States planted under some form of conservation tillage was 67.3 Mha in the year 2000. Of this, 20.6 Mha, or 31%, was under no-till (Table 15).

No-till farming in South America has also been widely adopted, especially in Brazil and Argentina (Table 16). Worldwide, about 48 Mha, or 3.5% of the world's cropland, was planted using a no-till system during 1998–1999, and the area under no-till increased to 5.1% in 2001–2002. Several chapters in this book address the adoption of no-till to sowing wheat following rice in South Asia. Principal advantages are the time saved between the harvesting of rice and the sowing of wheat, savings in fuel costs for seedbed preparation, and improvements in soil quality.

Conversion of plow-till to no-till for wheat, with continued intensive puddling for rice, may not result in significant improvements in soil structure and SOM concentration, especially if most of the wheat residue is removed for fodder and the rice residue is either burned or removed. Conversion from plow-till to

Table 15 Conservation Tillage in the United States (Mha)

Year	No-till	Conservation tillage	Reduced tillage	Total
		Cropland area under conservation tillage		
1990	6.8	22.8	28.7	58.3
1992	11.4	24.5	29.7	65.6
1994	15.7	24.5	29.6	69.8
1996	17.4	24.6	30.3	72.3
1998	19.3	24.9	31.6	75.8
2000	20.6	22.4	24.3	67.3

Sources: CTIC (2000); Köller (2003).

no-till for wheat production with no use of crop residue as mulch is not conservation tillage or no-till farming. Often rice straw is burned or removed because of the lack of appropriate seeding equipment that can cut through the residue and sow seed in a rough seedbed. The continued development and adaptation of drills for specific ecological conditions of South Asia remains a high priority.

Table 16 No-Till Farming in the World

Country	Area under no-till in the world (Mha)	
	1998–1999	2001–2002
Argentina	7.3	13.0
Australia	1.0	9.0
Bolivia	0.2	0.4
Brazil	11.2	17.4
Canada	4.1	4.1
Chile	0.01	0.13
Mexico	0.05	0.05
Paraguay	0.8	1.3
Spain	—	1.0
Uruguay	0.05	0.05
United State of America	19.3	22.4
Others	1.0	1.6
Total	45.0	70.2
World cropland area	1379	1369
% world cropland planted to no till	3.3%	5.1%

Sources: FAO (1998); Derpsch (2001); Hernanz et al. (2002); Derpsch and Benites (2003).

B. Viable Alternative to the Rice–Wheat System

Intensive rice–wheat cropping has adversely affected soil quality. It has depleted the groundwater reserve in some regions, and it has raised the water table in others. It is important, therefore, to find viable economic alternatives. Growing other crops such as cotton, canola, sunflower, sugarcane, maize, and vegetables within a rotation system may be extremely important. Rice has a high water requirement, and finding water-efficient substitutes remains a high priority.

Identifying compatible species, which can be grown either as mixed crops or in relay crop combinations, is another important consideration. The benefits of mixed versus mono cropping depend on a wide range of interacting factors such as species, relative density, soil conditions, irrigation frequency and method, fertilizer type, and the rate. The goal is to identify species that have a "facilitative effect" and "asymmetric interference" (García-Barrios, 2003).

C. Agroforestry

Agroforestry is a mixed cropping system that involves growing annuals in combination with perennials. Growing wheat in combination with poplar is a successful agroforestry system in Punjab, India (Fig. 2). Appropriate tree species within a suitable agroforestry system can provide forest products such as timber, fuel, fodder, and industrial raw material such as pulp for paper. Trees can be grown at all levels of agroforestry intensity. Soil carbon sequestration may be another important environmental benefit of an appropriate agroforestry system, especially if fast-growing and leguminous tree species are used.

VI. RECONCILING FOOD SECURITY WITH ENVIRONMENT QUALITY

Benefits of Green Revolution technologies must be preserved for input- responsive soils and in areas of irrigated agriculture by using no-till farming, mixed cropping, agroforestry, and other viable alternatives to rice–wheat systems. Some land area devoted to rice cultivation must also be converted to cash crops and water-efficient production systems. Benefits of Green Revolution technologies must be extended to soils of low inherent fertility and in areas of rainfed agriculture by restoring degraded/depleted soils and adopting improved systems of soil and crop management, including conservation-effective measures and on-farm systems of water management. To meet the food demands of increasing populations, yields of all crops must be increased by 50% to 75% for the South Asian region. These yield increases are attainable even through the adoption of existing

Figure 2 Wheat growing under poplar plantation in Punjab, India.

technologies (Borlaug, 2002). In addition to improved varieties and genetically engineered crops with built-in resistance to diseases and pests, a wide range of soil and water management technologies is appropriate for the South Asian region (Table 17). These technologies require soil-site-specific adoption through on-farm and participatory research. Adoption of such technologies can increase production, improve water quality, and reduce pollution/contamination of surface and groundwater.

VII. EMISSION OF GREENHOUSE GASES FROM AGRICULTURAL ACTIVITIES IN SOUTH ASIA

Agricultural activities lead to emissions of three principal GHGs: CO_2, CH_4, and N_2O. Emission of CO_2 is caused by oxidation/mineralization of soil organic matter through plowing, low or unbalanced use of fertilizers, removal/burning of crop residues, and low or no rates of applications of biosolids including compost, green manure, and sludge, etc. Emission of CH_4 is caused by rice paddy cultivation (Aulakh et al., 2001a; Adhya et al., 2000; Garg et al., 2001), from use of biosolids and manure, biomass burning, and anaerobiosis in soil caused by

Table 17 Technological Options for Sustainable Management of Soil and Water Resources in South Asia

Natural resource	Technological option
Soil	Conservation tillage based on mulch farming and use of appropriate seeding equipment
	Integrated nutrient management based on balanced application of essential macro- and micronutrients, and adopt technologies to enhance fertilizer use efficiency
	A complete ban on burning crop residue, using topsoil for brick making, and broadcasting fertilizers
	Precision farming based on soil-specific nutrient management
Crop	Use of improved and genetically engineered varieties
	Adoption of mixed and relay cropping systems including agroforestry where feasible
	Conversion of some area under rice paddy to other crops, and developing direct seedling technique for rice cultivation
	Improving shelf life of perishable agricultural products through better storage and post-harvest technologies
Water	Avoid excessive use of irrigation water, and provide provision for drainage of excess water
	Replace flooding by furrow irrigation, subirrigation, and drip irrigation (fertigation)
	Enhance water-use efficiency by decreasing seepage (delivery) and evaporation losses
	Use good-quality water for irrigation
Climate	Minimize emission of greenhouse gases (CO_2, CH_4, N_2O) into the atmosphere
	Enhance terrestrial carbon sequestration in soil and biota
	Enhance formation of secondary carbonates especially in irrigated soils
	Decrease emission of N_2O by improving N use efficiency and minimizing volatilization

waterlogging. Methane emission from rice paddies can be reduced by using nitrification inhibitors (Bharati et al., 2000) and appropriate varieties (Wang et al., 1997). Fertilizers and manures are the principal causes of N_2O emission. Emissions of N_2O from irrigated rice fields can be decreased by using nitrification inhibitors (Majumda et al., 2000) and by appropriate crop residue management (Aulakh et al., 2001b). Improving nutrient use efficiency of fertilizers and manures is important in order to reduce N_2O emissions from soils.

REFERENCES

Adhya, T.K., Misra, S.R., Rath, A.K., Bharati, K., Mohanty, S.R., Ramakrishnan, B., Rao, V.R., Sethunathan, N. 2000. Methane efflux from rice-based cropping systems under humid tropical conditions of eastern India. *Agric. Eco. & Env* 79:85–90.

Aulakh, M.S., Wassmann, R., Rennenberg, H. 2001a. Methane emissions from rice fields, role of management and mitigation options. *Adv. Agron* 70:193–260.

Aulakh, M.S., Khera, T.S., Doran, J.W., Bronson, K.F. 2001b. Denitrification, N_2O and CO_2 fluxes in rice-wheat cropping systems as affected by crop residues, fertilizer N and legume green manure. *Biology & Fertility of Soils* 34:375–389.

Bharati, K., Mohanty, S.R., Padmavathi, P.V.L., Rao, V.R., Adhya, T.K. 2000. Influence of six nitrification inhibitors on methane production in a flooded alluvial soil. *Nut. Cycling Agroec* 58:389–394.

Borlaug, N.E. 2002. Feeding a world of 10 billion people: The miracle ahead. In:. Bailey R.(), Ed. Global Warming and Other Eco-Myths: How the Environment Movement Uses False Science to Scare Us to Death. Forum: Prima Publishing, New York, 29–59.

Bumb, B.L., Berry, J.T. 2002. Global and regional data on fertilizer production and consumption 1961/62–2000/2001: IFDC. Muscle Shoals, AL.

Conway, G, Toenniessen, G. 1999. Feeding the world in the 21st century. *Nature* 402: C55–C58.

CTIC. 1997. National survey of conservation tillage practices: Conservation Tillage Information Center. West Lafayette, IN.

CTIC. 2000. Conservation Tillage Information Center: National Crop Residue Management Survey.

Derpsch, R., Benites, J. 2003. Situation of conservation agriculture in the world. IInd World Congress on Conservation Agriculture. 11–15 August 2003, Iguassu Falls, Parana, Brazil.

Derpsch, R. 2001. Conservation tillage, no tillage and related technologies. Proc. 1st World Congs. Conserv. Agric., Madrid, October 1–5, 2001, Conservation Agriculture: A Worldwide Challenge Vol I:161–170.

Engelman, R., P., LeRoy 1995. Conserving Land: Population and Sustainable Food Production: Population Action International. Washington, DC.

FAO. 1982. Production Supporting Capacities of Land in the Developing World: FAO. Rome. Italy.

FAO. 1990. Water and Sustainable Agricultural Development. Mar del Plata Action Plan for the 1990s: FAO. Rome. Italy.

FAO. 1994. Land Degradation in South Asia: Its Severity, Causes, and Effects upon People. World Soil Resources Report 78: FAO. Rome. Italy.

FAO. 2000. Production Yearbook: FAO. Rome. Italy.

Garcia-Barrios, L. 2003. Plant–plant interactions in tropical agriculture. In:. Vandermeer J.H., Ed. Tropical Agroecosystems: CRC Press. Boca Raton, FL, 11–58.

Gardner-Outlaw, T., Engelman, R. 1997. Sustaining Water, Easing Scarcity: A Second Update: Population Action International. Washington, D.C..

Garg, A., Bhattacharya, S., Shukla, P.R., Dadhwal, W.K. 2001. Regional and sectoral assessment of greenhouse gas emissions in India. *Atm. Env* 35:2679–2695.

Hazel, P.B.R., Ramasamy, C. 1991. The Green Revolution Reconsidered: The Impact of High Yielding Varieties in South Asia: Johns Hopkins Univ. Press. Baltimore.

Hernanz, J.L., Lopez, R., Navarrette, L., Sánchez-Girón, V. 2002. Long-term effects of tillage systems and rotations on soil structural stability and organic carbon stratification in semi-arid central Spain. *Soil & Tillage Res* 66:129–141.

Kaosa-Ard, M.S., Rerkasem, B. 2000. The Growth and Sustainability of Agriculture in Asia. Asian Development Bank. Manila, Philippines.

Köller, K. 2003. Techniques of soil tillage. In:. Titi A.E., Ed. Soil Tillage in Agroecosystems: CRC Press. Boca Raton, FL, 1–25.

Majumdar, D., Kumar, S., Pathak, H., Jain, M.C., Kumar, U. 2000. Reducing nitrous oxide emission from an irrigated rice field of North India with nitrification inhibitors. *Agric. Eco. & Env* 81:163–169.

Pachauri, P.K., Sridharan, P.V. 1998. Looking back to think ahead. Green India 2047: TERI. New Delhi.

Pingali, P.L., Heisey, P.W. 1999. Cereal productivity in developing countries: Past trends and future prospects. CIMMYT Economic Paper 99–03: CIMMYT. Mexico.

Roy, A.H. 2003. Fertilizer needs to enhance production-challenges facing India. In:. Lal R., Hansen D., Uphoff N., Slack S., Eds. Food Security and Environmental Quality in the Developing World: CRC/Lewis Publishers. Boca Raton, FL, 53–68.

Stauffer, M., Sulewski, G. 2000. Asia's Potential for Fertilizer Use. The Fertilizer Institute Outlook 2001. Alexandria, VA.

Swaminathan, M.S. 2000. Science in response to basic human needs. *Science* 287:425.

Swarup, A., Manna, M.C., Singh, G.B. 2000. Impact of land use and management practices on organic carbon dynamics in soils of India. In:. Lal R., Kimble J.M., Stewart B.A., Eds. Global Climate Change and Tropical Ecosystems: CRC/Lewis Publishers. Boca Raton, FL, 261–281.

Triplett, G.B., Van, D.M., Doren, Jr., Schmidt, B.L. 1968. Effect of corn (*Zea mays*) stover mulch on no-tillage corn yield and water infiltration. *Agron. J*: 236–239.

Twyford, I. 1994. Fertilizer use and crop yields. Paper presented at the 4th National Congress of Soil Science Society. Islamabad. Pakistan 1992.

Vandermeer, J. 2003. Introduction. In:. Vandermeer J.H., Ed. Tropical Agroecosystems: CRC Press. Boca Raton, FL, 1–9.

Van Doren, D.M., Triplett, G.B. 1969. Mechanism of corn response to cropping practices without tillage. Ohio Agric. Res. & Dev. Center Circular 169. Wooster, OH.

Velayutham, M., Pal, D.K., Bhattacharyya, T. 2000. Organic carbon stock in soils of India. In:. Lal R., Kimble J.M., Stewart B.A., Eds. Global Climate Change and Tropical Ecosystems: CRC/Lewis Publishers. Boca Raton, FL, 71–95.

Wang, B., Neue, H.U., Samonte, H.P. 1997. Effect of cultivar difference on methane emission. *Agric. Eco. & Env* 62:31–40.

3
Food Security and Environmental Sustainability

David Pimentel
Cornell University
Ithaca, New York, U.S.A.

Food security depends on an abundance of food, an even distribution, and an ability to purchase food, as well as political stability. Unfortunately, in the world today about 3 billion people are malnourished because of shortages of calories, protein, several vitamins, iron, and iodine (WHO, 2000). People can die because of shortages of any one of these nutrients. The 3 billion malnourished is the largest number of people ever affected in history.

Today more than 6.3 billion humans inhabit our planet (PRB, 2003). Based on current rates of increase, the world population is projected to double to more than 12 billion in less than 50 years (PRB, 2003). At a time when the world population continues to expand at a rate of 1.3% per year, adding more than a quarter million people daily, providing adequate food becomes an increasingly severe problem. Conceivably, the numbers of malnourished will reach 5 billion in future decades.

Reports from the FAO of the United Nations and the U.S. Department of Agriculture, as well as numerous other international organizations, further confirm the serious nature of the global supply. For example, the per capita availability of world cereal grains, which make up 80% of the world's food supply, has been declining for nearly two decades (Fig. 1). These shortages are not reflected in the price of cereal grains, because the poor people cannot afford to purchase grains despite the fact that prices are relatively low (Pimentel and Pimentel, 1996).

Malnourished people are more susceptible to numerous diseases, like malaria, tuberculosis, schistosomiasis, and AIDS. The World Health Organization reports that more than 2.4 billion people are infected with malaria, 2 billion are

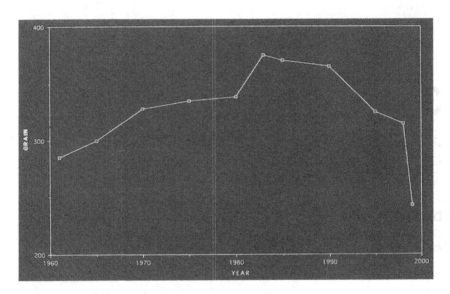

Figure 1 Cereal grain production per capita in the world from 1961 to 2000. (From FAO, 1961–2000. *Quarterly Bulletin of Statistics*, 1–13).

infected with tuberculosis, 600 million are infected with schistosomiasis, and 40 million are infected with AIDS (NIAID, 2002).

Thus as the world population continues to expand, greater pressure than ever before is being placed on all basic resources that are essential for food production and protection from diseases. Unfortunately, while the human population grows exponentially, food production only increases linearly. Furthermore, degradation of land, water, energy, and biological resources vital to agricultural sustainability continues unabated (Pimentel and Pimentel, 2003).

I. AGRICULTURAL RESOURCES

More than 99.7% of the world's food supply comes from the land, while less than 0.3% is from oceans and other aquatic habitats (FAO, 1998). Thus the continued production of an adequate food supply is directly dependent on ample quantities of fertile land, fresh water, energy, and natural biodiversity. Obviously, as the human population grows, the requirements for all these resources will escalate, because each person added to the population requires land, water, energy, biological resources. Even if these resources are never completely depleted and are managed in a sustainable manner, on a per capita basis, their supply will decline significantly because they must be divided among more and more people.

A. Land

Throughout the world fertile cropland is being lost from production at an alarming rate. True for all cropland, it is illustrated by the diminishing amount of land now devoted to cereal grains (Lal et al., 2002). Soil erosion by wind and water as well as overuse are responsible for the loss of about 30% of the world cropland during the past 40 years (WRI, 1994; Pimentel et al., 1995). Once fertile soil is lost, it takes about 500 years or more to replace a mere 25 mm of fertile soil. For crop production, at least 150 mm of soil is required.

The impact of wind erosion is illustrated by a photograph taken by NASA showing soil being blown from Africa toward the United States and South America (Fig. 2). Soil erosion by wind and water exposes the soil organic matter to rapid oxidation and thus contributes to the climate change and global warming (Lal et al., 1999).

Most replacement for eroded and unproductive agricultural land is coming from cleared forestland and marginal land. The need for more cropland accounts for more than 60% of the world's deforestation (Myers, 1994). Despite such land replacement strategies, world cropland per capita is declining and now stands at

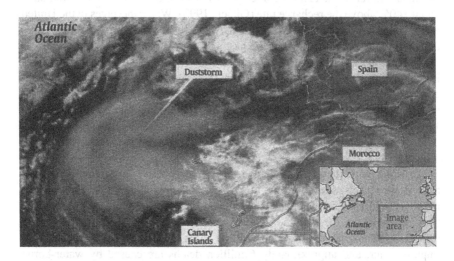

Figure 2 Cloud of soil from Africa being blown across the Atlantic Ocean. (Imagery by SeaWIFS Project, NASA/Goddard Space Flight Center and ORBIMAGE, 2000.)

only 0.25 ha per capita, or about 50% of the 0.5 ha per capita considered the minimum for the production of a diverse diet similar to that of the United States and Europe (Lal and Stewart, 1990). China now has only 0.08 ha per capita, or about 15% of the 0.5 ha per capita that is the accepted minimum.

B. Water

Rainfall and its collection in rivers, lakes, and vast underground aquifers provide the water needed by humans for their survival, agriculture, and diverse activities. Fresh water is critical for all vegetation, including crops. All use and transpire massive amounts of water during the growing season. For example, a hectare of corn, producing about 8,000 kg/ha, will transpire more than 5 million liters of water during one growing season (Pimentel et al., 1997). This means that about 10 million liters of water must reach each hectare of corn during the season. In total, agricultural production consumes more fresh water than any other human activity. Specifically, about 70% of the world's freshwater supply is consumed—that is, used up by agriculture—and thus is unavailable for other uses (Postel, 1999).

Water resources are under great stress as populous cities, states, and countries increase their withdrawal of water from rivers, lakes, and aquifers every year. For example, by the time the Colorado River reaches Mexico, it has almost disappeared before it finally trickles into the Gulf of Cortes. Also, the great Ogallala aquifer in the central United States is suffering an overdraft rate that is about 140% above the recharge rate (Gleick, 1993). Water shortages in the United States and elsewhere in the world already are reflected in the 10% per capita decline in crop irrigation during the past 10 years (Brown, 1997).

To compound the water problem, about 40% of the world population lives in regions that directly compete for shared water resources (Gleick, 1993; Gleick et al., 2002). In China, for example, more than 300 cities already are short of water, and these shortages are intensifying as Chinese urban areas expand (WRI, 1994). Competition for water resources among individuals, industries, regions both within and between countries is growing throughout the world community (Gleick, 1993; Gleick et al., 2002).

Along with the quantity of water, its purity also is important. Diseases associated with impure water and unsanitary systems rob people of their health, nutrients, and livelihood. These problems are most serious in developing countries, where about 90% of the diseases can be traced to a lack of pure water (WHO, 1992). Worldwide, about 4 billion cases of disease are contracted from impure water, and approximately 6 million deaths are caused by water-borne disease each year (Pimentel et al., 1998). Furthermore, when a person is stricken with diarrhea, malaria, or other serious disease, from 5%–20% of an individual's food intake is used by the body to offset the stress of the disease, further diminishing the benefits of his/her food (Pimentel et al., 1998).

In India, little clean water is available. For example, out of 3,120 cities in India, only 8 have full treatment of their waste sewage (Pimentel et al., 1998). According to the World Health Organization, virtually all of India's rivers are polluted (Zubrzycki, 1997). Typically in most developing countries, 95% of wastewater is dumped into rivers and lakes without treatment. A recent estimate indicates that more than 1 billion people throughout the world do not have access to clean water (Leshner, 2002).

Disease and malnutrition problems appear to be particularly serious, in the Third World where poverty and poor sanitation are endemic (Shetty and Shetty, 1993). The number of people living in urban areas is doubling every 10 to 20 years, creating other environmental problems, including a lack of water and sanitation, increased air pollution, and increased food shortages. For these reasons, the potential for the spread of disease is great in urban areas (*Science*, 1995).

C. Energy

Energy from many sources but especially fossil energy is a prime resource used in world food production (Pimentel et al., 2002). Nearly 75% of the fossil energy used each year throughout the world is consumed by populations living in developed countries. Of this, about 17% is expended in the production, processing, and packaging of food products (Pimentel and Pimentel, 1996). In particular, the intensive farming technologies characteristic of developed countries rely on massive amounts of fossil energy for fertilizers, pesticides, and irrigation and for machines that substitute for human labor. In contrast, developing countries use fossil energy primarily for fertilizers and irrigation to help maintain yields, rather than to reduce human labor inputs (Giampietro and Pimentel, 1993; Pimentel et al., 2002).

Because fossil energy is a finite resource, its depletion accelerates as populations expand and their food requirements increase. The United States is now importing 61% of its oil, and the projection is that the United States will be importing 90% or more in the next 20 years (BP, 2001; USBC, 2001). Consider that at present, the United States is importing more than 50% of its oil. To sustain its energy-based activities, U.S. oil imports will have to increase in future decades, further worsening the U.S. trade imbalance. The cost of fuel also will increase. The impact of price increases already is a serious problem for developing countries, where the relatively high price of imported fossil fuel makes it difficult, if not impossible, for poor farmers to power irrigation as they try to sustain needed harvests (Pimentel et al., 1997a).

Worldwide, per capita use of fertilizer has declined about 23% during the past decade (TFI, 1997). Supplies of fossil energy for fertilizers and irrigation show a significant decline, and this trend can be expected to continue. Furthermore, the current decline in per capita use of fossil energy, caused by the decline in oil supplies and increasing prices, is generating direct competition between developed and developing countries for fossil energy resources.

1. Energy Inputs in Crop Production

Cereal grains provide more than 80% of the world's food (Pimentel et al., 1999). Corn is one of the world's major grain crops (FAO, 1997). Under favorable soil, rainfall, and temperature conditions, corn is one of the most productive crops per unit area of land. An analysis of the energy inputs and yield (Table 1) suggests that the high yields of intensive corn production are in part related to the large inputs of fertilizers and pesticides. For every kcal of energy invested, about 4 kcal of corn are produced in this intensive system.

Producing corn employing irrigation requires enormous amounts of water plus the expenditure of energy to pump and apply the water (Postel, 1999). For example, a corn crop grown in an arid region requires about 1 meter of irrigated water per hectare and costs about $1,000 just for the water (Table 2). More than 13 million kcal of fossil fuel is required to pump and apply this water. Thus, for every kcal of fossil energy invested, only 1.5 kcal of corn are produced in this intensive crop production system.

D. Pests and Pesticides

Worldwide, an estimated 67,000 different pests attack and damage agricultural crops (Pimentel, 1997). In general, less than 10% are considered major pests. In most instance, the pests specific to a particular region have moved from feeding on native vegetation to feeding on crops that were introduced into the region (Pimentel, 1997). It should be noted that 99% of all crops are introduced species in each country (Pimentel, 2002).

Despite the yearly application of about 3 million metric tons of pesticides used in all countries in the world (Table 3), and at an expense of about $40 billion for the purchase and application of pesticides, pests in the world destroy more than 40% of potential food production (Pimentel, 1997). Worldwide, pesticide use has increased in total quantities from 1950 to 2000 (Table 4). If the toxicity of the newer pesticides is taken into account starting in 1980, the increased equivalent toxicity would be 10 to 20 times higher in the United States and world. Instead of applying many early pesticides at kilogram dosages per hectare, the newer pesticides are being applied at only gram dosages.

Worldwide, insect pests cause about a 15% loss, plant pathogens 12%, and weeds 13% (Pimentel, 1997). In the United States, crop losses are nearly the same as in the world, insect pest losses are 13%, plant pathogens 12%, and weeds

Table 1 Energy Inputs and Costs of Rainfed Corn Production per Hectare in the United States

Inputs	Quantity	kcal × 1000	Costs
Labor	6.2 hr[q]	250[f]	$62.00[h]
Machinery	55 kg[a]	1,018[e]	103.21[m]
Diesel	90 L[b]	900[e]	23.40[r]
Gasoline	56 L[b]	553[e]	14.60[r]
Nitrogen	148 kg[c]	2,738[s]	81.40[i]
Phosphorus	53 kg[c]	219[g]	12.72[i]
Potassium	57 kg[r]	186[g]	17.67[i]
Lime	699 kg[c]	220[e]	14.00[n]
Seeds	21 kg[a]	520[e]	74.00[c]
Herbicides	2.1 kg[c]	210[e]	21.00[j]
Insecticides	0.15 kg[c]	15[e]	6.00[l]
Electricity	13.2 kWh[b]	34[e]	2.38[k]
Transportation	222 kg[d]	268[e]	66.60[o]
Total		7,131	$498.98
8,590 kg yield[p]		30,924	
		kcal input: output = 1 : 4.3	

[a] Pimentel and Pimentel (1996).
[b] USDA (1991).
[c] USDA (1997).
[d] Goods transported include machinery, fuels, and seeds that were shipped an estimated 1,000 km.
[e] Pimentel (1980).
[f] It is assumed that a person works 2,000 hr per year and utilizes an average of 8,100 liters of oil equivalents per year.
[g] FAO (1999).
[h] It is assumed that farm labor is paid $10 per hour.
[i] *Soil Fertility Guide* (2002).
[j] It is assumed that herbicide prices are $10 per kg.
[k] Price of electricity is 7¢ per kWh (USBC, 1998).
[l] It is assumed that insecticide prices are $40 per kg.
[m] Hoffman et al. (1994).
[n] Assumed to be 2¢ per kg (Clary and Haby, 2002).
[o] Transport was estimated to cost 30¢ per kg.
[p] USDA (2001).
[q] NASS (1999).
[r] Diesel and gasoline assumed to cost 26.5¢ per liter.
[s] An average of energy inputs for production, packaging, and shipping per kg of nitrogen fertilizer from FAO (1999), Duffy (2001), and Fertilizer (2002).

Table 2 Energy Inputs and Costs of Corn Production per Hectare in the United States Using Pump Irrigation

Inputs	Quantity	kcal × 1000	Costs
Labor	6.2 hrs[q]	250[f]	$62.00[h]
Machinery	55 kg[a]	1,018[e]	103.21[m]
Diesel	90 L[b]	900[e]	23.40[t]
Gasoline	56 L[b]	553[e]	14.60[t]
Nitrogen	148 kg[c]	2,738[y]	81.40[i]
Phosphorus	53 kg[c]	219[g]	12.72[i]
Potassium	57 kg[c]	186[g]	17.67[i]
Lime	699 kg[c]	220[e]	14.00[n]
Seeds	21 kg[a]	520[e]	74.00[c]
Irrigation	100 cm[s]	13,400[r]	1000.00[u]
Herbicides	2.1 kg[c]	210[e]	21.00[j]
Insecticides	0.15 kg[c]	15[e]	6.00[l]
Electricity	13.2 kWh[b]	34[e]	2.38[k]
Transportation	222 kg[d]	268[e]	66.60[o]
Total		20,531	$1498.98
	8,590 kg yield[p]	30,924	
		kcal input: output = 1 : 1.51	

[a] Pimentel and Pimentel (1996).
[b] USDA (1991).
[c] USDA (1997).
[d] Goods transported include machinery, fuels, and seeds that were shipped an estimated 1,000 km.
[e] Pimentel (1980).
[f] It is assumed that a person works 2,000 hr per year and utilizes an average of 8,100 liters of oil equivalents per year.
[g] FAO (1999).
[h] It is assumed that farm labor is paid $10 per hour.
[i] *Soil Fertility Guide* (2002).
[j] It is assumed that herbicide prices are $10 per kg.
[k] Price of electricity is 7¢ per kWh (USBC, 1998).
[l] It is assumed that insecticide prices are $40 per kg.
[m] Hoffman et al. (1994).
[n] Assumed to be 2¢ per kg (Clary and Haby, 2002).
[o] Transport was estimated to cost 30¢ per kg.
[p] USDA (2001).
[q] NASS (1999).
[r] Batty and Keller (1980).
[s] USDA (1997).
[t] Diesel and gasoline assumed to cost 26.5¢ per liter.
[u] Irrigation for 100 cm of water per hectare costs $1,000 (Larsen et al., 2002).
[v] An average of energy inputs for production, packaging, and shipping per kg of nitrogen fertilizer from FAO (1999), Duffy (2001), and Fertilizer (2002).

Table 3 Estimated Annual Pesticide Use for 1950, 1980, and 2000 for the United States and World (million metric tons)

Region	1950	1980	2000
United States	0.2	0.5	0.5
World	0.4	2	3

12%. It should be pointed out that crop loss estimates in each country are based on the "cosmetic standards" that exist in that particular country. For example, most of the fruits and vegetables sold in the India and Guatemala markets would not be salable in the U.S. market. It is estimated that from 10% to 20% of the pesticide used in the United States is wasted, achieving impractical, high cosmetic standards in fruits and vegetables.

E. Fertilizer Use

After World War II, commercial nitrogen, phosphorus, and potassium fertilizers started to be applied to U.S. and world croplands. Initially the amounts were extremely small. In 1945, for example, only about 8 kg of nitrogen was applied to U.S. corn hectares (Pimentel and Pimentel, 1996) (Table 5). The quantity of nitrogen fertilizer applied to U.S. corn increased to about 150 kg/ha by 1985 and continues near this rate today (Table 5). Commercial phosphorus and potassium

Table 4 Estimated Annual Pesticide Use

Country/region	Pesticide use (10^6 metric tons)
United States	0.5
Canada	0.2
Europe	1.0
Other, developed	0.5
Asia, developing	0.3
China	0.2
Latin America	0.2
Africa	0.1
Total	3.0

Table 5 Quantities of Commercial Fertilizers Applied to U.S. Corn Land per Hectare Starting in 1945 up to 2000

Fertilizers	1945	1954	1964	1975	1985	2000
Nitrogen kg	8	30	55	111	152	147
Phosphorus kg	8	13	36	56	58	63
Potassium kg	6	20	28	62	75	55
Corn yield kg	2,132	2,572	4,265	5,143	7,400	8,000

Sources: Pimentel et al. (1990, 2002).

fertilizer nutrients also increased from 6 to 8 kg/ha to about 60 kg/ha today (Table 5).

Total fertilizer use in the United States increased from 10 million metric tons in 1950 to about 18 million tons in 2000 (Table 6). The increase in the use of world fertilizer during this same time period has been significantly greater than in the United States, growing in use from 30 million tons in 1950 to about 140 million tons today (Table 6).

The use of commercial fertilizers eliminated the need for 2 hectares to produce 1 hectare of corn. One hectare in a legume to produce nitrogen nutrients for corn production was no longer needed. Of course, large quantities of fossil energy are necessary to produce fertilizers, especially nitrogen fertilizer. For example, the production of 1 kg of nitrogen fertilizer requires about 18,690 kcal, or slightly more than the energy in 2 liters of gasoline (FAO, 1999).

Also, to illustrate how rapidly agricultural technology has changed once commercial fertilizers were discovered and produced, the energy inputs just for the nitrogen fertilizer today per hectare is greater than the total fossil energy inputs in producing corn in 1945. Corn yields have increased dramatically since 1945, when corn yields were only about 1,500 kg/ha. Corn yields today in the United States are nearly 9,000 kg/ha (USDA, 2001).

Worldwide, about 140 million metric tons of fertilizer are applied to crops (Table 6). These fertilizers are vital to crop production. Even with all the nitrogen,

Table 6 Estimated Annual Fertilizer Use for 1950, 1980, and 2000 for the United States and World (million metric tons)

Region	1950	1980	2000
United States	10	17	18
World	30	110	140

phosphorus, potassium, and other nutrients applied in world agriculture, the World Health Organization (2000) reports that more than 3 billion people are malnourished in the world. In addition, the Food and Agricultural Organization (FAO, 1961–2000) reports that per capita grain production has been declining for nearly two decades. In addition to shortages of cropland and irrigation water, per capita fertilizer use during the past decade has declined by more than 20% (Pimentel et al., 1999).

II. RICE–WHEAT SUSTAINABILITY ISSUE

No one knows exactly why problems exist with the rice–wheat production system in Asia. To improve the sustainability of the rice–wheat system, the following is suggested: (1) reduce soil erosion to about 1 t/ha/yr; (2) reduce water runoff, and this would be achieved in part by reducing soil erosion, (3) increase the soil organic matter by increasing the biomass production; (4) increase the diversity of crops grown in the region; and (5) increase the addition and use of livestock manure in the cropping system. Any one of these factors may not solve the sustainability problem, but the combination of all five factors would help improve the sustainability of the rice–wheat system.

III. BEEF PRODUCTION

Producing 1 kg of beef requires larger inputs of land, water, and energy than producing 1 kg of vegetable protein (Pimentel, 2003). In the United States, feedlot beef requires about 13 kg of grain and 30 kg of forage per 1 kg of beef produced (Table 7). This much feedlot beef requires approximately 43,000 liters of water (Table 7). On range land, about 200 kg of forage are required to produce 1 kg of beef or more than 200,000 liters of water per 1 kg of beef (Table 7).

Fossil energy inputs for beef production are also significant. About 57 kcal of fossil energy are required to produce 1 kcal of beef protein when produced under feedlot conditions (Table 7). Under range condition the energy required is

Table 7 Resources Inputs per Kilogram of Beef Production

Production system	Grain (kg)	Forage (kg)	Water (liters)	Energy (kcal)/ (1 kcal beef protein)
Feedlot beef	13	30	43,000	57
Range beef	0	200	200,000	20

about one third less than that of feedlot beef, or 20 kcal of fossil energy per 1 kcal of been protein (Table 7).

IV. BIODIVERSITY LOSS

A productive and sustainable agricultural system, indeed the quality of human life, also depends on maintaining the integrity of natural biodiversity that exists on earth. Often small in size, diverse species serve as natural enemies to control pests, help degrade wastes, improve soil quality, fix nitrogen for plants, pollinate crops and other vegetation, and provide numerous other vital services for humans and their environment (Pimentel et al., 1997b). Consider that in New York State on a bright sunny day in July, the wild and other bees pollinate an estimated 1,000,000 trillion blossoms so essential for the production of fruits and vegetables and other plants. Humans have no technology to substitute for this task and many of the other contributions provided by the estimated 10 million species that inhabit the earth.

The rapidly expanding human population and the expansion of diverse activities throughout the world disturb natural and managed ecosystems and thereby reduce biodiversity. The current extinction rate of species ranges from approximately 1,000 to 10,000 times higher than their natural extinction rate. If this worldwide trend continues, by the middle of this century as many as 2 million species worldwide will become extinct.

Diverse plants, wild and cultivated, provide food for humans, animals, and microbes. Although approximately 90% of the world food supply comes from 15 plant and 8 animal species, several thousand other plant species are also eaten by humans.

V. FOOD DISTRIBUTION

Some people assume that market mechanisms and international trade are effective insurance against future food shortages. However, when all nations reach the biological and physical limits of domestic food production, food importation will no longer be a viable option for any country, because at that point food importation by wealthy countries could only be sustained by starvation of the poor in other countries. In the final analysis, the existing biological and physical resource constraints regulate and limit all food production systems.

These concerns about the future are supported by two observations. First, most of the 183 nations of the world are now dependent on food imports. Most of these imports are cereal grain surpluses produced only in those countries that now have relatively low population densities, where intensive agriculture is practiced and where surpluses are common. For instance, the United States, Canada,

Australia, and Argentina provide about 80% of the cereal exports on the world market (USBC, 2001). This situation is expected to change when the U.S. population doubles in the next 70 years (USBC, 2001). Then based on this projection, instead of exporting cereals and other food resources, these foods will have to be retained domestically to feed approximately 570 million hungry Americans. The United States, along with other exporting countries, will cease to be a food-exporting country.

In the future, when the four major exporting countries retain surpluses for home use, Egypt, Jordan, and countless other countries in Africa and Asia will be without food imports that are basic to their survival. China, which now imports many tons of food, illustrates the severity of this problem. If, as Brown (1995) predicts, China's population increases by 500 million beyond their present 1.3 billion and their soil erosion continues unabated, it will need to import 200 to 400 million tons of food grains each year starting in 2050. This minimal quantity is equal to more than the current grain exports of *all* the exporter nations mentioned earlier (USBC, 2001). Based on realistic trends, sufficient food supplies probably will not be available for import by China or any other nation on the international market to import by 2050 (Brown, 1995).

VI. TECHNOLOGY

Over time, technology has been instrumental in increasing industrial and agricultural production, improving transportation and communications, advancing human health care, and overall improving many aspects of human life. However, much of its success is based on the availability of the natural resources of the earth.

In no area is this more evident than in agricultural production. No known or future technology will be able to double the world's arable land. Granted, technologically produced fertilizers are effective in enhancing the fertility of eroded croplands, but their production relies on the diminishing supply of fossil fuels.

The increase in the size and speed of fishing vessels has not resulted in increases in per capita fish catch (Pimentel and Pimentel, 1996). To the contrary for example, in regions like eastern Canada, overfishing has become so severe that about 80,000 fishermen have no fish to catch, and the entire industry has been lost (W. Rees, University of British Columbia, personal communication, 1996).

Consider also the world supplies of fresh water that are available must be shared by more individuals, and for increased agriculture, and for industry. No available technology can double the flow of the Colorado River — the shrinking groundwater resources in vast aquifers cannot be refilled by human technology. Rainfall is the only supplier of fresh water in the world.

Certainly, improved technology will help increase food production. This includes the more effective management and use of resources, but it cannot produce an unlimited flow of those vital natural resources that are the raw material for sustained agricultural production. Where is this technology and why has it not been employed now that cereal grain production per capita has been declining for nearly 20 years and continues to decline (Fig. 1)?

Biotechnology has the potential for some advances in agriculture, provided its genetic transfer ability is wisely used. However, biotechnology that started more than 20 years ago has not stemmed the decline in per capita food production during the past 20 years (Fig. 1). Currently, about 40% of the research effort in biotechnology is devoted to the development of herbicide resistance in crops (Paoletti and Pimentel, 1996). This technology does not increase crop yields, but it will increase the use of chemical herbicides and the pollution of the environment.

VII. CONCLUSION

Strategies for global food security must be based first and foremost on the sustainable conservation and management of the land, water, energy, and all biological resources required for food production. Our stewardship of world resources will have to change. The basic needs of all people must be brought into balance with the life-sustaining natural resources. The conservation of these resources will require the coordinated efforts of all individuals and countries. Once these finite resources are exhausted, they cannot be replaced by human technology. Along with this, more efficient and environmentally sound agricultural technologies must be developed and put into practice to support the continued productivity of agriculture (Pimentel and Pimentel, 1996).

Unfortunately, none of these conservation measures will be sufficient to ensure adequate food supplies for future generations unless the growth in the human population is simultaneously curtailed. Several studies have confirmed that to enjoy a relatively high standard of living, the optimum human population should be less than 200 million for the United States and less than 2 billion for the world (Pimentel et al., 1999; Smail, 2002). This harsh projection assumes that from now until such an optimum population is achieved, *all* strategies for the conservation of soil, water, energy, and biological resources are successfully implemented and an ecologically sound, productive environment is maintained. The lives and livelihood of future generations depend on what the present generation is willing to do now for future generations.

REFERENCES

Batty, J. C., Keller, J. 1980. Energy requirements for irrigation. In: Handbook of Energy Utilization in Agriculture. Pimentel D., Ed: CRC Press. Boca Raton, FL, 35–44.

BP. 2001. British Petroleum Statistical Review of World Energy. British Petroleum Corporate Communications Services. London.

Brown, L.R. 1995. Who Will Feed China?: W.W. Norton. New York.

Brown, L.R. 1997. The Agricultural Link: How Environmental Deterioration Could Disrupt Economic Progress: Worldwatch Paper 136, Worldwatch. Washington, DC.

Clary, G.M., Haby, V.A. 2002. Potential for Profits from Alfalfa in East Texas: http://ruralbusiness.tamu.edu/forage.alfcop.pdf (9/2/2002).

Duffy, M 2001. Prices on the Rise: How Will Higher Energy Costs Impact Farmers?: http://www.ag.iastate.edu/centers/leaopold.newsletter/2001leoletter/energy.html–(9/3/2002).

FAO. 1961–2000. Quarterly Bulletin of Statistics, 1–13.

FAO. 1997. Quarterly Bulletin of Statistics. Food and Agriculture Organization: United Nations. Rome. Italy.

FAO. 1998. Food Balance Sheet: http://armanncorn:98ivysub@faostat.fao.org/lim..ap.pl? (10/10/1998).

FAO. 1999. Agricultural Statistics: hhtp://apps.fao.org/cgi-bin/nph-db.pl? subset-agriculture, Food and Agriculture Organization, UN (11/22/1999).

Fertilizer. 2002. Fertilizer Use and Abuse. Land Use and Environmental Change in the Thompson-Okanagan: http://royal.okanagan.bc.ca/mpidwin/agriculture/fertilizer.html (9/3/2002).

Giampietro, M., Pimentel, D. 1993. The Tightening Conflict: Population, Energy Use, and the Ecology of Agriculture. In:. Grant L., Ed: Negative Population Forum, Negative Population Growth, Inc. Teaneck, NJ.

Gleick, P.H. 1993. Water in Crisis: Oxford University Press. New York.

Gleick, P.H., Burns, W.C.G., Chalecki, E.L., Cohen, M., Cushing, K.K., Mann, A.S., Reyes, R., Wolff, G.H., Wong, A.K. 2002. The World's Water: Island Press. Washington, DC.

Hoffman, T.R., Warnock, W.D., Hinman, H.R. 1994. Crop Enterprise Budgets, Timothy-Legume and Alfalfa Hay, Sudan Grass, Sweet Corn and Spring Wheat Under Rill Irrigation, Kittitas County, Washington: Farm Business Reports EB 1173. Pullman, Washington State University.

Lal, R., Stewart, B.A. 1990. Soil Degradation: Springer-Verlag. New York.

Lal, R., Hansen, D., Uphoff, N., Slack, S. 2002. Food Security & Environmental Quality in the Developing World: CRC Press. Boca Raton, FL.

Lal, R., Follett, R.F., Kimble, J., Cole, C.V. 1999. Managing U.S. cropland to sequester carbon in soil. *Journal of Water and Soil Conservation* 54(1):374–381.

Larsen, K., Thompson, D., Harn, A. 2002. Limited and Full Irrigation Comparison for Corn and Grain Sorghum: http://www.Colostate.edu/depts/prc/pubs/pl_pub.pdf (9/2/2002).

Leshner, A 2002. Science and sustainability. *Science.* (*August 9*) 297:897.

Myers, N. 1994. Tropical deforestation: Rates and patterns. In:. Brown K., Pearce D.W., Eds. The Causes of Tropical Deforestation: UBC Press. Vancouver, British Columbia, 27–41.

NASS. 1999. National Agricultural Statistics Service: http//usda.mannlib.cornell.edu (8/30/2002).

NIAID. 2002. HIV/AIDS Statistics. National Institute of Allergy and Infectious Diseases. Washington, DC: http://www.niaid.nih.gov/factsheets/aidsstat.htm (2/15/2002).

Paoletti, M.G., Pimentel, D. 1996. Genetic engineering in agriculture and the environment. *BioScience* 46(9):665–673.

Pimentel, D., Pimentel, M. 1996. Food, Energy and Society: Colorado Press. Niwet, CO.

Pimentel, D., Pimentel, M. 2003. World population, food, natural resources, and survival: World Futures (in press).

Pimentel, D., Dazhong, W., Giampietro, M. 1990. Technological changes in energy use in U.S. agricultural production. In: Gliessman S. R., Ed. Agroecology: Springer-Verlag. New York, 305–321.

Pimentel, D., Tort, M., D'Anna, L., Krawic, A., Berger, J., Rossman, J., Mugo, F., Doon, N., Shriberg, M., Howard, E.S., Lee, S., Talbot, J. 1998. Ecology of increasing disease: population growth and environmental degradation. *BioScience* 48:817–826.

Pimentel, D., Harvey, C., Resosudarmo, P., Sinclair, K., Kurz, D., McNair, M., Crist, S., Sphpritz, L., Fitton, L., Saffouri, R., Blair, R. 1995. Environmental and economic costs of soil erosion and conservation benefits. *Science* 267:1117–1123.

Pimentel, D., Houser, J., Preiss, E., White, O., Fang, H., Mesnick, L., Barsky, T., Tariche, S., Schreck, J., Alpert, S. 1997a. Water resources: Agriculture, the environment, and society. *BioScience* 47(2):97–106.

Pimentel, D., Bailey, O., Kim, P., Mullaney, E., Calabrese, J., Walman, F., Nelson, F., Yao, X. 1999. Will the limits of the Earth's resources control human populations?. *Environment, Development and Sustainability* 1:19–39.

Pimentel, D., Doughty, R., Carothers, C., Lamberson, S., Bora, N., Lee, K. 2002. Energy inputs in crop production: Comparison of developed and developing countries. In: Lal L., Hansen D., Uphoff N., Slack S., Eds. Food Security & Environmental Quality in the Developing World: CRC Press. Boca Raton, FL, 129–151.

Pimentel, D 1980. Handbook of Energy Utilization in Agriculture: CRC Press. Boca Raton, FL.

Pimentel, D 1997. Techniques for Reducing Pesticides: Environmental and Economic Benefits: John Wiley. Chichester. UK, 444.

Pimentel, D 2003. Livestock Production and Energy Use: Elsevier (in press).

Pimentel, D., Wilson, C., McCullum, C., Huang, R., Dwen, P., Flack, J., Tran, Q., Saltman, T., Cliff, B. 1997b. Economic and environmental benefits of biodiversity. *BioScience* 47(11):747–757.

Pimentel, D 2002. Biological Invasions: CRC Press. Boca Raton, FL, 369.

Postel, S. 1999. Pillar of Sand: Can the Irrigation Miracle Last?: W.W. Norton. New York.

PRB. 2003. World Population Data Sheet. Population Reference Bureau. Washington, DC. Science. 1995. Cities as disease vectors. Science 270 (editorial).

Shetty, P.S., Shetty, N. 1993. Parasitic infection and chronic energy deficiency in adults. *Supplement to Parasitology* 107:S159–S167.

Smail, J.K. 2002. Confronting a surfeit of people: Reducing global human numbers to sustainable levels: An essay on population two centuries after Malthus. *Environment, Development and Sustainability* 4(1):21–50.

Soil Fertility Guide. 2002: http://www.gov.nf.ca/agric/pubfact/Fertility/Fertiguide.htm-2k (7/20/2002).

TFI. 1997. The Fertilizer Institute: http://www.tfi.org/Statistics/worldfertuse.asp (8/16/2002).

USBC. 2001. Statistical Abstract of the United States. 1993. 200th ed. U.S. Bureau of the Census: U.S. Government Printing Office. Washington, DC.

USDA. 1991. Corn-State. Costs of Production. U.S. Department of Agriculture, Economic Research Service: Economics and Statistics System. Washington, DC. Stock#94018.

USDA. 1997a. 1997 Census of Agriculture. U.S. Department of Agriculture: http://www.ncfap.org (8/28/2002).

USDA. 1997b. Farm and Ranch Irrigation Survey 1998. 1997 Census of Agriculture Volume 3, Special Studies, Part 1.

USDA. 2001. Agricultural Statistics: USDA. Washington, DC.

WHO. 1992. Our Planet, Our Health: Report of the WHO Commission on Health and Environment: World Health Organization. Geneva.

WHO. 2000. Malnutrition Worldwide: http://www.who.int/nut/malnutrition_worldwide.htm (7/27/2000).

WRI. 1994. World Resources 1994–95: World Resources Institute. Washington, DC.

Zubrzycki, J. 1997. Pollution of rivers in India reaches a crisis. October 29, 1997: The Christian Science Monitor (International). Boston, MA.

4

Historical Development of No-Till Farming

Rattan Lal
The Ohio State University
Columbus, Ohio, U.S.A.

Agriculture first evolved around 10 to 12 millennia ago, in one of the fertile valleys of the Tigris and Euphrates rivers in West Asia. From there it spread to other hydraulic civilizations in the valleys of the Nile, Indus, Yangtze, and other rivers (Hillel, 1991).

Plowing is one of the oldest agricultural practices, and the plow is an ancient soil preparation tool. Despite its history, plowing is still among the least scientifically understood practices in terms of its impact on soil and environment quality and on the sustainable use of natural resources. The first farmers simply scattered seeds on the ground. Eventually some ancient farmers improvised a digging stick to place the seed in the soil to protect against birds and rodents and the vagaries of climate. The digging stick was later replaced by a range of "tillage" tools to create a loose soil layer and eradicate weeds. Sometime between 5000 and 4000 B.C., Sumerian and other civilizations developed simple tools to place and cover seed in the soil and to eradicate weeds (Troeh et al., 1980). Despite their diverse shape and forms, most tools evolved from a common ancestor, "a common stick in the form of a handle with a short, recurvous-pointed spur" (Keen, 1931). This was used to open a shallow furrow. Indeed, archaeological evidence of such ancient tillage tools is found in Mesopotamia about 3000 B.C. and in the Indus Valley about 2500 B.C.. We learn from the Old Testament (I Samuel 13:20) that the Israelites "went down to the Philistines to sharpen every man his share, and his coulter and his axe, and his mattock" (Kellogg, 1938). The Hindu scripture *Yajur Ved*, written in Sanskritic poetry between 1000 and 500 B.C. states, "let the plowshare turn up the furrow slice in happiness, air and sun nourishing the

55

earth and water elements." According to legend, the heroine of the Hindu epic *Ramayana*, Sita, was born when King Janak of the Ayodhya kingdom along the River Ganges was advised by pundits to plow his field with a silver plow so as to break a long drought.

I. TILLAGE IMPLEMENTS

The generic term "tillage" encompasses a broad range of techniques of physical manipulation of the soil specifically designed to prepare a seedbed that optimizes edaphological conditions for seed germination, seedling establishment, and crop growth. Important edaphological conditions include soil water and temperature regimes, soil air and aeration, seed–soil contact, nutrient availability, porosity and pore size distribution, and minimal incidence of pests.

A wide range of tillage tools has been developed over the millennia to prepare a seedbed (Söhne, 1992; Köller, 2003). The *ard* was invented long before the "plow." The term "*ard*" is derived from the Old Norse word *ardr*, which is related to the Latin *Aratrum* (Glob, 1951; Sach, 1968; Lerche, 1994) and takes its name from the fact that it "works" (*arat*) the soil (White, 1967). An *ard* is essentially a paddle-shaped digging tool without a coulter or a moldboard, initially pulled by humans. It was essentially a "scratch" tool.

Then around 3000 B.C. came the earliest of the human- or animal-pulled implements (Bowen, 1961; Lerche, 1994, 1995; Hillel, 1998; Fowler, 2002). There are two main types of ard. One has an oblique share and head, and it can penetrate deep into the soil. The first type is also called the *Døstrup ard*, named after the Døstrup bog/marsh in Jutland where a Bronze Age specimen was discovered. The other has an almost horizontal sole designed for shallow plowing and preparing a friable/crumbly tilth. The second type is also called *Triptolemos ard* after a Greek god and hero (Leser, 1931; Glob, 1951; Sach, 1968; Lerche, 1994). The ard pushes the soil to both sides without inversion, and the pointed share produces a narrow V-shaped furrow.

Plowing was used as a tool to eradicate weeds and conserve water in Palestine and Greece in the 9th century B.C. (Vergil, cited by Semple, 1928). The fields were regularly allowed to lie fallow in alternate years, and the idle field was kept constantly under cultivation (Theophrastus, cited by Semple, 1928). Three or four plowings were considered imperative. Homer frequently mentions "the thrice-plowed fallow" (Hesoid, cited by Semple, 1928). Plowing at that time was not only to be thorough, but Theophrastus and Cato considered it the essence of good tillage. The Romans considered that grain land had to be rich and well tilled for two feet down. Hence, the fallow was plowed deep (Columella, cited by Semple, 1928). The earliest Greek literature show that Odysseus in the Isle of the Cyclops observed "the fresh plowed field of rich tilth and wide where the upturned soil shows black" (*Odyssey*, IX, cited by Semple, 1928).

The ard was later fitted with a seed funnel and used as a drill. It is still widely used in South Asia and is called *Desi* or native plow. It is also used in Ethiopia and throughout the West Asia and North Africa region (Figs. 1, 2, and 3). The ard eventually evolved into the well-known "Roman plow" (White, 1967; Herrmann, 1985). Vergil gave the earliest description of the Roman plow (*Georgica* 1:61, 169–175, cited by Browne, 1944), "while still growing in the woods, an elm is bent with great strength and subdued so as form a plow-beam and is made to assume the shape of the curved plow. To this plow-beam a pole stretching out eight feet, two moldboards, and share-beam with two ridges are fitted at the base."

The moldboard plow was introduced into England by Tentonic and Scandinavians during the 5th to 10th centuries. The Saxon exploitation of the moldboard plow during this era is considered one of the four major events in England, which set off chain reactions and have radically affected the course of history. The other three events included the Black Death (1349), the introduction of turnips and clover as farm crops in the late 18th century, and the opening up of the American prairies in the late 19th century.

Jethro Tull operated a farm near Hungerford, England, during the early 18th century (1674–1741). He described different tillage implements used in

Figure 1 An ard or desi plough used in South Asia in 1995.

Figure 2   An ard being used to plow the field in southern India in 1998.

Britain and elsewhere in Europe in his book *The New Horse Houghing Husbandry* (Hillel, 1998). He believed, imaginatively but erroneously, that an objective of plowing was to pulverize the soil grains into small particles so that they can be ingested by plant roots.

The ard was eventually replaced by a plow that had an iron share and coulter and was steered by one or two stilts. The plow replaced the ard in Europe during the early Middle Ages. The Roman plow evolved in the later part of the 18th century into a rough inverting plow. A moldboard was attached, always to its right side, to lift and turn the furrow slice. In the United States, the earlier designs and history of plows is presented by Lewton (1943). The national museum has a collection of 14 plows designed since 1740. One of the plows is a John Deere Steel plow designed in 1838 at Grand Detour, Illinois, from a broken steel circular saw (Lewton, 1948). A moldboard plow was first designed by Jefferson in 1784. It was patented by Charles Newbold in 1796 and marketed as a "cast iron plow" in the 1830s by a blacksmith named John Deere (AAVIM, 1993; Köller, 2003). The plow consisted of a beam carrying: (1) a vertical knife or coulter to cut the side of the furrow away from the unplowed land; (2) a plow share to cut the bottom of the furrow and raise it from the subsoil; and (3) a moldboard to turn and invert the furrow slice. Indeed, the west was conquered

Figure 3   An Ethiopian ard carried by a farmer to his field near Addis Ababa in 1989.

by the plow. Farmers plowed their way to the west. The basic equipment for the prairie breaker was the plow (Rogin, 1931). The prairie breaker was comprised of a heavy, 14-foot-long wooden beam, and sometimes supported by a set of wheels at the forward end (Fig. 4). Located at the rear end of the beam was a massive steel or iron plow share weighing 25 to 60 kg that cut a furrow 40 to 75 cm wide and 5 to 15 cm deep (Hall, 1973). This gradual transition from a digging stick to moldboard plow is illustrated in Fig. 4.

The objectives of soil tillage are somewhat different for temperate and tropical climates. In temperate regions, fall plowing with a moldboard plow leaves the soil surface rough with large clods and numerous ridges and furrows. This rough surface undergoes a series of wet/dry and freeze/thaw cycles, "mellowing" or crumbling to become friable. During the spring, the mellowed soil is refined into a friable tilth by use of cultivators and harrows, which also eradicates weeds. Xenophon, around 400 B.C., recommended spring plowing (by ard) because "the land is more friable then" (Tisdale and Nelson, 1966). The loose soil surface dries and thus warms up rapidly during the spring. In the tropics, however, the soil at the onset of heavy seasonal rains (monsoons) following a long dry season is parched and extremely hot. Therefore, the objective of seedbed preparation in the tropics is to conserve water in the root zone, lower soil temperature, and

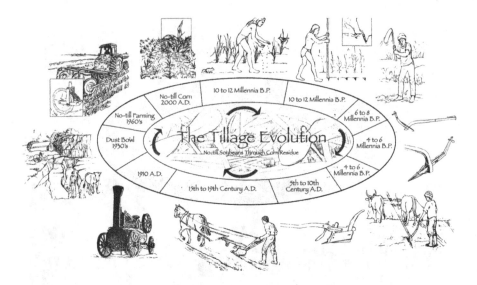

Figure 4   A chronological development of tillage and tools and practices.

suppress weeds. The suppression of weeds by plowing under the live biomass and residues reduces competition with crops for available water and nutrients, and it may also reduce the incidence of infestation by insects and fungus. Nonetheless, soil-inverting implements designed for temperate climates are not necessarily suited to edaphic conditions in the tropics, especially harsh climates characterized by intense storms.

II.   THE ENVIRONMENTAL CONSEQUENCES OF TILLAGE

The advent of the ard and plow greatly expanded the ability of agriculture to meet the food needs of growing populations. Cropland area in the world expanded dramatically over three centuries between 1700 and 2000, going from 265 million hectares (Mha) in 1700, to 537 Mha in 1850, 913 Mha in 1920, 1,170 Mha in 1950, and 1,500 Mha in 1980 (Richards, 1990; Myers, 1996; FAO, 1996). By 2000, it was estimated to have decreased about 10% to 1,360 Mha in 2000 (FAO, 2000).

Since their primitive invention between 5000 and 4000 B.C., various simple tools have been used to turn over, mix, and pulverize the soil. They have been greatly altered over time to suit the soil-specific needs for mechanized farm operations. Tillage implements include the plow to invert the soil, and cultivators and harrows to pulverize, mix, and stir soil clods. In addition, rollers are sometimes used to consolidate loose tilth, and hoes and broad shares to create soil mulch.

Soil can now be tilled to a deeper depth, more quickly and more intensively than ever before. It must be noted that tillage and cultivation are unlike any natural disturbance to soils. When repeated every season to a depth of 15 to 20 cm, tillage renders the soil in a state of unstable equilibrium. Some have called this drastic perturbation "rape of the earth" (Jacks, 1939).

The benefits of plowing on weed control and stand establishment are immediate, but it also has numerous long-term adverse impacts on soil quality and environments. Two major consequences of plowing-induced soil loosening, pulverization, and burial of weeds and crop residue are (1) exposure of the soil to the impact of raindrops and blowing wind exacerbating the risks of accelerated soil erosion by water and wind, and (2) oxidation of humus due to increased ease of microbial decomposition of organic matter.

The scientific basis of plowing has been questioned since the middle of the 20th century because of its negative long-term environmental consequences. The perceived benefits of cultivation for soil water conservation and weed control have been increasingly questioned and have often been found to be marginal or missing. Furthermore, alternative chemical weed control measures have been developed since the middle of the 20th century which make it possible to either completely eliminate the need for plowing or to greatly reduce its frequency and intensity.

A.   Soil Degradation, Erosion, and Desertification

Increased expansion of plow-based agriculture brought increased soil degradation, accelerated soil erosion, and desertification with it. The process of soil degradation is set in motion by crusting and by compaction of loosened soil that results from raindrop impact and from human, animal, and vehicular traffic.

Plowing and turning under crop residues and weed biomass accentuate the risks of accelerated erosion by water on sloping land and wind on flat terrain. Vergil was one of the earliest Roman writers to mention the destruction caused by wind erosion. Thus, Vergil doubted the value of plowing when soil had been baked to a crust (Vergilius Maro, cited by Browne, 1944). Vergil also realized the destruction caused by floods and runoff. Vergil recommended contour plowing and stone terraces for controlling erosion. Indeed, accelerated erosion is analogous to "cancer" of the land. Its disastrous consequences destroyed ancient civilizations including the Phonecians (Eckholm, 1976), Mesopotamia (Lowdermilk, 1939), ancient kingdoms of Lydia and Sardis in present-day Turkey (Beasley, 1972), the Harappan-Kalibangan culture in the Indus Valley (Singh, 1982), and the Mayan civilization in Central America (Olsen, 1981).

Soil erosion has been "the quiet crisis" even in modern times (Brown and Wolf, 1984). Pimentel et al. (1995) have estimated the global economic loss by water erosion at hundreds of billions of U.S. dollars per year due both to on-site and off-site effects. The on-site effects of erosion on crop yields are not this great, however, especially for commercial farming of North America and Europe, and especially for those now practicing conservation tillage (den Biggelaar et al., 2003a, b).

B. Water Quality

Transport of chemicals from agricultural land leads to a major loss of costly inputs as well as the pollution of natural waters. Sediment load, accumulation of plant nutrients, and transport of pesticides in runoff all contribute to pollution of surface water. Preferential transport of nitrogen in runoff and that of phosphorus in sediments leads to eutrophication of surface waters (Sharpley and Halvorson, 1994; Owens, 1994). Concentrations of plant nutrients and pesticides in water runoff are related to the amount and kind of sediments. The transport of sediments and of sediment-borne pollutants is exacerbated by plowing and other tillage operations.

C. Soil Erosion and the Greenhouse Effect

Soil erosion leads to preferential removal of the lighter soil fractions by water runoff and blowing wind. The light fractions include soil organic matter and unaggregated clay and silt-sized particles. Erosion causes breakdown of aggregates and exposure of hitherto encapsulated/protected soil organic matter. Assuming that gross erosion by water is 75 billion Mg, the amount of erosion-induced transport of soil carbon is 4 to 6 Pg per annum (Lal, 2003). With 20% emission due to mineralization of the displaced carbon, erosion-induced emissions may be 0.8–1.2 Pg C/year. Regardless of the exact amount emitted, accelerated erosion has a strong impact on the global C cycle and the atmospheric concentration of CO_2 (Lal, 2003).

III. EVOLUTION OF NO-TILL FARMING IN THE UNITED STATES

No-till farming *per se* is nothing new. It is exactly what most resource-poor farmers and shifting cultivators do (Figs. 5 and 6). Vergil was the first person to recognize the danger of plowing in causing erosion (Browne, 1944). In the context

Figure 5   A small landholder around Arush, Tanzania, prepares seedbed manually with minimal soil disturbance.

of modern agriculture, however, the usefulness of plowing and other tillage operations as the main means of weed control and of seedbed preparation has long been questioned (Russell and Keen, 1941). The "Dust Bowl" catastrophe of the 1930s in the Midwestern United States was a turning point (Steinbeck, 1939). A cloud of dust rising up to 4,500 m high obscured the sun in May 1934, all the way from the Texas Plains up through the Dakotas and from Montana to the Ohio Valley. On May 12, 1934, dust sifted through the windows of the White House and covered President Roosevelt's desk. The effect of prolonged drought in the Midwestern and Great Plains states was exacerbated by excessive tillage leaving the ground bare and unprotected against persistent winds. Contrary to the teaching of Prophet Isaiah, the "plow share had done more damage than the sword."

Subsequent to the Dust Bowl phenomenon, the rationale for plowing was questioned, even prior to the availability of herbicides (Faulkner, 1943). Faulkner claimed that "no one has ever advanced a scientific reason for plowing." Louis Bromfield, another Ohioan and the creator of "Malabar Farm," also sought to demonstrate that conventional methods of soil cultivation were incompatible with environmental stability.

During the late 1940s and early 1950s, soil scientists, agronomists, and agricultural engineers started conducting research on weed control. Availability

Figure 6   A traditional shifting cultivator in Burkina Faso, West Africa, where the people sow their crops without plowing and store the harvest on a tree to safeguard against animals and pests.

of 2,4-D and other herbicides after World War II provided farmers with a viable alternative to plowing and tillage for seedbed preparation. A major breakthrough in weed control occurred during the 1950s and 1960s with the development of atrazine, simizine, and cyanazine, which were used as postemergence herbicides. Complete replacement of tillage with herbicides was made possible by the development of Paraquat by ICI in England during the early 1960s. These contact (Paraquat) and residual (S-triazines) herbicides made it possible to sow directly into the killed sod with an appropriate drill (Hood et al., 1963, 1964; Van Doren, 1967; Phillips and Young, 1972). The development of Roundup and Roundup-Ready crops by Monsanto during the 1990s revolutionalized the concept of seedbed preparation and tillage. These herbicides, along with development of appropriate seeding drills, made no-till farming a "tillage revolution."

Main features of no-till farming include the following:

1.   Crop residue mulch is maintained on the soil surface,
2.   Presowing mechanical seedbed preparation is eliminated,
3.   Weeds and any cover crop are controlled by herbicides (chemical) or slashed mechanically,

   4.   An appropriately designed drill is used to sow seed in an untilled and rough seedbed covered with crop residue mulch, and

   5.   All secondary tillage operations for weed control are replaced by postemergent herbicides.

Numerous variants of no-till farming represent different types and intensities of presowing tillage. Collectively, these methods of seedbed preparation are called "conservation tillage." By definition, conservation tillage is any method of seedbed preparation that leaves at least 30% of the soil surface covered by crop residue mulch. Since it was initiated in the late 1960s, conservation tillage was practiced on about 37% of the U.S. cropland during 2000 (Table 1). Different variants of no-till farming include the following (Fig. 7):

> *No-till*: Any method of sowing a crop without any presowing tillage (with or without the use of herbicides to kill weeds or a cover crop). No-till sowing is called sod seeding when crops are sown into a sod produced by application of herbicides on a cover crop. Direct drilling, a term commonly used in Europe, implies sowing through the residue of a previous crop (Allen, 1981).
>
> *Minimal till*: Any method of seedbed preparation for which the preplanting mechanical tillage is limited. Most involve sowing after chisel or disc operation, or on a semipermanent ridge-furrow system, or following a strip tillage (Fig. 5).

Advantages of no-till and conservation tillage systems include

    Erosion control
    Water conservation
    Improvements in water quality

Table 1   Cropland Area Sown by Conservation Tillage in the United States

Year	Land area under conservation tillage (Mha)	% of planted cropland area
1968	2.4	2.0
1970	4.1	3.4
1975	7.3	5.6
1980	15.8	10.9
1985	38.5	27.8
1990	29.7	26.1
1995	40.0	35.5
2000	43.0	36.5

Sources: Scherfz (1988); CTIC (2000).

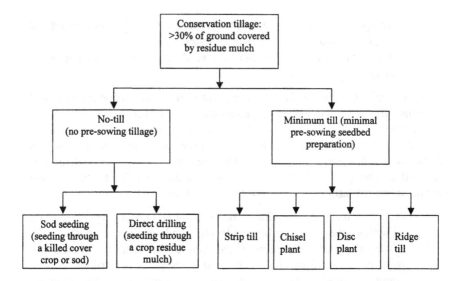

Figure 7   Variants of conservation tillage methods.

> Savings of time, fuel, labor, and money for seedbed preparation
> Increase in soil organic matter content, and soil carbon sequestration
> Double cropping potential.

No-till and conservation tillage systems are essentially mulch farming techniques. Benefits of mulch farming are well documented in Europe and elsewhere (Jacks et al., 1955). Mulch farming has also been supported by the "organic farming" movement (Fukuoka, 1978). Independent of the research done in the United States, direct drilling experiments were also conducted in Europe, especially in the United Kingdom, during the early 1960s (Hood et al., 1963, 1954; Jeater and Mcilvenny, 1965; Jeater and Laurie, 1966). However, adoption of conservation tillage has not been as widespread in Europe as in the United States and Canada (Kuipers, 1970).

IV.   EVOLUTION OF CONSERVATION TILLAGE IN THE TROPICS

The problem of soil erosion and other degradative processes is more severe in the tropics than in the temperate climate of North America and Europe due to different climatic and soil factors. Results from several soil and crop management experiments conducted in the tropics during the 1950s and 1960s indicated the potential of no-till and conservation tillage methods for soil and water conservation. In the Belgian Congo (Zaire), Muller and De Bilerling (1953) reported that upland rice (*Oryza sativa*) yielded more on unplowed than on plowed soil. In Colombia, Ospina and Nel (1957) reported that plowing caused severe losses by erosion. Pereira et al. (1954, 1958) initiated experiments on tillage methods and cover crops in East Africa and documented the beneficial effects of cover crops in terms of improved soil structure and fertility.

In northern Nigeria, Lawes (1962) observed significantly higher infiltration rates for plots with mulch than for plots on which the crust had been broken mechanically. Jurion and Henry (1969) identified benefits from mulch farming at several locations in Zaire. Kannegieter (1967, 1969) first used the term "zero tillage" in Africa when describing the results of his experiments conducted in Ghana. He documented that short-term Pueraria fallow, combined with mulch tillage, provided effective soil protection against erosion, improved soil moisture storage, and produced satisfactory yields of maize (*Zea mays*).

Long-term no-till experiments initiated at the International Institute for Tropical Agriculture (IITA) in Ibadan, Nigeria, quantified the benefits of no-till farming with crop residue mulch in terms of soil and water conservation, lower soil temperature, and better soil structure and crop yields (Lal, 1973, 1974, 1976a, b, 1989, 1995) (Fig. 8). Despite a long history of research, adoption of conservation tillage or no-till farming in Africa continues to be a major challenge. The failure to adopt no-till farming in Africa may be due primarily to socio-economic factors and weak institutions.

Considerably more progress has been made in spreading no-till farming in South America, especially in Brazil and Argentina. In Costa Rica, Shenk and Sanders (1981) reported that no-till systems produced more maize grain yield than plow till and suffered significantly less damage by insects and pathogens. A series of long-term no-till experiments was initiated in Paraná by researchers from GTZ and EMBRAPA (Sidiras et al., 1982, 1983, 1984, 1985a, b; Sidiras and Roth, 1985; Derpsch et al., 1985, 1986). They obtained satisfactory yields of wheat (*Triticum aestivum*) and soybeans (*Glycine max*) and better soil structure with no-till than plow-till methods of seedbed preparation. Because of the research information and farmers' initiatives, no-till farming in Brazil is practiced on more than 17 Mha of cropland (Landers, 2001; Calegari, 2002; Köller, 2003; and Chapter 2). In contrast to Africa, there have been strong institutional support and enthusiastic responses by farmers to no-till opportunities, leading to widespread adoption of no-till farming in Brazil, Argentina, and elsewhere in South America.

V. NO-TILL FARMING IN RICE

Upland rice can be grown using a no-till system, similar to that used for any other row crop, provided that weeds are successfully controlled. Upland rice has

Figure 8 A no-till research plot of soybeans sown through the residue of maize in 1972 at IITA, Ibadan, Nigeria.

a lower germination, a poorer stand, and lower grain yields when grown under no-till rather than a plowed-based system (Ogunremi et al., 1986a, b). Poor seed–soil contact, damage to young seedlings by birds, and nutrient imbalances in the soil are major factors. Transplanting of rice seedlings through the dead mulch of a cover crop, although labor-intensive, may be a satisfactory method of obtaining a good crop stand. In western Cameroon, however, Ambassa-Kiki et al. (1984) reported no differences in rice grain yield among no-till, minimum-till, and plow-till methods when rice was grown on Vertisols with a high moisture retention capacity. The mean grain yields and soil bulk densities of the 6–11-cm layer were 6,240 kg/ha and 1.46 g/cm^3 for minimum-till, and 5,934 kg/ha and 1.56 g/cm^3 for the plow-till, respectively. Hydromorphic soils with low infiltration rates give satisfactory yields of lowland irrigated rice with an unpuddled or no-till system (Rodriguez and Lal, 1985; Lal, 1986). Finding viable alternatives to puddling is a high priority. In northern India, Ghildyal (1971) reported that compacting a sandy soil with a roller produced as good a yield of flooded rice as puddling.

For rice grown on a sandy hydromorphic soil in western Nigeria, Rodriguez and Lal (1985) reported that mean grain yields and soil bulk density of the 6–11-cm layer were 6,240 kg/ha and 1.46 g/cm^3 for no-till, 6,332 kg/ha and 1.61

g/cm^3 for minimum-till, and 5,934 kg/ha and 1.56 g/cm^3 for the plow-till, respectively. Also, for hydromorphic soils with low infiltration rates, satisfactory yields of lowland irrigated rice have been reported with an unpuddled or no-till system (Rodriguez and Lal, 1985; Lal, 1986).

Lowland rice is a key crop in Southeast Asia. The conventional method of seedbed preparation is to puddle the soil (Fig. 9) when it is almost saturated in order to destroy soil structure and decrease percolation. The puddling system has many disadvantages, including: (1) high energy and labor costs; (2) the time needed for seedbed preparation when the turnaround time is limited for double cropping; and (3) low yields of the many upland crops due to poor soil structure. Sharma and DeDatta (1986) evaluated the use of a no-till system for lowland rice. They reported satisfactory yields in some seasons but not in others. Other researchers have also reported satisfactory yields in the Philippines (Mabbayad and Buensosa, 1967; DeDatta and Karim, 1974; Sharma and DeDatta, 1986).

VI. NO-TILL FARMING IN SOUTH AND SOUTHEAST ASIA

Experiments on no-till farming in South and Southeast Asia were initiated in the early 1980s. Blevins (1984) described the potential for minimum-till in Bangla-

Figure 9 Puddling the soil with buffalo-driven tillage implements (a) in the Philippines, and (b) in Sri Lanka. Soil puddling destroys soil structure, decreases infiltration, and facilitates transplanting.

Figure 9 Continued.

desh and observed a promising potential for many soils in India. Thamburaj et al. (1980) reported that cassava tuber yield in Madras was satisfactory when no-till was combined with the use of residue mulch. Wijewardene (1982) reported that a range of upland crops can be grown profitably in Sri Lanka with a no-till system.

Conservation tillage systems have been tried in northern Thailand, where Ratchdawong et al. (1984) reported satisfactory crop yield and less runoff and erosion on no-till cropping compared with twice-plowed plots. Higher yields were obtained with residue mulch than without it. Research in Indonesia has also shown the usefulness of no-till farming for erosion control and for producing satisfactory crop yields. Suwardjo et al. (1984) reported that mechanical tillage including plowing reduced structural stability. They therefore recommended minimum-till with mulching for soils prone to erosion. Satisfactory crop yields were obtained with no-till systems for maize, groundnuts (*Arachis hypogea*), mung beans (*Phaseolus aureus*), and upland rice.

VII. NO-TILL FARMING IN THE RICE–WHEAT SYSTEM

Application of no-till farming to the rice–wheat system of South Asia began in the mid- and late- 1990s. The rationale for no-till sowing of wheat following rice includes a reduction in the turnaround time between crops, savings in fuel, and water conservation. However, the no-till practiced in rice–wheat system in South Asia is different from conservation tillage systems used in North America, Brazil, and Argentina. First, little if any crop residue mulch is left on the field (Figs. 10, 11, and 12). Most of the rice straw is either removed or burned (Fig. 13). Thus, wheat sown in an unplowed soil does not benefit from any residue mulch for water conservation, soil temperature moderation, soil organic matter improvement, or soil faunal activity enhancement.

One reason why farmers remove or burn rice straw is the lack of a seeding drill that can cut through the wet residue and establish good seed–soil contact for the following wheat. Another reason is that seedbed preparation for transplanted rice sown in the summer is done by the conventional method of soil puddling (Fig. 9). Soil structure is deliberately destroyed to reduce seepage losses. Therefore, no-till wheat is sown in a soil with poor soil structure without residue mulch. Under these circumstances, environmental benefits of the classic no-till system are not fully realized.

Figure 10 The wheat residue is removed from the field to the threshing floor where it is stored for use as fodder (photo in 2000 near PAU, Ludhiana).

Figure 11 Whatever wheat residue is left, it is heavily grazed. Intense grazing also causes soil compaction (photo in 2000 near PAU, Ludhiana).

VIII. SOIL CARBON SEQUESTRATION, CARBON TRADING, AND CLEAN DEVELOPMENT MECHANISMS

One of the principal ecological benefits of no-till farming is enhancement of the soil carbon (C) pool. In the context of C sequestration, there are two advantages: decrease in emission, and increase in the terrestrial carbon pool (Fig. 14).

Conversion from plow-till to no-till may decrease the use of fossil fuel. For example, the average fuel consumption is 49.4 l/ha for plowing, 31.3 l/ha for chiseling, 28.4 l/ha for discing, 25.2 l/ha for ridge planting, and only 13.4 l/ha for no-till farming (Köller, 2003). Similar information is available from other sources (Stout, 1984; USDA, 1977). Conversion from plow-till to no-till can save a total of 30 to 35 l/ha of fuel. In addition, no-till also has the potential for saving labor hours. Properly implemented, no-till farming may also reduce losses of fertilizer and water and thus can enhance use efficiency of these inputs. Adoption of no-till thus has many energy-saving benefits that include (1) less fuel consumption due to reduced field operations, (2) less power required due to better soil structure, (3) less time and labor required, (4) less fertilizer use due to increase

Figure 12   The cattle dung is often not returned to the field and is used as cooking fuel (photo in 2000, near Karnal, India).

in fertilizer use efficiency, and (5) less irrigation used due to increased water-use efficiency.

Some activities associated with no-till may require more energy than plow-till. Examples of these activities are (1) weed control with herbicides, (2) higher incidence of insects and pathogens where natural controls have been reduced or not restored, (3) higher seeding rate, and (4) special seeding equipment. Frye and Phillips (1981) and Poincelot (1986) estimated that fuel use for all operations listed above was 71.6 l/ha for conventional plowing, 65.0 l/ha for chisel till, 53.4 l/ha for disc plant, and 38.4 l/ha for no-till. These data may be soil-site-specific and need to be obtained for local specific conditions.

In addition to avoiding emission, adoption of no-till farming also enhances the terrestrial C pool through C sequestration. Increase in biomass production through better soil quality is an important consideration. The biotic C sequestration in plants may be a short-term gain. However, a fraction of the net primary production (NPP) converted into humus or soil organic matter has a long residence time. The rate of soil organic carbon (SOC) sequestration is lower in warm and arid regions than in cool and humid regions, and the base rates may be enhanced by judicious management of crop residue. The more the residue is returned to

Figure 13 Secondary salinization is a severe problem in Punjab, Pakistan (photo in 1997 near Gujranwala, Pakistan).

the soil, the greater is the SOC sequestration. The rates of SOC sequestration in semi-arid and subhumid tropics/subtropics may be 50 to 250 kg/ha/yr (Lal, 1999, 2001) though any net assessment requires long-term experiments.

Secondary salinization is a major problem in South Asia (Figs. 13, 15; see Chapter 5). Growing salt-tolerant (halopytic) plants can improve above- and below-ground biomass production and increase SOC concentration. Singh (1989) observed that growing *Prospis juliflora* increased the SOC pool in alkaline soils of northwest India. Singh et al. (1994) reported that growing salt-tolerant species increased SOC concentration in the top 30-cm depth by 2 to 4 times (0.24% to 0.93%). The SOC pool increased from about 10 Mg C/ha to 30 to 45 Mg C/ha over an 8-year period. Reclaimed soils could then be grown to rice–wheat system. Similar results were obtained by Garg (1998).

Biofuels are increasingly recognized as a feasible alternative to fossil fuel (Abelson, 1995; Giampietro et al., 1997). In this context, biofuel is any type of solid, liquid, or gaseous fuel that can be produced from biomass substrates and that can be used as substitute for fossil fuel. Short-duration woody plants (poplar, etc.) could lead to substantial amounts of fossil C offset. They may also substitute for use of crop residue and animal dung as household fuel and allow them to be

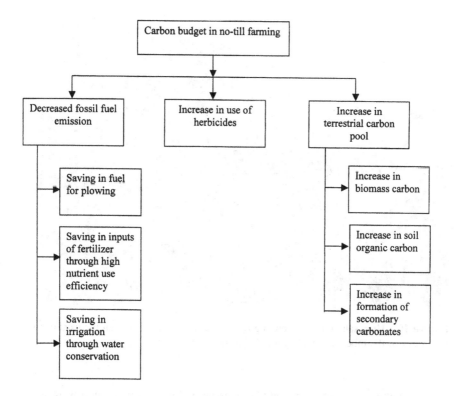

Figure 14 Terrestrial carbon sequestration upon conversion from plow-till to no-till farming.

used as mulch or manure, and as soil amendments. Indiscriminate burning of these biosolids causes severe air pollution (Pachauri and Sridharan, 1998).

There is also a potential of sequestration of soil inorganic carbon (SIC) through formation of secondary carbonates, especially in irrigated ecosystems. In systems of partial or complete soil leaching, carbonates are sequestered primarily through movement of HCO_3^- into groundwaters, especially if irrigation water is not saturated with carbonates. Formation of secondary carbonates can be enhanced by use of biosolids such as green manure, compost, farmyard manure, and mulches. The rate of carbonate formation may range from 50 to 200 kg/ha/y (Lal, 1999). It is important to assess these rates, especially in irrigated soils of South Asia.

Implementation of C trading through the Clean Development Mechanism (CDM) under the Kyoto Protocol will require development and strengthening of

Figure 15 High salt concentration in groundwater is apparent from salt incrustation on the bricks of the tube well in Haryana, India (photo in 2000).

databases about rates of terrestrial C sequestration by different processes as outlined herein. While terrestrial C sequestration is a major challenge in the tropics, the environmental and economic benefits can be enormous. The revised Kyoto Protocol has incorporated two specific clauses regarding soil/terrestrial carbon sequestration:

1. Countries are allowed to subtract from their industrial C emissions some increases in C sequestered in sinks such as forests and soil.
2. Countries are allowed to trade emission allowances that can reduce abatement costs. The UNFCCC/Kyoto Protocol recognizes soil C sinks provided that the rate of SOC sequestration, and the cumulative magnitude can be verified by standard procedures. Thus, terrestrial carbon sequestration may be credited under the Clean Development Mechanism (CDM, Article 12 of the Kyoto Protocol) and emission trading (Article 17) or joint implementation activities (Article 6).

Terrestrial C sequestration provides new opportunities to restore degraded soils and ecosystems through C trading in the international market. This opportunity can be realized through the development of databases on rates of soil and biotic C sequestration in principal soils, land uses, and ecoregions.

IX. CONCLUSION

Evaluation of tillage systems has gone a full circle over 10 or 12 millennia since the dawn of settled agriculture. The transition from hunter-gatherer societies to agrarian societies occurred by raising crops using no-till methods in which seeds were either scattered on the ground or put in a shallow hole made with a digging stick. However, the expansion and intensification of agriculture occurred because of the evolution of the ard, and later a moldboard plow pulled by draught animals. The animal-drawn plow was a major factor in horizontal expansion of agriculture (Miller, 2000).

Agricultural intensification during the second half of the 20th century was achieved through mechanization. However, excessive and indiscrete tillage led to accelerated erosion and degradation of soil resources with attendant adverse impacts on water quality and emission of greenhouse gases into the atmosphere. Concerns about sustainable management of soil and water resources resulted in the search for alternatives to plowing for weed control and seedbed preparation.

There is a widespread awareness about the need for reducing or completely eliminating presowing seedbed preparation. In addition to reducing fuel use, the adoption of no-till farming is also known to result in soil C sequestration, which can mitigate climate change. World soils have lost 66 to 90 Pg of C due to plowing and the attendant erosion and mineralization of soil organic matter (Lal, 1999). Conversion of plow-till to no-till can enhance soil organic matter. We have gone a full circle by developing no-till farming techniques. However, their use and promotion must be objectively validated, rational, and site-specific. They will contribute most by being part of larger strategies of sustainable natural resource management respecting agroecological principles.

ACKNOWLEDGMENT

The artwork in Fig. 4 was drawn by Mary Hoffelt, FAES, OSU.

REFERENCES

AAVIM. 1993. Fundamentals of No-Till Farming. American Association for Vocational Instructional Material. Athens, GA.

Abelson, P.H. 1995. Renewable liquid fuels. *Science* 268:955.

Allen, H.P. 1981. Direct Drilling and Reduced Cultivations: Farming Press Ltd.. Suffolk. U.K.

Ambassa-Kiki, R., Aboubakar, Y., Boulama, T. 1984. Compartement physique et mécanique du sol en relation avec l'utilisation de matériel agricole lourd. Institut de la Recherche Agronomique (IRA). Garoua, Cameroon.

Beasley, R.P. 1972. Erosion and Sediment Pollution Control, First Edition: Iowa State Univ. Press. Ames, IA.

Blevins, R.L. 1984. Potential for minimum tillage in Bangladesh. Bangladesh Agricultural Research Council, International Agricultural Development Service, Dhaka.

Bowen, H.C. 1961. Ancient Fields: A Tentative Analysis of Vanishing Earthworks and Landscapes. British Association for the Advancement of Science, London.

Brown, L.R., Wolf, E. 1984. Soil erosion: Quiet crisis in the world economy. Worldwatch Paper 60. Worldwatch Institute, Washington, DC.

Browne, C.A. 1944. Elder John Leland and the mammoth Cheshire cheese. *Agric. History* 18. (publius Vergilius Maro, Georgica 1. In the Georgicas and Ecolognes of Virgil): 61, 169–175.

Calegari, A. 2002. The spread and benefits of non-till agriculture in Paraná State. Brazil. In: Agroecological Innovations. Uphoff N., Ed: Earthscan Publications. London, 187–202.

CTIC. 2000. National survey of conservation tillage practices. Conservation Tillage Information Center, West Lafayette, IN.

DeDatta, S.K., Karim, A.A.S.M.S. 1974. Water and nitrogen economy of rainfed rice as affected by soil puddling. *Soil Sci. Soc. Am. Proc* 38:515–518.

den Biggelaar, C., Lal, R., Wiebe, K., Breneman, V. 2003a. The global impact of soil erosion on productivity. I. Absolute and relative erosion-induced yield losses. *Adv. Agron. (in press)*.

den Biggelaar, C., Lal, R., Wiebe, K., Eswaran, H., Breneman, V. 2003b. The global impact of soil erosion on productivity. II. Effects on crop yields and production over time. *Adv. Agron. (in press)*.

Derpsch, R., Sidiras, N., Roth, C.H. 1986. Results of studies made from 1977–1984 to control erosion by cover crops and no-tillage techniques in Parana, Brazil. *Soil and Tillage Research* 8:253–263.

Eckholm, E.P. 1976. Losing Ground, Norton, New York.

FAO. 1996. The Production Yearbook. Rome. Italy.

FAO. 2000. The Production Yearbook. Rome. Italy.

Faulkner, E.H. 1943. Plowman's Folly: University of Oklahoma Press. Norman, OK.

Fowler, P. 2002. Farming in the First Millennium AD: British Agriculture between Julius Caesar and William the Conqueror: Cambridge Univ. Press. Cambridge.

Frye, W.W., Phillips, S.H. 1981. How to grow crops with less energy. In: Cutting Energy Costs. Hayes J., Ed. The 1980 USDA Yearbook, U. S. Department of Agriculture. Washington, DC.

Fukuoka, M. 1978. The One-Straw Revolution: An Introduction to Natural Farming: Rodale Press. Emmaus, PA.

Garg, V.K. 1998. Interaction of tree crops with a sodic soil environment: potential for rehabilitation of degraded environments. *Land Degradation and Development* 9: 81–93.

Ghildyal, B.P. 1971. Soil and water management for increased water and fertilizer use efficiency for rice production. *Proc. Int. Symp. Soil Fert. Eval.* New Delhi.

Giampetro, M., Ulgiati, S., Pimentel, D. 1997. Feasibility of large-scale biofuel production. *Bioscience* 147:587–600.

Glob, P.V. 1951. Ard og plow i Nordens oldtid (Ard and Plough in Pre-historic Scandinavia). Jysk arkaeologisk selskab I, Aarhus.

Hall, T.G. 1973. Wilson and the food crisis: Agricultural price control during World War I. *Agric. History* 47:25–56.

Herrmann, K. 1985. Pflügen, Säen, Ernten. Landarbeit und Landtechnik in der Geschichte: Rowohlt Taschenbuch Verlag Gmbh. Reinbek.

Hillel, D. 1991. Out of the Earth: Civilization and the Life of the Soil: The Free Press. New York.

Hillel, D. 1998. Environmental Soil Physics: Academic Press. San Diego.

Hood, A.E.M., Jameson, N.R., Cotterell, R. 1963. Destruction of pasture by paraquat as a substitute for ploughing. *Nature 4869, 23 Feb*:748.

Hood, A.E.M., Jameson, N.R., Cotterell, R. 1964. Crops grown using paraquat as a substitute for ploughing. *Nature 4869, 14 March*:1070–1072.

Jacks, C.V. 1939. The Rape of the Earth: Faber & Faber. London.

Jacks, C.V., Brind, W.D., Smith, R.W. 1955. Mulching. Commonwealth Bureau of Soil Sci. Technical Communication 49, Harpenden. U.K.

Jeater, R.S.L., Mcilvenny, H.C. 1965. Direct drilling of cereals after use of paraquat. *Weed Research* 5:311–318.

Jeater, R.S.L., Laurie, D.R. 1966. Comparison of rates of paraquat prior to direct-drilling cereals. *Weed Research* 6:332–337.

Jurion, F., Henry, J. 1969. Can Primitive Farming Be Modernized? I.N.E.A.C. Hors Series, O.C.D, Brussels. Belgium.

Kannegieter, A. 1967. Zero cultivation and other methods of reclaiming Pueraria fallowed land for food crop cultivation in the forest zone of Ghana. *Tropical Agriculturist (Sri Lanka)*:123.

Kannegieter, A. 1969. The combination of short-term Pueraria fallow and zero cultivation and fertilizer application: Its effects on a fallowing maize crop. *Tropical Agriculturist* 3 and 4:125.

Keen, B.A. 1931. The Physical Properties of the Soil: Longmans, Green & Co.. London.

Kellogg, C.E. 1938. Soil and Society. In "Soils and Men." Yearbook of Agriculture. USDA. Washington, DC, 863–886.

Köller, K. 2003. Techniques of soil tillage. In: Soil Tillage in Agroecosystems. Titi A.L., Ed: CRC Press. Boca Raton, FL, 1–25.

Kuipers, H. 1970. Historical notes on zero tillage concept. *Neth. J. Agric. Sci* 18:219–294.

Lal, R. 1973. Effects of methods of seedbed preparation and time of planting of maize in western Nigeria. *Exp. Agric* 9:303–313.

Lal, R. 1974. Soil temperature, soil moisture and maize yield from mulched and unmulched tropical soil. *Plant and Soil* 40:129–143.

Lal, R. 1976a. No-tillage effects on soil properties under different crops in western Nigeria. *Soil Sci. Soc. Amer. Proc* 40:762–768.

Lal, R. 1976b. Soil erosion on Alfisols in western Nigeria. I. Effects of slope, crop rotation and residue management. *Geoderma* 16:363–375.

Lal, R. 1986. Effects of six years of continuous no-till or puddling systems on soil properties and rice (Orryza sativa) yield of a loamy soil. *Soil and Tillage Research* 8: 181–200.

Lal, R. 1989. Conservation tillage for sustainable agriculture: Tropics vs. temperate environments. *Adv. Agron* 42:85–197.

Lal, R. 1995. Tillage systems in the tropics: Management options and sustainability implications. FAO Soil Bulletin 71. Rome. Italy.

Lal, R. 1999. Soil management and restoration for carbon sequestration to mitigate the
accelerated greenhouse effect. *Prog. Env. Sci* 1:307–326.

Lal, R. 2001. World cropland soils as source or sink for atmospheric carbon. *Adv. Agron*
71:145–191.

Lal, R. 2003. Soil erosion and the global carbon budget. *Env. Intl. (in press)*.

Landers, J.L. 2001. Zero-tillage development in tropical Brazil. FAO Agricultural Services
Bulletin 147. Rome. Italy.

Lawes, A.D. 1962. Rainfall conservation and yield of cotton in northern Nigeria. Samaru
Research Bulletin 16: Ahmadu Bello Univ.. Samaru. Nigeria.

Lerche, G. 1994. Ploughing Implements and Tillage Practices in Denmark from the Viking
Period to About 1800: Experimentally Substantiated. Commission for Research on
the History of Agricultural Implements and Field Structures 8. Royal Danish Academy of Sciences and Letters: Poul Kristensen. Herning.

Lerche, G. 1985. Radiocarbon datings of agricultural implements. In "Tools and Tillage"
1868–1995. Revised calibrations and recent additions. Tools and Tillage 4:172–205.

Leser, P. 1931. Entstehung und Verbreitung des Pfluges. Münster in W. (reproduced 1971
by the International Secretariat, Copenhagen).

Lewton, F.L. 1943. Notes on the old plows in the United States National Museum. *Agric.
History* 17:62–64.

Lowdermilk, W.C. 1939. Conquest of the Land Through 7000 Years. USDA-ARS Bulletin
99, Washington, DC.

Mabbayad, B.B., Buensosa, I.A. 1967. Tests on minimum tillage of transplanted rice.
Philippine Agriculture 5:541–555.

Miller, L.R. 2000. Horse-drawn Plows and Plowing. Small Farmer's Journal, Sisters, OR.

Muller, J., De Bilerling, O. 1953. Les methods culturales indigenes sur les sols equatoriaux
de plateau. *I.N.E.A.C. Belge Bull. Inform* 2:21–30.

Myers, W.B. 1996. Human Impact on Earth: Cambridge University Press. Cambridge.
U.K.

Ogunremi, L.T., Lal, R., Babalola, O. 1986a.. Effects of tillage methods and water regimes
on soil properties and yield of lowland rice from a sandy loam soil in southwest
Nigeria. *Soil and Tillage Research* 6:223–234.

Ogunremi, L.T., Lal, R., Babalola, O. 1986b. Effects of tillage and seeding methods on
soil physical properties and yield of upland rice for an Ultisol in southeast Nigeria.
Soil and Tillage Research 6:305–324.

Olson, G.W. 1981. Archeology: Lessons on Future Soil Use. *Journal of Soil and Water
Conservation* 32:130–132.

Ospina, V., Nel, P. 1957. La agriculture moderna y la conservacion de les recurses naturales.
Agri. Trop 13:163–180.

Owens, L.B. 1994. Impacts of soil nitrogen management on the quality of surface and
sub-surface water. In: Soil Processes and Water Quality. Lal R., Stewart B.A., Eds:
CRC/Lewis Press. Boca Raton, FL, 137–162.

Pachauri, R.K., Sridharan, P.V. 1998. Looking Back to Think Ahead: Green India 2047.
Tata Energy Research Institute, New Delhi. India.

Pereira, H.C., Jones, P.A. 1954. A tillage study in Kenya. II. The effects of tillage practices on structure of soil. *Emp. J. Expl. Agric* 23:323–331.

Pereira, H.C., Wood, R.A., Brzostowobi, H.W., Hosegood, P.H. 1958. Water conservation by fallowing in semi-arid tropical East Africa. *Emp. J. Expl. Agric* 26:213–228.

Phillips, S.H., Young, H.M. 1972. No-Tillage Farming. Reiman Associate. Milwaukee, WI.

Pimentel, D., Harvey, C., Resosudarmo, P., Sinclair, K., Jurz, D., McNair, M., Crist, S., Shpritz, L., Fitton, L., Saffouri, R., Blair, R. 1995. Environmental and economic costs of soil erosion and conservation benefits. *Science* 267:1117–1123.

Poincelot, R.P. 1986. Toward a More Sustainable Agriculture: AVI Publishing Co.. Westport, CT.

Ratchdawong, S., Boonchee, S., Ryan, K.T., Brigatti, J. 1984. Conservation Farming Systems in Northern Thailand. Land Development Division, Bangkok. Thailand.

Richards, J.F. 1990. Land transformation. In: The Earth as Transformed by Human Action: Global and Regional Changes in Biosphere over the Past 300 Years. Turner B.L., Clark W.C., Kates R.W., Richards J.F., Mathews J.T., Myers W.B., Eds: Cambridge University Press. Cambridge. U.K., 163–178.

Rodriguez, M., Lal, R. 1985. Growth and yield of paddy rice as affected by tillage and nitrogen levels. *Soil & Tillage Res* 6:163–178.

Rogin, L. 1931. The Introduction of Farm Machinery in Its Relation to the Productivity of Labor in the Agriculture of the United States in the Nineteenth Century: Univ. California Press. Berkeley.

Russell, E.W., Keen, B.A. 1941. Studies in soil cultivation. X. The results of a six-year cultivation experiment. *J. Agric. Sci* 31:326–347.

Sach, F. 1968. Proposal for classification of pre-industrial tilling implements. *Tools & Tillage* 1(1):3–27.

Schertz, D.L. 1988. Conservation tillage: An analysis of acreage projections in the United States. *Journal of Soil and Water Conservation* 43:256–258.

Semple, E.C. 1928. Ancient Mediterranean agriculture (citing: Odyssey, IX, 136; Vergil, Georgicas, I, 42–49, 71–72, Theophrastus, De Causis, III, Chap. X, 1; Xenophon, Oeconomicus XVI: 1–15, Hesoid, Theogony XVIII, 542; Odyssey X, 127; Odyssey XIII, 32).

Sharma, P.K., DeDatta, S.K. 1986. Physical properties and processes of puddle rice soils. *Adv. Soil Sci* 5:139–178.

Sharpley, A.N., Halvorson, A.D. 1994. The management of soil phosphorus availability and its impact on surface water quality. In: Soil Processes and Water Quality. Lal R., Stewart B.A., Eds: CRC/Lewis Press. Boca Raton, FL, 7–90.

Shenk, M.D., Saunders, J.L. 1981. Vegetation management systems for crop production in tropical regions of Central America: Costa Rica. In: No-tillage Crop Production in the Tropics. Akobundu I.O., Deutsch A.B., Eds. Plant Protection Center: Oregon State Univ.. Corvallis, 3–85.

Sidiras, N., Derpsch, R., Mondardo, A. 1983. Effect of tillage systems on water capacity, available moisture, erosion and soybean yield in Paraná, Brazil. In: No-tillage Crop Production in the Tropics. Akobundu I.O., Deutsch A.E., Eds. IPPC: Oregon State Univ.. Corvallis, 154–165.

Sidiras, N., Vieira, M.J. 1984. Compartamento de um Latossolo Roxo distrofico, compactado pelas rodes do trator na semeadura. *Pesquisa Agropecuaria Brasileira* 19: 1285-1293.

Sidiras, N., Roth, C.H. 1985. Measurements of infiltration with double-ring infiltrometers and a rainfall simulator as an approach to estimate erosion by water under different surface conditions on an Oxisol. 10th ISTRO Conf. 8-12 July 1985, Guelph, Ontario, Canada.

Sidiras, N., Henklain, J.C., Derpsch, R. 1982. Comparison of three different tillage sytems with respect to aggregate stability, the soil and water conservation and the yield of soybean and wheat on an Oxisol. Proc. 9th ISTRO Conf, Osijek, Yugoslavic, 537-544.

Sidiras, N., Vieira, S.R., Roth, C.H. 1985a. Determinacao de algumas caracteristicsas fisicas de um latossolo roxo distrofico sob plantio directo e preparo convencional. *Revista Brazileira de Ciencia do Solo* 8:265-268.

Sidiras, N., Heinzman, F.X., Kahnt, G., Roth, C.H., Derpsch, R. 1985b. The importance of winter crops for controlling water erosion and for the summer crops on two Oxisols in Paraná, Brazil. *Zeitschrift für Acker- und Pflanzenbau* 155:205-214.

Singh, B. 1989. Rehabilitation of alkaline wasteland on the Gangetic alluvial plains of U.P., India through afforestation. *Land Degradation and Rehabilitation* 1:305-310.

Singh, G., Singh, N.T., Abrol, I.P. 1994. Agroforestry techniques for rehabilitation of degraded salt-affected lands in India. *Land Degradation and Rehabilitation* 5: 223-242.

Singh, H.P. 1982. Management of desertic soils. In: Review of Soil Research in India. ICAR, New Delhi. India, 676-699.

Söhne, W. 1992. Bodenbearbeitungs und Erntetechnik: Ein historischer Abri von Anbeginn bis heute: DLG-Verlag. Frankfurt.

Steinbeck, J. 1939. The Grapes of Wrath: New York: Penguin.

Stout, B.A. 1984. Energy Use and Management in Agriculture: Breton Publishers. North Scituate, MA.

Suwardjo, H., Abdurachman, A., Sutono, A. 1984. Effect of mulch and tillage on soil productivity of a Lampung Red Yellow Podsolic, Indonesia. *Pembr. Pen. Tanah dan Pupuk* 3:12-16.

Thamburaj, S., Shanmugavelu, K.G., Muthukrishnan, C.R., Vijaykumar, M. 1980. Studies on no-tillage in tapioca. Natl. Seminar Tuber Crops Prod. Techn: Tamil Nadu Agric. Univ. Coimbatore. India.

Tisdale, S.L., Nelson, W.L. 1966. Soil Fertility and Fertilizers, Third Edition. Macmillan, New York.

Troeh, F.R., Hobbs, J.A., Donahue, R.L. 1980. Soil and Water Conservation for Productivity and Environmental Protection: Prentice-Hall Inc.. Englewood Cliffs, NJ.

USDA. 1977. A Guide to Energy Savings for the Field Crop Producer. U. S. Department of Agriculture, Washington, DC.

Van Doren, D.M. 1967. Changes in seed environment due to tillage. In: Proc. Conf. on Tillage for Greater Crop Production: ASAE Publication 168, 5-9.

White, K.D. 1967. Agricultural Implements of the Roman World, Cambridge.

Wijewardene, R. 1982. Conservation Farming for Small Farmers in the Humid Tropics. IITA/Sri Lanka Program. Colombo. Sri Lanka.

5

Agroecological Thoughts on Zero-Tillage: Possibilities for Improving Both Crop Components of Rice–Wheat Farming Systems with Rice Intensification

Norman Uphoff
Cornell University
Ithaca, New York, U.S.A.

The progress reported and evaluated at this conference on the use of zero-tillage (ZT) practices in the Indo-Gangetic Plains (IGP) of South Asia is very welcome in a region where it is crucial to maintain advances in agricultural productivity while also conserving the natural resource base on which further advances depend. It should not be surprising that ZT practices offer South Asian farmers the same kinds of benefits that their counterparts in Brazil and other countries in Latin America's Southern Cone have gotten from ZT (Calegari, 2002).

In the spirit of interdisciplinary discussion and exploration, I suggest here some ways in which zero-tillage systems in South Asia might be further strengthened by making some changes in the agronomic practices for growing irrigated rice as part of the rice–wheat farming systems (RWS) that are benefiting from ZT. I am not an agroecologist by training or by profession, but rather by co-optation, being a social scientist who has worked with agricultural scientists over the past 30 years on a variety of problems of agricultural and rural development. I have been learning from them and with them how to improve rural livelihoods and food security in different parts of the world. I was drawn into the field of agroecology through my involvement with the System of Rice Intensification (SRI) developed in Madagascar (Uphoff, 1999, 2002a). My approach to agronomic subjects is inductive, prompted by observations and a need to explain how this system could give such remarkable increases both in yield and in the productivity of land, labor, capital, and water as all factors of production.

This chapter raises some questions about future directions for ZT development and complements that by Duxbury (this volume), which seeks to expand consideration beyond ZT to address fundamental constraints to productivity and sustainability in the rice–wheat farming system, particularly considering soil microbiology. This chapter also reports briefly on experience with ZT in Brazil, the country where such practices were first widely utilized. This has been one of the most dramatic and important cases supporting the application of agroecological principles.* Zero-tillage has been viewed usually as a change in mechanical practices that has beneficial impacts on soil physics and thus plant growth. In agroecological terms, it is a promising approach to plant, soil, and water management that has potential positive consequences for soil biology and whole ecosystems.

I. THE SYSTEM OF RICE INTENSIFICATION

Zero-tillage practices are being introduced in South Asia within the rotational rice–wheat farming systems that dominate large areas of northern India, parts of Pakistan and Bangladesh, and the terai in Nepal. The relevance of the SRI developed in Madagascar during the 1980s and now starting to be adopted in other countries, already more than 15, could be twofold for ZT systems. We have found that SRI methods can raise yields by 50% to 100% or even more, without requiring the use of new seeds or any purchased inputs. These higher yields are achieved with reduced use of water for the rice crop, which with SRI, in contrast to current irrigation practices, is not grown under continuously flooded conditions.

With reference to the rice–wheat systems that are our focus in this volume, the methods used for SRI can mitigate or eliminate many of the problems that have been identified by contributors to this volume from India, Pakistan, Bangladesh, and Nepal as threatening the productivity and sustainability of rice–wheat systems in the IGP region (see chapters from Bhuiyan, Gil, Harrington, Hobbs, Pathik, Samra, and Singh):

> the *need to raise productivity per unit of land* given declining availability of arable land per capita in the region;

---

* Most of my acquaintance with ZT has come through the chapter written by Ademir Calegari (2002), one of the researchers who has contributed most to the understanding and spread of ZT in the Paraná state of Brazil, for a volume that I edited with papers and conclusions coming from a Bellagio conference on sustainable agriculture (Uphoff, 2002b). This book gives an overview and analyses, amplified by a dozen case studies, of productive agroecological approaches to agriculture in Asia, Africa, and Latin America for the 21st century.

degraded soil physical properties for the wheat crop due to puddling of
fields for the preceding rice crop;

dropping water tables as a result of heavy irrigation demand, particularly
for rice;

low soil organic matter;

pollution of ground- and surface water from insecticides and chemical
fertilizers;

pest and disease pressures, including soilborne pathogens;

lodging of grain;

rising costs of production, especially for energy inputs; and most important,
declining factor productivity for the rice–wheat system as a whole.

Harrington and others show how ZT practices can contribute to solving a number
of these problems. As will be shown here, so can SRI. Duxbury suggests that a
focus on ZT should be expanded to address other factors that constrain rice–wheat
farming systems, bringing in other, complementary practices such as soil solarization to increase seedling health. Similarly, the plant, soil, water, and nutrient
practices associated with SRI can minimize many of the problems and constraints
listed above if used in conjunction with ZT and practices suggested by Duxbury
(this volume).

1. By growing rice under intermittently wetted conditions rather than continuous flooding during the vegetative growth period, SRI methods for the rice part
of RWS could provide *more favorable soil conditions for the following wheat
crop*. There could also be other advantages with SRI for the RWS system.* But
these are not the focus of this paper.

2. The SRI experience raises interesting questions about what are *optimizing
agronomic management practices for plant growth*. Possibly some of the ways

---

* Soil chemical and physical properties can be improved with SRI methods for rice as upper horizons
are less reduced and will have better structure for the subsequent wheat crop. The soil after an SRI
crop would be more suitable for wheat because with SRI, rice roots penetrate more deeply and
grow more voluminously. Perhaps more important would be soil biological changes as the rice root
growth with SRI can enhance soil organic matter directly by contributing more to soil C and
indirectly by promoting more diverse microbial populations in the soil. Use of compost rather than
NPK fertilizer adds to soil organic matter. Compost is not, however, a necessary part of SRI, as
these methods can improve yield also with fertilizer. Since SRI requires only about half as much
irrigation water for a season of rice production compared to conventionally flooded rice, the long-
term viability of RWS could be enhanced by these new methods. Especially where irrigation water
is drawn from a falling water table, this could be very important for RWS sustainability. Further,
with the productivity of water as well as of land, labor, and capital increased when SRI methods
are properly used, RWS should become more profitable, which would contribute to its sustainability.
An additional environmental benefit could be reduction in methane emissions from ending the
continuous flooding of rice paddies. A cross-national assessment of SRI potentials and limitations
can be found in Uphoff et al. (2002).

in which plants, soil, water, and nutrients are jointly managed within SRI to obtain higher yields can provide leads for better understanding and further improving ZT. This is the focus of the paper.

The System of Rice Intensification, developed by Fr. Henri de Laulanié through his 34 years of work with farmers in his adopted country of Madagascar (de Laulanié, 1993), changes some practices that farmers in Asia and elsewhere have used for growing irrigated rice for centuries, even millennia. There is demonstrable synergy among the new but simple practices discussed below. SRI was probably not discovered earlier because its practices appear to be risky, though they need not be. They are certainly counterintuitive, giving expression to the illogical principle that "less can be more"—fewer and smaller seedlings with reduced external inputs give higher production. Not everything in the world, especially in the biological realm, is logical or intuitive. SRI should not be rejected just because it appears "too good to be true." It must bear a greater burden of proof to gain acceptance, but logical *a priori* objections do not make the system false.

A. SRI Practices

1. SRI involves, first, transplanting *young seedlings*, 8–12 days old and certainly not more than 15 days old, compared to the 3–4 weeks that is the norm (deDatta, 1987). We find that early transplanting preserves genetic potential for massive tillering and root growth that is suppressed when transplanting occurs after the start of the fourth phyllochron and when other conventional practices for growing irrigated rice are followed.

2. When transplanting SRI seedlings, they are *widely spaced* in a square pattern, 25 × 25 cm or even farther apart if soil quality is good, and *one per hill*, rather than in clumps of 3 or 4, as is usual practice (deDatta, 1987). The seeding rate is 5–10 kg/ha compared with conventional rates 10 times higher, which can save as much as 100 kg of rice seed per hectare.

3. Transplanting should be done *quickly and carefully*, getting seedlings from the nursery, which is watered like a garden rather than kept flooded, into the field in 15 to 30 minutes, with minimum trauma to the roots and no desiccation. The root tips should be placed in the soil carefully so that they are pointing downward or are at least horizontal and not inverted upward. The one or two weeks of growth lost through conventional transplanting, where seedlings are jammed downward into flooded paddies, greatly reduce subsequent tillering.

4. During the vegetative growth phase, rice paddies are not kept constantly flooded. Instead, the *soil is kept unsaturated and aerated as much as possible* to avoid the die-back of roots occurring under hypoxic conditions. It is commonly

believed that rice is an aquatic plant.[*] However, although it can survive under continuously flooded conditions and even give some good results, there is evidence (e.g., Ramasamy et al., 1997) that rice does not perform its best under such conditions.

Because of the assumption that rice is well adapted to flooded conditions, three observable phenomena have been given little attention in the literature or in practice.

1. When soil is hypoxic, *roots remain fairly close to the soil surface* (the soil–water interface, where dissolved oxygen is available). About three fourths of irrigated rice plant roots are in the top 6 cm of soil 29 DAT (Kirk and Solivas, 1997), whereas under unflooded conditions, roots grow to depths of 30 to 40 cm (Barison, 2002).

2. The roots of rice plants, whether of varieties developed for irrigated or upland production, form air pockets (aerenchyma) when grown under flooded conditions. They do *not* form aerenchyma when grown in soil that is well drained.[†]

3. Moreover, under flooded conditions, by the time of panicle initiation, *about three fourths of rice roots degenerate* (Kar et al., 1974) and are no longer able to access nutrients for themselves and the plant. Because there is more accompanying senescence of leaves, reducing the contribution of photosynthesis, plants growing in saturated soil must depend more on nutrient uptake during their vegetative growth than if their roots and leaves remained intact and vigorous throughout their reproductive period as well.

The superficiality and degeneration of irrigated rice roots are widely known, but this morphology and "senescence" are considered to be natural rather than a consequence of conventional water management practices. They are regarded more as givens than as variables, and thus there is little research comparing rice grown under flooded versus unflooded conditions. Because upland (rainfed) rice, which is subjected to many stresses and is usually grown in poorer soils, has generally lower yields than paddy (irrigated) rice, the superiority of the latter has been assumed rather than tested, e.g., compared with lightly irrigated rice in moist but well-drained soil.

---

[*] "[Rice] thrives on land that is water-saturated, or even submerged, during part or all of its growth cycle" (DeDatta, 1987:43). "Most varieties maintain better growth and produce higher grain yields when grown in flooded soil than when grown in unflooded soil" (DeDatta, 1987:297–298).

[†] Puard et al. (1989) showed that an "irrigated" rice variety grown under unflooded conditions does not form air pockets (aerenchyma) in its roots to improve the supply of O_2 to its root cells as it does under flooded conditions. Kirk and Bouldin (1991) have described this disintegration of the rice root cortex to form aerenchyma as "often almost total," adding that "disintegration of the cortex [under saturated soil conditions] must surely impair the ability of the older parts of the root to take up nutrients and convey them to the stele." After panicle initiation, they add, "the main body of the root system is largely degraded and seems unlikely to be very active in nutrient uptake."

The superficiality and degeneration of irrigated rice roots appear, to some-one not trained in plant science, to be unfortunate anomalies bound to have some adverse effect on plant growth and performance. Little is made of these distortions of "normal" plant growth, however, because the prevailing scientific paradigm for crop improvement focuses on raising the *harvest index* (HI), the proportion of "total plant biomass" that ends up in the edible portion of the plant. This perspective pays little attention to roots and their function. Indeed, roots are not even included in calculations of "total plant biomass." In fact, a preoccupation with HI has led some agronomists to consider roots to be "a waste." Our experi-ence with SRI gives evidence as to why this neglect in crop science is so unfortu-nate, and why it should be avoided in our thinking about ZT.[*]

5. The other elements of SRI are not controversial, but their contributions to its synergy are important. When rice paddies are not kept continuously flooded during the vegetative growth phase, it is necessary to do more *weeding* to control the weeds that can grow. Herbicides could be used, but they do not contribute to soil aeration. For SRI, we recommend use of a mechanical weeder ("rotating hoe") that was developed by IRRI several decades ago. It not only removes weeds but churns up the top layers of the soil, aerating them. At least two weedings are recommended, starting 10–12 days after transplanting, and as many as 3 or 4 if possible before canopy closure makes further weeding difficult and unneces-sary. We have found that additional weedings can increase yield, other things being equal, probably not for their removal of weeds but because of the associated soil aeration.[†]

6. The NGO that has been promoting SRI most actively in Madagascar, Association Tefy Saina, with which CIIFAD has worked since 1994, does not consider the *use of compost* to be an essential characteristic of SRI. Tefy Saina regards compost more as an accelerator of production than as a requirement for SRI. Indeed, SRI was developed in the 1980s with the use of chemical fertilizer, when it was still subsidized in Madagascar; when its price shot up at the end of

---

[*] For evidence of this neglect, readers can consult the most widely cited text on rice by DeDatta (1987). In its chapter on the morphology, growth, and development of the rice plant (Chapter 5), out of 390 lines of text, only 8 are devoted to roots; among the more than 1,100 entries in the book's index, there is not a single entry on roots. DeDatta explained this deficiency by saying that there is a great paucity of research on roots in rice (personal communication), which is, unfortunately, correct.

[†] We have data from farmers' fields showing increases of 0.5 to 2.0 t/ha increase in yield, other things being equal, for each additional weeding beyond the minimum of two (Uphoff, 1999). Recent data from a researcher at the University of Madagascar found, on similar soil with comparable other practices, a yield of 3.75 t/ha with hand weeding, 5.3 t/ha with two rotating-hoe weedings, 6.1 t/ha with three rotating-hoe weedings, and 7.6 t/ha with four rotating-hoe weedings. With use of herbicide (2,4-D), the yield was 5 t/ha, and with herbicide plus one rotating-hoe-weeding 5.3 t/ha (Ndriantsoavina, unpublished data, 2002).

the decade, Fr. de Laulanié began working instead with compost, made from any available biomass, and got even better results.

The use of compost with the other SRI practices is encouraged because there is evidence from factorial trials of the greater benefits from using compost compared to NPK, especially for traditional varieties not bred for fertilizer-responsiveness (Tables 1, 2). As discussed below, compost can support greater diversity and dynamics of soil microbial populations, which appear to be a major influence increasing SRI yields. Processes such as biological nitrogen fixation (BNF) can be inhibited by application of chemical fertilizer (Van Berkum and Sloger, 1983).

B. SRI Results

To date, we have observed substantial increases in yield from these practices when used with any and all rice varieties, both traditional and improved (high-yielding). The practices apparently provide soil and other environmental conditions that are more favorable for rice plant growth and facilitate a different, more productive phenotypic expression of genotypic potential that has existed in rice plants for millennia.

One sees, for example, that with SRI management, varieties produce several times *more tillers* than observed with conventional management. Averages of 30 to 50 tillers per plant are not difficult to achieve, and the number can reach 70 or even more. While there are fewer plants per m^2 with SRI, there are as a rule more tillers per m^2 in total.

At the same time, there is *greater root growth* with SRI methods, which supports both more tillering and grain production. Measured root-pulling resistance per plant is about 5 times greater for SRI plants than for ones grown conventionally (Barison, 2002; Joelibarison, 1998).

With SRI, there is a *positive correlation between tillering and grain filling*, contrary to the previously reported *negative correlation* between the number of tillers per plant and the number of grains per panicle (Ying et al., 1998).

Specific yield results depend on numerous factors, as in any biological process. We have found that with SRI, yields often *increase over time*, even when they start off much higher than present averages. So the levels first reported with SRI often rise as farmers' skill with these practices increases and/or as soil quality improves as a result of those management practices.[*]

---

[*] It is possible that when soil conditions are more aerobic than in flooded rice paddies, over time there could be more problems with aerobic soilborne pathogens or nematodes. Paddies are kept superficially flooded during the reproductive phase after panicle initiation, maintaining 1–2 cm of water on the field, so SRI is not a fully anaerobic system, with the soil-water management objective being one of alternating wetting and drying.

To summarize yield results, without going into much detail, in Madagascar we find that on-farm yields average 8 or more t/ha with SRI practices under a wide variety of conditions, with somewhat higher yields at higher elevations with more day–night temperature contrast. This contrasts with a world average of 3.7 t/ha and in Madagascar with usual yields of about 2 t/ha, even though soils there are, in chemical terms, some of the poorest in the world (Johnson, 1994). Similar yields of 8–9 t/ha have been reported from China, Cuba, and Sri Lanka. Somewhat lower yields in the 5–6 t/ha range are reported from Cambodia, Myanmar, and the Philippines, but since this represents a two- to three-fold increase in their usual yield, farmers there consider this gain quite attractive. In Bangladesh, Indonesia, and Gambia, yields vary more, ranging between 5–9 t/ha (Uphoff et al., 2002).

Yield is not the most important consideration when discussing SRI, though it is the most dramatic. Some individual farmers have gotten yields in the 15–20 t/ha range with SRI methods in Madagascar and Sri Lanka. This sounds incredible, even though true. One should focus on gains in *productivity* since the objective is not to double or triple rice production, but rather to raise the productivity of land, labor, water, and capital devoted to rice production. In this way, SRI can support the *diversification* of agricultural production, which will enhance both its nutritional contribution and its profitability.

Madagascar and other countries presently devote too large a share of their land, labor, water, and capital to producing their staple food. They can never develop very far with staple-food productivity as low as it is today. SRI should enable them to move resources into higher-value production, with more income earned and more nutritional value. But this is enough said about SRI and its prospects and impacts as a system of production. The main concern of this volume is how to make ZT more profitable and have more sustainable agriculture in South Asia. To consider this, the following is a summary of what we have learned from the SRI experience about opportunities for agricultural development more generally in terms of two basic concerns:

1. More attention should be paid to the growth and functioning of *plant roots*, which are obviously crucial for plant performance. We know relatively little about how ZT contributes to the growth and functioning of roots. Everyone acknowledges that the plant sciences have devoted little attention to roots, and can give reasons for this neglect, but explanations and excuses do not remedy this major gap in our knowledge.

2. More attention should at the same time be paid to the contribution of *soil biology and microbiology* when evaluating and trying to improve agricultural systems. There has been some work on biological N fixation and mycorrhizal uptake of available soil P, but most soil science has been preoccupied with chemical and physical aspects of soil–plant–nutrient–water interactions, giving short shrift to biological processes in the soil. This gets justified more on pragmatic

than on theoretical grounds, although there is experimental evidence that should promote more interest than shown to date.[*] Biological work is difficult, costly, and often ambiguous. But we should be looking at ZT and its impacts on soil organic matter (SOM) not just in terms of what they can contribute *chemically* to plant nutrition and *physically* to plant growth, but what can they contribute to the *microbiological* dynamics that affect both chemical and physical aspects of the soil and contribute directly to plant growth.

With ZT, there is no question of different transplanting practices having a positive effect, but there could be benefits from *wide spacing* and particularly from maintaining *well-aerated soil* that would be relevant in ZT systems. Mycorrhizae play particularly important, though poorly understood, roles in plant growth, accessing phosphorus and other nutrients through hyphae as long as 10–12 cm and increasing the volume of soil accessed unaided by the host plant as much as 100-fold (Habte and Osorio, 2002). Being fungi, mycorrhizae do not grow under anaerobic conditions, and thus the benefits they can confer on rice are lost when paddies are kept flooded. Tillage has a definitely adverse effect on mycorrhizal populations that provide protection against various diseases and stresses (Johnson and Pfleger, 1992). That such practices should be considered and evaluated in a multivariate context can be seen from the SRI experience, which demonstrates the potential role of *synergy* among practices such as those that get combined in ZT.

II. EVIDENCE OF SYNERGY WITH AGROECOLOGICAL PRACTICES

Below are reported the results of two sets of carefully conducted factorial trials, both carried out by students in the Faculty of Agriculture at the University of Antananarivo for baccalaureate theses (*memoires de fin d'études*) (Rajaonarison, 2000; Andriankaja, 2001). The first trials ($N = 288$) were conducted in 2000 by Jean de Dieu Rajaonarison at the Centre de Baobab near Morondava on the west coast of Madagascar. This tropical area was known to have low soil fertility, but the likelihood of fewer pest and disease problems during the minor season made it attractive for conducting factorial trials there, since we were looking for any systematic but possibly small differences among varied combinations of factors.

These trials were conducted with **variety** as a first factor: high-yielding variety ($\times 2798$) versus traditional (*riz rouge*), with four conventional practices

---

[*] See, for example, the findings in Magdoff and Bouldin (1970), which show that mixing and alternating aerobic and anaerobic horizons of soil increases biological nitrogen fixation (BNF). The water management and weeding practices recommended for SRI have the effect of intermittently aerating the soil and rhizosphere.

compared with SRI alternatives to see what effect different combinations would have on yield and various yield components:

> **Water management**: saturated soils (SS) versus aerated soils (AS), the latter kept well-drained during the vegetative growth phase;
> **Age of seedling**: 20-day seedlings versus 8-day seedlings;
> **Plants per hill**: 3 versus 1;
> **Nutrients**: application of NPK fertilizer (NPK) in the amounts recommended by the national rice program versus compost (C), with a third option of no fertilization as a control.

A sixth factor of **spacing** was evaluated, with seedlings planted in a square pattern 25×25 cm versus 30×30 cm; both spacings were within the range recommended for SRI. There was no difference in yield between the two sets of trials (both $N = 144$), the mean yields being identical, so the data in Table 1 combine trials results from these two spacings. This means that yield reported for each of the combinations of practices represents the average for *six* replications, rather than the usual three.

Table 1 shows how under these soil and climatic conditions, the yield with four conventional practices (saturated soil, 16-day seedlings, three per hill, with NPK application), averaged 2.5 t/ha, combining results from both HYV and traditional varieties. Substituting one SRI practice for any one of the four conventional practices added an average of *half a ton* per hectare, and using any two SRI practices added on average a further *three quarters of a ton* to this; so did using three out of four SRI practices. Significantly, going from three to four, i.e., all SRI practices, added *almost two additional tons*.

Similar factorial trials ($N = 240$) were done the following year by Andry Andriankaja on farmers' fields in the village of Anjomakely, 18 km south of the capital Antananarivo on the high central plateau where soil and other growing conditions are more favorable with a more temperate climate. This time, variety was kept constant (all trials were with *riz rouge*), while **soil conditions** were varied as a factor (clay versus loam). There were fewer trial plots in this experiment because no control trials (without fertilization) were conducted on the poor (loam) soil plots.

Table 2 shows a similar set of relationships. The increases from adding one, two, or three SRI practices to the conventional set are greater—*between 1 and 1.6 tons per hectare* for each increment moving from conventional to SRI cultivation—but the biggest increase came when all SRI practices were used together—again, *almost 2 tons* per hectare. The patterns are similar for both better and poorer soil, with more than a tripling of yield achieved by SRI methods with both soil types. Tests of statistical significance showed the differences between sets of practices all significant in the 0.03 to 0.000 range.

Table 1 Factorial Trial Results, Morondava, 2000 Season (*N* shown in parentheses; SRI practices indicated in **boldface**)

| | Variety | | |
	HYV	Traditional	Average
Convetional practices			
SS/16/3/NPK	2.84 (6)	2.11 (6)	**2.48** (12)
+ 1 SRI practice			
SS/16/3/**C**	2.69 (6)	2.67 (6)	
SS/16/**1**/NPK	2.74 (6)	2.28 (6)	
SS/**8**/3/NPK	4.08 (6)	3.09 (6)	
AS/16/3/NPK	4.04 (6)	2.64 (6)	
	3.34 (24)	**2.67** (24)	**3.01** (48)
	[+0.50 t]	[+0.56 t]	[+0.53 t]
			+21.4%
+ 2 SRI practices			
SS/16/**1**/**C**	2.73 (6)	2.47 (6)	
SS/**8**/3/**C**	3.35 (6)	4.33 (6)	
AS/16/**1**/NPK	4.10 (6)	2.89 (6)	
AS/16/3/**C**	4.18 (6)	3.10 (6)	
SS/**8**/**1**/NPK	5.00 (6)	3.65 (6)	
AS/**8**/3/NPK	5.75 (6)	3.34 (6)	
	4.28 (36)	**3.24** (36)	**3.78** (72)
	[+0.94 t]	[+0.62 t]	[+0.78 t]
			+22.2%
+ 3 SRI practices			
SS/**8**/**1**/**C**	3.85 (6)	5.18 (6)	
AS/16/**1**/**C**	3.82 (6)	5.88 (6)	
AS/**8**/3/**C**	4.49 (6)	4.78 (6)	
AS/**8**/**1**/NPK	6.62 (6)	4.29 (6)	
	4.69 (24)	**4.28** (24)	**4.48** (48)
	[+0.41 t]	[+0.99 t]	[+0.70 t]
			+18.5%
All SRI practices			
AS/**8**/**1**/**C**	**6.83** (6)	**5.96** (6)	**6.40** (12)
	[+2.14 t]	[+1.68 t]	[+1.92 t]
			+42.9%

Source: Uphoff et al., 2002.

Table 2 Factorial Trial Results, Anjomakely, 2001 Season (*N* shown in parentheses; SRI practices indicated in **boldface**)

	Soil type		Average
	Clay	Loam	
Conventional practices			
SS/20/3/NPK	**3.00** (6)	**2.04** (6)	**2.52** (12)
+ 1 SRI practice			
SS/20/3/**C**	3.71 (6)	2.03 (6)	
SS/20/**1**/NPK	5.04 (6)	2.78 (6)	
SS/**8**/3/NPK	7.16 (6)	3.89 (6)	
AS/20/3/NPK	5.08 (6)	2.60 (6)	
	4.25 (24)	**2.83** (24)	**3.54** (48)
	[+1.25 t]	[+0.79 t]	[+1.02 t] +40.5%
+ 2 SRI practices			
SS/20/**1**/**C**	4.50 (6)	2.44 (6)	
SS/**8**/3/**C**	6.86 (6)	3.61 (6)	
AS/20/**1**/NPK	6.07 (6)	3.15 (6)	
AS/20/3/**C**	6.72 (6)	3.41 (6)	
SS/**8**/**1**/NPK	8.13 (6)	4.36 (6)	
AS/**8**/3/NPK	8.15 (6)	4.44 (6)	
	6.74 (36)	**3.57** (36)	**5.16** (72)
	[+2.49 t]	[+0.74 t]	[+1.62 t] +45.8%
+ 3 SRI practices			
SS/**8**/**1**/**C**	7.70 (6)	4.07 (6)	
AS/20/**1**/**C**	7.45 (6)	4.10 (6)	
AS/**8**/3/**C**	9.32 (6)	5.17 (6)	
AS/**8**/**1**/NPK	8.77 (6)	5.00 (6)	
	8.31 (24)	**4.59** (24)	**6.45** (48)
	[+1.57 t]	[+1.02 t]	[+1.29 t] +25.0%
All SRI practices			
AS/**8**/**1**/**C**	**10.35** (6)	**6.39** (6)	**8.37** (12)
	[+2.04 t]	[+1.80 t]	[+1.92 t] +30.0%

Source: Van Berkum and Sloger, 1983.

These data are presented to suggest that we approach and assess "no-till" strategies from an *agroecological* perspective, looking particularly at possible *interactions* among multiple factors: retention of soil moisture; the impacts of greater plant root growth on soil quality as well as on yields; increase in soil C (not lost by plowing); and promotion of soil biological activity. The latter can increase plant-available N through biological nitrogen fixation (BNF) (Baldani et al., 1997; Boddy et al., 1995), can enhance soil phosphorus supplies through P solubilization (Turner and Haygarth, 2001), and/or can contribute in other ways to soil enrichment and plant performance.

These effects can only be hypothesized at this stage, needing to be tested and established by proper experimental design and measurement. Such evaluation should not be on a once-and-for-all basis since the phenomena we are dealing with are biological processes, not just chemical ones. The effects of ZT should always be tracked over at least five years to establish the benefits and potentials of these alternative practices. Their effects are not just on plant nutrition from available nutrient pools but on dynamic processes that may increase (or decrease) these through soil biological activity.

One aspect of ZT that should be of special concern is its possible impact on *soil aeration*. With SRI, we think that this is of crucial importance not just for the O_2 that it provides to the soil and to the plant roots directly, but for its positive impact on microbial populations underground, on what can be called "subterranean biodiversity," including particularly mycorrhizae.

One of the positive contributions of tillage is to get air into lower horizons of the soil, although in the process, plowing exposes and loses soil C. There can be other losses from wind and water erosion that also offset the benefits of aeration. With tractor tillage there are some definitely adverse effects of soil compaction (plow pans) that accompany the soil mixing, as discussed in chapters by Morrison and Murray. This reduces soil aeration by altering the physical structure of the soil.

Tillage also adversely affects the populations of soil microfauna and macrofauna, particularly earthworms (Haynes, 1999). When abundant, macrofauna perform important soil aeration functions of their own. Soil fauna take up carbon and other minerals but also contribute these in organic form for subsequent mineralization; many symbiotically exist with root growth and death, being nurtured by plant exudates. Various bio-tillage agents such as earthworms and their burrowing predators enrich the soil in many ways. Possibly with ZT, the loss of soil aeration by not tilling is compensated for by the increased activity of macrofauna, offsetting the C and nutrient losses noted above. But this is something that needs to be investigated, neither assumed nor ignored.

The net effects of ZT on soil physical and biological properties, which in turn affect soil chemistry and nutrient uptake, should be assessed. Possibly some

of the beneficial effects reported for ZT can be explained by a better understanding
of these subsurface processes.

III. AGROECOLOGICAL PRECEDENTS: NO-TILL
AGRICULTURE IN BRAZIL

The ideas and concerns expressed here have been shaped by participation with
colleagues at a 1999 Bellagio conference on sustainable agriculture. Among the
cases considered was the introduction and spread of no-till cultivation in Brazil,
the best-documented example of successful ZT to date. Starting in the 1970s,
from the work of a single farmer, the area under no-till cultivation spread to
800,000 hectares throughout Brazil by 1985, and today it covers about 13.5 million hectares in that country.[*]

The main purpose of no-till practices in Brazil has been to curb soil erosion,
which had become a major problem by the 1970s due to widespread monoculture
that exposed soil to water and wind losses. A variety of plants are grown as
cover crops between cropping seasons. These are cut and used as mulch, with
succeeding crops planted into the residue, or they are incorporated into the soil
as green manure without deep plowing. The benefits of this are as follows:

> *Soil physical effects*—increased soil aggregate stability, and enhanced soil
> water infiltration rates due to greater soil porosity and aeration;
> *Soil chemical effects*—higher levels of N, P, K, Ca, Mg, and organic matter
> in the soil surface horizons, better soil organic C equilibrium, improved
> nutrient cycling and/or nitrogen fixation, plus decreases in Al toxicity;
> *Soil biological effects*—increased soil microbial populations and reduced
> populations of nematodes, also reduced impact of weeds due to changes
> in weed species and weed biomass.

These benefits are enhanced by well-designed systems of crop rotation, often
with leguminous crops, in cycles extending over two or three years. The larger
root systems encouraged by these practices in turn contribute to more soil organic
matter and better soil structure, improving soil quality and performance over time.
With no-till practices, particularly soil P is increased by the use of different winter
cover crops (see Figure 15.2 in Calegari, 2002).

Soybean yields have shown increases attributable to reduced tillage of 44%
to 56% compared to conventionally tilled systems over a 10-year period (see

---

* Information in this section comes from Calegari (2002), which gives extensive references to Brazilian research on ZT operations and results. In this volume, chapters by Benitez and   ιas also give
useful background on and insight into the development of ZT strategies and practices in Brazil

Figure 15.3 in Calegari, 2002), although the major gains for farmers have been in terms of lowered costs of production, which raised their net profits from agriculture. Benefits to the environment and society in terms of reduced soil erosion are hard to measure and assess, but EMBRAPA in 1991 estimated the potential savings from no-till technologies to be US$110 billion over a decade, calculated in terms of the value of the fertilizer equivalent of soil nutrients lost unnecessarily through erosion.

IV. DISCUSSION

This volume should enable many persons to gain a better understanding of the ZT farming systems being developed and diffused in South Asia and elsewhere. These should not be automatically classified as constituting a single kind of farming system, and the agroecological purposes that each type of ZT system serves should be carefully examined. The no-till farming systems spreading in Brazil, Paraguay, and Argentina were developed for different objectives and under different ecological conditions than those now found in India, Pakistan, Nepal, and Bangladesh. The multiplicity of ZT systems is stressed in Benites' contribution to this volume. Thus, ZP is not a technology but rather a challenging *category* of many practices—actually various *combinations of practices*—that serve a range of economic, ecological, and social purposes.

The System of Rice Intensification reported on above could be another component in the evolving rice–wheat farming systems within the region. Like ZT, it is also a combination of practices, many of which will vary according to soil conditions, labor availability, and other considerations, rather than being a single technology, to be adopted or not.

Some of SRI's practices or principles are already being demonstrated as valuable for the wheat component of RWS, particularly the planting of *single seedlings* with wide spacing and the growing of wheat on *raised beds* so that the rhizosphere is more aerated than with surface planting.[*] SRI, by reducing the inundation of fields during the rice phase of the system, could make for better growing conditions for the following wheat crop, giving higher yields and higher returns to land and labor while drawing down the water table less than with continuous flooding.

---

[*] These innovations were introduced by wheat-growing farmers on the Yaqui Valley of the Sonora state in Mexico starting in the 1980s. Their economic advantages are documented in Sayre (1997), although their agronomic effects are not analyzed there. It was found that a seeding rate of 15–25 kg/ha could produce as high a yield as 200 kg/ha, given that more widely spaced plants tillered more profusely and filled available field space.

We should know more about the short-term and long-term effects of ZT on soil organic matter. Such effects are attributable to the way that soil and water are managed, but also to the growth and functioning (and demise) of plant roots. Not only do roots contribute organic matter to the soil through their degeneration, but they also infuse the soil with carbohydrates and amino acids synthesized aboveground into the root zone in the form of *exudates*, thereby nourishing and increasing the populations of bacteria, fungi, protozoa, and other microorganisms in the soil. The amount of knowledge currently available on exudation in rice plants is minute (Wassman and Aulakh, 2000). The potential contributions of mycorrhizae are similarly little understood, largely because the continuously flooded growing environment for paddy rice has eliminated them as factors affecting productivity.

These several processes can most easily be (and should be) analyzed and evaluated in chemical terms, but more important will be to gain a better understanding of what is going on biologically—in, on, and around the roots. How can those beneficial processes be safeguarded and even accelerated? What are their implications for sustainable development in the region? Might ZT be mining the soil of nutrients, or building up the pool of nutrients available for plant growth through a process of microbiological "weathering" that can be much more rapid than the more commonly considered geochemical weathering processes?

Many grassland soils under diverse collections of plant species over thousands of years are often richer both chemically and biologically than other kinds of soil. Millennia of plant root–soil–water–nutrient–microbial interactions have transformed whatever parent material was available under grasslands into soils that are today some of the most capable of supporting a variety of plant species, particularly those of interest to us as crops. Accordingly, ZT, which moves agriculture toward more of a "grasslands" ecosystem, may serve to build up soil capabilities, at least for cereal crops, in ways that tillage practices—despite their other merits—have diminished. This suggests the importance of assessing zero-tillage not just in crop and soil science terms but also from an ecological (which includes historical) perspective.

REFERENCES

Andriankaja, A.H. 2001. Mise en evidence des opportunités de développement de la riziculture par adoption du SRI, et evaluation de la fixation biologique du l'azote. Mémoire de fin d'études, Ecole Supérieure des Sciences Agronomiques: University of Antananarivo. Antananarivo.

Baldani, J.I., Caruso, L., Baldani, V.L.D., S.R., Goi, Döbereiner, J. 1997. Recent advances in BNF with non-legume plants. *Soil & Biol Biochem* 29:911–922.

Barison, J. 2002. Nutrient-use efficiency and nutrient uptake in conventional and intensive [SRI] rice cultivation systems in Madagascar. M.S. thesis, Dept of Crop and Soil Sciences: Cornell University. Ithaca, NY.

Boddy, R.M., de Oliveira, O.C., Urquiaga, S., Reis, V.M., de Olivares, F.L., Baldani, V.L.D., Döbreiner, J. 2001. Biological nitrogen fixation associated with sugar cane and rice: Contributions and prospects for improvement. *Plant & Soil* 174:195–209, 1995.

Calegari, A. 2002. The spread and benefits of no-till agriculture in Paraná State, Brazil. In: Uphoff N., Ed. Agroecological Innovations: Increasing Food Production with Participatory Development. London, Earthscan, 187–202.

DeDatta, S.K. 1987. Principles and Practices of Rice Production. Malabar, FL: Robert Krieger.

de Laulanié, H. 1993. Le système de riziculture intensive malgache. *Tropicultura (Brussels)* 11:110–114.

Habte, M., Osorio, N.W. 2002. Mycorrhizas: Producing and applying arbuscular mycorrhizal inoculum: Overstory #102. http://agroforester.com/overstory/ovbook.html.

Haynes, R.J. 1999. Size and activity of the soil microbial biomass under grass and arable management. *Biol & Fert Soils* 30:210–216.

Joelibarison, R.J. 1998. Perspective de développement de la region de Ranomafana: Les mechanismes physiologiques du riz sur de bas-fonds: Cas du SRI. Mémoire de fin d'études, Ecole Supérieure des Sciences Agronomiques: University of Antananarivo. Antananarivo.

Johnson, B.K. 1994. Soil survey. In: Final Report for the Agricultural Development Component of the Ranomafana National Park Project in Madagascar. Raleigh, NC: Soil Science Department, North Carolina State University, 5–12.

Johnson, N.C., Pfleger, F.L. 1992. Vesicular-arbuscular mycorrhizae and cultural stresses. In Bethlenfalvay G.J., Linderman R.G., Eds. Mycorrhiza in Sustainable Agriculture. Madison, WI: Soil Science Society of America:71–79.

Kar, S., Varade, S.B., Subramanyam, T.K., Ghildyal, B.P. 1974. Nature and growth pattern of rice root system under submerged and unsaturated conditions. Il, Riso (Italy) 23, 173–179.

Kirk, G.J.D., Bouldin, D.R. 1991. Speculations on the operation of the rice root system in relation to nutrient uptake. In: F.W.T. Penning de Vries et al., eds., Simulation and Systems Analysis for Rice Production. Wageningen: Pudoc, 195–203.

Kirk, G.J.D., Solivas, J.L. 1997. On the extent to which root properties and transport through the soil limit nitrogen uptake by lowland rice. *Eur J Soil Sci* 48:613–621.

Magdoff, F.R., Bouldin, D.R. 1970. Nitrogen fixation in submerged soil-sand-energy material media and the aerobic–anaerobic interface. *Plant & Soil* 33:49–61.

Puard, M.P., Coucha, P., Lasceve, G. 1989. Etude des mecanismes d'adaptation du riz aux contraintes du milieu I: Modification de l'anatomie cellulaire. *L'Agron Trop* 44:156–173.

Rajaonarison, J.D. 2000. Contribution a l'amelioration des rendements de 2ème saison de la double riziculture par SRI sous experimentations multifactorielles: Cas des sols sableux de Morandava. Mémoire de fin d'études, Department of Agriculture: University of Antananarivo. Antananarivo.

Ramasamy, J.D., ten Berge, S.H.F.M., Purushothaman, S. 1997. Yield formation in rice in response to drainage and nitrogen application. *Field Crops Research* 51:65–82.

Sayre, K.D., Moreño Ramos, O.H. 1997. Applications of Raised-Bed Planting Systems to Wheat. Wheat Program Special Report No. 31. Mexico, CF: CIMMYT.

Stoop, W., Uphoff, N., Kassam, A.. A review of agricultural research issues raised by the System of Rice Intensification (SRI) from Madagascar: Opportunities for improving farming systems for resource-poor farmers. *Agricultural Systems* 71:249–274.

Turner, B.L., Haygarth, P.M. 2001. Phosphorus solubilization in rewetted soils. *Nature* 411, May 17:258.

Uphoff, N 1999. Agroecological implications of the System of Rice Intensification (SRI) from Madagascar. *Envir, Dev & Sustainability* 1:297–313.

Uphoff, N 2002a. Opportunities for raising yields by changing management practices: The system of rice intensification in Madagascar. In: Uphoff N., Ed. Agroecological Innovations: Increasing Food Production with Participatory Development. London: Earthscan, 145–161.

Uphoff, N., Ed. 2002b. Agroecological Innovations: Increasing Food Production with Participatory Development. London: Earthscan.

Uphoff, N., Fernandes, E.C.M., Yuan, L.P., Peng, J., Rafaralahy, S., Eds. 2002. Assessments of the System of Rice Intensification, Proceedings of an International Conference on the System of Rice Intensification. Sanya, China. April 1–4, 2002. Ithaca, NY: Cornell International Institute for Food, Agriculture, and Development.

Van Berkum, P., Sloger, C. 1983. Interaction of combined nitrogen with the expression of root-associated nitrogenase activity in grasses and with the development of N_2 fixation in soybean (Glycine max L. Merr.). *Plant Phys* 72:741–745.

Wassman, R., Aulakh, M.S. 2000. The role of rice plants in regulating mechanisms of methane emissions: A review. *Biol. Fertil. Soils* 31:20–29.

Ying, J., Peng, S., He, Q., Yang, H., Yang, C., Visperas, R.M., Cassman, K.G. 1998. Comparison of high-yield rice in tropical and subtropical environments, I: Determinants of grain and dry matter yields. *Field Crops Res* 57:71–84.

6

Problems and Challenges of No-Till Farming for the Rice–Wheat Systems of the Indo-Gangetic Plains in South Asia

Peter R. Hobbs
Cornell University
Ithaca, New York, U.S.A.

Raj Gupta
CIMMYT India Office
Pusa, New Delhi, India

The Rice–Wheat Consortium (RWC) is a CGIAR ecoregional program that has the following goal:

> *Strengthen existing linkages and partnerships with national research programs (NARSs), other international centers, advanced institutions and the private sector working in the region to develop and deploy more efficient, productive and sustainable technologies for the diverse rice–wheat production systems of the Indo-Gangetic Plains (IGP) so as to produce more food at less cost and improve livelihoods of those involved with agriculture and eventual decrease poverty.*

The consortium was established in 1994 to deal with the rice (*Oryza sativa L.*)–wheat (*Triticum aestivum L.*) (RW) farming systems practiced extensively in the Indo-Gangetic Plains (IGP) and Himalayan midhills of South Asia. These RW systems are one of the most important cropping systems in South Asia for food production, and millions of farmers depend on it for their livelihoods. Since almost as much area is devoted to RW systems in China, they represent one of the major farming systems in the world. This chapter provides information on various aspects of these systems in relation to no-till farming to acquaint readers with background information for better understanding the other contributions to this volume.

I. THE INDO-GANGETIC PLAINS

The IGP is a large ecoregion in South Asia with some of the most productive farmland in the world because of its fertile soils, favorable climate, and the availability of irrigation water (Fig. 1). The IGP occupies nearly onesixth of the total geographical area of the Indian subcontinent and is home to nearly 42% of the total population of 1.3 billion of South Asia. The population is increasing at about 2% per year, meaning that 26 million more mouths need to be fed each year. The RW systems are practiced on nearly 13.5 million ha in South Asia (Ladha et al., 2000) with another 12 m ha in China. In South Asia, these systems are used on almost one sixth of the cultivated area, and they produce more than 45% of the region's total food. Demand for rice and wheat is expected to grow at 2.5% per year over the next 20 years. Meeting this demand will become increasingly more difficult as the per capita rice–wheat growing area has already shrunk from 1200 m^2 in 1961 to less than 700 m^2 in 2001. Future food production growth will have to come from sustainable and profitable yield increases.

The IGPs are a relatively homogeneous ecological region in terms of vegetation but can be subdivided into five broad transects — the Trans IGP (shown in Fig. 1 as region 1 in Pakistan and region 2 in the Indian Punjab and Haryana); the Upper IGP (region 3); the Middle IGP (region 4); and the Lower IGP, part of region 4 in eastern India, and region 5 in Bangladesh (Fig. 1). These transects have been delineated based on factors such as the increasing demand for food, the development of ground- and surface water, infrastructure for inputs and research institutions, variables driving agricultural development, factors that affect the expansion of the rice–wheat area, and productivity constraints.

The climate of the IGP ranges from semi-arid in the west to subhumid in the east, with a distinct wet monsoon summer season and a dry, cool winter season. This allows rice and wheat to be grown in a double cropping pattern within one calendar year, rice in the summer and wheat in the winter. Temperatures can exceed 45°C in the summer, as well as frost in some areas in the winter. Evapotranspiration exceeds precipitation in the dry season and even in some parts in the wet season (especially in the semi-arid areas of the west), making supplemental irrigation necessary for crop production. The IGP was developed with extensive canal irrigation systems using water storage reservoirs in the Himalayan midhills. Canal irrigation is supplemented with tubewell water, and most of the rice–wheat areas are irrigated or partially irrigated.

Twenty percent of the soils are alluvium derived as a result of the deposition of the Indus and Ganges river systems. Many of the soils are alkaline in pH,

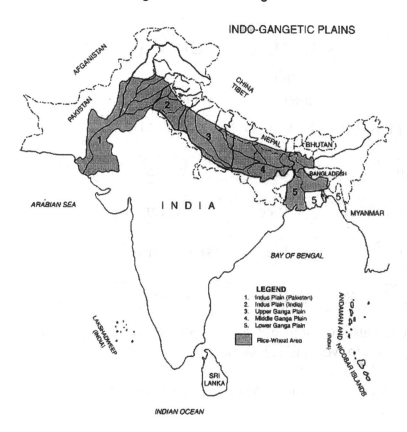

Figure 1 The rice–wheat areas of the Indo-Gangetic Plains by transect.

although acid soils are also present in the piedmont and some floodplains with acidic geology. Soils range in texture from loamy sands to silty clay loams.

The rice–wheat cropping system does not grow only rice and wheat. The cropping patterns are many and varied, with at least two and sometimes three or more crops grown in any one calendar year (Fig. 2). The triple and more intensive cropping patterns are found in transects 4 and 5 in the east, where average temperatures are warmer than in the west. There are many fields where farmers grow continuous rice–wheat, but in many cases, rotations break this continuous cereal system. Sugarcane, for example, is used in rotation with rice and wheat in parts of the IGP where it occupies the land for two or more years before returning to rice and wheat.

The high population density of the region and the fact that agriculture is the main form of employment and income mean that farm size is relatively small.

R-W Systems Calendar

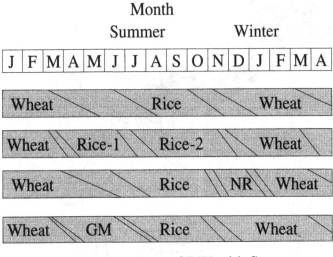

There are also rotations of RW with Sugarcane
NR = potato, vegetable, legume, oilseed

Figure 2   The rice–wheat systems calendar for the Indo-Gangetic Plains.

There are some larger farmers in the area, and some of these use tenants to crop their land. Many farmers crop less than one hectare of land, and many of these have to rely on off-farm employment to feed their families. Small landholders also lease out extra land to farm.

Mechanization has increased gradually from west to east over the last 30 years. Small four-wheeled tractors (35 hp mainly) in NW India and Pakistan do most of the land preparation, with farmers who do not own tractors renting them from service providers. In the east, many farmers still rely on animal power for land preparation. This is also changing since it is becoming expensive to keep a pair of bullocks year-round just for this purpose. Two-wheeled tractors are becoming popular on small farms in Bangladesh. There is also the issue of status, as the use of tractors gives higher status than the use of animals.

Many of the other farm operations like rice transplanting, weeding, fertilizer application, harvesting, and threshing are done manually. However, combine harvesters are becoming popular in NW India and Pakistan, and wheat threshers have spread from west to east over the past 20 years. Labor constraints, especially at key times like planting and harvesting, are gradually becoming significant, affecting timeliness of operations as industrialization grows. Many younger people shun the drudgery of farming and leave the rural areas for the cities in search of employment. Mechanization is therefore bound to increase to fill this power gap.

II. TRADITIONAL TILLAGE PRACTICES IN THE IGP

The most common practice for establishing rice in the rice–wheat systems of South Asia involves puddling—plowing soils when they are saturated—before transplanting rice seedlings. Puddling benefits rice by reducing water percolation and controlling weeds. However, puddling also results in degraded soil physical properties, particularly for finer-textured soils, and it subsequently creates difficulties in terms of providing good soil tilth for wheat. It also promotes the formation of a plow pan, which affects rooting depth in the next crop. This conflicting soil management situation is unique to the rice–wheat cropping system.

The implement most frequently used for preparing the soil with tractor power is either a nine-tine cultivator or a disk harrow. Deep plowing with a moldboard or disk plow is rare. With animal power a wooden plow with a metal tip is used. Although tractors allow land to be prepared more rapidly for wheat after rice, farmers still make many passes of plowing and planking: Six to eight passes with the plowing implement are common, usually followed by planking (leveling of the field using a wooden plank). Table 1 illustrates the time and number of operations for plowing in several areas of the IGP based on diagnostic surveys conducted at selected locations.

Table 1 Data on Tillage and Crop Establishment from Diagnostic Surveys of Selected Rice–Wheat Cropping Systems, South Asia

Location	Area planted late (%)	Turnaround time (days)	Average number of passes with plow
Punjab, Pakistan	40	2–10	2–10 (6)
Pantnagar, India	35	15–20	5–12 (8)
Faizabad, India	25	20–45	5–12 (6)
Haryana, India	25	15–35	4–12 (8)
Bhairahawa, Nepal	40	15–35	4–8 (6)

Note: Late planting is defined as wheat planted after the first week of December.
Source: Data for Punjab, Pakistan, from Byerlee et al. (1984); for Pantangar, India, from Hobbs et al. (1991); for Faizabad, India, from Hobbs et al. (1992); for Haryana, India, from Harrington et al. (1993b); and for Bhairahawa, Nepal, from Harrington et al. (1993a).

III. NO-TILL FARMING IN THE IGP AND ITS IMPORTANCE FOR TIMELY PLANTING AND GOOD PLANT STANDS

Late planting is a major problem in most rice–wheat areas of South Asia, except for the Indian Punjab (Fujisaka et al., 1994). To improve the productivity of the rice–wheat system, the wheat crop must be planted at the optimal time. The typical response of wheat to different dates of planting in South Asia, shown in Figs. 3 and 4, makes the optimum date some time at the end of November, with a linear decline in yield of 1%–1.5% per day after that (Ortiz-Monasterio et al., 1994). Although the slope of the line varies by variety and year, all show a decline regardless of whether they are short- or medium-duration varieties.

Late planting not only reduces yield but also reduces the efficiency of the inputs applied to the wheat crop. Nitrogen responses are much flatter in plots planted late compared to those planted at an optimal time (Saunders, 1990). This means that applying more nitrogen cannot compensate for the decline in yield due to late planting.

The reasons for late planting of wheat in the rice–wheat system are many (Fig. 5). An obvious cause of late planting is the late harvest of the preceding rice crop. Sometimes a short-duration, noncereal third crop will be planted after rice. In some part of the IGP, farmers grow long-duration, photosensitive, high-quality *basmati* rice that matures late. Farmers prefer to grow *basmati* despite its lower yields because of its high market value, good straw quality (livestock feed), and lower fertilizer requirements. *Basmati* varieties cannot be readily replaced

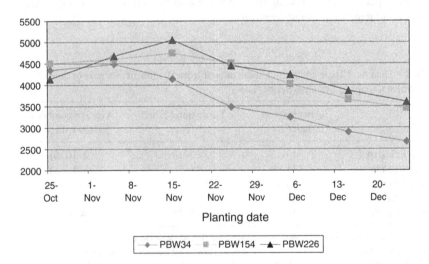

Figure 3 The effect of planting date on wheat yields by variety, PAU, Ludhiana, India.

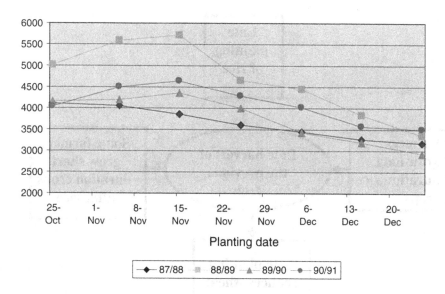

Figure 4 The effect of planting date on wheat yield, 1987–1991.

by a shorter-duration rice variety and thus late rice harvest results. However, in other areas, high-yielding, modern rice varieties are grown, and their planting dates can be manipulated so that wheat planting is not delayed. This is the case in the Indian Punjab, where modern rice varieties are planted early and harvested in early October, and the wheat is planted by the end of October or early November.

The other major cause of late wheat planting is the long turnaround time between rice harvest and wheat planting (Fig. 5). Long turnaround time can be caused by many factors, including excessive time for tillage, soil moisture problems (soil too wet or too dry), lack of animal or mechanical power for plowing, and the priority that farmers place on threshing and handling the rice crop before preparing land for wheat.

In addition to the problem of late planting of wheat is also the problem of poor germination and plant stands. The majority of farmers in IGP plant wheat by broadcasting the seed into plowed land and then incorporating the seed by another plowing. Part of the reason for this is residue management problems in fields following rice. The loose straw and stubbles are raked and clog the seed drills. Broadcast seed results in seed placement at many different depths and into different soil moistures with variable germination as a result.

The problems of late planting and poor plant stand have been addressed in the RWC by promoting various resource-conserving tillage and crop establish-

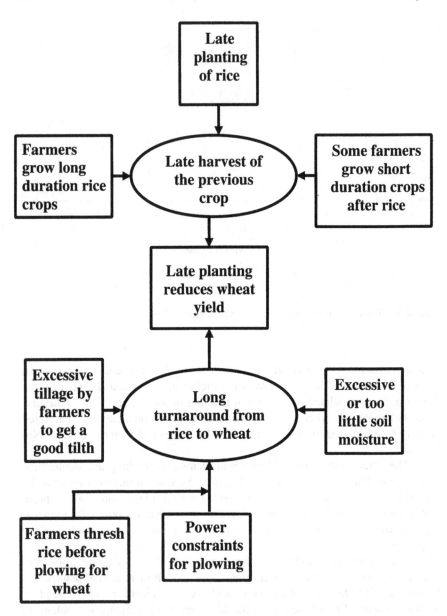

Figure 5   The most common causes of late wheat planting following rice harvest.

ment techniques described below. No-till farming is now gaining popularity in all five transects of the IGP as a way to overcome the above timing and plant-stand issues as seen in other chapters.

IV. NO-TILL TECHNIQUES PROMOTED IN THE RWC

A. Surface Seeding

This is the simplest no-till system being promoted (Fig. 6). In this tillage option, wheat seed is placed onto a saturated soil surface without any land preparation. This is a traditional farmer practice for wheat, legumes, and other crops in parts of eastern India, Nepal, and Bangladesh. Wheat or other crop seed is broadcast either before the rice crop is harvested (relay planting) or after harvest. One of the major advantages of surface seeding is that no equipment is needed, and any farmer can easily adopt this practice.

Promotion of surface seeding for planting wheat has been done for several years in areas where the soils are fine-textured and drain poorly, and where land preparation is difficult and often results in a cloddy tilth. These are areas that are mostly in rice-fallow but where sufficient moisture is available to grow a crop and where land preparation is very difficult and costly. It is one technology that can increase crop acreage in the future. The key to success with this system is having the correct soil moisture at seeding time. Too little moisture results in poor germination, and too much moisture can cause the seed to rot. A saturated soil is best. The seeds germinate into the moist soil and roots follow the saturation

Figure 6 Seed sprouting on surface of the soil after seeding.

fringe as it drains down the soil profile. The high soil moisture reduces soil strength and thus eliminates the need for tillage. Once the seed germinates and the root extends into the soil, the root can follow the saturation fringe and still get sufficient oxygen for growth. A good crop may even be possible without any additional irrigation.

As long as soil moisture can be manipulated, surface seeding is also success-ful on coarse-textured soils. There must be enough surface moisture to germinate the seed (soaking the seed before sowing can help), and soil strength at the root penetration stage must be low (moisture-dependent) to allow root growth. This may require an early, light irrigation on coarse-textured soils. Some farmers who relay wheat into the standing rice crop place the cut rice bundles on the ground after harvest. This allows the rice to dry but also acts as mulch, keeping the soil surface moist and ensuring good wheat rooting. In China, where surface seeding is also practiced, farmers apply cut straw to mulch the soil, reduce evaporative losses of moisture, and control weeds. The standing stubble also protects the young seedlings from birds. However, relay planting can be done only if the amount of soil moisture is correct for planting at this stage.

B.   No-Till with Inverted-T Openers

No-till is another resource-conserving technology (RCT), where the seed is placed into a soil slit made by a seed drill without prior land preparation (Fig. 7). This technology is presently being tested throughout the IGPs, and it has made substan-tial gains in NW India and Pakistan. It is estimated that 300,000 ha of no-till wheat were grown in 2002 in the IGP (Fig. 8). In eastern zones, equipment adapted to animal power and two-wheeled hand tractors (Fig. 9) is being devel-oped. However, farmers in this region are even using four-wheeled equipment through contractual services.

The basis for this technology is inverted-T opener equipment (Fig. 10) that was developed and imported from New Zealand to Pakistan by CIMMYT in 1983. This coulter and seeding system places the seed into a narrow slot made by the inverted-T as it is drawn through the soil by the four-wheeled tractor (Fig. 10). The coulters can be rigid or spring-loaded depending on the design and cost of the machine. This type of seed drill works very well in situations where there is little surface residue after rice harvest. This usually occurs after manual har-vesting.

Using a combine to harvest wheat is becoming popular among farmers in northwestern India and Pakistan. In this combine system, loose straw and residue are commonly left after harvest (Fig. 11). The inverted-T opener does not work as well where combines are used, since the opener acts as a rake for the loose straw. Farmers presently burn residues to overcome this problem (Fig. 11) of loose stubble whether they use no-till or the traditional system. Since the RWC

Figure 7   No-till seeder made in Ludhiana in India.

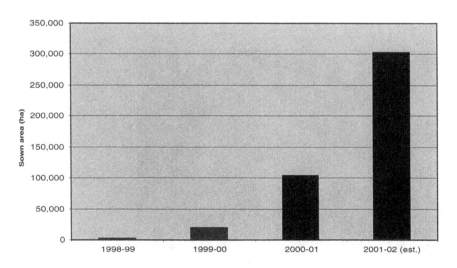

Figure 8   Trends in wheat no-tillage area rice–wheat systems in India and Pakistan.

Figure 9   Modified no-till drill with inverted T openers for two-wheeled hand tractors.

discourages the practice of burning because of its environmental and air pollution impacts, future strategies will look at alternative machinery and techniques to facilitate planting into the residue. Leaving the straw as mulch on the soil surface has not been given much thought in Asian agriculture until now. However, results from rainfed systems and some preliminary results elsewhere in Asia suggest

Figure 10   The Inverted-T coulter used for zero-tillage in the IGP.

Figure 11   Surface residue after rice combine harvesting and burning of the residue.

that this may be very beneficial to the establishment and vigor of crops planted with no-till (Sayre, 2000). Studies are needed to explore the benefits and long-term consequences of this approach.

C.   Reduced Tillage

Many farmers are reluctant at first to accept that no-till will work. It goes against their experience with farming. Many farmers first try reduced tillage using the same no-till drill described above. In other situations, carryover weeds from the rice crop create problems. A reduced tillage system may be more feasible since few farmers have the skills or equipment for efficient herbicide application and effective weed control. The following is a brief description of what the RWC is promoting in the IGP on this topic.

The Chinese have developed a seeder for their 12-horsepower, 2-wheeled diesel tractor that prepares the soil and plants the seed in one operation. This system consists of a shallow rotovator followed by a six-row seeding system and a roller for compaction of the soil (Fig. 12). Funding from the Department for International Development (DFID, UK) and CIMMYT made it possible to import several tractors and implements from Nanjing, China, into Nepal, Pakistan, and India in the early 1990s, where they have been tested over the past several years

Figure 12 The Chinese seeder for the two-wheeled hand tractor.

with positive results. In Bangladesh, farmers are using more than 200,000 hand tractors from China in their agriculture.

Soil moisture is a critical factor in this reduced tillage system. The rotovator fluffs up the soil, which then dries out faster than with normal land preparation. The seeding coulter does not place the seed very deep, so soil moisture must be high during seeding to ensure seed germination and root extension before the soil dries appreciably. Modifying the seed coulter to place the seed a little deeper would help correct this problem.

The main drawback of this technology is that the tractor and the various implements are not easily available, and access to spare parts and maintenance is a major issue. It would help if the private or public sector in South Asian countries could import this machinery or develop a local manufacturing capability. As it becomes more costly to keep and feed a pair of bullocks year-round, more farmers in the region are turning to significantly cheaper tractorized/mechanized options of land preparation. One of the benefits of this tractor is that it comes with many options for carrying out other farm operations; it can function also as a reaper, rotary tiller, and moldboard plow; it can also drive a mechanical thresher, winnowing fan, or pump. However, most farmers are attracted to the tractor because it can be hitched up to a trailer and used for transportation. For small-scale farmers who cannot afford their own tractors, custom hiring is a common alternative.

Engineers are experimenting with removing some of the blades that rototill the soil. In this way, a strip of soil rather than the whole area is cultivated. This reduces the power needs and costs and makes it easier for farmers to manage the tractor. In India, a four-wheeled tractor version of this ''strip-tillage'' machinery is available in the Punjab (Fig. 13).

Most farmers in Bangladesh do not have the Chinese seeder attachment and instead broadcast seed and fertilizer and then rototill to incorporate. This is popular because it saves time and cost and gives good results. Engineers are also substituting the smaller rotary tines found in the Chinese seeder onto the rototiller that is often provided with the two-wheeled tractor. They will also change the speed gear so that the rototiller can be used to incorporate seed and fertilizer faster with a finer tilth. Interestingly, many farmers prefer this system because they often want to plant at night and cannot see the seed drop in the Chinese seeder.

D. Bed Planting

Bed planting is another RCT practice being promoted (Fig. 14). It has many advantages in the RW system, especially in regard to water saving, mechanical weeding possibilities, fertilizer placement, better grain production, and less lodging. When this system is combined with no-till or reduced tillage, the permanent beds may be preferred more by farmers since bed-making costs are reduced. This is the latest RCT activity being researched within the RWC. For this system to be attractive, a permanent bed system would have to be successful for rice as

Figure 13 A strip till drill blade on machinery made in Punjab, India.

Figure 14   A bed former cum planter from Punjab, India.

well as wheat even if it provides better drainage and better conditions of growth for nonrice crops in the wet season. Data in the past two years suggest that rice beds is a feasible technology. Rice can be grown either by transplanting or by direct seeding, although the latter system has weed problems that need to be addressed. Farmers in India and Pakistan have obtained up to 9 t/ha of rice on beds and saved 60% of the water needed to grow normal rice. This is confirmed by monitoring farmer fields where this practice is being adopted. Yields might be further enhanced by SRI practices discussed in other chapters in this volume.

V.   WHY IS NO-TILL FARMING GAINING POPULARITY IN SOUTH ASIAN RW SYSTEMS?

Subsequent chapters give examples of the benefits of no-till farming in the IGPs, which are enumerated below. The following set of benefits is based on observations and discussions with scientists, farmers, and other stakeholders:

1.   No-till significantly reduces costs of production. Farmers estimate this at about 2,500 rupees/ha ($60/ha), mostly due to using less diesel and less labor. Farming has become uneconomic for many farmers as prices for inputs rise and the prices received for products decline. No-till allows farmers to make a profit.

2. Less tillage means less tractor wear and tear, and less maintenance expense.
3. Earlier, more timely planting is the main reason for the additional yields obtained. No-tilled plots can be planted closer to the optimum date for planting wheat, and the resulting higher yield means more income.
4. The seed and fertilizer drill used with no-till improves germination and plant stands over traditional broadcasting systems and improves fertilizer efficiency through better placement. Less water is used in no-till, and therefore there is less leaching of nitrogen.
5. Because weeds in wheat germinate only after temperatures drop below critical levels, no-till results in less weed emergence because less soil is disturbed. Fewer weeds mean less need for herbicide and lower cost. This is very important in areas where weeds have developed herbicide resistance to commonly available chemicals.
6. It takes less time for water to flow across the field in no-till compared to conventional-till plots for the first irrigation. That means farmers can save on water applications and, just as important, this reduces waterlogging and yellowing of the crop.
7. If surface residue management is combined with no-till, there is no need to burn the residue. This reduces air pollution as well as GHGs emitted into the atmosphere.
8. No-till reduces diesel consumption by 60–80 liters per hectare. This not only saves the foreign exchange for imports but also reduces GHG emissions significantly.
9. With no-till cultivation, anchored residues are left standing. Data show that this promotes the population of beneficial insects since it provides a good habitat for their survival and this reduces insect damage. This effect is enhanced if burning is stopped.
10. No-till means less carbon oxidation during plowing and possibly enhanced carbon sequestration, especially if residue management is good. However, we are not sure what effect the subsequent puddled rice crop has on the soil organic matter dynamics. This is being studied by monitoring farmer fields over time where no-till has been accepted.

VI. PROBLEMS INHIBITING WIDESPREAD ADOPTION OF NO-TILL FARMING

The main constraint to accelerating the adoption of no-till wheat farming in RW systems is people's mindset and the ability to make farmers aware of the technology and benefits. Examples abound of farmers being ridiculed when they first tried this technology. However, once the crop germinated, other farmers wanted to know how they achieved the results and were convinced that it would work, although many waited until seeing the harvest before making a final conclusion. It is not only by seeing but also by doing that a farmer is convinced. Since his livelihood is dependent on farming, he is not going to change practices unless he can see and do for himself. The RWC is promoting a participatory research approach where we provide the equipment to the farmer so he can experiment with the technology and come to his own conclusions. This is a major change in paradigm compared to the traditional top-down and researcher-demonstration model. Another mechanism that works is to identify service providers, convince them of the utility of no-till, and then provide training. They will then seek business and promote the technology.

Another issue is availability of suitable, good-quality equipment. Most of the no-till drills in the region are made by small-scale manufacturers in simple workshops. In order for the technology to be available to millions of farmers who grow RW, many more machines are needed. This scaling-up is hampering adoption as farmers accept this technology. We are encouraging service providers to help with this issue. Many farmers do not own tractors and get their land plowed on contract. When a service provider gets a no-till drill, he can plant at least 200 acres of no-till per season by contract and maximize the use of each drill. This benefits the farmer and the tractor owner. With a charge of 250 rupees an acre for planting no-till, the operator can pay for the drill in one year. The cost of the drill is about $400. This system means that any farmer, resource-poor or-medium, who does not have a tractor can benefit from this technology.

There is the issue of resistance from those farmers who own tractors and make money by plowing land on contract. One can try to handle this by encouraging these tractor owners to buy a no-till drill and continue to be service providers.

VII. FUTURE RESEARCH ISSUES

Several issues need to be addressed by site-specific research. These are as follows:

1. There is still a need for better no-till equipment that provides more uniform results. This is especially important in areas where combine harvesting leaves loose residues on the field. This will require involving local manufacturers in the program and introducing new, innovative designs.
2. Research is needed on the total farming system. Would the benefits of no-till wheat be better if rice soils were not puddled? Data suggest this may be so, and this could receive benefits from the system of rice intensification (SRI).

3. Monitoring long-term consequences of no-till cultivation is needed to make certain that soil and biotic factors (insects, diseases, weeds, rodents, etc.) are not going to become a future problem.
4. The social issues of no-till farming need to be assessed. Who benefits from the adoption of no-till, and who loses? What happens to the livelihoods of the poor if no-till becomes a common practice?
5. More research is needed on the role of no-till practices on soil organic matter and on soil properties.
6. There is a need to effectively document the effects of no-till cultivation on GHG emissions and its possible impact on global warming.
7. No-till establishment of other crops including rice, legumes, and maize is possible. What happens if both rice and wheat are no-tilled?
8. Permanent bed-planting systems need more research to determine if they are sustainable systems for a whole range of cropping.

VIII. CONCLUSIONS

No-till farming is becoming very popular in the rice–wheat systems of South Asia as farmers struggle to make farming profitable. It provides an opportunity for farmers to grow more food at reduced cost and thus improve their livelihoods. It also has several important benefits in regard to the environment. Several farmers have raised no-till wheat for the past four years without any problems and, in fact, report higher yields. Overcoming age-old prejudices about ''more tillage giving better crops'' and changing the mindsets of farmers is an important problem. A lack of available, good-quality machinery has also hampered and slowed adoption. However, this is being corrected. A ''tillage revolution'' is underway in the IGPs that will help maintain food security, improve farmer welfare, and possibly have some demonstrable effect on poverty. However, these gains will be futile unless South Asia can more rapidly reduce its massive population growth.

REFERENCES

Byerlee, D., Sheikh, A., Aslam, M., Hobbs, P.R. 1984. Wheat in the Rice-based Farming System of the Punjab: Implications for Research and Extension. PARC/CIMMYT Wheat Paper No. 84. Islamabad. Pakistan: Pakistan Agricultural Research Council (PARC) and CIMMYT.

Fujisaka, S., Harrington, L.W., Hobbs, P. R. 1994. Rice–wheat in South Asia: System and long-term priorities established through diagnostic research. *Agricultural Systems* 46:169–187.

Harrington, L.W., Fujisaka, S., Hobbs, P.R., Adhikary, C., Giri, G.S., Cassaday, K. 1993a. Rice–Wheat Cropping Systems in Rupandehi District of the Nepal Terai: Diagnostic

Surveys of Farmers' Practices and Problems, and Needs for Further Research. Mexico. D.F.: CIMMYT.

Harrington, L.W., Fujisaka, S., Morris, M.L., Hobbs, P.R., Sharma, H.C., Singh, R.P., Chaudhary, M.K., Dhiman, S.D. 1993b. Wheat and Rice in Karnal and Kurukshetra Districts, Haryana, India: Farmers' Practices, Problems and an Agenda for Action. Mexico. D.F.: Haryana Agricultural Unuversity, Indian Council for Agricultural Research, CIMMYT, and the International Rice Research Institute.

Hobbs, P.R., Hettel, G.P., Singh, R.P., Singh, Y., Harrington, L.W., Fujisaka, S. 1991. Rice–Wheat Cropping Systems in the Terai Areas of Nainital, Rampur, and Pilibhit Districts in Uttar Pradesh, India: Sustainability of the Rice–Wheat System in South Asia. Diagnostic Surveys of Farmers' Practices and Problems, and Needs for Further Research. Mexico. D.F.: CIMMYT, ICAR, and IRRI.

Hobbs, P.R., Hettel, G.P., Singh, R.K., Singh, R.P., Harrington, L.W., Singh, V.P., Pillai, K.G. 1992. Rice–Wheat Cropping Systems in Faizabad District of Uttar Pradesh, India: Exploratory Surveys of Farmers' Practices and Problems, and Needs for Further Research. Mexico. D.F.: ICAR, NDUAT, CIMMYT, and IRRI.

Ladha, J.K., Fischer, K.S., Hossain, M., Hobbs, P.R., Hardy, B. 2000. Improving the productivity of rice–wheat systems of the Indo-Gangetic Plains: A synthesis of NARS-IRRI partnership research. IRRI Discussion Paper No. 40. Los Baños: IRRI.

Ortiz-Monasterio, J.I., Dhillon, S.S., Fischer, R.A. 1994. Date of sowing effects on grain yield and yield components of irrigated spring wheat cultivars and relationships with radiation and temperature in Ludhiana, India. *Field Crops Research* 37:169–184.

Saunders, D.A. 1990. Crop management research summary of results. Wheat Research Centre Monograph 5. Nashipur. Bangladesh: Wheat–Rice Consortium.

Sayre, K.D. 2000. Effects of tillage, crop residue retention and nitrogen management on the performance of bed-planted, furrow irrigated spring wheat in northwest Mexico. Presented at the 15th Conference of the International Soil Tillage Research Organization; July 2–7, 2000;. Fort Worth. Texas. USA.

7

Opportunities and Constraints for Reduced Tillage Practices in the Rice–Wheat Cropping System*

J.M. Duxbury, J.G. Lauren, and M.H. Devare
Cornell University
Ithaca, New York, U.S.A.

A.S.M.H.M. Talukder, M.A. Sufian, and A. Shaheed
Wheat Research Center, Nashipur
Dinajur, Bangladesh

M.I. Hossain
Bangladesh Agricultural Research Station
Rajshahi, Bangladesh

K.R. Dahal
Institute of Agriculture and Animal Science
Rampur, Nepal

J. Tripathi and G.S. Giri
Regional Agriculture Research Station
Bhairahawa, Nepal

C.A. Meisner
CIMMYT
Dhaka, Bangladesh

* Research reported in this chapter was supported by US-AID through the Soil Management CRSP.

I. INTRODUCTION

Intensive tillage of soil is traditionally practiced in the rice–wheat cropping system of the Indo-Gangetic Plains (IGP) region of South Asia, including puddling of soils for paddy rice production and multiple passes over fields to prepare seed beds for wheat and other crops in the upland phase of the rotation. Intensive tillage has degraded soil physical structure and maximized losses of soil organic matter. No-tillage and reduced-tillage practices offer opportunities to improve the physical condition of soils and to increase soil organic matter levels, possibly to levels similar to those in the soils prior to agricultural development. For these benefits to be realized, no-/reduced-tillage practices will have to be applied across the system.

No-tillage for wheat (*Triticum aestivum L.*) is rapidly being adopted in the mechanized areas of northwest India and Pakistan, with 300,000 ha planted this way in 2002 (Hobbs and Gupta, 2002); however, rice production practices remain unchanged. Adoption of no-tillage in wheat will probably increase wheat yields on a regional basis because it enables timely planting, which minimizes yield reductions associated with excessive heat during the grain filling period (Ortiz-Monasterio et al., 1994; Hobbs and Gupta, 2002). However, this practice does not, by itself, address the suboptimal soil physical condition that will likely still constrain wheat yields, especially on heavier textured soils. Several studies have shown reduced yields of wheat when soil is puddled for rice (*Oryza sativa L.*) and wheat is planted by no-tillage at the same time as that with conventional tillage (Singh et al., 1994; Aggarwal et al., 1995; Singh et al., 1998; Bajpai and Tripathi, 2000; Srivastava et al., 2000). The challenge then is to develop no-/reduced-tillage practices that are effective for the rice phase of the rotation and that benefit both rice and wheat (upland crop) phases of the rotation. This issue needs to be addressed from a long-term cropping systems perspective as crop productivity and soil-quality gains through improved soil management will not be immediately realized.

II. NO/REDUCED TILLAGE MANAGEMENT OPTIONS FOR THE RICE–WHEAT SYSTEM

The rice–wheat system was developed to optimize the soil environment for rice—at least as far as water retention is concerned—rather than for wheat. An alternative would be to optimize the soil environment for wheat and to grow rice without puddling soils, and possibly without flooding soils. Until recently, this approach has been rejected by rice agronomists on the basis that such a system would lead to high yield penalties with rice and increased problems with weeds. Nevertheless, several technologies such as raised beds and surface seeding offer opportunities to address no-/reduced tillage at a system level. Raised beds and the system of rice intensification (SRI; N. Uphoff, paper in these proceedings) also offer the possibility of using less water, which is important as groundwater resources are being used nonsustainably in the western part of the IGP and are contaminated with arsenic in parts of the eastern IGP.

III. PERMANENT RAISED BEDS

Raised beds are widely used in agriculture in developed countries and have proven to be an excellent option for wheat (Limon-Ortega et al., 2000a, b). Permanent raised beds may also offer good possibilities for the rice–wheat system. With reshaping, these would allow soil aggregation to be rebuilt over time, providing deeper rooting and better air and water relationships in the soil. Particularly attractive are the possibilities that water use can be reduced, as furrow irrigation is more efficient than flood irrigation, and that fertilizer N recovery can be increased for both rice and wheat by banding nitrogen into the soil between two rows on a bed. The latter is especially important for rice where fertilizer N recovery is generally poor (20%–40%) due to the difficulty of avoiding losses by denitrification and ammonia volatilization in flooded soil (Peoples et al., 1995).

Experiments to compare crop productivity and other parameters on beds and on flat land have been established at two sites in Bangladesh (Nashipur and Rajshahi) on soils that differ in texture. The soil at Nashipur is a sandy loam and that at Rajshahi is a silty clay loam. The beds were established by hand with similar dimensions (15 cm high, 75 cm furrow to furrow) to those created by the FIRBS bed formerly developed at Punjab Agricultural University (see Rice–Wheat Consortium web site at http://www.rwc-prism.cgiar.org/rwc/bed_planting.asp). Two rows of plants, 20 cm apart, were planted on the beds for all crops. Conventional spacing and plant densities were generally used on the flat. However, at the Rajshahi site, 30-day-old rice seedlings planted 3 to 4 per hill were used in 2000, whereas single 10-day-old seedlings were used in 2001. The spacing between rice hills/plants within rows was 15 cm. The experiments are currently in their second cropping cycle of an annual rotation of rice–wheat–mung bean (*Vigna radiata L.*). They include three N levels (50%, 100%, and 150% of recommended N) and two placement methods (band and broadcast). The effect of planting on beds on crop yields across all N treatments is shown for the two sites in Fig. 1. To date, yields of all three crops have been higher on the bed compared to the flat at both sites.

The response of rice and wheat to nitrogen inputs is shown in Fig. 2. Wheat yield responses on the bed and on the flat were essentially parallel, suggesting similar recoveries of fertilizer N in both cases, but better use of soil N for crops on the beds. Some divergence in response curves between the bed and the flat is seen for monsoon season rice at both sites for both band and broadcast treat-

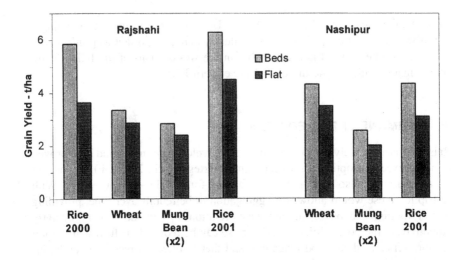

Figure 1 Crop yields on permanent raised beds and conventional flat field production systems in a rice–wheat–mung bean annual rotation at two locations in Bangladesh.

ments. Rice yields were higher at Rajshahi than Dinajpur, in part because of droughty conditions at the latter site (sandy soil and low rainfall) and perhaps also because of a modified SRI approach using young, single-seedling transplants at Rajshahi. The yield of 7.1 t/ha from beds with N at the highest level (180 kg N/ha) is high for the monsoon season rice.

Water use in wheat beds was 50% to 60% of that normally used in flood irrigation. Rice grown during the monsoon season may or may not require irrigation, depending on rainfall patterns. Heavy rainfall after the first month of the initial rice crop at the Rajshahi site kept the field flooded for most of the remaining season. The beds were never flooded during the second-year rice crop, and about half the irrigation water was used compared to conventional paddy rice. Rice yields on the beds were higher in both years, even with this difference in water regimes.

The advantage of the raised beds on crop yield will likely increase over time for wheat and mung bean as the soil physical condition improves. However, it is not clear what will happen with rice if the soils should gradually become more permeable. Perhaps compacting the bottom of the furrows would reduce downward movement of water while maintaining lateral infiltration rates, partially maintaining the water retention advantage of the paddy. Compaction of furrows will occur naturally with permanent beds as they control traffic over the field, a feature that is also essential for an effective reduced-tillage system.

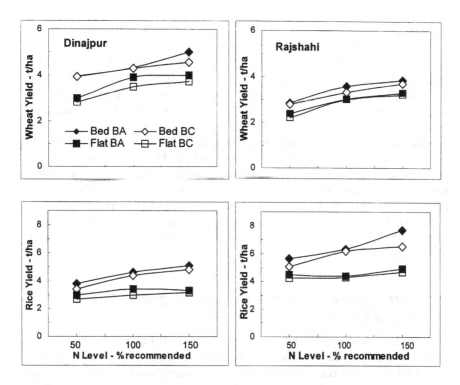

Figure 2 Effect of N rate on yields of rice and wheat on raised beds and on the flat at two sites in Bangladesh, 2001.

The bed system is most easily undertaken by farmers with tractors and large fields and is more difficult for resource-poor farmers with small land holdings. Farmers in the Kathmandu valley in Nepal make raised beds by hand for wheat production on heavy soils, although this is a special case as religious reasons prevent the use of animals for plowing. Nevertheless, the farmers are willing to invest labor into making beds and should consider keeping them for rice or maize (*Zea mays L.*) rather than breaking them down for paddy rice.

IV. SURFACE SEEDING

Another no-tillage option that is very good for small fields is surface seeding on flat land (Ahmad and Chatha, 1992; Hobbs and Giri, 1997), which requires no machinery and minimal labor. Surface seeding has been promoted for timely sowing of wheat following rice on clay soils that were normally left fallow after rice because of the difficulty of preparing a seed bed for wheat. Surface seeding can be done after rice harvest or as a relay seeding into standing rice. We have shown that surface seeding can also be successful on light-textured soils for both rice and wheat, using a preseeding irrigation and straw mulch to aid germination and stand establishment. The mulch can subsequently be removed if it is needed for other purposes such as feed or fuel. However, leaving the straw will suppress weeds, increase soil organic matter levels, and quickly improve soil physical condition at the soil surface, with subsequent benefits to aeration and water infiltration. Leaving the mulch in place significantly increased ($p < 0.05$) yields of both rice and wheat by 0.5–1.2 t/ha in the two cycles of the experiment to date (Table 1). The increase in rice yields with mulch may be due to reduced losses of N by ammonia volatilization as the decomposing mulch releases CO_2 and lowers floodwater pH.

V. INTERACTIONS BETWEEN TILLAGE, SEEDING METHOD, AND PUDDLING OF SOIL

A rice–wheat rotation experiment was established at Bhairahawa, Nepal, to determine whether deep tillage would improve wheat yields. It also included comparison of direct seeded rice without puddling with transplanted rice with puddling in order to determine effects of no puddling on rice and wheat yields. No effect of tillage or crop establishment method was found on rice yields over four years. For wheat, the experiment included two wheat establishment methods: using the seed drill attachment to the Chinese hand tractor and surface seeding. Using the

Table 1 Yields of Rice and Wheat in a No-Tillage Surface Seeding Experiment with and Without Straw Mulch (6 t/ha) at Rampur, Nepal

Crop and year	Crop yield (t/ha)	
	+Mulch	−Mulch
Wheat 1999–2000	4.5a	3.3b
2000–2001	4.0a	3.5b
Rice 2000	4.2a	3.7b
2001	3.4a	2.9b

Different letters within a row indicate a significant difference at $p < 0.05$.

drill required that the soil be sufficiently dry to use the equipment, which meant a delay in planting compared to surface seeding in years where the soils were wet at planting time. Consequently, wheat yields were significantly higher with surface seeding compared to using the drill (Fig. 3). Similarly, deep tillage prior to rice significantly increased wheat yield compared to the normal tillage practice (Fig. 3), presumably due to deeper rooting of wheat. Tillage by rice-planting method interactions also showed that the lowest wheat yields were associated with puddling of soil (Fig. 3, treatment NT-TPR) and that deep tillage or simply avoiding puddling of soil ameliorated this yield reduction.

VI. MAJOR CONSTRAINTS TO CROP PRODUCTIVITY IN THE RICE–WHEAT SYSTEM

The most important issue to consider with regard to use of no-/reduced-tillage practices is the extent to which they address the major constraints in productivity

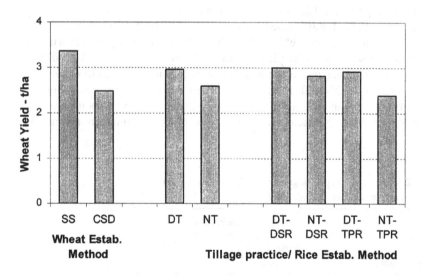

Figure 3 Effect of tillage and wheat and rice crop establishment methods on mean wheat yields at Bhairahawa, Nepal, 1997–2001. SS = surface seeding; CSD = Chinese seed drill; DT = deep tillage (ripping to 50 cm); NT = normal tillage; DSR = direct seeding without puddling; and TPR = transplanting with puddling.

with the cropping system. As previously mentioned, the practices currently being used for wheat production in northwest India successfully address yield reductions associated with late planting. They do not address the question of yield constraints due to poor soil physical condition. Furthermore, diagnostic surveys conducted with soil solarization have shown that poor soil biological health is the major constraint to crop productivity in the rice–wheat system in Bangladesh and Nepal. The particular pathogens and pests involved are still being identified. How tillage and other soil management practices interact with soil biological health is therefore also of paramount importance. A soil solarization study superimposed onto the tillage experiment at Bhairahawa illustrates this point. Without soil solarization, deep tillage significantly ($p < 0.05$) improved wheat yield (Table 2). However, with soil solarization, yields were substantially increased in both tillage treatments and the effect of deep tillage was eliminated. The deep tillage treatment probably partially addresses the soil biological health problem by promoting deeper rooting, but none of this was necessary when the primary constraint was addressed.

A strategy that may combat poor soil biological health is the use of seedlings that are free from infectious pathogens and nematodes. Such "healthy seedlings" can be produced by a combination of seed treatment with the biocide vitavax-200 and solarization of soil in the rice nursery to control seed- and soilborne pathogens, respectively. Rice yields were increased up to 45% when treated seedlings are transplanted into flooded soils. Results obtained by farmers are shown in Fig. 4, which compares yields with seedlings from nurseries with and without solarization of soil. The mean yield for all 25 farms was 5.2 t/ha with seedlings from nurseries on solarized soil compared to 4.0 t/ha with seedlings from conventional nurseries. In this trial, 16 of the 25 farmers were able to increase their rice yield between 30% to 40% just by solarizing nursery soils. We hypothesize that the healthy seedling approach is successful because aerobic pathogens and nema-

Table 2 Effect of Deep Tillage and Soil Solarization on Yields of Wheat at Bhairahawa, Nepal, 1999

	Yield (t/ha)	
Tillage treatment	Nonsolarized soil	Solarized soil
Conventional	3.0a	5.2c
Deep tillage	4.0b	5.1c

Different letters indicate a significant difference at $p < 0.05$.

todes are stressed in the flooded main field and have difficulty infecting rice plants. In contrast, pathogens and nematodes continue to negatively affect growth and reduce yields in plants that are already infected. One implication of this hypothesis is that soil biological health would be a greater problem when rice is grown more aerobically than in the paddy, such as with raised beds or with the SRI method.

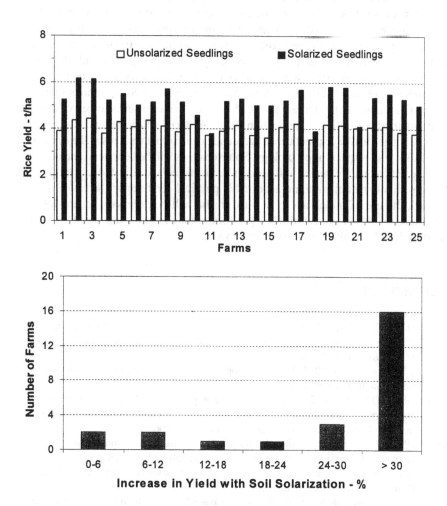

Figure 4 Effect of soil solarization in the rice nursery on monsoon (aman) season rice yields from 25 farm trials in the Dinajpur district, Bangladesh, 2001.

VII. FUTURE RESEARCH

Reduced-tillage practices are likely to become increasingly important management options in the rice–wheat cropping system. As research and practice move
forward, we urge that much attention be given to identifying and overcoming the
major biophysical constraints to crop productivity in the rice–wheat cropping
system. Only with this knowledge will we be able to have a sustainable system
and a rational basis for choosing between different tillage options.

Of the various tillage options, permanent raised beds appear to hold the
most promise as they have the potential to improve productivity of all crops in
the rotation while simultaneously reducing water use and improving crop recovery
of fertilizer N. Reduced-tillage practices should also be explored with the SRI
method of rice production, which is also a system that saves water and grows
rice more aerobically. The possibility that raised bed and SRI methods of rice
production are likely to increase the need for P fertilization needs to be evaluated
as do methods of fertilizer application for no-/reduced-tillage practices.

Changes in tillage practice will gradually cause other changes in the system,
especially with respect to soil biology. Some changes will be beneficial and others
may be detrimental; changes will need to be identified and adjusted to over time.
It is also likely that maintaining a desirable physical condition at the soil surface
through the use of mulch and controlled traffic will be necessary for the long-
term success of no-/reduced/tillage. The former may pose a constraint for small
farmers who have alternative uses for straw.

REFERENCES

Aggarwal, G.C., Sidhu, A.S., Sekhon, N.K., Sandhu, K.S., Sur, H.S. 1995. Puddling and
    N management effects on crop response in a rice-wheat cropping system. *Soil and
    Tillage Research* 36:129–139.
Ahmad, S., Chatha, A.A. 1992. New technology for wheat production in rice-wheat crop-
    ping system. *J. Agric. Res., Lahore* 30:335–339.
Bajpai, R.K., Tripathi, R.P. 2000. Evaluation of non-puddling under shallow water tables
    and alternative tillage methods on soil and crop parameters in a rice–wheat system
    in Uttar Pradesh. *Soil and Tillage Research* 55:99–106.
Hobbs, P.R., Giri, G.S. 1997. Reduced and zero-tillage options for establishment of wheat
    after rice in South Asia. In: Braun H.J., et al., Eds. Wheat: Prospects for global
    improvement, Developments in Plant Breeding 6: Kluwer Academic Publishers.
    Dordrecht, The Netherlands, 455–465.
Hobbs, P.R., Gupta, R.K. 2002. Resource Conserving Technologies for Wheat in
    Rice–Wheat Systems. In: Ladha J.K., et al., Eds. Improving the productivity and
    sustainability of rice–wheat systems: Issues and impact: ASA, Spec. Publ. 00. ASA
    Madison. WI (in press).

Limon-Ortega, A., Sayre, K.D., Francis, C.A. 2000a. Wheat nitrogen use efficiency in a bed planting system in Northwest Mexico. *Agron. J.* 92:303–308.

Limon-Ortega, A., Sayre, K.D., Francis, C.A. 2000b. Wheat and maize yields in response to straw management and nitrogen under a bed planting system. *Agron. J.* 92: 295–302.

Ortiz-Monasterio, J.I., Dhillon, S.S., Fischer, R.A. 1994. Date of sowing effects on grain yield and yield components of irrigated spring wheat cultivars and relationships with radiation and temperature in Ludhiana, India. *Field Crops Res.* 37:169–184.

Peoples, M.B., Freney, J.R., Mosier, A.R., Bacon, P.E. 1995. Minimizing gaseous losses of nitrogen in Nitrogen Fertilization in the Environment: Marcel Dekker. New York, 565–602.

Singh, P., Aipe, K.C., Prasad, R., Sharma, S.N., Singh, S., Singh, P. 1998. Relative effect of zero and conventional tillage on growth and yield of wheat (Triticum aestivum) and soil fertility under rice (*Oryza sativa*)–wheat cropping system. *Indian J. Agron.* 43:204–207.

Singh, P.K., Singh, Y., Kwatra, J. 1994. Effect of tillage and planting management on yield and economics of rice–wheat cropping system. *Agricultural Science Digest, Karnal* 14:41–43.

Srivastava, A.P., Panwar, J.S., Garg, R.N. 2000. Influence of tillage on soil properties and wheat productivity in rice (*Oryza sativa*)–wheat (*Triticum aestivum*) cropping system. *Indian J. Agric. Sci* 70:207–210.

8

No-Tillage Farming in the Rice–Wheat Cropping Systems in India

Ram Kanwar Malik, A. Yadav, S. Singh, and P. K. Sardana
Haryana Agricultural University
Hisar, India

Peter R. Hobbs
Cornell University
Ithaca, New York, U.S.A.

Raj Gupta
CIMMYT
New Delhi, India

I. INTRODUCTION

In the 1950s, rice (*Oryza sativa L.*) and wheat (*Triticum aestivum L.*) accounted for only one third of the grain production in India. This proportion rose to more than 70% by 1989 (Maklin and Rao, 1991). One third of the irrigated rice and half of the irrigated wheat in South Asia come from the rice–wheat cropping system. Rice–wheat systems occupy 24 million ha of cultivated land in the Asian subtropics. In South Asia, the system occupies about 13.5 million ha (10 million in India, 2.2 million in Pakistan, 0.8 million in Bangladesh, and 0.5 million in Nepal), extending across the Indo-Gangetic floodplain into the Himalayan foothills (Ladha et al., 2000). Rice–wheat systems cover about 32% of the total rice area and 42% of the total wheat area in these four countries, and they account for one quarter to one third of the total rice and wheat production (Hobbs and

Morris, 1996; Huke *et al.*, 1994a, b; Woodhead *et al.*, 1994a, b). In the first decade after the introduction of improved rice and wheat varieties, grain production grew rapidly, propelled by increased cropped area as well as by higher yields. More recently, however, the area devoted to rice and wheat has stabilized, and further area expansion seems out of the question—in fact, cropped area may even decline in the years to come (Hobbs *et al.*, 1998). The need for higher yields has come at a time when evidence is accumulating that growth in rice and wheat yields has started slowing down in the high-potential agricultural areas of northwestern India and Pakistan. There is evidence of declining partial and total factor productivity* (PFP or TFP) (Hobbs and Morris, 1996; Ali and Byerlee, 2000; Murgai, 2000). A diagnostic study of constraints in rice–wheat cropping systems was conducted by CCS Haryana Agricultural University, Hisar, CIMMYT, IRRI, and ICAR and showed that a high population of *Phalaris minor*—a serious weed of wheat in the rice–wheat cropping system—and a decline in soil productivity were two major constraints of the system (Harrington *et al.*, 1992). Other causes of decline in the TFP of the rice–wheat cropping system throughout the region included declines and changes in soil organic matter (SOM), a gradual decline in the supply of soil nutrients causing nutrient (macro and micro) imbalances due to inappropriate fertilizer applications, a scarcity of surface and groundwater, and in some places poor water quality (Paroda *et al.*, 1994). At the time of the diagnostic survey conducted in 1991 (Harrington et al., 1992), scientists at Haryana Agricultural University were gathering evidence that herbicide (e.g., isoproturon) resistance had developed in *Phalaris minor*, and it was confirmed in 1992–1993 (Malik and Singh, 1993, 1995). These studies provided the scientific community with a framework for setting research priorities (Hobbs, 1994). The resource conservation technology requirements differed in various areas of the Indo-Gangetic Plains. Herbicide resistance and problems with lowering water tables were major factors in Haryana, India (Malik *et al.*, 1998). To answer these sustainability issues, work began on no-tillage and bed planting in wheat in collaboration with the Rice–Wheat Consortium for Indo-Gangetic Plains, CIMMYT-Mexico and ACIAR-Australia in 1997.

II. APPROACH

Many scientists tried no-tillage on experimental farms without any tangible output and ended their experiments before they could demonstrate anything on farmers' field. Until 1998, results in Haryana showed that no-tillage was not a successful

---

* The PFP or TFP is the measure of grain output divided by the quantity of some inputs or all inputs taken together, respectively.

technology and was only an option for late sown wheat (DWR, 1996–1997; 1997–1998). By using a farmer's participatory approach in 1996, HAU scientists provided a significant breakthrough for no-till acceptance by farmers. Farmers were primarily interested in the cost reduction of cultivation and increased yields. The approach involved farmers' experimenting with the technology and avoiding and identifying potential problems. This approach provided information from different field conditions at selected sites within the rice–wheat cropping tract of Haryana. Ten one-acre demonstrations were performed with farmers and HAU scientists in 1996–1997. In 2001–2002, the area under zero tillage in Haryana had increased to approximately 101,000 hectares (Fig. 1). No-till also helps reduce carbon emissions (reduced diesel use) and improves input use efficiency while raising system productivity and farm level profits. The Rice–Wheat Consortium organized traveling seminars for scientists, farmers, and manufacturers in 1998–1999 and 1999–2000 to assess the success of zero tillage in northwest India. The adoption rate of zero tillage in Haryana (India) and Punjab (Pakistan) (Gill *et al.*, 2000) was basically a result of a strong farmers' participatory approach.

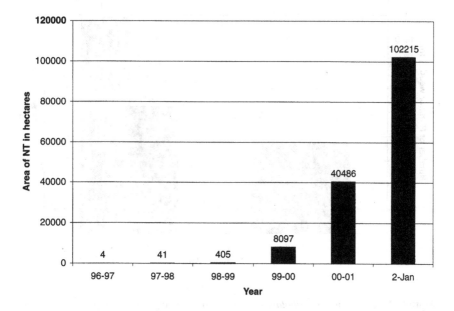

Figure 1 Area growth under no-tillage from 1996–1997 to 2001–2002 in Haryana State, India.

III. OPPORTUNITIES FOR IMPROVING WHEAT PRODUCTIVITY

Based on 1998–1999 government statistical data, the average yields of rice and wheat in Punjab, Haryana, Uttar Pradesh, Bihar, and West Bengal were 7.5, 6.1, 4.5, 3.3, and 4.4 t/ha, respectively. Most of this yield variation was due to the time of sowing. Punjab sows wheat the earliest followed by Haryana, western UP, eastern UP, and Bihar. Farmers in eastern UP and Bihar have the ability to increase wheat productivity by adopting no-till to get timely planting. In states like Punjab and Haryana, no-till not only helps to get timely planting but also solves second-generation problems such as herbicide resistance in *Phalaris minor* (Fig. 2). However, herbicide use is still needed. Monitoring data over several years shows a change in the spectrum of weed flora in favor of broad leaf weeds especially where wheat planting is done by the end of October (Yadav *et al.*, 2002). No-till technology will maintain and improve the sustainability of the rice–wheat system.

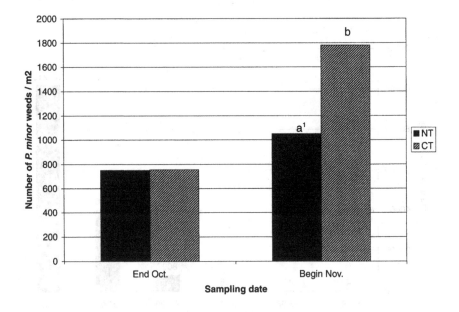

Figure 2 Population of *Phalaris minor* before sowing wheat on farmer fields in 1998–1999 in Haryana State in India. Means compared at the same date followed by different letters are significantly different at $P = 0.05$ using DMRT. NT = no-tillage and CT = conventional tillage.

Over the past two years, the wheat yields in the rice–wheat cropping system in these two states have increased by advancing sowing to the third and fourth weeks of October. Data in Fig. 3 and 4 show that the no-till advantage in the rice–wheat cropping system in Haryana was more pronounced in the early and timely planted wheat than the late planted wheat. Similarly, data in Fig. 5 show that no-till improved productivity of wheat in Bihar (Malik et al., 2000). It also indicates that tillage responses are more sensitive to planting time than any other factor. Provided farmers maintain their tillage reforms, they should continue to grow more rice and wheat. There also seems to be a good opportunity to introduce no-till in transplanted rice (Piggin et al., 2002). Once tillage is conducted on a systems basis, benefits will be even greater. No-till helps solve the problem of late planting and excessive costs of production in wheat, but if rice can be grown without puddling, the total system productivity would be even greater. There is a strong need to coordinate research in a multidisciplinary mode to take advantage of expertise and develop better recommendations.

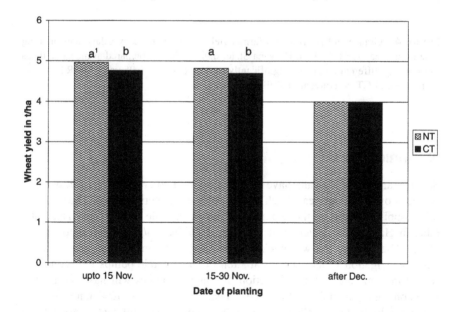

Figure 3   Grain yield of wheat with no-tillage and conventional tillage on farmer fields sown at different times in 1997–1998 in Haryana State, India. [1]Means compared at the same dates followed by different letters are significantly different at $P = 0.05$ using DMRT. NT = no-tillage and CT = conventional tillage.

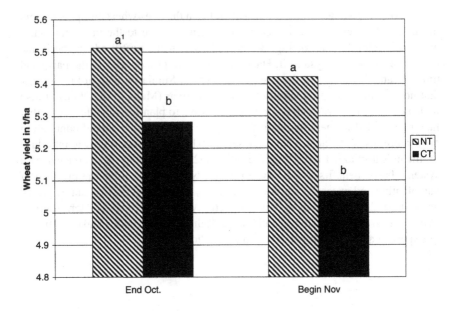

Figure 4 Grain yield in wheat on farmer fields as influenced by tillage and planting time in the year 1998–1999 in Haryana State, India. [1]Means compared at the same dates followed by different letters are significantly different at $P = 0.05$ using DMRT. NT = no-tillage and CT = conventional tillage.

IV. IRRIGATION

Several farmers in Haryana have started growing two crops of rice followed by wheat in one calendar year. Such practices have planners worried. The effect of this cropping pattern on the depletion of the water table is very high. The water table in Haryana is depleting at the rate of 20–30 cm/year (Harrington et al., 1992). The drop in the water table has escalated the installation and pumping costs of irrigation water. The installation of submersible pumps is becoming a reality in some areas. Such depletion of water resources will have an adverse effect on the sustainability of the rice–wheat system. Data show that no-till saves water, facilitates timely planting, and reduces costs. In no-till, water passes faster over the soil compared to tilled soil and so less water is used in the first irrigation. No-till wheat can also benefit from the residual moisture by planting immediately after harvesting of rice without a presowing irrigation. Data from farmer surveys show that no-till wheat uses 15% to 25% less water than normal tilled wheat

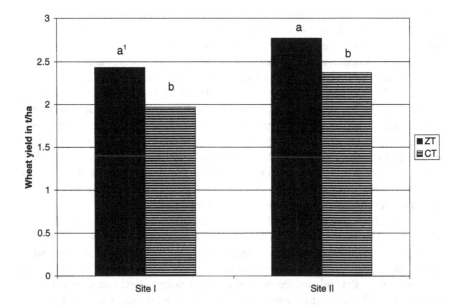

Figure 5 No-tillage compared with conventional tillage at 2 sites in Phulwari Sharif in Bihar State, India. [1]Means compared at the same sites followed by different letters are significantly different at $P = 0.05$ using DMRT. NT = no-tillage and CT = conventional tillage.

(Malik et al., 2000). Savings also result from less fuel needed for pumping water and less wear on the tube-well so farmers improve profits.

V. SOIL QUALITY

The term "soil quality" includes soil productivity and its environmental moderating capacity and received considerable attention in the 1990s (Doran *et al.*, 1994; Karlen *et al.*, 1997). Scientists, extension agencies, and even policy makers have taken the rice–wheat cropping system for granted. Now that evidence suggests that TFP of the system is failing (Harrington *et al.*, 1992), how should we respond? Prolonged dependence on high-yielding varieties for yield growth is unlikely. Intensification of the cropping system has increased productivity but has had negative effects on soil quality. However, the soils of the Indo-Gangetic Plains (IGP) have the capacity to feed the present and the future populations provided soils are restored and improved (Lal, 2000). Adoption of no-till is a solution to soil-quality issues and maintaining food security in the region. According to Lal (2000), among the principal global concerns of the 21st century are soil degradation and its mismanagement in relation to soil organic carbon. The effects of these soil productivity issues are closely linked with global warming and the dramatic increase in the concentration of atmospheric carbon dioxide (Etheridge et al., 1996) and other greenhouse gases (GHGs) (Harvey et al., 2000). Today, high synthetic fertilizer use is linked to soil productivity. Efficient use of fertilizer is essential to reduce unnecessary GHG emissions like methane and nitrous oxides. Strategies are also needed to increase sequestration of carbon. No-till offers a strategy to do this by using surface mulch and retaining anchored stubbles. This will help improve long-term soil organic matter quality and quantity. This carbon and the inherent carbon pool will also provide food for micro-organisms and lead to improved biological health and microbial diversity.

Long-term sites maintained at farmers' fields in Teek, Uchani, and Nangla villages in Haryana revealed an improvement in the NPK and organic matter status in soil with no-till (Kumar et al., 2002). Similarly, long-term tillage sites at Wooster, Ohio, also revealed that organic matter, nutrients, and soil enzymes accumulated at the soil surface but decreased at deeper depths in no-tilled plots (Dick et al., 1991). At another site in Coshocton, Ohio, the earthworm population increased in no-till soils (Dick, 1983). Most of these studies indicate that zero-tillage can be continuously practiced with yield and soil advantages in corn (Zea Mays L.) and soybean (Glycine max L.) especially in well-drained soil with benefits increasing over time (Dick et al., 1991). Improvement in soil quality should boost productivity of the cropping system in the long run. At present, farmers in the IGP are using an increased amount of fertilizer; therefore, improvements in soil productivity and fertilizer efficiency will increase profits and sustainability. No-till can help achieve this goal. When phosphorus (P) is not mixed into the soil profile with no-till, it rapidly accumulates in the surface layer as a horizontal band (Dick et al., 1991). Since the sowing of no-till wheat is done in relatively moister soil than conventional till, the efficiency of P is likely to be more in no-till. The dark green color of wheat leaves observed in no-till confirms this possibility attributable to improvements in nitrogen availability. The socioeconomic audit of no-till done in Haryana in 2001–2002 (Malik et. al., 2002) has shown that there is no yellowing of wheat leaves after the first postsowing irrigation compared to conventional till, possibly because of the efficient use of fertilizer applied at sowing and in some cases after the first irrigation. The emission of gases into the atmosphere due to plowing and deforestation is increasing at the rate of 0.5%/year for CO_2, at the rate of 0.75%/year for methane, and at the rate of 0.25%/year for N_2O (IPPC, 1996). The concentration of CO_2 has increased due to soil cultivation, biomass burning, fuel burning, and the use of nitrogenous fertilizers (Lal, 2000). The no-till area has been targeted to increase from about 0.1 million ha in 2001–2002 to 1 million ha under the NATP project of ICAR that ends in

2004. Grace *et al.* (2000) have indicated that current land preparation practices for wheat after rice involve as many as 12 tractor passes. Changing to a no-till system on 1 hectare of land would save up to 98 liters of diesel and approximately 1 million liters of irrigation water. This represents about a quarter ton fewer CO_2 emissions per hectare. These benefits increase dramatically if extended across a portion of the region's 13.5 million hectares of rice–wheat. Adoption of no-till on only 5 million hectares would represent a savings of 5 billion cubic meters of water each year. Planners need to accelerate the adoption of no-till throughout the IGP. International research groups including CIMMYT (News briefing in Hague in October 2001), ACIAR-Australia (Vincent and Quirk, 2002), and Haryana Agricultural University, India, see no-till in India as the beginning of a new tillage revolution after the Green Revolution based on new varieties introduced in 1966. The constraints to the adoption of this technology in India are being resolved through a World Bank-aided NATP project under the Indian Council of Agricultural Research. Four core themes, including tillage and crop establishment, nutrient management, water management, and integrated pest management, will be further strengthened through this project and accelerate the adoption of resource conservation technologies in the Indo-Gangetic Plains of South Asia.

VI. GAINS FROM NO-TILL

A survey was conducted immediately after wheat harvesting in 2001 with 100 farmers from 17 villages practicing no-till, sometimes in combination with conventional till. The villages covered in the survey have fertile soils and good irrigation facilities, allowing farmers to grow an intensive rice–wheat rotation. The salient findings of the survey are given below:

1. Wheat grown at farms with 4 and 3 years of no-till had significantly higher yield than those with 1 year of no-till (Fig. 6).
2. Per hectare profitability on no-till farms was higher than conventionally tilled farms (Fig. 7).
3. Most of the adopters have an education level of middle or high school. The highest level of education in most families is high school. They do not have any association with institutions such as Panchayat, cooperative society, etc. It means that adopters of no-till are not influential and affluent persons of the village (material possession score = 19.5 out of a maximum of 44).
4. Only 15% of the adopters have tractors of 35 hp, 12% have tractors more than 35 hp, 25% have trolleys, and only 1% owns a combine. A seed drill is owned by only 7%, while a no-till drill is owned by 17% of the sample.

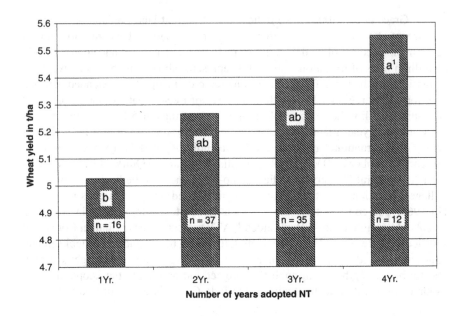

Figure 6   Impact of no-tillage on wheat yield on farmer fields where no-tillage has been adopted for 1–4 years. [1]Means followed by different letters are significantly different at $P = 0.05$ using DMRT. (From Malik et al., 2002).

5. The operational land holding that includes leased land was 75% with small farmers (8.01 acres), 17% with medium farmers (23.38 acres), and 8% with large farmers (40 acres). If categorization is considered on the Rural Credit Survey basis (standard in most socioeconomic studies), the corresponding figures are 13, 45, and 42. This indicates that small landowners are benefiting from this technology.

6. Farmers' response to change was rated as moderate in the survey. Most farmers keep updating their information but do not try all the new methods. About 67% of farmers reported that they see the results of a neighbor before trying out a new practice.

7. Farmer attitude toward no-till was observed to be highly positive. The reasons attributed for this attitude were high profitability of the technology (96%), increase in the yield of wheat (81%), saving diesel (98%), saving money (98%), saving water in the first irrigation (85%), saving water in the subsequent irrigation (79%), crop does not turn yellow after first irrigation (98%), and simplicity of the technology

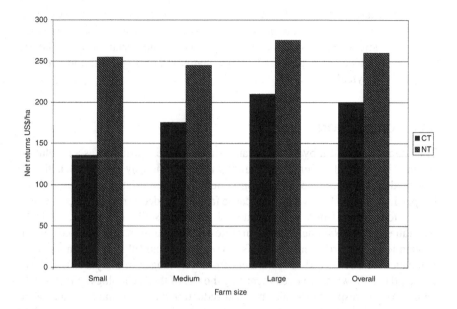

Figure 7 Profitability (net returns) to no-tillage compared to conventional tillage on farmer fields of different farm size in Haryana in 2001–2002. All no-tillage plots are significantly different from the conventional tilled plots when compared at each farm size.

(96%). The farmers advocated strong involvement of government agencies to promote no-till (95%). Ninety-seven percent of the farmers strongly agreed that no-till will be successful.

8. Knowledge of no-till was observed to be moderately high. About 86% of farmers had no problem in no-till with anchored stubbles of rice. About 58% of farmers reported that emergence of wheat is 1 or 2 days earlier than conventional sowing. About 96% agreed that the *Phalaris minor* population is less in no-till. About 95% of farmers reported that water does not remain stagnant after first irrigation, and 68% of farmers strongly agreed that lodging is less of a problem in no-till.

9. Constraints in no-till: Farmers felt no technical constraints in adoption of no-till. However, financial constraints such as lack of adequate manpower from state extension agencies, extension literature, attention from mass media, credit facilities, money to buy a new machine, and costs of other inputs were a problem.

10. Farmers felt highly satisfied with no-till because it gave them a sense of achievement; a merit for promotion; technical feasibility; reduced stress, chances for custom hiring services; appreciation by neighbors; reduced cost, time, and energy; status and prestige; and improvement in yield.

VII. CONCLUSION

The adoption of no-till by South Asian farmers can be considered as a revolution because it is part of a significant paradigm shift in the way wheat production is managed in the rice–wheat cropping system. The accelerated adoption did not happen in isolation. It was mainly because farmers needed to reduce costs of wheat production. Joint efforts by international institutions like CIMMYT, ACIAR-Australia, and the National Agricultural Research Systems have identified several economic and environmental payoffs from adopting no-till in India. In a survey conducted in 2000–2001, 100 farmers from 19 villages in Haryana, who have accepted no-till, were surveyed regarding the use of this technology. Ninety-eight percent of the respondents said that cost reduction and yield improvement were the main reasons for accepting this technology. Wheat yields in Haryana had problems due to herbicide resistant weeds and soil productivity issues in northwest India. Due to no-till, the profitability of growing wheat has increased. In the last two years, acceleration of no-till has become a major preoccupation of scientists. In addition to cutting costs, no-till also encourages farmers to plant wheat early. This will result in higher yields in India. After the success achieved with no-till, scientists have now identified other opportunities in crop establishment techniques for the rice–wheat cropping system as a whole.

REFERENCES

Ali, M., Byerlee, D. 2000. Productivity growth and resource degradation in Pakistan's Punjab: Policy Working Paper. World Bank. Washington, DC (in process).

Anonymous. 1998. Research Highlights 1996–98. Directorate of Research: CCS Haryana Agricultural University. Hisar, 28.

Dick, W.A. 1983. Organic carbon, nitrogen and phosphorus concentrations and pH in soil profiles as affected by tillage intensity. Soil Sci. Soc. Am. J 47:47–102.

Dick, W.A., McCoy, E.L, Edwards, W.M., Lal, R. 1991. Continuous application of no-tillage in Ohio soils. Agron. J 83:83–65.

Doran J.W., Coleman D.C., Bedzicek D.F., Stewart B.A., Eds. 1994Defining Soil Quality for a Sustainable Environment: SSSA Spec. Publ. 35. Madison, WI.

DWR. 1997. Directorate of Wheat Research, Annual Report 1996–97

DWR. 1998. Directorate of Wheat Research, Annual Report 1997

Etheridge, D.M., Steele, L.P, Langenfelds, R.L., Francey, R.J. 1996. Natural and anthropogenic changes in atmospheric CO_2 over the last 1000 years from air in Antarctic ice and firn. *J. Geophys. Res* 101:101–4115.

Gill, M.A., Kahlown, M., Choudhary, M.A, Hobbs, P.R. 2000. Evaluation of Resource Conservation Technologies in Rice–Wheat System of Pakistan. Water and Power Development Authority Report. Lahore. Pakistan.

Grace, P.R., Jain, J.C, Harrington, L.W. 2000. Environmental concerns in rice–wheat systems. Proceedings of an International Workshop on "Developing an action program for farm level impact in rice–wheat systems of the Indo-Gangetic Plains." Sept. 25–27 2000. New Delhi. India.

Harrington, L.W., Morris, M.L, Hobbs, P.R., Singh, V.P, Sharma, H.C., Singh, R.P, Chaudhary, M.K., Dhiman, S.D. 1992. Wheat and rice in Karnal and Kurukshetra districts, Haryana, India: Exploratory survey report. CCS Haryana Agricultural University: Indian Council of Agricultural Research, CIMMYT, and IRRI.

Harvey, L.D.D. 2000. Global Warming: The Hard Science: Prentice Hall. U.K..

Hobbs, P. R. 1994. Rice–wheat system in South Asia. Paper presented at a symposium on "Sustainability of Rice–Wheat System in Northwest India" held at CCS Haryana Agricultural University, Regional Research Station. Karnal on May 7–8, 1994.

Hobbs, P.R., Morris, M.L. 1996, Meeting South Asia's future food requirements from rice–wheat cropping systems: Priority issues facing researchers in the post green revolution era: NRG Paper 96–01. CIMMYT. Mexico, D.F..

Hobbs, P.R., Giri, G.S, Grace, P. 1998. Reduced and zero-tillage options for the establishment of wheat after rice in South Asia. Rice–Wheat Consortium Paper Series 2: RWC. New Delhi. India, 1–18.

Huke, E., Huke, R., Woodhead, T. 1994a. Rice–wheat atlas of Bangladesh: IRRI/CIMMYT/BARC. Los Baños. Philippines.

Huke, R., Huke, E., Woodhead, T. 1994b. Rice–wheat atlas of Nepal: IRRI/CIMMYT/NARC. Los Baños. Philippines.

IPCC. 1996. Climate change 1995. Working Group 1, IPCC: Cambridge University Press. Cambridge. U.K..

Karlen, D.L., Mausbach, M.J, Doran, J.W, Cline, R.G, Harris, R.F., Schuman, G.E. 1997. Soil Quality: A concept, definition and framework for evaluation. *Soil Sci. Soc. Am. J* 61:61–4.

Kumar, V, S, Yadav, Singh, A., Malik, R.K, Hobbs, P.R. 2002. Studies on the effect of zero tillage in wheat on physico-chemical properties of soil in comparison conventional tillage. In. Malik R.K., Balyan R.S., Yadav A., Pahwa S.K., Eds. Herbicide resistance management and zero tillage in rice-wheat cropping systems: proccedings of an international workshop March 4–6, 2002. Hisar. India: CCS Haryana Agricultural University, 172–176.

Ladha, J. K., Fischer, K.S, Hossain, M., Hobbs, P.R, Hardy, B. 2000. Improving the productivity and sustainability of rice–wheat systems of the Indo-Gangetic Plains: A synthesis of NARS-IRRI partnership research: IRRI Discussion Paper No. 40. IRRI. Los Baños. Philippines.

Lal, R. 2000. Controlling greenhouse gases and feeding the globe through soil management. University distinguished lecture presented on February 17, 2000, at Wexner Center for the Arts: The Ohio State University. Columbus, Ohio, 1–36.

Maklin, M.C., Rao, M.V. 1991. Rice and wheat production and sustainability of the irrigated rice-wheat cropping system in India: World Bank Internal Report. New Delhi.

Malik, R.K., Singh, S. 1993. Evolving strategies for herbicide use in wheat: Resistance and integrated weed management. *In* Proceedings of the International Symposium on Integrated Weed Management for Sustainable Agriculture. Hisar. India, Nov. 18–20: Indian Society of Weed Science. Hisar. India, 225–238.

Malik, R.K., Singh, S. 1995. Littleseed canary grass (*Phalaris minor*) resistance to isoproturon in India. *Weed Technology* 9:9–419.

Malik, R.K., Gill, G., Hobbs, P.R. 1998. Herbicide Resistance—a Major Issue for Sustaining Wheat Productivity in Rice–Wheat Cropping Systems in the Indo-Gangetic Plains: Rice-Wheat Consortium Paper Series 3. New Delhi. India, 1–32, RWC.

Malik, R.K., Mehla, R.S, Singh, B.K 2000. Conservation tillage technologies and farmers' participatory research and extension approaches in Haryana—a case study. Proceedings of an International Workshop on developing an action programme for farm level impact in rice-wheat systems of the Indo-Gangetic Plains. Sept. 25–27 2000. New Delhi. India.

Malik, R.K., Yadav, A., Singh, Samar, Malik, R.S, Balyan, R.S., Banga, R.S, Sardana, P.K., Jaipal, S., Hobbs, P.R., Gill, G., Singh, Samunder, Gupta, R.K, Bellinder, R. 2002. Herbicide resistance management and evolution of zero tillage—a success story. Hisar. India: CCS Haryana Agricultural University, Research Bulletin.

Murgai, R. 2000. The Green Revolution and the productivity paradox: Evidence from the Indian Punjab: World Bank Working Paper. New Delhi. India.

Paroda, R.S., T., Woodhead, R.B, Singh 1994. Sustainability of rice–wheat production systems in Asia: FAO-Rapa Publication 1994/11.

Piggin, C.M., Garcia, C.O, Janiya, J.D. 2002. Establishment of irrigated rice under zero and conventional tillage systems in the Philippines. In. Malik R.K., Balyan R.S., Yadav A., Pahwa S.K., Eds. Herbicide resistance management and zero tillage in rice-wheat cropping systems: proceedings of an international workshop, March 4–6 2002. Hisar. India: CCS Haryana Agricultural University, 190–195.

Vincent, D., Quirke, D. 2002. Controlling *Phalaris minor* in the Indian Rice–Wheat Belt. ACIAR Impact Assessment Series No. 18. ACIAR. Canberra. Australia: Available at the following Web site: www.aciar.gov.au/publications/db/abstract.asp?-pubsID = 523.

Woodhead, T., Huke, R., Huke, E. 1994a. Rice–wheat atlas of. Pakistan. IRRI/CIMMYT/PARC. Los Baños. Philippines.

Woodhead, T., Huke, R., Huke, E., Balababa, L. 1994b. Rice–wheat atlas of India: IRRI/CIMMYT/ICAR. Los Baños. Philippines.

Yadav, Ashok, Malik, R.K, Banga, R.S., Singh, Samar, Chauhan, B.S, Yadav, D.B., Murthi, Ram, Malik, R.S. 2003. Long-term effects of zero tillage on wheat in rice–wheat cropping system. In. Malik R.K., Balyan R.S., Yadav A., Pahwa S.K., Eds. Herbicide resistance management and zero tillage in rice–wheat cropping systems: Proccedings of an international workshop, March 4–6 2002. Hisar. India: CCS Haryana Agricultural University, 158–161.

9

No-Till Option for the Rice–Wheat Cropping System in the Indo-Gangetic Plains of India: Some Experiences

J. S. Samra
ICAR
New Delhi, India

D. K. Painuli
IISS
Bhopal, India

I. INTRODUCTION

Sequential cropping of rice and wheat on the same field came into prominence in the irrigated Indo-Gangetic Plains (IGPs) of India after the mid-1960s to provide food security to a burgeoning population. Since then it has occupied a dominant place in livelihood gathering, employment generation, poverty alleviation, and environmental impacts. Rice (*Oryza sativ L.*) and wheat (*Triticum aestivum L.*) constituted more than 75% of the total food grain production of India during the past five years. The majority of the 10.5 million ha under the rice–wheat system of India (Woodhead et al., 1994) is concentrated in Punjab, Haryana, Western Uttar Pradesh, the *Tarai* region of Uttaranchal, Bihar, and West Bengal. The total area of 13.5 mha of India, Pakistan, Bangladesh, and Nepal account for 25% of the rice and 33% of the wheat production in the South Asian region (Ladha et al., 2000). About 32% of the total rice and 42% of the total wheat area of South Asia is seeded in the rice–wheat system (Hobbs and Morris, 1996).

Soils of the IGPs are mainly riverine alluvial except a small part of West Bengal where laterite, lateritic, and red-yellow soils also occur. The soil varies

from coarse to fine texture, calcareous, moderately saline/alkaline to slight/
strongly acidic, and low- to moderate-base saturation.

Tillage has been an important operation for seedbed preparation, moisture
conservation, water management, weed control, improving nutrient supply, ensuring biological activities, and maintenance of soil physical environment as reviewed by Unger (1990). Tillage constitutes about 25%–30% of the total cost of
cultivation because of its above-said multifunctionalities. Tillage is dynamically
complex and has generally responded to demands of diversification in crops/
varieties/inputs/farming systems, availability of improved machinery, cost-effec-
tiveness, production of chemical/biological weedicides, resource use efficiency,
conservation, and market forces. Cultivation of the rice–wheat system has also
witnessed a whole range of tillage options varying from no-till (NT) in eastern
parts to input intensive puddling for rice transplanting in the western zone over
time. Compaction of the subsurface and plow-layer formation due to puddling
under the rice–wheat cropping system has been reported (Sur and Prihar, 1989;
Tripathi, 1992). Deep plowing to alleviate puddling-related subsurface compact-
ness, prepuddling cultivation, reduced till, NT, bed planting, and residue handling
of combine-harvested crops have all experienced research activity from time to
time. No-till research was triggered in Australia in the 1960s, Kentucky in 1962,
New Zealand in 1972, and Brazil during the 1970s after the commercial release
of Paraquat herbicides by the British Company ICI. Argentina also initiated trials
on NT in the 1970s, which were mainly promoted by Monsanto for its Glyphosate
herbicide. No-till versus conventional till (CT) for continuous corn (*Zea mays*
L.) and the corn–soybean (*Glycine max* L.) system is being compared in long-term
experiments for 40 years in two rainfed agro-ecologies by the Ohio Agricultural
Research and Development Center at Wooster and Hoytville (Dick, 2002). Results
of another 18 years of continuous experimentation demonstrated several positive
effects including accumulation of organic matter/nutrients, enhanced biological
activities, better management of herbicides, avoidance of surface sealing, less
erosion, and better trafficability under NT (Tebrugge and During, 1999). No-till
was estimated to be used on 59 million hectares in the world during 2000–2001.
No-till and residue management are being practiced mostly in rainfed conditions
to prevent soil erosion and to conserve rainwater for aerobic cropping systems
of soybean, corn, wheat, and other crops. Tillage operations for the anaerobic
rice and aerobic wheat system under irrigated situations are unique and, comparatively, inadequately experimented until recently. Issues of conserving/improving
soil biodiversity, air pollution due to excessive use of fossil fuels, emission of
greenhouse gases by residue burning, pollution of natural resources by weedicides, and carbon sequestration are the latest dimensions of these tillage operations. Realizing higher total/partial factor productivity by improving efficiency
of inputs is another concern in order to be globally competitive through cost-
effective technologies in the new liberalized trade regime. Soil type, socio-eco-

Table 1 Tillage and Crop Establishment from Diagnostic Surveys of Selected Areas Under Rice–Wheat Cropping System

Location	Turnaround time (days)	Average number of passes with plow	Late planted area (%)[a]
Pantnagar	15–20	5–12 (8)	35
Faizabad	20–45	5–12 (6)	25
Haryana	15–35	4–12 (8)	25

[a] Late planting is defined as wheat seeded after the first week of December.
Sources: Hobbs et al. (1991) for Pantnagar; Hobbs et al. (1992) for Faizabad; and Harrington et al. (1993) for Haryana.

nomic condition of the farmers, availability of farm machinery, regional concerns, and irrigation water availability are unique to a region. Therefore, tillage is expected to be situation-specific to different agro-ecologies.

Wet tillage (puddling) for rice followed by dry tilled wheat is the most used land preparatory practice of the Indo-Gangetic Plains. Puddling for rice planting compacts soil below a 10-cm depth and requires a long turnaround time (Table 1) and more tillage operations/energy to prepare a good seedbed leading to delays in wheat sowing (Hobbs et al., 1997; Mehla et al., 2000; Sharma et al., 1984). Delay in wheat sowing after mid-November is expected to reduce productivity at the rate of 35 kg ha^{-1} day^{-1} (Ortiz-Monasterio et al., 1994). This paper endeavors to make arguments on some of the above-mentioned tillage-related issues with specific reference to the unique irrigated rice–wheat system of the Indo-Gangetic Plains.

II. RICE TILLAGE

Puddling in rice is generally preferred in the irrigated western plains for the control of weeds, to reduce water percolation losses, realize better root growth, accomplish higher water-use efficiency (Table 2), and improve the availability of redox potential sensitive nutrients. Interestingly, rice yield under nonpuddled conditions was also quite high in this case (5.5 t/ha). Even prepuddling tillage has been attempted to improve efficacy of the puddling practice by different investigators (Gajri et al., 1999; Sarkar and Rana, 1999). Although puddling is common for wetland rice, it alters the soil physicochemical environment and is labor- and capital-intensive. Data from major rice growing areas illustrated that it deteriorated soil structure below the 5-cm soil layer (Tables 3 and 4), the magnitude of which increased as intensity of puddling and the number of cropping

Table 2 Effect of Seedbed Preparation with Different Implements on Average Yield of Rice and Water-Use Efficiency in Rice–Wheat Rotation Grown for 4 Years in Haryana

Puddling treatments in rice	Rice yield ($Mg\ ha^{-1}$)	Water-use efficiency in rice ($kg\ ha^{-1}cm^{-1}$)
No puddling	5.55	49.00
Puddling by country plow	6.30	57.07
Puddling by bullock-drawn disc harrow	5.99	54.28
Puddling by bullock-drawn puddler	6.44	58.01

Source: Faroda (1992).

seasons increased (Aggawal et al., 1995; Balloli et al., 2000). Relatively lower bulk density, higher hydraulic conductivity, and resistance to penetration in the 0–5-cm layer shown in Table 3 were attributed to high rice root activity in this restricted zone. Lack of deeper penetration of rice roots limits exploitable volume of soil, increases fertilizer requirements, and promotes volatilization losses. The adverse effect of puddling in rice manifests itself through lower productivity of wheat in the cropping sequence (Aggarwal et al., 1995; Chatterjee and Khan, 1977) and may even require an occasional expensive deep chiseling.

In view of the adverse effects of puddling, alternative tillage for planting of rice in the rice–wheat system is being experimented with in a participatory

Table 3 Bulk Density, Hydraulic Conductivity, and Resistance to Penetration of Different Layers at Harvest of Wheat in a Soil of Punjab Under Rice–Wheat (RW) and Corn–Wheat (CW) Rotation

Depth (cm)	Bulk density ($g\ cm^{-3}$)		Hydraulic conductivity ($cm\ h^{-1}$)		Resistance to penetration at mass water content of			
					8%		14%	
	RW	CW	RW	CW	RW	CW	RW	C
0–5	1.53	1.60	0.17	0.11	13.5	12.5	9.0	10.0
5–10	1.65	1.62	0.08	0.14	17.0	16.0	11.0	10.0
10–15	1.69	1.63	0.06	0.13	29.5	27.5	20.5	19.0
15–20	1.81	1.68	0.02	0.04	48.0	31.5	30.5	20.0
20–25	1.73	1.64	0.02	0.08	29.8	25.0	10.0	6.0
25–30	1.65	1.64	0.06	0.08	30.0	22.5	7.5	5.6

Source: Sur and Prihar (1989).

Table 4   Soil Bulk Density After Wheat Harvest in a Long-term
Rice–Wheat Rotation Experiment in a Silty Clay Loam Soil at
Pantnagar, Utteranchal

	Average bulk density (g cm^{-3})	
Depth (cm)	Initial	Nine years after rice–wheat cropping
0–7	1.34	1.38
15–22	1.41	1.50
25–32	1.48	1.58
40–57	1.52	1.53
55–62	1.50	1.51

Source: Tripathi (1992).

approach with farmers. Bed planting of rice consumes less water in northern
Australia (Borrell et al., 1997). Ninety-five percent of wheat is being seeded on
beds in the Yaqui Valley of northern Mexico and is saving 25% of the irrigation
water requirements; it is recently being evaluated in India (Meisner et al., 1992).
Band placement of nitrogen (N) in the crop grown on beds is possible with new
drills and improves the efficiency of N utilization, protein content of wheat grains,
and reduced environmental hazards associated with broadcasting (Dhillon et al.,
2000). Conservation farming for high input efficiency is an important research,
development, and extension agenda for Pakistan, Nepal, Bangladesh, China, Iran,
Turkey, Sudan, and several CIS republics.

III. NO-TILLAGE

A diagnostic survey done in 2001 in the rice–wheat growing belt of the Indian
Punjab by the Punjab Agriculture University (PAU), Ludhiana, revealed that the
majority of farmers (75%) carried out more than four tillage operations (2 to 3
with disc harrow and 1 to 2 with a 9-tyned cultivator) for seedbed preparations
in wheat. Their main intentions were to control weeds and get higher germination
(unpublished).

Alternative tillages for wheat including NT are being evaluated to reduce
time of operation, lower cost of cultivation (Table 5), reduce consumption of
diesel (Table 6), save water (Mehla et al., 2000; Mehta et al., 2001), and prevent
the adverse effect of late sowing. The time used with machinery was reduced by
3.7 times and cost of operation by 3.9 times by practicing NT (Table 5). Similarly,
the efficiency of tillage energy was improved by 3.4 times without much loss in

Table 5 Time and Cost of Operation Under Different Tillage Options for Wheat in a Clay Loam Soil at Kaul (Kurukshetra), Haryana

Tillage treatments	Time taken (h ha^{-1})	Operation cost (Rs. ha^{-1})
Conventional	7.71	657
No-tillage (Pantnagar no-till drill)	2.10	166

Source: CIAE (1999).

yield under NT (Table 6). Farmer participatory demonstrations in six districts of Haryana showed higher wheat yields under NT compared to CT even when seeded at the same time (Table 7). The mean grain yields of wheat planted under NT, conventional tillage timely sown (CTT), and conventional tillage delayed sown (CTD) conditions were 4,252, 4,016, and 3,533 kg ha^{-1}, respectively. Thus the difference in yield under NT and CTD treatments ($Y_{ZT} - Y_{CTD}$) was 17% on a regional basis. On an average, productivity of NT wheat was higher by 5.5% when compared with the CTT system ($Y_{ZT} - Y_{CTT}$). The increase in productivity was ascribed primarily to enhanced fertilizer and water-use efficiency and a significant reduction in weed population (Table 8). *Phalaris minor* population was about 4.6 times less in NT as compared to CT. The yield benefit of 13% under NT compared to CT (four harrowing followed by broadcasting of seed + tillage + planking) was also observed by Sharma et al. (1984) at Kaul (Haryana). In that situation, NT advanced sowing of wheat by 4 to 5 days and was beneficial to the productivity.

Table 6 Energy Requirement, Grain Yield, and Tillage Energy Efficiency Under Different Land Preparation Tillage for Wheat in Rice–Wheat System at IARI, New Delhi

Tillage treatment	Yield (t ha^{-1})	Energy requirement (MJ ha^{-1})	Tillage energy/efficiency (MJ t^{-1})
No-tillage	3.12	785	251.8
Chiseling	3.21	2040	643.1
Rototilling	3.25	1800	554.0
Chiseling plus rototilling	3.63	3045	858.9
Disc harrowing	2.90	1390	482.7
Chiseling plus harrowing	3.07	2645	843.4

Source: Srivastava et al. (2000).

Table 7 Effect of Tillage Practices and Sowing Time on Yield of Wheat in Haryana

Districts	Growing season (year)	No of trials	Average sowing time difference (days) (NT-CTD)	Grain yield (kg ha^{-1})		
				NT	CTT	CTD
Kaithal	1997–98	04	14	4,407	3,829	3,583
	1998–99	32	14	4,897	4,605	4,288
Karnal	1997–98	17	12	4,391	3,850	3,465
	1998–99	32	14	4,616	4,553	4,273
Sonepat	1998–99	07	13	3,489	3,270	2,059
Panipat	1998–99	06	13	3,800	4,000	3,583
Ambala	1998–99	02	13	3,825	3,638	2,997
Kurukshetra	1998–99	32	13	4,593	4,383	4,016
Mean				4,252	4,016	3,533
% decrease				—	5.5	17.0

NT = no-tillage; CTT = conventional tillage timely sown; CTD = Conventional tillage delayed sown.
Source: Mehla et al. (2000).

Table 8 Effect of Tillage Practices and Sowing Time on *Phalaris minor* Population in Haryana

Districts	Growing season (year)	No. of trials	Average sowing time difference (days) (NT-CTD)	*Phalaris minor* population (plants m^{-2})	
				NT	CTT & CTD
Kaithal	1997–98	04	14	129	560
	1998–99	32	14	114	550
Karnal	1997–98	17	12	103	438
	1998–99	32	14	110	473
Sonepat	1998–99	07	13	75	333
Panipat	1998–99	06	13	89	440
Ambala	1998–99	02	13	73	379
Krukshetra	1998–99	32	13	97	444
Mean				98.7	452.1

NT = no-tillage; CTT = conventional tillage timely sown; CTD = conventional tillage delayed sown.
Source: Mehla et al. (2000).

Table 9 Relative Economics and Energetics of Various Tillage Options for Wheat in Rice–Wheat Sequence[a]

Parameters	Rotary	NT	FIRB	CT
Cost of equipment (Rs)	42,000	13,000	16,000	13,000
Yield grain, q/ha	60.5	54.9	50.5	54.7
Straw, q/ha	104	95	82	93
Gross income, Rs/ha	40,069	36,331	33,257	36,189
Cost of production, Rs/ha	11,906	11,675	12,450	13,101
Cost of sowing, Rs/ha	398	167	1605	1593
Net return, Rs/ha	28,163	24,656	20,807	23,088
Energy requirement, MJ/ha	31,056	30,576	31,504	33,653
Specific energy, MJ/kg	1.88	2.05	2.38	2.28

[a] The cost of production does not include land rent, interest on working capital, risk factor, and management charges.

NT = no-tillage; FIRB = furrow irrigated raised bed; CT = conventional tillage.

Brar et al. (1986), based on their experience under Punjab conditions, observed that wheat can be grown without preparatory tillage on coarse- to medium-texture soil economically. A large number of trials on farmer fields were also carried out by the Directorate of Wheat at Karnal, and analysis showed the minimum cost of production was with NT (Table 9). However, the highest net return and productivity were realized from a single rotavator cum seeding operation on their sandy-loam soils. This is yet to be tried in other agro-ecologies and soils. Moreover, the rotavator requires higher initial investment and more horsepower (Chauhan et al., 2001).

Table 10 Grain Yield and Root Length Density in Different Soil Depths at Maturity of Wheat Under Different Tillage Treatments at GBPUA&T, Pantnagar

Parameter	Soil depth (cm)	CT	NT
Yield (q ha^{-1})	—	42.45	33.90
RLD (mm mm^{-3})	0–10	6.8	5.6
	10–20	3.4	2.3
	20–30	2.3	1.5
	30–40	1.8	1.3
	40–50	1.2	0.8

Source: Bajpai and Tripathi (2000).

Table 11 Grain Yields and Total Water-Use Efficiency of Wheat Under Different Tillage Treatments in Rice–Wheat System at PDCSR, Modipuram[a] and BCKV, Gayeshpur[b]

Tillage treatment	Grain yield[a] (Mg ha^{-1}) Modipuram	Grain yield[b] (Mg ha^{-1}) Gayeshpur	Water-use efficiency[b] (kg ha^{-1}mm^{-1})
Conventional	4.50	2.97	5.51
No	3.24	1.76	3.38

Sources: [a]PDCSR (1999); [b]Sarkar and Rana (1999).

However, the yield advantage under NT has not been observed in some other field trials (Tables 10, 11, 12). It should be mentioned here that some of these NT trials did not use a suitable seed drill and were planted manually. This plus the fact that NT plots were planted at the same date and moisture as the CT plots probably account for the different results. Therefore, results need to be interpreted with great caution. Better physical condition under CT (Tables 13, 14) including reduced mechanical resistance/soil strength in the soil profile encouraged better root growth (Table 15) and thus improved the utilization of nutrients and profile water. High weed infestation under NT (Table 16) also favored higher yield under CT. However, recent NT trials on farmer fields indicate that weed germination is significantly less with this practice than conventional systems because of less soil disturbance (Malik et al., 2002). Initial higher yield under CT disappeared during subsequent years in some cases. In fact, NT led to improved soil physical conditions. Decay of roots and leaves and undisturbed channels in the uncultivated soil supplemented by the build-up of micro as well as meso fauna help, but this process can take some years. Permanent trials in Wooster, Ohio, showed NT under two rotations was superior, whereas at the Hoytville site, it was better only in the wheat–soybean rotation 13 years later.

Table 12 Physical Condition of the Soil as Affected by Tillage Methods

Tillage method	Shear tester value (rad)	Bulk density (g cm^{-3})	Water content (%)	Bearing capacity of soil (kg m^{-2})
No	2.002	1.58	22.4	3.6
Conventional	1.279	1.50	26.3	2.5

Source: Dhiman et al. (2001).

Table 13 Effect of Tillage Treatments on Physical Properties of a Deep, Loamy
Alluvial Soil at Wheat Sowing and on Grain Yield of Wheat

Treatment	IR (cm h^{-1})	BD (g cm^{-3})	MWD (mm)	Grain yield (Mg ha^{-1})
No-tillage	0.8	1.57	–	1.99
Rotavator once and sowing	1.3	1.47	14.50	2.51
Rotavator twice and sowing	1.5	1.44	13.16	2.62
Cross-plowing by cultivator and sowing	0.5	1.54	25.93	2.30
Victory plow once plus Desi plow twice and sowing	1.0	1.39	20.28	2.39
Cultivator once and sowing and again one cultivator	1.2	1.43	31.10	2.21
Desi plow four times and sowing	0.8	1.56	27.23	2.33

Source: RAU (1995).
IR = infiltration rate; BD = bulk density; MWD = mean weight diameter.

Table 14 Soil strength in Top 30-Cm Soil in Three Tillage Systems at Wheat Seeding
Time at PAU, Ludhiana

Soil depth (cm)	Cone index (M Pa)			
	No-tillage	Conventional tillage	Deep tillage	LSD (*P* 0.05)
0–5	0.75	0.46	0.20	0.25
5–10	1.65	1.03	0.43	0.33
10–15	1.95	1.92	1.11	0.43
15–20	2.69	2.75	1.61	0.59
20–25	2.78	2.75	1.59	0.62
25–30	2.71	2.75	1.56	0.71

Source: Gajri et al. (1992).

Table 15 Tillage Effects on Root Length Density Profiles of Wheat in Rice–Wheat System at PAU, Ludhiana

Depth (cm)	Root length density ± SD (cm cm^{-3})		
	No-tillage	Conventional tillage	Deep tillage
1985–1986 (58 DAS)			
0–30	0.81±0.18	0.83±0.44	0.67±0.10
30–60	0.07±0.02	0.15±011	0.20±0.06
60–90	—	0.03±0.03	0.07±0.02
90–120	—	—	0.04±0.02
1985–1986 (85 DAS)			
0–30	1.23±0.16	1.38±0.11	1.27±0.34
30–60	0.34±0.02	0.40±0.07	0.58±0.24
60–90	0.27±0.01	0.35±0.17	0.24±0.09
90–120	0.14±0.04	0.16±0.06	0.19±0.06
120–150	0.04±0.02	0.06±0.03	0.15±0.07
150–180	—	—	0.03±0.02
1986–1987 (60 DAS)			
0–30	0.30±0.15	0.41±0.19	0.40±0.18
30–60	0.14±0.11	0.20±0.12	0.24±0.12
60–90	0.05±0.03	0.07±0.05	0.15±0.08
90–120	—	0.04±0.03	0.10±0.08
120–150	—	0.01±0.01	0.04±0.04

Source: Gajri et al. (1992).

Table 16 Effect of Tillage Methods on Density and Dry Weight of Weeds

Tillage method	Density (m^{-2})	Total dry weight	
		Phalaris minor	Broad leaf weeds
No	160(72)	12	14.9
Conventional	135(2)	8	0.8

Figures in the parentheses are the population of *Phalaris minor* emerged in the first flush, that is, along with wheat.
Source: Dhiman et al. (2001).

IV. BED PLANTING

The yellowing of lower leaves and significant reduction in chlorophyll and wheat yield due to temporary flooding/anaerobic conditions have been reported from India by Bandhopadhayay and Dutt (1983), Sharma and Swaroop (1988), and Gill et al., (1992, 1993). The flowering stage and the early tillering stage were the most sensitive to the stress of excessive moisture or inadequate aeration as compared to crown root initiation and the grain-filling growth phases. Several wheat varieties responded significantly differently to temporary flooding or partially anaerobic conditions. Stagnation of water due to excessive irrigation or rainfall, especially irrigation followed by rains or accumulation of water in a part of an inadequately leveled field, is a common phenomenon in impermeable soils (heavy-textured, sodic, or with a clay/calcic pan or continuously rice puddled fields). Management of this kind of temporary stress by bed planting or other strategies is important for precision farming. Bed planting is also being evaluated in the rice–wheat system where the groundwater level is falling, herbicide resistance in weeds is developing, and conservation farming is assuming significance. The system is supposed to improve water- and fertilizer-use efficiency, reduce seed rate, allow mechanical control of weeds, and decrease lodging (Hobbs et al., 1997). The ridge and furrow bed system for growing wheat, rice, soybean, pigeonpea, black gram, linseed, chickpea, safflower, and other crops in deep Vertisols in a 1,000-mm rainfall region was experimented with for 16 years in Jabalpur (M.P.) (Annon, 2002). Wheat, soybean, and blackgram yield improved by more than 2 times, sesamum by 6 times, and pigeonpea by 12 times, with still further higher productivity in mixed cropping. The rice crop failed to mature in 2 rainfall-deficit seasons out of 16 years, with an average productivity of 2,980 kg/ha for the remaining 14 years. The payback period of the additional cost of making permanent beds was 1 to 2 years. The benefit cost ratio ranged from 1.10 to 1.74 depending on cropping system and kind of mixed crops, and financing by banks was attractive. Preliminary results of participatory research demonstra-

Table 17 Average Grain Yield of Wheat Planted with Different Tillage Systems in Farmers' Participatory Trials in Haryana During 1998–2000

Tillage	Number of trials	Grain yield (mg ha^{-1})
Conventional[a]	23	5.28
Bed planting[a]	23	5.73
No-tillage[b]	134	5.52

[a] Seeding dates comparable. Mean yield of two years.
[b] No-till wheat planted earlier than conventional method.
Sources: Singh (2000); HAU, Uchani (personal communication).

tion and extension of bed planting in the Indo-Gangetic Plains were quite encouraging (Table 17). However, it involves significantly higher tillage for bed formation and thus may not be considered a reduced or NT system unless the bed is made permanent. Many issues of seeding or planting of rice on permanent beds in this cropping sequence are yet unresolved. Desirable aerobic rice varieties for the bed systems are not available, and chlorosis of existing varieties due to deficiency of iron, zinc, and manganese has been reported. Besides, it needs to address some emerging issues of crop establishment, physical condition of soil, management of weeds, some insects (termites), water-use efficiency, and nutrients.

V. SYSTEM PRODUCTIVITY

In a rice–wheat system perspective, it is pertinent to consider both the crops and their management practices together and not in isolation since tillage interactions exist between the two crops in the sequence. Many studies have, however, recommended NT for wheat without including the yield of rice in the system. In a silty clay loam soil of the *Tarai* (Uttaranchal, India), puddling for rice in combination with CT for wheat was the most productive (10.14 Mg ha^{-1}) tillage system. The direct-drilled no-puddled rice in combination with NT wheat was the least productive (8.77 Mg ha^{-1}) tillage system (Bajpai and Tripathi, 2000). Findings of Sarkar and Rana (1999) on a fine loam Fluventic Ustochrept of West Bengal were similar. Overall analysis of the entire sequence is relatively less debated in the literature. However, these data need to be diagnosed with caution since it is important to use the correct management for these new technologies in order to get successful results. That means using the correct equipment, planting at the optimal soil moisture, and controlling weeds, and taking advantage of optimal planting dates is necessary before valid comparisons can be made against conventional practices.

VI. EFFECTS OF NO-TILL IN WHEAT ON SOIL PROPERTIES AND MANAGEMENT

No-till farming is a unique system with different kinds of physicochemical and biological processes involved. Decay of roots left behind as channels in the profile and activities of earthworms, ants, and other fauna improve porosity. The impact of NT on biodiversity of soil microbes and associated processes is inadequately understood. Some of the available information on soil properties is summarized below.

A. Soil Physical Factors

A field investigation revealed that NT soil contained 17.4% less water than the CT when measured one week after irrigating the field (Dhiman et al., 2001). This kind of moisture distribution pattern may prove beneficial in heavy-textured or reclaimed alkali soil where wheat becomes yellow after the first irrigation due to anaerobic conditions. These researchers also reported an increase in the bulk density of the soil from 1.50 g cm^{-3} in CT to 1.58 g cm^{-3} in NT plots. The value of bearing capacity and strength of the soil also increased by about 50% under the NT practice. These changes may reduce soil loss in runoff water due to lesser splash erosion. Similar findings of Dixit and Gupta (2001) also indicate that adoption of NT practices in wheat may alter the physical condition of the soil in many ways. These changes are long-term in developing and also need to be assessed with care. Malik et al. (2002) have data that show the physical properties of NT are improved after four years of adoption.

In Haryana, farmers generally practice one to two dry plowings with a disc harrow after the harvest of the wheat crop. The depth of penetration of the disc harrow in the soil was almost double (8.3 cm) in the CT over NT (Dhiman et al., 2001). Thus, the requirement of tillage operations for the preparation of the soil would be increased for the next crop of rice in NT plots of wheat as compared to CT. This may nullify the saving of tillage operations done before the sowing of wheat in NT plots.

B. Weed Flora

Research in NT was triggered by the commercial release of herbicides like Paraquat by the ICI Company and Glyphosate by Monsanto since weed management is an important issue of productivity. There is a growing recognition that *Phalaris minor* has developed herbicide resistance and is affecting wheat production adversely over a large area in northwest India. Incidence of *Phalaris minor* germination was less under NT (Mehla et al., 2000) because of less soil disturbance and lower temperature at germination time experienced by the weed seed below 5-cm soil depth (Mehta et al., 2001). However, Dhiman et al. (2001) reported that the total density of *Phalaris minor* was 18% higher in NT than CT, and these differences may be due to sowing time. Besides this, the density of the first flush of this weed that germinated along with the wheat crop was 24 times higher in the crop sown with the NT machine. Dhiman et al. attributed this to the seed redistribution pattern of *Phalaris minor* in the soil after puddling of the soil during the *kharif* season. They further argued that 60%–70% of the seeds in the upper 5-cm soil layer accumulated in the uppermost half-centimeter layer of the puddled soil. This preferential emergence of *Phalaris minor* in the first flush along with the wheat crop in NT may lead to development of cross-resistance in *Phalaris minor* against the newly recommended herbicides, such as Fenoxaprop and Clodinafop. They also reported that after the harvesting of wheat, the population of perennials, particularly *Paspalum distinchum* and *Cynodon dactylon*, was conspicuously higher in NT fields than the fields sown with the CT. However, other authors have shown the opposite effect with less *Phalaris minor* in NT fields (Malik et al., 2002). They also reported that by using NT combined with some of the new wheat herbicides, the problem of herbicide-resistant weeds could be overcome (Vincent and Quirke, 2002). Apart from the high cost of new weedicides, increased dependence on chemicals may not be desirable for various economic, biodiversity, and environment reasons. In some IGP areas, excessive use of herbicides decreased soil microbial properties (Yadav et al., 1998). The impact of herbicides on the dynamics of soil biology, meso fauna, and contamination of soil/water/air and its second generation consequences are inadequately known. The use of chemically synthesized weedicides is also debarred in organic farming.

C. Insect, Pest, and Disease

Rice residues harbor rice stem borers and if residues are not plowed, the larvae have a greater chance of survival and harm the next rice crop. However, data from Inayatullah (1989) show that this is not very critical except maybe in stemborer hot spots. Their data show that although the stemborer numbers are high just after rice harvest, by irrigating and applying N, the rice residue decomposes and exposes the larvae to predacious insects. Data from Jaipal et al. (2002) also show that the diversity of beneficial insects in NT fields is significantly higher than in CT fields. This led to effective biological control of the stemborers. Also, diseases such as the leaf blights (*Helminthosporium spp*) are more likely to proliferate on crop residues (Hobbs *et al.*, 1997). On the other hand, residues are expected to promote greater dynamics of meso- and macrofauna, which in turn modify several physicochemical properties of soil.

D. Microbial Activity

Reduced tillage causes only a slight incorporation of surface residues and soil disturbance. In NT, the only soil disturbance is that associated with seed placement. However, CT mixes the soil greatly and consequently stimulates microbial activity reflected by higher census counts and increased rate of soil respiration. Inversion and fragmentation of surface residues fosters a zone of intense microbial activity at the plow-sole depth. In contrast, reduced and NT foster microbial activity at or near the soil surface (Paul and Clark, 1996). Chaudhary et al. (2002) recently reported as follows: "Thus it can be concluded that no-tillage which resembles the undisturbed system is definitely having an improvement in soil quality as being evidenced by the quantity and quality of soil biological health parameters.'' Thus, introducing a change in the tillage system may alter the distribution of the microbes, and its consequences to the rice–wheat system need to be documented.

E. Residue Burning

Burning crop residues is most prevalent in the rice–wheat growing areas of the IGP where combine harvesting is becoming popular for both rice and wheat. The latest estimates revealed that 95% of wheat and 92% of the rice crop are combine-harvested in the Indian Punjab. The anchored stubbles do not interfere with the NT operation whereas loose straw thrown in small windrowed piles by the combine harvesters poses a real problem for present NT drills and CT fields. The farmers are invariably burning this detached straw, losing nutrients from the system, and causing environmental air pollution. The temperature of burning rice straw is about 700°C, which leads to loss of carbon (C), release of greenhouse gases, and killing of beneficial soil insects and microorganisms. Herbicide performance is also impaired due to its adsorption on the C particles or ash, and it creates a health and environmental hazard in the air. The area under combine harvesting is extending, and residue management problems will be a major issue to resolve although NT and permanent-bed planting may offer a solution to this problem. Since the present NT and strip-tillage drills available in India operate best in clean and leveled fields, new machinery will be needed to overcome this problem. A drill that will allow crops to be planted into residue is needed. Once that is available, the loose straw can be chopped and spread evenly on the ground by the combine and can be used as a mulch instead of burning. This should have positive effects on surface soil's physical and biological properties. Paddy straw is relatively moist at harvest, is nonbrittle, contains high silica content, and cannot be shredded easily, making this a challenge to agricultural engineers. This development may also create other issues for the research agenda of rice–wheat cultivation. If the size of straw is too small, it may float in the irrigation water, collect in low-lying areas, and may even smother the growing wheat. On the other hand, uniform spread of the chopped straw on the surface may produce tremendous beneficial effects of mulching depending on the agro-ecological situation. Once the straw is evenly spread, it should be easier to design a drill to plant into this loose residue.

VII. CONCLUSIONS

NT of wheat after rice is being adopted by farmers at a very fast rate.

The presently recommended practice of NT of wheat after puddled rice only covers part of the possibilities within a systems perspective. In the long run, it is desirable to NT of both rice and wheat for maximum benefit, which calls upon direct seeding of rice, especially in the irrigated system where transplanting is well entrenched.

Bed planting may be adopted by farmers provided the beds are permanent in nature and suitable seeding/planting practices especially for rice and residue management are generated.

Adequate data on C sequestration by NT, residue management, and biological status of soils are not available for the Indo-Gangetic Plains.

Residual dynamics of weedicides being used in NT and CT and its contaminating implications are not properly analyzed from the viewpoint of organic foods.

REFERENCES

Aggarwal, G.C., Sidhu, A.S., Sekhon, N.K., Sandhu, K.S., Sur, H.S. 1995. Puddling and N management effects on crop response in a rice–wheat cropping system. *Soil Till. Res.* 36:129–139.

Annon, H.S. 2002. Raised–sunken bed technology for rainfed Vertisol of high rainfall areas. Technical Bulletin Painuli D.K., Tomar S.S., Tenbe G.P., Sharma S.K., Eds: IISS. Nabi-Bagh, Berasia Road, Bhopal, (M.P.), India.

Bajpai, R.K., Tripathi, R.P. 2000. Evaluation of non-puddling under shallow water tables and alternative tillage methods on soil and crop parameters in a rice–wheat system in Uttar Pradesh. *Soil Till. Res.* 55:99–106.

Balloli, S.S., Rattan, R.K., Garg, R.N., Singh, G., Krishnakumari, M. 2000. Soil physical and chemical environment as influenced by duration of rice-wheat cropping system. *J. Indian Soc. Soil Sci.* 48:75–78.

Bandyopadhyay, B.K., Dutt, S.K. 1983. Problem of yellowing of winter crops after heavy rainfall or irrigation in coastal saline soil. *J. Indian Soc. Soil Sci.* 31:521–526.

Borrell, A., Garside, A., Fukai, S. 1997. Improving efficiency of water use for irrigated rice in a semi-arid tropical environment. *Field Crops Res.* 52:231–248.

Brar, S.S., Sandhu, H.S., Dhaliwal, J.S. 1986. Scope and limitations of minimum tillage under an intensive system of cropping in Punjab (India). *In:* Portch S. et al., Ed. International Symposium on Minimum Tillage, Feb. 26–27, 1986, Dhaka, Bangladesh, 133–140.

Chatterjee, B.N., Khan, S. 1977. Growth of wheat established with different amounts of tillage after the harvest of direct seeded and transplanted rice. *Indian J. Agri. Sci.* 48:654–659.

Chauhan, D.S., Sharma, R.K., Tripathi, S.C., Kharub, A.S., Chhokar, R.S. 2001. New paradigms in tillage technologies for wheat production. Res. Bull No. 8: Directorate of Wheat Research. Agarsain Marg, Karnal, India.

Chaudhary, A., Choudhary, R., Kalra, N. 2002. Impact of no-tillage practices on soil biological health. Division of Environmental Sciences Report: IARI. New Delhi, India.

CIAE. 1999. Annual Report (June 1997–May 1998) and Half Yearly Report (June–November 1998) of the Ad-hoc Project Increasing Productivity of Rice–Wheat Cropping System by Mechanizing the Operations of Tillage and Crop Establishment: CIAE. Bhopal, Madhya Pradesh, India.

Dhillon, S.S., Hobbs, P.R., Samra, J.S. 2000. Investigations on bed planting system as an alternative tillage and crop establishment practice for improving wheat yields sustainably. *In* Proc. 15th Conf. Int. Tillage Res. Org, Fort Worth, Texas, U.S.A., July 2–7, 2000.

Dhiman, S.D., Om, H., Kumar, H. 2001. Advantages and limitations of no-tillage in wheat. *Indian Farming* 51(6):8–10.

Dick, W. 2002: Ohio State University, Wooster, OH, USA (personal communication).

Dixit, J.S., Gupta, R.S.R. 2001. Effect of tillage on soil physical properties under paddy-wheat cropping system. *Ann. Pl. Soil Res.* 3:65–69.

Faroda, A.S. 1992. A decade of agronomic research in rice–wheat system in Haryana. *In:* Pandey R.K., et al., Ed. Rice–Wheat Cropping System: Project Directorate for Cropping Systems Research. Modipuram, Meerut, 233–238.

Gajri, P.R., Arora, V.K., Prihar, S.S. 1992. Tillage management for efficient water and nitrogen use in wheat following rice. *Soil Till. Res.* 24:167–182.

Gajri, P.R., Gill, K.S., Singh, R., Gill, B.S. 1999. Effect of pre-planting tillage on crop yields and weed biomass in a rice-wheat system on a sandy loam soil in Punjab. *Soil Till. Res.* 52:83–89.

Gill, K.S., Quadar, A., Singh, K.N. 1992. Response of wheat (*Triticum aestivium*) genotopes to sodicity in association with waterlogging at different stages of growth. *Indian J. Agri. Sci.* 62:124–128.

Gill, K.S., Quadar, A., Singh, K.N. 1993. Effect of interaction of flooding and alkalinity at various growth stages on grain yield of wheat (*Triticum aestivum*). *Indian J. Agri. Sci.* 63:795–802.

Harrington, L.W., Fujisaka, S., Morris, M.L., Hobbs, P.R., Sharma, H.C., Singh, R.P., Chaudhary, M.K., Dhiman, S.D. 1993. Wheat and Rice in Karnal and Kurukshetra Districts, Haryana, India: Farmers' Practices, Problems and an Agenda for Action. Haryana Agricultural University (HAU), Indian Council for Agricultural Research (ICAR), CIMMYT, and the International Rice Research Institute (IRRI). Mexico City.

Hobbs, P.R., Hettel, G.P., Singh, R.P., Singh, Y., Harrington, L.W., Fujisaka, S. 1991. Rice–Wheat Cropping Systems in the Tarai Areas of Nainital, Rampur, and Pilibhit Districts in Uttar Pradesh, India: Sustainability of the Rice–Wheat System in South Asia. Diagnostic Survey of Farmers' Practices and Problems, and Needs for Further Research: CIMMYT. Mexico City.

Hobbs, P.R., Hettel, G.P., Singh, R.K., Singh, R.P., Harrington, L.W., Singh, V.P., Pillai, K.G. 1992. Rice–Wheat Cropping Systems in Faizabad District of Uttar Pradesh, India: Exploratory Surveys of Farmers' Practices and Problems, and Needs for Further Research: CIMMYT. Mexico City.

Hobbs, P.R., Morris, M.L. 1996. Meeting South Asia's future food requirements from rice-wheat cropping systems: Priority issues facing researchers in the post-Green Revolution era. NRG Paper 96-01: CIMMYT. Mexico City.

Hobbs, P.R., Giri, G.S., Grace, P. 1997. Reduced and no-tillage options for the establishment of wheat after rice in South Asia. RCW Paper No. 2. Rice–Wheat Consortium for the Indo-Gangetic Plains. New Delhi, India.

Inayatullah, C., Ehsan-ul-Haq, C., Ata-ul-Mohsin, C., Rehman, A., Hobbs, P.R. 1989. Management of rice stem borers and the feasibility of adopting no-tillage in wheat. PARC Mimeograph: NARC. Islamabad, Pakistan.

Jaipal, S., Singh, S., Yadav, A., Malik, R.K., Hobbs, P.R. 2002. Species diversity and population density of macro-fauna of rice–wheat cropping habitat in semi-arid subtropical north-west India in relation to modified tillage practices of wheat sowing. Pp. 166–171. *In.* Malik et al. R.K., Ed. Herbicide resistance management and no-tillage in the rice-wheat cropping system. *Proc. Internat. Workshop*, Hissar, India, March 4–6, 2002: CCSHA Univ. Hissar, India.

Ladha, J.K., Fischer, K.S., Hossain, M., Hobbs, P.R., Hardy, B. 2000. Improving the productivity of rice–wheat systems of the Indo-Gangetic Plains: A synthesis of NARS-IRRI partnership research. IRRI Discussion Paper No. 40: IRRI. Los Baños, Philippines.

Malik, R.K., Yadav, A., Singh, S., Malik, R.S., Balyan, R.S., Banga, R.S., Sardana, P.K., Jaipal, S., Hobbs, P.R., Gill, G., Singh, S., Gupta, R.K., Bellinder, R. 2002. Herbicide resistance management and evolution of no-tillage: A success story. Research Bulletin: Haryana Agricultural University. Hisar, India.

Mehla, R.S., Verma, J.K., Gupta, R.K., Hobbs, P.R. 2000. Stagnation in the productivity of wheat in the Indo-Gangetic plains: No-till-seed-cum-fertilizer drill as an integrated solution: Rice–Wheat Consortium Paper Series 8: RWC. New Delhi, India.

Mehta, A.K., Singh, R., Hansra, B.S. 2001. No-tillage sowing of wheat—a boon for farmers. *Intensive Agriculture* 38:21–23.

Meisnerz, C., Acevedo, E., Flores, D., Sayre, K., Ortiz-Monasterio, I., Byerlee, D. 1992. Wheat production and grower practices in the Yaqui Valley, Sonora, Mexico. Wheat Special Report No. 6: CIMMYT. Mexico.

Ortiz-Monasterio, J.I., Dhillon, S.S., Fischer, R.A. 1994. Date of sowing effects on grain yield and yield components of irrigated spring wheat cultivars and relationships with radiation and temperature in Ludhiana, India. *Field Crops Res.* 37:169–184.

Paul, E.A., Clark, F.E. 1996. Soil microbiology and biochemistry, 2nd ed.: Academic Press. Boston, MA, USA.

PDCSR. 1999. Annual Progress Report (July 1998–March 1999) of the Ad-hoc Project on Alternate Tillage and Residue Management Strategies in Rice-wheat Cropping System. Publ: Project Directorate for Cropping Systems Research. Modipuram, U.P., India.

RAU. 1995. Tillage requirement in rice–wheat cropping system. Pre-workshop meeting at Wheat Directorate, Karnal on Aug. 16–17, 1995: Dept. of Agronomy, RAU. Pusa (Samastipur), Bihar, India.

Sarkar, S., Rana, S.K. 1999. Role of tillage on productivity and water use pattern of rice-wheat cropping system. *J. Indian Soc. Soil Sci.* 47:532–534.

Sharma, D.N., Jain, M.L., Sharma, S.K. 1984. Evaluation of no-tillage and conventional tillage systems. *AMA* 15:14–18.

Sharma, D.P., Swarup, A. 1988. Remove excess water for high wheat productivity in alkali soil. *Indian Farming* 38:18–19.

Srivastava, A.P., Panwar, J.S., Garg, R.N. 2000. Influence of tillage on soil properties and wheat productivity in rice (*Oryza sativa*)–wheat (*Triticum aestivum*) cropping system. *Indian J. of Agril. Sci.* 70:207–210.

Sur, H.S., Prihar, S.S. 1989. Soil management in rice-based cropping systems in Punjab. *In*: Physical Environment of Rice Eco-systems. Indian Society of Agrophysics. Agricultural Physics Division: IARI. N. Delhi, 37–44.

Tebrugge, F., During, R.A. 1999. Reducing tillage intensity—a review of results from a long-term study in Germany. *Soil & Tillage Res.* 53:15–28.

Tripathi, R.P. 1992. Physical properties and tillage of rice soils in rice–wheat system. *In* Pandey R.K., et al., Eds. Rice–Wheat Cropping System: Publ. Project Directorate for Cropping Systems Research. Modipuram, U.P., India, 53–67.

Unger, P.W. 1990. Conservation tillage systems. *Adv. Soil Sci.* 13:28–67.

Vincent, D., Quirke, D. 2002. Controlling *Phalaris minor* in the Indian Wheat Belt. ACIAR Impact Assessment Series 18: ACIAR. Canberra, Australia.

Woodhead, T., Huke, R., Huke, E., Balababa, L. 1994. Rice–wheat atlas of India: IRRI/CIMMYT/ICAR publication. Los Baños, Philippines.

Yadav, R.L., Prasad, K., Gangwar, K.S. 1998. Analysis of eco-regional production constraints in rice-wheat cropping system. PDCSR Bulletin No. 98-2: Project Directorate for Cropping System Research. Modipuram, U.P., India.

10

Principles and Practices of Tillage Systems for Rice–Wheat Cropping Systems in the Indo-Gangetic Plains of South Asia

Yadvinder-Singh
Punjab Agricultural University
Ludhiana, India

J. K. Ladha
International Rice Research Institute
Manila, Philippines

I. INTRODUCTION

For centuries, cultivation has been used to prepare the soil for crop establishment. The main purpose has been to control weeds, but cultivation also removes crop residue, reduces compaction, and prepares a "seedbed." Tillage is usually referred to as physical manipulation of soil properties for the express purpose of crop production (SSSA, 1987). Tillage management aims at achieving optimal range of edaphic factors for the two plant processes—seed germination and seedling emergence, and growth and proliferation of root systems of crops. Conventional tillage (CT) is known to enhance soil erosion, deplete soil organic carbon, and deteriorate soil structure. The intensive rice (*Oryza sativa L.*)–wheat (*Triticum aestivum L.*) cropping system of the Indo-Gangetic Plains (IGP) has thrown new challenges for devising tillage systems where the crops following in quick succession have divergent adaptive requirements. Tillage under intensive cropping systems has the additional challenge to ensure high water-, nutrient-, and energy-use efficiencies.

Concerns in the 1960s about erosion and fuel costs during the oil crisis, as well as development of broad-spectrum, nonresidual herbicides, led to interest in developing systems with less reliance on tillage in many parts of the world. Tillage is a major component of soil management influencing the performance of a crop production system. It also enhances the agricultural sustainability due its ability to modify soil properties and processes and changes soil quality. In many soils of the world, physical properties do not deteriorate and crop yields do not fall by reducing or omitting tillage, provided weeds are controlled effectively (Ehlers, 1984). In the long run, intensive tillage leads to a decrease in soil organic matter content (Gajri et al., 2002). The intensively tilled soils may become more sensitive to slaking, compaction, and erosion. The choice of any tillage practices should be based on prevailing environmental factors and specific crop and edaphic requirements. There is a need to rationalize or optimize soil tillage for achieving the twin objectives of increasing crop production and maintaining environmental quality.

The rice–wheat rotation is one of the largest agricultural production systems of the world, occupying about 13.5 million ha of the productive land in the IGP of South Asia (Bangladesh, India, Nepal, and Pakistan), and on approximately 10.5 M ha in China (Ladha et al., 2000). The rice–wheat cropping system satisfies the caloric needs of over one billion people within Asia. Major increases in area and production of rice and wheat occurred during the 1970s and 1980s as a result of large-scale adoption of Green Revolution technologies. Favorable prices for rice and wheat and input subsidies for fertilizers, electricity, and water make the rice–wheat system the most profitable and least risky option for farmers. Future food security for the region's expanding population may, however, be threatened by a range of difficulties, including stagnant or declining on-station and on-farm yields, slower growth in rice and wheat area, few options for expanding cropped area, and alarming signs of natural resource depletion and degradation since the 1990s (Ladha et al., 2002; Pathak et al., 2002). In addition, emerging indications suggest that the use efficiency of purchased inputs is going down, resulting in higher input use and lower farmer profit (Harrington et al., 1992). The declining resource-use efficiency in the IGP raises serious questions about both the sustainability and potential environmental consequences of current and future production systems. The causes of low yields of wheat after rice are often poor crop establishment and inferior root growth due to adverse physical conditions of the soil, which, in turn, are claimed to be caused by the puddling undertaken for rice (Sur et al., 1981; Boparai et al., 1992). The continuous rice–wheat cultivation has led to a build-up of pests and diseases. The major weed affecting wheat is *Phalaris minor*, which is normally controlled with the herbicide Isoproturon. But in some places this weed has developed resistance to the herbicide (Malik et al., 1998).

Other major threats to sustainability include groundwater depletion, waterlogging and salinity in some areas, declining soil fertility, and changes in the pest scenario. Pumping from groundwater has led to a serious depletion of groundwater in many parts of the IGP (Hira et al., 1998; Pingali and Shah, 1999). This was due to increased pumping of water for rice, which replaced maize in the previous maize–wheat systems grown on coarse-textured soils. There is a continuing need to improve the efficacy of irrigation water use in rice–wheat production because water presents a major production cost to most farmers. High water input in the cultivation of rice, particularly on coarse-textured soils, has raised the doubts of the long-term sustainability of the system. Most rice stubble is burned, causing atmospheric pollution (Gupta and Rickman, 2002), and soil organic matter content is low with consequences for nutrient cycling and soil structure. Alternative improved soil, water, stubble, and nutrient management approaches for the rice–wheat system in the IGP are clearly needed to increase input use efficiencies, to reduce costs, and to make the system more productive, resilient, and sustainable. This review focuses on the tillage and residue management options for the rice–wheat system in the IGP and on the effects of tillage practices on changes in soil properties in order to ensure sustainable food production and to increase the income of the farmers and meet food needs of the ever-increasing population in South Asia.

II. CHARACTERISTICS OF THE INDO-GANGETIC PLAINS

Although the IGP agro-ecological region is relatively homogenous, based on agroclimatic characteristics, soil type, and physiographic features, it is divided into five transects (Fig. 1): (1) Trans-Gangetic Plains or IGP Transect 1 and 2 (areas in Pakistan, and parts of Punjab and Haryana in India); (2) Upper-Gangetic Plains or IGP Transect 3 (most of Uttar Pradesh and parts of Bihar, India, and parts of Nepal); (3) Middle-Gangetic Plains or IGP Transect 4 (parts of Bihar, India, and parts of Nepal); and (4) Lower-Gangetic Plains or Transect 5 (parts of Bihar, West Bengal, India, and parts of Bangladesh) (Narang and Virmani, 2001). In addition, a substantial area in rice–wheat exists outside the IGP, in China, India, and Nepal. In India, the system includes hilly parts of Himachal Pradesh, and Uttranchal, and parts of Jharkhand, Madhya Pradesh, and Rajasthan. A considerable amount of rice–wheat farming is also practiced in hilly parts of Nepal. In China, the rice–wheat rotation is common in the provinces of Jiangsu, Zhejiang, Hubei, Guizhou, Yunnan, Sichuan, and Anhui. Descriptions of areas under rice–wheat in IGP and outside IGP including China are available elsewhere (Huke et al., 1994; Zheng, 2000; Narang and Virmani, 2001; Gupta et al., 2002; Ladha et al., 2002). The rice–wheat areas are located within subtropical to warm temperate climates characterized by cool and dry winters and warm and wet summers (Timsina and Connor, 2001; Ladha et al., 2002). Solar radiation decreases as we move from IGP Transect 1 to 5 in the rice season, whereas the trend is reversed in the wheat season (Pathak et al., 2002). The minimum temperature in

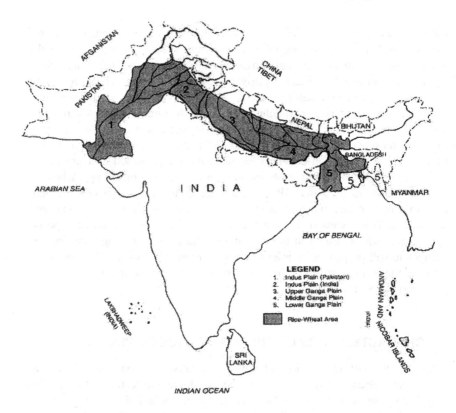

Figure 1 The rice–wheat areas of the Indo-Gangetic Plains by transect. (Adapted from Huke and Huke, 1992.)

rice and wheat increases as you move from IGP Transects 1 to 5. This is also true for maximum temperature in the wheat season, but during the rice season, the average maximum temperature is similar across all the IGP. Rainfall also follows a distinct pattern, increasing from Transects 1 to 5 of the IGP, with Transects 1 and 2 receiving only 650 mm of rainfall per annum and Transect 5 receiving over twice as much. Except for rainfall, the climatic conditions make the upper transects of the IGP more favorable for rice and wheat cultivation. Access to assured irrigation has alleviated the problem of low rainfall periods and made the zone (Transects 1 and 2) very productive. The less favorable climatic conditions and limited irrigation facilities are the major constraints to achieve higher yields in the lower transects (Transects 3, 4, and 5) of the IGP (Fig. 1).

The soils of the region are predominantly coarse-textured with smaller areas of silty clays and clays. The traditional rice fields have loam to clay loam soils, but the nontraditional rice soils are more porous (sands, loamy sands, and sandy loams). Currently, about 60% of the rice soils of Punjab, India, are coarse-textured. Lowland rice develops larger root systems in submerged and puddled soils than in upland soils. Because their root system is inherently shallow, seldom extending below 40 cm, lowland rice can be grown on shallow, unstructured soils with few large pores. Soils in relation to hydrological situations differ throughout the IGP. On the lower terraces with heavier poorly drained soils, wheat is less suited. The lowest of these may be flooded or wet all year. Middle terraces have somewhat lighter soils and fewer drainage problems. On upper terraces with coarser soils and good drainage, rice is less suitable. A unique feature of rice–wheat systems, particularly in Transects 1 and 2, is to grow long-duration, tall traditional aromatic rice (basmati). The basmati group of rice occupies more than 40% of the total area under the rice–wheat system in Haryana (India). In Pakistani Punjab, the basmati group also occupies at present 80% of the area under rice. Wheat sowing after basmati rice is generally delayed, resulting in poor crop yields. A feature of the rice–wheat system is that the amount of soil water remaining after the rice crop is usually adequate for the seeding of the following wheat crop.

Wheat needs a well-aerated soil and rooting medium with oxygen in waterfree soil pores for root respiration. In well-drained, loamy sand and sandy loam soils without restricting layers, wheat roots can grow up to two meters deep (Jalota et al., 1980). Upland crops like wheat do not grow well on soils with dense layers at a shallow depth. Because water comes to them intermittently, they depend on water stored in the root zone for survival between wettings. Deeper wetting and deeper root systems therefore are desirable.

III. CURRENT SOIL MANAGEMENT PRACTICES FOR THE RICE–WHEAT SYSTEM

There are two critical issues for soil management in rice–wheat systems. The first relates to physical management to minimize the disadvantages of puddling to the ensuing wheat and other nonrice crops. Another important issue is related to the decline in SOM and biological activity and the health of the soil under the CT system. In rice–wheat systems, new soil management techniques are required to provide suitable soil conditions for the growth of both crops. Rapid establishment of wheat before cracks are formed improves its germination, establishment, growth, and yield. Generally, tillage methods are dictated by the amount of crop residue remaining on the surface after planting (Gebhardt et al., 1985).

A. Soil Management in Rice

Tillage for rice consists of dry tillage after harvest of wheat (prepuddling tillage), wet tillage (puddling), and planking. Recent diagnostic surveys showed that in

Table 1   Tillage and Seed Bed Preparation Practices for Rice in Rice–Wheat System in Northwestern India

State	Percent NT farmers					
	Prepuddling tillage		Puddling tillage		Planking	
Number of tillage	2–3	>3	0–2	>2	0–1	>1
Punjab	72	28	84	16	48	52
Haryana	30	70	30	70	90	10
Tarai UP	67	33	64	36	46	54
Eastern UP	67	33	93	7	11	89

Source: P.R. Gajri, personal communication.

the Punjab (India), the majority of farmers perform two to three prepuddling and two puddling operations, and in Haryana farmers perform four to five prepuddlings and three to four puddling operations (Table 1). In the *tarai* region and eastern UP, the majority of farmers perform two to three prepuddling operations and two to four puddling operations. In eastern UP, 90% of the farmers do only two puddling operations. Four-wheel tractors are the major power source in the high potential areas. Even small landholders take advantage of this power source by renting tractors from large landholders in the village. Tractors will probably eventually replace animal power, as the cost of maintaining a bullock pair for a whole year becomes prohibitive. The use of animals is more prevalent in the eastern rice–wheat areas where yields are lower, although many farmers use a tractor for the first tillage operation and animals for planking and later plowing.

Puddling, or wet cultivation, is the traditional method of land preparation for establishing rice in Asia. It produces soft mud for easy transplanting of rice seedlings, reduces water percolation losses, provides anaerobic conditions that promote rice growth (Lal, 1985), lessens weed competition, and increases the availability of certain nutrients such as P, Mn, and Fe. However, puddling destroys soil aggregates, breaks capillary pores, disperses clay particles, decreases saturated hydraulic conductivity, reduces gaseous exchange, and lowers soil strength in the puddled layer (Sharma and De Datta, 1986). Greater retention and slower release of water prolong the period for which the puddled soils remain wet compared with the unpuddled soils (Prihar et al., 1985). Adachi (1992) concluded that in fine-textured soils, the percolation rate was reduced due to a blocking layer just below the puddled soil of fine dispersed particles. In medium-textured soils, puddling decreased the percolation rate by increasing the bulk density of the lower part of the puddled layer. In coarse-textured soils, reduction was mainly due to a clayey layer formed in the top 1–2 cm of the puddled layer. Repeated puddling leads to formation of a hardpan at shallow depth (Moorman and van Breemen, 1978; Aggarwal et al., 1995). The hardpan created by repeated puddling may restrict root growth and performance of nonrice crops.

A subsurface hardpan, which occurs at 10–40-cm depth in lowland rice soils, is characterized by higher bulk density, higher soil strength, lower total porosity, and lower water permeability than overlying and underlying soil horizons (Table 2). The magnitude of the effect depends, however, upon soil type. Farmers need to plow their fields many times to get a good seedbed for planting wheat after rice, which results in late planting of wheat. The breaking of the hardpan before planting of wheat may, however, adversely affect the growth of the next rice crop. Hasegawa (1992) provides an optimum range of values of selected soil physical properties for the plow pan. He suggests that mechanical impedance values higher than 1.2–1.5 MPa are likely to create root penetration problems. Likewise, stagnated water must be able to drain within one day after rain or irrigation. Relative gas diffusivity in the surface soil should be higher than 0.005 at the same time to ensure oxygen supply to plant roots.

Sur et al. (1981) studied the changes in soil physical properties after six years of cropping under a maize–wheat and rice–wheat system (Table 3). Puddling in rice increased soil bulk density and reduced saturated hydraulic conductivity relative to that after maize in the 10–20-cm soil layers. Total root mass of wheat was 48% more in maize–wheat than in rice–wheat. The thick hardpan developed from long-term puddling and soil strength increased rapidly as the soil dried, restricting root growth and yield of the succeeding wheat or other nonrice crops in Asia (Kirchhof and So, 1990; Boparai et al., 1992; Aggarwal et al., 1995). This also led to a range of potential problems including water deficit in the

Table 2 Comparison of Soil Physical Properties of the Plow Layer and Plow Pan of a Typical Lowland Rice Soil in the Taihu Lake Region of China

Soil property	Plow layer	Plow pan
Bulk density (Mg m^{-3})	1.20	1.35
Total porosity (% v/v)	53.5	44
Aeration porosity (% v/v)	15	2
Ks (um s^{-1})	120	0.19
Diameter of pores (%)		
>0.2 mm	22	11
0.2–0.05 mm	4	2
0.05–0.005 mm	8	6
<0.005 mm	66	81

Source: C. yun-Sheng (1983).

Table 3 Effect of Rice–Wheat and Maize–Wheat Cropping Systems on Soil Physical Properties After Six Cropping Cycles on a Sandy Loam Soil at Ludhiana, India

Soil depth (cm)	Bulk density (Mg m^{-3})		Hydraulic conductivity (cm h^{-1})	
	Rice–wheat	Maize–wheat	Rice–wheat	Maize–wheat
0–5	1.53 b[a]	1.60 a	0.17 a	0.11 a
5–10	1.65 a	1.62 a	0.08 b	0.14 a
10–15	1.69 a	1.63 b	0.06 b	0.13 a
15–20	1.81 a	1.68 b	0.019 b	0.043 a
20–25	1.73 a	1.64 b	0.026 b	0.086 a

[a] Values in the same row followed by the same letter do not differ significantly at $P = 0.05$ by Duncan's multiple-range test.
Source: Sur et al. (1981).

absence of frequent irrigation and the inability to extract nutrients and water below the hardpan.

B. Soil Management in Wheat

In wheat, farmers perform up to 12 tillage operations for land preparation in the IGP. In Punjab (India), with coarse-textured soils, a minimum of two operations is most common and in other areas, minimum operations are four to five (Table 4). Turnaround time between rice harvest and wheat sowing is especially challenging in lower terraces, or in the eastern parts of the IGP. Poor drainage can delay wheat sowing even when the rice harvest is timely. In irrigated areas of the IGP, there is a short window period available for land preparation and sowing of wheat, particularly after long-duration "basmati" rice, which is harvested from November to mid-December. It normally takes 2 to 3 weeks for rice fields to become workable for land preparation due to antecedent moisture. Late planting of wheat after rice also occurs due to excessive tillage. In the rice–wheat system of South Asia, each day of delay in wheat planting after mid-November results in a 1%–1.5% loss in yield (or 30 to 45 kg grains ha^{-1}) (Ortiz-Monasterio et al., 1994; Regmi et al., 2002). Repeated tillage to get a good seedbed for wheat increases costs. Another factor hindering optimum wheat yields in rice fields is the sowing of seed as broadcast onto the poor seedbeds with clods, which causes poor crop establishment. No-till (NT) seed drills allow seeds to be placed evenly at the right depth and the plant stand is more uniform. But drills cannot be used if the rice residues have been incorporated and not properly decomposed; otherwise, the drill clogs. In NT, loose residues are a problem, because they again clog the drills. Under such situations, farmers opt for broadcasting of seed, which

Table 4 Data on Tillage and Crop Establishment for Wheat from Diagnostic Surveys of Selected Rice–Wheat Cropping Systems in South Asia

Location	Area planted late[a] (%)	Turnaround time (days)	Number of passes with plow (average)
Punjab, Pakistan	40	2–10	2–10 (6)
Pantnagar, India	35	15–20	5–12 (8)
Faizabad, India	25	20–45	5–12 (6)
Haryana, India	25	15–35	4–12 (8)
Bhairahawa, Nepal	40	15–35	4–8 (6)

[a] Late planting is defined as wheat planted after the first week of December.
Source: Adapted from Hobbs (2001).

results in seed placement at many different depths and into different soil moistures, resulting in variable germination (Hobbs, 2001). The Rice–Wheat Consortium for the IGP is working on machinery that can plant into loose residues.

IV. CROP RESIDUE MANAGEMENT PRACTICES

Throughout the IGP there is little retention of crop residues in the field—they are either burned in the field or harvested for fuel, animal feed, or bedding. Combine harvesting of rice and wheat has become a common practice in the western parts of the IGP. It leaves behind both anchored (40–50 cm high) and loose crop residues in wind rows, which hinders the planting of the following crop. In the Punjab and Haryana provinces of India, about 90% and 45% of the area under rice are harvested by machine, respectively, and rice residue is either completely burned after use of a shredder or partially burned having a burned component of 45% (Table 5). Under wheat the machine-harvested area is 82% in Punjab (India), 30% in Haryana, 55% in Tarai UP, and 31% in eastern UP. In the machine-harvested wheat fields, a chaff maker has recently been introduced in Punjab and Haryana to collect the chaff for fodder, leaving only about 15%–20% residue in the field, which is usually disposed by burning. Stubble burning is the easiest and cheapest way of residue disposal.

Residues are an important source of nutrients and organic matter and should be utilized efficiently. Burning causes air pollution, leading to global warming and health concerns. Burning also results in loss of nutrients and active soil C, leading to increased costs and soil structure decline (Angus et al., 1998). It reduces the number and activity of soil fauna and flora. On the other hand, burning is perceived to improve weed, insect, and disease control. Retention of crop residues improves organic carbon content, water-stable aggregates, bulk density, and hy-

Table 5 Current Area Harvest by Machine and Straw Management Practices (% of Total Area) in the IGP of India

Method of harvest/straw burning	Punjab	Haryana	Tarai UP	Eastern UP
Rice				
Machine harvest	91	45	55	27
Complete burn	55	37	55	5[a]
Partial burn	36	8	—	22
Wheat				
Machine harvest	82	30	52	31
Complete burn	10	18	55	31
Partial burn	72	12	—	—

[a] Straw incorporated.
Source: P. R. Gajri, personal communication.

draulic conductivity and reduces runoff (Wilhelm et al., 1986; Yadvinder-Singh et al., 2000). Residue management influences soil temperature, moisture conservation, erosion and machinery performance, and weed and pest management. Not only the quantity but also the distribution and type of residue are important. An uneven distribution results in poor performance of planters, herbicides, and overall effects on crop productivity. High loads of rice straw hinder the performance of drills during wheat seeding.

Residue incorporation can increase N tie-up and release of phytotoxins causing harmful effects on succeeding crops. Lynch (1973) reported that under certain conditions, substances toxic to cereal seedlings are produced by cereal residues that decay near the seedlings. These findings will assume greater importance when cereals are grown immediately after cereals and with minimal cultivation. However, there is no direct evidence of deleterious effects due to phytotoxins released during rice straw decomposition on the growth and yield of wheat in the IGP. There is a need to study the effect of surface retention of crop residues on the growth and yields of crops in the rice–wheat system, particularly in the absence of puddling. A recent review by Gajri et al. (2002) indicates the beneficial effects of surface mulch on weed control, physical properties of soil, water infiltration, and crop yields, particularly under dryland cropping. However, surface mulching reduced seedling emergence and corn yields in a temperate climate (Kaspar et al., 1990). If the straw is plowed in or is given time to decompose near the soil surface before the next crop is sown, the risk of damage may be lower. In an NT system, residues are left on the surface and decompose slower than when they are incorporated into the soil. Heavy loads of residues on the surface cause trouble in the use of machinery for planting. For efficient straw management under direct drilling, chopping and uniform spreading as mulch with suitable modifications in combine harvesters may be needed. Although retention of straw is generally known to improve soil-quality parameters, often the improvements are not translated into significant increases in crop yield (Yadvinder-Singh, personal communication). This is because (1) a long-term timeframe is needed to exhibit positive effects and (2) some negative effects of straw offset the positive effects. Nevertheless, straw retention in the field is most desirable to avoid negative environmental effects of burning.

V. TILLAGE SYSTEMS

Tillage systems to counter ecological constraints to plant growth depend on the soil type, climate, residue status, and crop and management factors. In addition, tillage systems are affected by socioeconomic factors such as farm size, availability of labor, power sources, and technical skill. Tillage systems can broadly be divided into two categories: those where crop residues are removed or burned before tillage and those where residues are left in the field. In the absence of residue in the field, the tillage system is either CT or NT.

When the residues are left on the surface, the percent of surface covered with residue at planting time constitutes the basis for further distinction (CT, reduced tillage, conservation tillage). Conservation tillage can further be classified as mulch tillage, ridge tillage, and NT.

With the development of chemical weed, disease, and pest control, the role of CT is being questioned. In the last 50 years, CT and its modifications have been developed and are increasingly applied in the temperate region of North America and Europe, and some countries of the tropics, with the overall objective of reducing soil erosion and energy expenditure in agriculture. It has been stated that adoption of CT on a global scale must address the issues of crop residue management, herbicide use for weed control, and planting and fertilizer application machinery (Frye and Lindwall, 1986).

The major issue is adapting drills to sow into high levels of loose residues. Strategies include chopping and spreading of straw during or after combining or use of disc-type trash drills. Other issues in the adoption of an NT system with residue on the soil surface include delayed drying and warming of soils after wet, cold winters and increased incidence of diseases and pests (Carter, 1994; Kirkby, 1999). Replacement of CT with NT or conservation tillage is believed to result in substantial C sequestration (Lal, 1997b).

The prevention of soil structure degradation such as formation of a hardpan caused by puddling in rice–wheat systems is achieved through several "best management practices." These options include reduced puddling to slow the formation of a hardpan, or deep tillage to destroy them to stimulate rooting for optimum growth of the following nonrice crop. Alternatives to puddling for rice include dry seeding on the flat (De Datta, 1986) or on beds (Cooten and van Borrell, 1999). Controlled traffic is another form of reduced tillage to prevent soil structure degradation. The aim of controlled traffic is to keep the wheels of all field equipment always in the same paths, year after year. All in-field operations like sowing, spraying, harvesting, and so forth are more efficient with controlled traffic, as there is reduced overlap (Blackwell, 1998).

A. No-Tillage System

Several terms have been used for NT, such as no-till (NT), direct seeding, direct seeding with residues, conservation tillage, and conservation agriculture. Conservation tillage implies that tillage has been reduced compared to the present level of CT (even to a level without any tillage) and also that crop residues are left on the soil surface. The discovery of broad-spectrum, nonresidual herbicides stimulated NT research in the early 1960s (Hood et al., 1964; Taylor, 1967). It was concluded from these studies that excellent wheat and barley crops could be grown without tillage by using Paraquat for weed control. Studies in many countries have shown that many crops can be established on uncultivated soil. These suggested that the primary function of plowing was weed control and showed that omission of tillage in weedfree situations did not result in any yield reduction (Christian and Ball, 1994; Riley et al., 1994). Considering the additional potential savings in machinery, labor, fuel, and soil, this is remarkable,. Allen (1981) noted from the studies carried out in the U.K. that there was little difference in yield compared to plowing where sufficient N was applied, and yields of second- and third-year NT crops were often higher than in the first year.

The potential advantages of NT practices for wheat after rice include the opportunities to reduce the production costs and minimize or even reverse the probable long-term detrimental effects on sustainability resulting from the excessive tillage being used. No-till practices in wheat can save up to 45–60 liters/ha of diesel fuel by saving up to 10 tillage passes normally used to plant the wheat crop after flooded puddled rice. Irrigation water savings in NT also average about 20% including costs for pumping, and many annual weeds like *Phalaris minor* appear to be less prolific under NT (Mehla et al., 2000). Since soil disturbance is minimal with NT, few weed seeds are exposed to germination. Apart from the above, NT has several environmental benefits. It reduces the pollution of waterways with agrochemicals due to increased water infiltration and lesser use of agrochemicals. By reducing the consumption of fossil fuel and the amount of organic matter that is transformed into carbon dioxide, greenhouse-gas emissions are also reduced (Grace et al., 2002). It is clearly a new production system for the rice–wheat system of the IGP of South Asia, which is a step toward increasing sustainability. No-till has problems with the loose residues left on the soil after combine harvesters have been used. Currently farmers burn or remove the residues of rice. Research is underway to resolve this problem in South Asia.

Alternative tillage practices that reduce costs, raise productivity and bring environmental benefits need to be evaluated for the rice–wheat-based cropping systems in the IGP. The NT seed drills work very well in situations where there is little surface residue after the rice harvest. This usually occurs after manual harvesting or burning of residue after combine harvesting of rice. The inverted-T may not work well where combines are used and loose rice straw is left in the field, since the opener acts as a rake for the loose straw. In this case other NT coulter designs will be needed. Inherent soil characteristics may cause different degrees of constraints to successful no-till drilling. Soils with low clay content have been identified as those likely to be problematic with NT and require deep plowing to 30-cm depth. It is common that conversion from CT to an NT system with residue retention requires several crop cycles before potential advantages/disadvantages begin to become apparent (Phillips and Phillips, 1984). Because of its many environments and rainfall patterns, the IGP regions of South Asia present a formidable challenge for the development of a permanent NT package for the rice–wheat system.

1. No-Till Wheat

For increased and sustained production of wheat, it is essential to adopt an integrated approach that helps in timely planting, reduces weeds, and improves water- and nutrient-use efficiencies. No-till reduces the turnaround time between the rice harvest and wheat sowing and allows planting closer to the optimal dates. Timely sowing of wheat contributes to enhanced yields and water- and nutrient-use efficiency. The drill also ensures seed and fertilizer placement at the desired depth and in the right quantities. No-till improves soil physical properties and prevents the formation of plow pans. Additionally, NT is known to reduce the *Phalaris minor* infestation in wheat (Hobbs, 2001; Mehla et al., 2000). In CT wheat, yellowing of plants is common after the first irrigation at 25 to 30 days after seeding due to temporary waterlogging, particularly on fine-textured and partially reclaimed soils. Wheat planted with the NT system is unlikely to be constrained by waterlogging due to better percolation and drainage rates. Farmers do not need to apply presowing irrigation to NT plots, and the amount of irrigation required for the first irrigation at the crown root stage is less in NT than in CT. The use of the NT technique in Punjab (Pakistan) as compared with CT, resulted in saving 21%–63% (mean 38%) of irrigation water during the first irrigation after wheat sowing and 15%–20% in the subsequent irrigations (Gill et al., 2000).

Studies conducted in India and Pakistan on the relative performance of CT and NT (residue removed) systems under an assured supply of irrigation water and optimum nutrient supply conditions showed that NT wheat was either comparable or better although sometimes lower than CT (Table 6). In general, yields with NT are higher than CT when the advantage of advanced sowing is considered and when NT wheat is planted at the optimal moisture content. Results of field trials conducted in Haryana, India, showed that wheat can be seeded with the NT system both for timely and late sown conditions (Mehla et al., 2000). Tomar and Sharma (1998) conclude that in "swell-shrink" Vertisols, rice should be grown by direct seeding in rows and NT or minimum tillage should be adopted in the following wheat for obtaining maximum benefits.

Studies conducted in the Sichuan province of China showed that wheat planted after rice by NT increased the yield by 7%–10% over that planted with CT (Table 6). The main reason ascribed for high yields with NT was that it maintained good soil structure. Bacon and Cooper (1985) reported from Australia that NT with surface retention of rice stubble produced significantly higher yield of dryland wheat than with incorporation or burning.

Studies have also shown reduced yields of wheat when planted by NT as compared to CT under constrained resources (Chaudhary et al., 1991; Gajri et al., 1992; Singh et al., 1998; Srivastava et al., 2000; Bajpai and Tripathi, 2000) (Table 6). This could be due to improper moisture conditions causing poor crop establishment in NT wheat and inefficient fertilizer management due to nonavailability of a proper seed-cum-fertilizer drill. No-till plots normally require higher moisture content at wheat seeding than under CT.

In most of the studies on NT in wheat, crop residues were either removed or burned. A recent field study in Punjab, India, showed that rice straw (7 Mg ha^{-1}) retained as mulch on the surface of NT plots caused deleterious effects on the initial wheat growth, which might be due to low soil temperatures (1.5–2.5°C), and N immobilization and/or phytotoxins released during straw decomposition (Table 7). The amount of dry biomass of wheat 92 days after planting was, however, significantly higher in NT and straw–mulch plots than that on CT–straw-burned plots. Grain yield was, however, not significantly affected by tillage and residue management.

2. No-Till Rice

Lowering the drudgery of the traditional transplanting approach by evaluating new establishment techniques and doing away with the puddling of soils as a way to reduce soil degradation and increase system productivity are important aspects in the cultivation of rice in the IGP. Water for agriculture, especially for irrigated rice production, is becoming increasingly scarce and is likely to decline per capita over the next two decades (Kim et al., 2001). In many areas of NW India and Pakistan, water tables are declining rapidly as more water is removed than recharged (Hira et al., 1998). More efficient use of water in rice ecosystems

Table 6 Grain Yield (Mg ha^{-1}) of Wheat Under NT and CT Systems in Rice–Wheat Cropping Systems of South Asia

Reference	No-till[a]	CT	Location and soil type
Roy and Sarkar (1993)	3.53 a	3.00 b	Jamalpur, Bangladesh; sandy loam soil, irrigated wheat
	3.13 a	3.10 a	Sandy loam soil, rainfed wheat
Brar and Kumar (2000)	4.42 a	4.35 a	Ludhiana, India; sandy loam soil, average for 14 years of studies
Aslam et al. (1993)	3.89 a	3.53 b	Pakistan; loam to clay loam soils, mean for 34 locations
	3.68 a	2.56 b	Late sowing under CT by an average of 24 days
Mehla et al. (2000)	4.56 a	4.32 a (timely sown) 3.93 b (late sown)	Haryana, India; mean of 132 trials
Shafiq et al. (1995)	2.44 a	2.45 a	Pothwar, Pakistan; clay loam
Sandhu (1980)	3.45 a	3.30 a	Punjab, India; sandy loam to loam, mean for 12 locations
Verma et al. (1988)	2.67 a	2.86 a	Ranchi, Bihar (India); silt loam, mean for 2 years
Verma et al. (1989)	2.92 a	2.93 a	Ranchi, Bihar (India); clay loam, timely sown wheat
	2.06 a	2.17 a	Late sown wheat
Bhardwaj and Singh (1998)	4.62 a	4.55 a	Uttaranchal, India; sandy loam soil, mean for 5 years
Gajri et al. (1992)	2.87 b	3.57 a	Ludhiana, India; sandy loam soil, mean for 3 years
Chaudhary et al. (1991)	3.41 b	4.02 a	Ludhiana, India; sandy loam soil, mean for 3 years
Singh et al. (2001)	3.67 b	3.97 a	Delhi, India; sandy clay loam, mean for 3 years
Yonglu et al. (2000)	6.26 a	5.86 b	Sichaun Province of China; average of 40 on station trials
	4.97 a	4.53 b	Average of 52 on-farm trials

[a] Values in the same row followed by the same letter do not differ significantly at $P = 0.05$ by Duncan's multiple-range test.

Table 7 Effect of Tillage, Rice Straw Management and Fertilizer N on Wheat Growth at 52 and 92 Days After Sowing and Grain Yield on a Sandy Loam Soil During 2001–2002 at Ludhiana, India

| | Dry matter (g m^{-2})[a] | | | | Grain yield (Mg ha^{-1})[a] | |
| | 52 days | | 92 days | | | |
Tillage/straw management	No N	120 kg N ha^{-1}	No N	120 kg N ha^{-1}	No N	120 kg N ha^{-1}
Conventional till—straw burned	103 a	191 a	165 b	440 b	1.86 a	4.83 a
No-till—straw burned	91 a	182 a	209 a	479 ab	2.33 a	4.73 a
Conventional till—straw incorporated	100 a	207 a	176 b	511 a	2.17 a	5.19 a
No-till—straw retained after chopping	63 b	121 b	223 a	539 a	2.21 a	4.93 a

[a] Values in the same column followed by the same letter do not differ significantly at $P = 0.05$ by Duncan's multiple-range test.
Source: Yadvinder-Singh, personal communication.

is of critical importance in view of the increased demand to meet the food needs of a growing world population.

Some alternate systems such as NT, bed planting, laser leveling, and dry seeded rice can help improve water-use efficiency. In recent years, the need to increase productivity against the background of rising labor costs and water shortage have led to considerable increases in rice area under direct seeding, particularly in southeast Asia (Pandey and Velasco, 1999; Kim et al., 2001). Tillage studies conducted in western Nigeria on different soil types indicated that for soils of heavy texture, NT can be adopted successfully for both seeded and transplanted rice (Maurya and Lal, 1979).

Dry seeded rice is less labor-intensive than puddled, transplanted rice and no puddling can benefit the ensuing wheat crop. The dry seeded crop can be raised as aerobic rice with no flooding throughout the crop cycle or as a semidry rice with flooding after the crop has reached a certain stage. This system will, however, extend the main field time by about 10 to 15 days for the crop to mature. In addition, weeds are a major problem in dry-seeded rice, especially under no flooding after crop establishment. Thus, more emphasis is needed on water management during crop establishment to ensure good stands and effective weed control in direct seeded rice. Water used for puddling can be saved, but percolation losses may increase in nonpuddled soil. Thus, the total water balance needs to be evaluated on different soil types (Hobbs et al., 2002). Long-term trials of dry-seeded rice should evaluate the sustainability of this system by looking at yield trends, evolution of weeds and other pests, water retention, water- and nutrient-use efficiency, and changes in soil quality over time.

One possible subsystem of dry-seeded rice could be NT rice. In this system, fields are irrigated, and weeds are allowed to germinate and are then controlled by plowing or with nonselective herbicides such as Glyphosate. An NT drill is then used to sow the rice seed. Since the soil is less disturbed, fewer weeds may germinate. NT can be successfully adapted to both direct seeded and transplanted rice in soils of fine texture. Perhaps, ploughing after 5 to 6 years may be necessary during the dry season (wheat) to rid the soil of any harmful effects due to NT.

Rice can be grown under unflooded conditions without sacrificing yield (De Datta, 1986; Garside et al., 1992). Studies carried out at PAU, Ludhiana (India), have revealed that optimum management of irrigation water in rice, which consists of keeping water standing in the field for two weeks after transplanting and subsequently irrigation water is applied intermittently two days after the ponded water has filtered into the soil, saved 65 cm of water without any reduction in yield relative to the conventional practice of continuously ponding of water (Sandhu et al., 1980). It has further been shown that irrigation to rice could be terminated about 14 to 17 days before harvesting, saving about 60 cm of water and also substantial time gained for wheat seedbed preparation following the rice harvest (Sandhu et al., 1982). In direct seeded rice, irrigation water application every four days reduced the water amount by about 21 cm without any reduction in yield as compared to irrigation every 3 days (H.S. Uppal, personal communication). Mondal and Islam (1991) found no significant yield reduction in rice raised on a clay-loam soil among continuous standing water, continuous saturated conditions, and saturation to field capacity. But saving in water under continuous saturated conditions, and saturation to field capacity over continuous standing water was 26% and 46%, respectively.

Research on NT rice is very limited at present in the IGP of South Asia. Preliminary results at IRRI suggest that a satisfactory lowland rice crop can be established using NT seeders mounted on hand tractors (Piggin et al., 2001; Table 8 here). Effective weed management can be obtained with Glyphosate herbicide and proper water management. Conventional-till, transplanted rice requires four to five tillage operations and seedling propagation and transplanting. NT requires one spray and one seeding operation. The economic advantage of NT rice, relative to CT rice, will depend on the relative prices of herbicides, labor, and fuel. The emission of greenhouse gases, particularly methane, will be markedly reduced under NT cultivation of rice. Sharma and De Datta (1986) showed that the grain yield of rice was lower under NT than under CT, and the decrease was more on clay-loam than on clay soil with a low percolation rate. Mahata et al. (1990) and Singh et al. (2001) have reported lower yields of NT rice than puddled trans-planted rice. On a sandy loam soil at Pusa, Bihar (India), Sharma and Pandey

Table 8 Effect of Tillage on Grain Yield of Direct-Seeded Rice Under Manual Weeding Conditions at IRRI, Philippines

	Grain yield (Mg ha^{-1})	
Treatment	1998	1999
No-till—Chinese seeder	3.5	4.2
No-till—split seeder	3.2	4.5
Conventional till—direct dry seeded	3.8	5.2
Conventional till—direct wet seeded	4.2	4.0
LSD 0.05	0.7	—

Source: Piggin et al. (2001).

(1992), however, observed no significant differences in the initial rate of water loss, hydraulic conductivity, and grain yield of rice between CT and NT rice after CT wheat. The lower rice yields obtained under NT than under CT in some of the above studies may be due to improper seed establishment and ineffective weed control.

A few studies have examined the effect of different rice establishment methods combined with different wheat establishment methods on total system productivity. Recent experiments in rice–wheat sites in Nepal and Pantnagar (India) reveal that rice yields were not reduced by direct seeding in nonpuddled fields compared with puddled transplanted rice (Table 9). The transplanted rice

Table 9 Grain Yield of Rice and Wheat with and Without Puddling at Three Sites in South Asia

Location	Soil type and duration of study	Treatment	Mean yield (Mg ha^{-1})		
			Rice[a]	Wheat[a]	Total[a]
Pantnagar, India	Silt loam, 3 yrs	Puddled	6.1 a	4.1 b	10.2 a
		Unpuddled	5.6 a	4.6 a	10.2 a
Pantnagar, India	Sandy loam, 6 yrs	Puddled	5.6 a	3.9 a	9.5 a
		Unpuddled	5.3 a	4.0 a	9.3 a
Bhairahawa, Nepal	Silty clay loam, 2 yrs	Puddled	5.3 a	3.1 b	8.4 b
		Unpuddled	5.4 a	3.4 a	8.8 b

[a] Means followed by the same letter in a column are not significantly different by Duncan's multiple-range test.
Source: Hobbs et al. (2002).

had a higher yield in one year on both the sites at Pantnagar due to a higher population of brown plant hopper in direct seeded rice than in the transplanted plots. Wheat yield (averaged across all the years) was significantly higher in plots where soils were not puddled in silt loam soil at Pantnagar and silty clay loam at Bhairahawa and were not different in sandy loam soil at Pantnagar (Table 9). There were no differences in total grain yield per year in any of the three locations irrespective of the treatment. Tillage practices for sowing wheat had no significant effect on yield when averaged across all years on two soils at Pantnagar (Hobbs et al., 2002). But the wheat yields were significantly higher under NT than under CT on a silty loam soil at Bhairahawa, Nepal. Both treatments were planted on the same day. However, some differences were noted between the years. In drier years, NT was significantly worse than CT plots due to less germination, but in wet years it performed better. The study suggested that no-tillage is equal to CT on sandy loam soil when planted on the same day. But when the correct soil moisture level is present at planting, NT can do better than CT. No-till needs higher soil moisture at planting than CT plots.

From a long-term study, Sharma et al. (2000) report that in the rice–wheat system on a Vertisol, mean grain yield of direct seeded, nonpuddled rice was significantly lower than that of puddled transplanted rice. Grain yield of the following wheat was similar under direct seeded, nonpuddled and puddled transplanted rice (Table 10). The total grain yields under the rice–wheat system were, however, similar under the two tillage systems. Likewise, grain yields of both rice and wheat were not affected by tillage practices in wheat (Table 10).

In permeable soils, leaching losses of water and N can be substantial in unpuddled and flooded rice. These losses can be drastically curtailed by compacting the soil and/or by using slow-release N fertilizers. Humphreys et al. (1992)

Table 10 Effect of Tillage Management in Rice–Wheat System on the Grain Yields of Rice and Wheat on a Clay Loam Soil (Data Averaged for 8 Years)

Treatment to rice	Rice yield (Mg ha^{-1})[a]	Wheat yield (Mg ha^{-1})[a]
Nonpuddled direct-seeded rice	3.6 b	3.1 a
Puddled transplanted rice	4.1 a	2.8 a
Treatment to wheat		
No-tillage	3.9 a	3.1 a
Conventional tillage	3.8 a	3.0 a

[a] Values in the same column followed by the same letter do not differ significantly at $P = 0.05$ by Duncan's multiple-range test.
Source: Sharma et al. (2000).

reported that compaction with a vibrating roller reduced the percolation rate in rice from 17 mm to 4 mm/day on a permeable soil. Another N management issue is the leaching loss of NO_3–N from the soil during puddling operations, especially after incorporating a green manure. In nonpuddled DSR, this soil N is less prone to leaching loss and would be utilized by the young rice crop. The superiority of nonpuddling over puddling in rice, and of direct seeded over transplanted rice, over a range of soil types and their effect on following wheat require further investigation. Direct seeded rice, either on beds or on the flat, would require different cultivars and very different nutrient, water, and weed management strategies than for transplanted rice.

B. Bed Planting Systems

The terms "bed," "raised beds," and "ridges" are used interchangeably in the literature. Bed farming is a system where the crop zone and traffic lanes are distinctly and permanently separated. Soil is moved from the furrows and added to the crop zone, raising its surface level. The furrows serve as irrigation channels, drains, and traffic lanes. In bed-planting systems, wheat or other crops are planted on raised beds. The use of beds or ridge tillage for the production of nonrice crops was pioneered in the poorly drained heavy clay soils of the irrigation areas of southern NSW about 25 years ago (Maynard, 1991; Tisdall and Hodgson, 1990). Now, virtually all of the nonrice summer crops, and considerable areas of winter crops, are grown on beds of variable width, typically 1.6–2 m. Hobbs (2001) and Limon-Ortega et al. (2000a) have listed the following advantages for bed planting in rice–wheat systems:

1. Management of irrigation water is improved and simpler.
2. Bed planting facilitates irrigation before seeding and thus provides an opportunity for weed control prior to planting.
3. Weeds can be controlled mechanically, between the beds, early in the crop cycle in wheat.
4. Wheat seed rates are lower.
5. Band application of fertilizers in the bed at planting followed by banding of side dress N at critical growth stages after emergence can significantly improve N use efficiency and enhance grain quality.
6. Herbicide dependence is reduced in wheat, and hand weeding and roguing are easier.
7. Less lodging occurs in wheat.
8. Higher yields of wheat could be obtained on soils where drainage problems exist.
9. Less emission of greenhouse gases, particularly CH_4 with rice on beds, occurs.

Tisdall and Hodgson (1990) found a higher yield for crops grown on ridges compared with those on the flat due to better soil aeration. Beds on a poorly drained clay soil improved drainage and N fertilizer efficiency and enhanced wheat growth (Sweeney and Sisson, 1988). Beds could greatly improve drainage for crops sown after rice, improving their establishment and yield. In the Yaqui Valley of Mexico, about 90% of the farmers plant wheat and many other crops on beds. Farmers growing wheat on beds obtain about 8% higher yields with nearly 25% fewer operational costs as compared to those still planting convention-ally on the flat (Aquino, 1998). This system is now being assessed for suitability in South Asia. Initial studies in Punjab and Haryana (India) showed that wheat could be successfully raised on beds (two rows per bed; bed size 67.5 cm; 37.5-cm top and 30-cm furrow) as an alternative to CT. Two of the major constraints on higher yields in northwest India and Pakistan are weeds and lodging. Both can be reduced in bed-planting systems. Studies show that *Phalaris minor* is less prolific on dry tops of raised beds than on the soil found in conventional flat planting, and reduced incidence of weeds in bed planting means less use of herbicides (S.S. Dhillon, personal communication).

Results from Punjab, India, show that there are no significant differences between flat- and bed-planted systems in wheat (S.S. Dhillon, personal communi-cation), which means that yield was not sacrificed by moving to a bed system. Average yield of wheat in 32 farmers' participatory trials in Haryana, India, was higher (5.73 Mg ha^{-1}) under bed planting compared to under the conventional flat (5.28 Mg ha^{-1}) system (Gupta et al., 2002). However, farmers are particularly pleased with the water savings they obtained from the bed system.

Currently the only option the farmers have is to destroy the beds after the harvest of the bed-planted wheat crop using several tillage operations before transplanting of rice. New beds are formed for planting the next wheat crop. Research is therefore underway to keep the beds permanent by growing rice in this system. In this way, benefits in terms of reduced costs of production would be enhanced.

C. Permanent Raised Beds for the Rice–Wheat System

Researchers are working to develop a permanent raised-bed culture for rice, wheat, and other crops to improve the system productivity in irrigated rice–wheat of the IGP. A bed-planting system has more benefits when beds are "perma-nent"—that is, when they are maintained over the medium to long term and not broken down and reformed for every crop (Hobbs, 2001). In this system, wheat is harvested and straw is retained or burned. Passing a shovel down the furrows reshapes the beds. If the rice could be grown on beds, it may significantly increase water-use efficiency and reduce land preparation costs. Research in farmers' fields has shown that rice can be grown on beds, transplanting seedlings or direct seeded, making this system feasible for rice–wheat regions. The permanent raised-bed cropping system gave improved water, nitrogen, and phosphorus economies, energy savings, greater timeliness of operations, and improved soil structure of the crop zone due to reduced soil compaction (Garside et al., 1992). In Missouri, furrow-irrigated rice used approximately half the amount of water compared to flooded rice (Tracy et al., 1993). Farmers in Haryana and Punjab, India, use a raised-bed machine to prepare raised beds and furrows for rice. Dry rice seeds are drilled in rows on the beds or seedlings are transplanted on the beds. Preliminary results of eight on-farm trials conducted in U.P. and Haryana, India, and one at PAU, Ludhiana, India, showed that grain yields were slightly lower for transplanted rice on raised beds than on flats, but the irrigation water requirement was significantly lower in the case of the former (Table 11). Direct seeded rice on beds, however, produced significantly lower yield compared with transplanted rice on flats. In this experiment both direct seeded and transplanted rice were planted on the same date. Poor plant stand and ineffective weed control may be the other reasons for lower yields under DSR. Phosphorus and iron management in rice under a permanent raised-bed system is likely to be more important than rice transplanted on flats.

Changing from flat to bed layouts alters the hydrology of the system and the transport and transformation of nutrients. The water moves horizontally from the furrow into the bed, upward toward the bed surface driven by evaporation and capillarity, and downward driven largely by gravity. Permanent beds can enhance horizontal and reduce vertical infiltration. In some situations, solutes are transported to the surface of the beds, creating problems such as salinization in

Table 11　Effect of Tillage and Crop Establishment on Grain Yield and Water Use by Rice in Rice–Wheat System During 2001 in the IGP of India

Tillage and crop establishment	Experiment 1 (Gaziabad, India)		Experiment 2 (Ludhiana, India) (sandy loam soil)
	Grain yield (Mg ha^{-1})	Total irrigation time (hr)	Grain yield (Mg ha^{-1})[a]
Dry seeded rice—beds	5.0	153	6.8 c
Transplanted rice—beds	5.6	146	7.9 b
Conventional-till transplanted rice—flats	5.3	250	8.3 a

[a] Values in the same column followed by the same letter do not differ significantly at $p = 0.05$ by Duncan's multiple-range test.
Sources: R. K. Gupta (personal communication) (Experiment 1); Yadvinder-Singh and P. R. Gajri (personal communication) (Experiment 2).

the seed zone, and reducing the availability of nutrients to the crop. Planting rice on beds would be similar to dry-seeded rice on nonpuddled soils. Rice planted on permanent beds can be NT and fertilizer can be placed at 10-cm depth to reduce ammonia volatilization losses of N. Further DSR, either on raised or flat beds, would allow mechanization and different nutrient, water, or weed management strategies than for transplanted rice.

Limon-Ortega et al. (2000b) and Sayre (unpublished data) studied the permanent raised-bed system for irrigated wheat in the Yaqui Valley of Mexico. They observed small yield differences between the treatments for the first five years, but after five years the treatment with permanent beds with full residue retention after chopping had the highest yields (Table 12). When a part of the residues was removed (leaving about 30% in the field), yields were reduced and burning was by far the worst practice. Yields with CT beds with residues incorporated were similar to permanent beds with full residue retention, but operational costs for permanent beds were about 25% less than CT. The yield advantage with permanent beds seems to be associated with gradual improvements in physical, chemical, and biological properties of soil when tillage has been reduced and crop residues retained.

There is an urgent need to conduct systematic studies to quantify total water balances including quantification of different components such as evaporation from exposed raised beds, percolation, and seepage losses under different water and soil management systems in a range of soil types. Interrow compaction of furrows on coarse-textured soils may be a useful option for reducing nutrient and water losses through leaching by reducing their permeability. In bed systems, the controlled traffic in the furrow also leads to targeted compaction that may be useful in water management.

Table 12 Effect of Tillage and Residue Management on Wheat Grain Yield (Mg ha^{-1}) Under a Long-Term Permanent Raised-Bed System in Maize–Wheat rotation at Obregon, Mexico

	Average grain yield	
Treatment	1993–1997	1998–2001
Conventional till beds, residue incorporated	6.3 a	7.6 a
Permanent beds—residue burned	6.2 a	6.6 c
Permanent beds—residue removed	6.4 a	7.3 b
Permanent beds—residue retained	6.5 a	7.7 a

Sources: Adapted from Limon-Ortega et al. (2000a) and K.D. Sayre, unpublished data.

VI. TILLAGE EFFECTS ON SOIL PHYSICAL PROPERTIES

Productivity-related soil physical parameters include soil depth, available soil water storage, movement of air and water into and within the soil, transmission of water to the growing plants, and soil resistance to root penetration. These parameters are the manifestations of soil structural properties, like bulk density, pore size distribution, stability and continuity of pores, soil aggregation, and sealing and crusting of soil. Limited information is available in the literature on long-term changes in various soil properties as affected by NT under rice–wheat systems in the IGP. Most of the data discussed in the following sections pertain to studies conducted mainly on cropping systems other than rice–wheat in North America, Europe, Australia, and Asia. The same principles will, however, be applicable to the rice–wheat systems of South Asia.

A. Bulk Density and Porosity

One soil property that is nearly always altered by tillage operations is bulk density; most changes in the soil's physical environment are mediated through bulk density. The magnitude and direction of change in this property depend upon the antecedent soil properties, type and intensity of tillage operation, and duration of tillage system. Both pore space and size of pores are affected by tillage. From an 18-year-study in Germany, Tebrugge and During (1999) reported that under NT while bulk density of the upper soil layer increased, macropores, total porosity, and saturated hydraulic conductivity decreased compared with CT. Because of greater earthworm population and activity, the number of biopores was much greater in the NT system than in the CT system. Bulk density of the 0–15-cm layer and saturated hydraulic conductivity were higher after unpuddled direct seeded rice than after puddled transplanted rice (Pandey et al., 2000). After three years of continuous rice–wheat cropping, bulk density was less in unpuddled plots and particularly where soil was tilled for wheat (Table 13). The bulk density of subsurface layer was quite high in all plots.

The pore system in NT plots is often more continuous than that under CT plots because of earthworm activity and old root channels and vertical cracks between beds especially in clayey soils (Cannell, 1985). Yonglu et al. (2000) reported that bulk density in the surface 0–15-cm layer of NT plots after rice was more suitable for normal growth of wheat compared to that of CT. No-till plots kept the soil capillary and pore distribution intact. The capillary porosity was 4.1%–5.5% higher on NT than on CT plots, but noncapillary porosity became lower than that of CT soil. The optimum porosity for a crop will depend on the water holding capacity, the soil water content at tillage, or the soil water content during crop growth.

Table 13 Soil Physical Properties Under Different Tillage Practices in Surface and Subsurface Soil After Three Years of Rice–Wheat Cropping on a Silt Loam Soil at Pantnagar, India

Treatment		Bulk density $(Mg\ m^{-3})$ in soil depth of[a]		Hydraulic conductivity$(mm\ h^{-1})$ in soil depth of[a]		Infiltration rate $(mm\ h^{-1})$[a]
Rice	Wheat	0–7 cm	12–19 cm	0–7 cm	12–19 cm	
DS	NT	1.49 a	1.70 a	2.08 b	1.48 b	1.40 b
DS	CT	1.43 b	1.67 a	3.08 a	2.47 a	2.35 a
TR	NT	1.54 a	1.74 a	1.76 b	1.29 b	1.23 b
TR	CT	1.46 b	1.69 a	2.44 a	2.01 a	1.92 a

[a] Means followed by the same letter in a column are not significantly different by Duncan's multiple-range test. DS = direct seeding, TR = transplanted, NT = no-tillage, CT = conventional tillage.
Source: Hobbs et al. (2002).

B. Soil Aggregation

Tillage also influences stability and size of aggregates through changes in organic carbon content of the soils. Douglas and Goss (1982) observed higher values of mean weight diameter of water stable aggregates in the surface layer of silt loam soil in NT than in moldboard tillage. Differences in aggregate stability between NT and CT decreased with increasing clay content of soils. Boparai et al. (1992) reported that puddling in rice reduced the mean weight diameter of soil aggregates as compared to that after maize. Tillage will likely have greater effects on soil aggregate stability when residues are retained than when residues are removed or burned.

C. Water Transmission Properties

Changes in number, continuity, and connectivity of pores caused by tillage and traffic are reflected in hydraulic characteristics of a soil (Boone, 1988). Increased porosity associated with a decrease in bulk density will increase the amount of water held at high soil water potentials. Radford et al. (1995) measured a 28% increase in plant available soil water at sowing with NT, linked to both a fourfold increase in earthworm numbers and significantly greater soil macropores in the 1.5–3-mm size.

Pandey et al. (2000) reported that hydraulic conductivity was significantly higher after nonpuddled direct-sown rice than after puddled rice. Tillage, which provides water-stable aggregates at the surface and soil pores open to the surface, is effective in achieving favorable infiltration. Zheng (2000) reported better drainage of NT than CT plots under the rice–wheat system. After three years of continuous rice–wheat cropping, hydraulic conductivity and the infiltration rates were higher in plots where soil was not puddled (Table 13). Tillage for wheat sowing (CT) significantly increased both hydraulic conductivity and the infiltration rate. More weed growth was observed early in the direct seeded rice crop than in the puddled treatment, in which higher infiltration rates made it more difficult to maintain standing water, which helps check weed growth (Hobbs et al., 2002).

The effects of tillage on the soil physical environment bring out soil water status as a key link that influences the response of thermal, aeration, and mechanical impedance regimes to tillage. This has led to the development of the composite concept, for example "nonlimiting water range" (Letey, 1985) or "least-limiting water range" (De Silva et al., 1994). This single parameter describes a range of water content, which is based on the limitations imposed by aeration and mechanical impedance of soil.

D. Soil Aeration

Tillage influences the soil aeration status characterized in terms of air-filled porosity of soil, the diffusion coefficient of a soil, or O_2 concentration in soil air. An increase in the number of large pores in the disturbed layer causes rapid drainage and restores adequate air-filled porosity faster after saturation by rain or irrigation. Dowdell et al. (1979) reported that the average O_2 concentration in NT and plowed soils during a wet winter averaged 10.2% and 7.2%, respectively. This effect was possible due to a system of large pores and channels, which developed in the NT plots but was destroyed by plowing. Aeration may become limiting for root growth in soils of low permeability or those with a relatively hard or impervious layer at shallow depths. Waterlogging for more than one day generally occurs in rice fields after heavy irrigation or rainfall due to development of a compact layer, which restricts drainage. In this case, the wheat crop suffers due to deficiency of O_2.

E. Soil Temperature

Tillage-induced changes in surface roughness and residue cover influence soil temperature by modifying the reflectance of incidental solar energy and turbulent exchange at the surface. Alternations in different soil physical properties associated with tillage affect thermal conductivity and heat capacity of a soil. Decrease in bulk density and reduced particle contact associated with soil loosening with tillage reduce thermal conductivity and volumetric heat capacity (Arshad and Azooz, 1996). Decrease in reflectance with surface roughness of tilled soil along with reduced heat capacity increases soil temperature in the surface layers. Tillage effects on soil temperature are confounded by the amount of residues left on the soil surface. Arshad and Azooz (1996) reported that mean diurnal temperature at 5-cm depth during the first three weeks after planting barley on a silt loam soil in northern British Columbia was 11.2°C in NT against 13.3°C in CT. This affect was ascribed to higher volumetric heat capacity and thermal conductivity caused by greater soil wetness under the NT system. In the tropics and subtropics, crop residues associated with tillage tend to moderate superoptimal soil temperature. For example, in Nigeria, NT reduced maximum temperature at the 5-cm depth by 11°C in maize and 9°C in soybean two weeks after planting (Lal, 1976). In central India, Sharma (1991) observed that maximum temperature in growing corn remained higher in NT compared to moldboard or cultivator tilled soil, regardless of residue mulch. However, residue mulch kept the soil cooler in all tillage treatments. In NT wheat, surface residues lowered maximum soil temperature by 1.5° to 2.0°C measured 10 to 45 days after seeding as compared to no mulch (B.-S. Yadvinder-Singh, personal communication). At significantly sub- or superoptimal temperatures, temperature modifications caused by tillage will have a strong influence on crop growth. When the soil temperature is nonlimiting and the soil is very wet, insufficient aeration is the primary risk, which increases with increasing soil temperature.

F. Mechanical Impedance and Root Growth

Tillage-induced changes in bulk density and particle reorientation affect soil mechanical impedance to growing roots or emerging seedlings. The mechanical resistance will affect the volume of rooted soil and the distribution of the roots in the volume. In somewhat drier situations, water is becoming limiting and crop water availability strongly decreases due to the high mechanical resistance, which restricts rooting depth. The properties of the soil in contact with and in the immediate vicinity of the emerging seed determine the uptake of water and O_2. High-strength soil layers may be formed at shallow depths with the use of heavy machinery for field operations. Generally, tilled soils have lower cone index values than NT system or untilled soils, particularly down to the depth of tillage. Gajri et al. (1992) observed that in a sandy loam soil under rice–wheat culture for 10 years, the cone index at field capacity wetness in the top 10-cm layer was 1.20 MPa in NT wheat plots and 0.75 MPa in 10-cm deep CT plots (Table 14). But there were no differences in soil strength in layers below 10 cm. The critical values of penetration resistance at which root growth is inhibited vary from 1.5 to $\geq$5 Mpa.

Physical property changes in the soil profile resulting from a particular tillage system affect the amount, size, and pattern of crop roots. Root growth may be affected due to increased mechanical strength under NT in wheat and/or

Table 14 Effect of No-Tillage (NT) and Conventional Tillage (CT) on Soil Strength and Root Length Density of Wheat at 58 Days After Seeding in a Loamy Sand Soil at Ludhiana, India

Soil depth (cm)	Soil strength (MPa)[a]		Soil depth (cm)	Root length density (cm cm^{-3})[a]	
	NT	CT		NT	CT
0–10	1.20 a	0.75 b	0–30	0.81 a	0.83 a
10–20	2.32 a	2.34 a	30–60	0.071 b	0.15 a
20–30	2.75 a	2.75 a	60–90	0.00 b	0.03 a

[a] Values in the same column followed by the same letter do not differ significantly at $P = 0.05$ by Duncan's multiple-range test.
Source: Gajri et al. (1992).

rice. Rice root growth may be better under unpuddled NT transplanted or direct seeded rice. In somewhat drier situations, water is becoming limiting and crop water availability strongly decreases due to a high mechanical resistance, which restricts rooting depth. In the absence of tillage, soil strength was increased, which may restrict root growth of wheat (Mahata et al., 1990). Poor root growth can reduce yields if nutrient and water uptake is hindered by inadequate rooting density.

Chaudhary et al. (1991) reported that NT, in the absence of crop residues, adversely affected the root length and density of wheat grown after puddled rice due to higher bulk density and soil strength. Lower grain yields of wheat on sandy loam soil under NT may be due to poor root growth and low N uptake. Gajri et al. (1992) reported that the effect of NT on root growth of wheat was more pronounced in the early growing season (Table 14) and decreased with time. Shallow rooting as in the NT system has not been a deterrent to high yield where residue is on the surface and soil water is not limiting throughout the growing season (Sprague and Triplett, 1986). It is the stability and continuity of the channels caused by earthworms and related biological activity and macropores created by decaying roots in the NT soil that favor deep rooting system development.

In conclusion, tillage influences soil physical properties by altering the soil water retention and hydraulic conductivity of soil, which govern the apportioning of rain and applied irrigation water into runoff and infiltered water. No-till systems with residues retained on the surface are highly efficient in reducing soil loss by wind and water erosion and thus maintain soil quality. Surface residues in NT will reduce surface sealing and the process of crust formation, and soil strength. There is overwhelming evidence that bulk density of the soil surface increases with conversion from CT to NT. But organic matter level, water-stable aggregates, and the number and continuity of large biopores are greater under NT than in CT. The least-limiting water range in NT is lower in NT than in CT. However, water movement into and within soil as indicated by infiltration rate, hydraulic conductivity, and preferential flow in NT is normally equal to or greater than that in CT.

VII. TILLAGE EFFECTS ON SOIL CHEMICAL PROPERTIES

The effect of tillage on soil chemical properties will depend upon soil, climate, cropping system, amount of residue, management practices, and time since initiation of the tillage system (Lal, 1997a).

A. Soil Organic Matter

Soil organic matter (SOM) plays an important role in nutrient supply and aggregate stability. Changes in frequency and intensity of tillage can influence decomposition of SOM and crop residues and cause differential distribution in the soil. Conservation tillage results in stratification of organic matter with a high concentration at the surface and a sharp decline with depth (Alverez et al., 1995). The increase in organic C under NT is generally in the top 0–7.5-cm soil layer with little difference below in NT and CT systems (Wander et al., 1998). Crop residues incorporated into the soil decompose several times faster than those lying on the soil surface. Generally, a tillage system, where residue is left at or near the surface and soil disturbance is reduced such as in NT, results in a higher organic C than when residue is incorporated into soil as in CT.

Plow tillage exposes organic C in inter- and intra-aggregate zones and is immobilized by microbial tissue by rapid oxidation because of improved availability of substrate and O_2 (Jastraw et al., 1996). In contrast to CT, conservation tillage keeps the soil cooler, wetter, less oxidative, and more acidic and reduces soil–residue contact (Rice et al., 1986). These conditions reduce the decomposition rate compared with that under CT. The effect of tillage on SOM is also dependent on the initial C level and the time after initiation of tillage. In soils with low organic matter, conservation and NT practices usually increase organic matter, which may remain constant or decrease with CT (Rasmussen and Collins, 1991). Only limited information is available on the long-term effects of tillage with or without residues on SOM content under the RW system. Yonglu et al. (2000) reported a greater amount of SOM in the surface layers under NT than under CT after five years of rice–wheat cropping. The increased level of SOM under NT was due to a slower rate of decomposition of rice stubble and roots. Tillage systems and residue incorporation will likely have a relatively smaller effect on SOM under subtropical and tropical environments than that in temperate regions due to greater rates of organic matter decomposition.

B. Nitrogen Dynamics

Tillage considerably influences the transformation and availability of N in soils (Silgram and Shepherd, 1999). Tillage generally leads to a temporary increase in soil mineral N as it enlarges the pool of C substrate being made available to support greater microbial activity. Tillage exposes microsites within soil aggregates where organic matter was previously protected from microorganisms and their enzymes. Nitrogen mineralization is generally lower in NT than in CT, and differences between the two can be as much as 65 kg N ha^{-1} (Silgram and Shepherd, 1999). However, these differences may be for a short term and may become similar in the long term (Radford et al., 1995; B.-S. Yadvinder-Singh, personal communication). The decreased rates of N mineralization during fallow periods under NT and the resulting lower soil test values will lead to a greater requirement of fertilizer N than CT (about 20 kg ha^{-1}). Tillage effects on N mineralization depend on the nature of the residue, the extent of soil disruption, and the cropping system and are strongly dependent on the duration the field has been under a particular tillage system (Staley et al., 1988).

Since there is a greater organic matter content in the surface layer and fresh residues with wide C:N ratios are located at or near the soil surface in the NT system relative to CT, there is a likelihood of temporary N immobilization. While greater N immobilization might cause short-term decreases in yield and N fertilizer recovery, the long-term benefits in terms of higher crop yields and lower N losses are likely to be better. Although the rate constant for organic N mineralization may continue to be greater in plowed soils, gradual accumulation of a larger organic N pool may compensate for reduced mineralization rate in the NT system on a long-term basis. In the rice–wheat system in the Sichuan province of China, Yonglu et al. (2000) reported a higher amount of total N (0.164%) under NT than that under CT (0.121%) in the upper 15-cm soil layer. Soils recently brought under NT will typically have higher fertilizer N demand during the first few years due to the initial immobilization process (Fox and Bandel, 1986). Immobilization of N within surface residue may have a positive effect on the subsequent crop because N remains near the root zone.

With NT, there is a greater probability of loss of fertilizer N, particularly in wheat, by denitrification and volatilization because of wetter soil, a larger denitrifying population, and the fact that fertilizer N may not be incorporated (Sprague and Triplett, 1986). The longer the rainfree period following application, the greater the likelihood and magnitude of N losses via NH$_3$ volatilization where urea is broadcast at the soil surface. Therefore, with NT seed-cum-fertilizer drill N losses will be markedly reduced over surface broadcast of fertilizers under NT.

The use of slow-release fertilizers and urea supergranules in direct seeded rice on beds may be advantageous. Greater leaching losses of NO_3 are expected from nonpuddled direct seeded rice plots, particularly on coarse-textured soils. But, leaching losses of NO_3 are likely to be substantial during puddling for transplanted rice. Greater N losses coupled with lower N availability in a soil can result in lower yields in NT soils than in CT soils at the same N rate, particularly at low N rates. Gajri et al. (1992) reported that 136 kg N ha^{-1} is needed for producing 95% of relative wheat yield on NT plots as compared to 102 kg N ha^{-1} under CT. Perhaps an inefficient N application method used in the above study was responsible for the low N use efficiency in NT wheat.

C. Phosphorus Availability

Tillage alters the concentration and distribution of P in soils. Greater amounts of SOM could increase P availability by increasing the organic pool and by shielding adsorption sites on clay minerals under CT (Schomberg et al., 1994). Because of less soil disturbance under NT, relatively immobile P tends to accumulate in areas close to the surface of application. This may reduce P fixation by soil colloids as continuous surface application saturates the P fixation sites and thus makes it more available to crops. Plant roots may be able to absorb more P due to a greater rate of P diffusion in NT with high soil water content than under the CT system, particularly in wheat. Increased organic matter content, acidification, and higher water content will increase the P availability under the NT system. Phosphorus and iron availability may also be an important issue on beds. On one hand, Willett and Higgins (1978) suggested that P requirements for field crops increase after flooded rice, and thus Garside et al. (1992) suggested that, with rice grown on beds, increases in P requirements for field crops will be small relative to those following flooded rice. On the other hand, P availability to rice on beds is likely to be reduced due to the aerobic soil conditions and may lead to P deficiency in rice, particularly on coarse-textured soils. Yadvinder-Singh et al. (1988) reported an increase in the availability of P in soils under waterlogged conditions relative to a field-capacity moisture regime. The tillage system also influences the amount and distribution of K and other exchangeable cations.

D. Micronutrient Availability

Franzluebbers and Hons (1996) reported greater accumulation of Mn and Zn in the surface layers with NT than with CT. Extractable Fe and Cu were lower under NT compared to CT in the surface 0–5-cm layer, but greater under NT at 5–60 cm with N fertilization. Greater water-filled pore space under NT compared with CT may create reduced soil conditions leading to higher extractable Fe and Mn.

Complexation of Cu and Fe by organic matter under NT may reduce their availability in soil. Long-term effects of NT without residues on nutrient availability have not been sufficiently investigated to permit its comparison with CT.

Nayyar et al. (2001) reported that Fe deficiency in rice occurs in upland or highly percolating coarse-textured soils because less mobilization of Fe^{2+} as the desired degree of reduction does not occur. However, the magnitude of the difference in the availability of P and Fe between bed and ponded flat layouts will depend on the extent and duration of saturation in the beds (Garside et al., 1992) and soil chemical properties.

VIII. TILLAGE EFFECTS ON SOIL BIOLOGY

Tillage affects the biological properties of a soil through its effect on soil mixing and loosening, crop residue status, and hydrothermal regime (Carter, 1986). Tillage effects on biological properties of a soil also depend on soil texture, quality of residue, previous management and history, time of the year, and time since the tillage system was initiated. These properties have a strong bearing on nutrient cycling, pesticide degradation, and soil structure development. Accumulation of organic matter at the soil surface in conservation tillage systems may lead to greater biological activity and microbial biomass compared with plowed soil (Doran, 1980; Hoffman et al., 1991). In a per-humid region, no-tillage for three years increased microbial biomass C and N in the 0–5-cm layer by 26%–28% compared to a plow-based system (Carter, 1986). Doran (1980) reported that the numbers of aerobic microorganisms, denitrifying microorganisms, facultative anaerobes, and certain soil enzyme activities appeared to be stratified in the surface of NT soils. Nondisturbance of soil under NT system generally stimulates earthworm growth. Carter (1991) reported an increase of 140%–160% in earthworm biomass in rotary-harrowed, and NT compared with CT on a sandy loam soil. The effects of tillage systems on earthworm population under the rice–wheat cropping sequence are likely to be small because anaerobic conditions during the rice phase inhibit the growth and activity of earthworms. This, however, needs to be investigated under nonpuddled rice–wheat systems. Furthermore, earthworm populations are not abundant in alkaline soils with low organic matter levels, especially when there are little surface residues.

IX. SUMMARY AND CONCLUSIONS

The vast majority of area under the rice–wheat system is characterized by the use of intensive tillage, often with crop residue removal or burning and inefficient irrigation water management by flood irrigation. Excessive tillage, especially when residues are removed or burned, contributes to degradation of soil productivity and diminishes input use efficiency. New technologies that reduce input use, such as NT, nonpuddled rice, and planting on permanent raised beds, can help in the transition to a more sustainable rice–wheat production system. Crop responses to tillage are highly variable depending on soil conditions, climate, implements used, and agronomic management. The conservation tillage practices offer opportunities to improve the physical conditions of the soil and to increase SOM levels, which are very low in the soils under the rice–wheat system in the IGP. However, these benefits will only be realized if NT or minimum tillage is applied across the system rather than in wheat only. No- or reduced tillage typically leads to soil conditions that differ markedly to those under more conventional arable systems. NT is advantageous in reducing energy inputs, allowing greater flexibility due to the reduced workload, and improving soil and water conservation.

With the development of improved NT seed-cum-fertilizer drills, NT and bed planting in wheat have shown encouraging results in reducing production costs and increasing farmer profits in recent years in many parts of the IGP. Studies indicate that NT with proper weed control is feasible for lowland rice. Development of new technologies to seed rice and wheat using reduced or NT combined with retention of crop residues on the soil surface requires the development of new, appropriate machinery. There is a need to develop soil- and site-specific tillage systems not only to sustain production of rice and wheat but also to conserve the natural resources.

Permanent raised beds are likely to emerge as a possible solution for increasing the productivity and sustainability of rice–wheat systems in the IGP. Fertilizer N recovery can be increased for both wheat and rice by banding of N into soil between the crop rows on a bed, and it is especially important for rice where fertilizer N recovery is generally poor. It is likely that the dynamics of water and N and other nutrients on beds with furrow irrigation will be complex due to the effect of the geometry of the system on water and N fluxes, and the occurrence of alternate wetting and drying of the soil on the beds. Bed planting provides new opportunities and challenges for stubble management in rice–wheat systems, which need to be addressed. Detailed investigations of permanent raised-bed systems for growing rice and wheat, with the goals of determining their suitability for a range of agro-ecological situations and for identifying optimal water, nitrogen, and stubble management for improving the productivity, resource use efficiency, and sustainability of rice–wheat systems in the IGP, are needed.

Many issues need further investigation under the NT or permanent raised-bed systems as follows:

1. How can weeds and crop residues be managed?
2. Which pests and diseases might increase, and what are the effects on the following wheat?
3. What physical conditions are affected, and what are the implications of changes?

4. Will fertilizer requirements remain the same?
5. What changes in machinery should be considered?
6. Can small-plot results be replicated to a field scale? The results of field experiments will provide evaluation of a limited range of agronomic management options, soil, and climatic conditions. Tillage experiments will be more valuable when they provide indispensable input data for ecophysiological models, which integrate separate effects.
7. What are the effects on water and nutrient (N) balance and dynamics?
8. What long-term trends might be expected on important soil types and different groundwater conditions and under different agro-ecological regions of the rice–wheat system in the IGP?

To date, there is limited information on suitable wheat and rice cultivars and species for NT with improved root anchorage and penetration ability through hard soil layers and improved yield potential. Scant information is available in the literature on changes in various soil properties as affected by NT under the rice–wheat systems in the IGP.

REFERENCES

Adachi, K. 1992. Effect of puddling on rice soil properties: Softness of puddled soil and percolation. *In* Murty V.V.N., Koga K., Eds. Proc. Intl. Workshop on Soil and Water Engineering for Paddy Field Management, Asian Institute of Technology. Bangkok. Thailand, 220–231.

Aggarwal, G.C., Sidhu, A.S., Sekhon, N.K., Sandhu, K.S., Sur, H.S. 1995. Puddling and N management effects on crop response in a rice-wheat cropping systems. *Soil Till. Res* 36:129–139.

Allen, H.P. 1981. Direct drilling and reduced cultivation. Ipswich. (U.K.): Farming Press Ltd.

Alverez, R., Diaz, R.A., Barbero, N., Santanatogila, O.J., Blotta, L. 1995. Soil organic carbon, microbial biomass, and CO_2–C production from three tillage systems. *Soil and Tillage Res* 33:17–28.

Angus, J., Poss, R., Kirkegaard, J. 1998. Long-term benefits of stubble. Australian Grain, April–May 1998, 29–30.

Aquino, P. 1998. The adoption of bed planting of wheat in the Yaqui Valley, Sonora, Mexico. Wheat Special Report No. 17a. CIMMYT. Mexico.

Arshad, M.A., Azooz, R.H. 1996. Tillage effects on soil thermal properties in semi-arid cold region. *Soil Sci. Soc. Am. J* 60:561–567.

Aslam, M., Majid, A., Hashmi, N.I., Hobbs, P.R. 1993. Improving wheat yield with rice–wheat cropping system of the Punjab through no-tillage. *Pakistan J. Agri. Res* 14:8–11.

Bacon, P.E., Cooper, J.L. 1985. Effect of rice stubble and fertilizer management techniques on yield of wheat sown after rice. *Field Crops Res* 10:241–250.

Bajpai, R.K., Tripathi, R.P. 2000. Evaluation of nonpuddling under shallow water tables and alternative tillage methods on soil and crop parameters in a rice–wheat system in Uttar Pradesh. *Soil Tillage Res* 55:96–106.

Bhardwaj, A.K., Singh, Y. 1998. Tillage options for high productivity in rice–wheat system. In Proc. First Intl. Agronomy Congress, Extended Sumaaries. Indian Society of Agronomy, Indian Council of Agricultural Research. New Delhi, 474–475.

Blackwell, P. 1998. Customized controlled traffic farming systems instead of standard recommendations or "Tramlines ain't tramlines." *In* Tullberg J.N., Yule D.F., Eds. Second National Controlled Farming Conf. University of Queensland. Australia.

Boone, F.R. 1988. Weather and other environmental factors influencing crop responses to tillage and traffic. *Soil Till. Res* 11:283–324.

Boparai, B.S., Yadvinder-Singh, Sharma, B.D. 1992. Effect of green manuring with *Sesbania aculeata* on physical properties of soil and on growth of wheat in rice–wheat and maize–wheat rotations in semiarid region of India. *Arid Soil Res. Rehabil* 6: 135–143.

Brar, S.S., Kumar, S. 2000. Experience with no-tillage in Indian Punjab. *In* Proc. Intl. Conf. on Managing Natural Resources for Sustainable Agricultural Production in the 21st Century, Extended Summaries 3, 313–314.

Cannell, R.Q. 1985. Reduced tillage in northwest Europe—A review. *Soil Till. Res* 5: 129–177.

Carter, M.R. 1986. Microbiology biomass as an index for tillage-induced changes in soil biological properties. *Soil Till. Res* 7:29–40.

Carter, M.R. 1991. Evaluation of shallow tillage for spring cereals on fine sandy loam II. Soil physical, chemical and biological properties. *Soil Till. Res* 21:37–52.

Carter, M.R. 1994. A review of conservation tillage strategies for humid temperature regions. *Soil Till. Res* 31:289–301.

Chaudhary, M.R., Khera, R., Singh, C.J. 1991. Tillage irrigation effects on root growth, soil water depletion and yield of wheat following rice. *J. Agril. Sci. (Cambridge)* 116:9–16.

Christian, D.G., Ball, B.C. 1994. Reduced cultivation and direct drilling for cereals in Great Britain. *In* Carter M.R., Ed. Conservation Tillage in Temperate Agroecosystems. Boca Raton, FL, (USA): Lewis Publishers, CRC Press, 117–140.

Cooten, D.E., van Borrell, A.K. 1999. Enhancing food security in semi-arid eastern Indonesia through permanent raised-bed cropping: A review. *Aust. J. Exptl. Agric* 39: 1035–1046.

De Datta, S.K. 1986. Technology development and the spread of direct-seeded rice in southeast Asia. *Exp. Agric* 22:417–426.

De Silva, A.P., Kay, B.D., Perfect, E. 1994. Characterization of the least limiting water range of soils. *Soil Sci. Soc. Am. J* 58:1775–1781.

Doran, J.W. 1980. Soil microbiological and biochemical changes associated with reduced tillage. *Soil Sci. Soc. Am. J* 44:765–771.

Douglas, J.T., Goss, M.J. 1982. Stability and organic matter content of surface soil aggregates under different methods of cultivation and in grassland. *Soil Till. Res* 2: 155–175.

Dowdell, R.J., Cress, R., Burford, J.R., Cannell, R.Q. 1979. Oxygen concentration in a clay soil after ploughing or direct drilling. *J. Soil Sci* 30:230–245.

Ehlers, W. 1984. The need for soil physics in tillage research. *Soil Till. Res* 4:1–3.

Fox, R.H., Bandel, V.A. 1986. Nitrogen utilization with no-tillage. In Phillips R.E., Phillips H.S., Eds. No-Tillage Agriculture. Van Nostrand, Reinhold, NY, 117–148.

Franzluebbers, A.J., Hons, F.M. 1996. Soil-profile distribution of primary and secondary plant available nutrients under conventional and no-tillage. *Soil Till. Res* 39: 229–239.

Frye, W.W., Lindwall, C.W. 1986. No-tillage research priorities. *Soil Till. Res* 8:311–316.

Gajri, P.R., Arora, V.K., Prihar, S.S. 1992. Tillage management for efficient water and nitrogen use in wheat following rice. *Soil Till. Res* 24:167–182.

Gajri, P.R., Arora, V.K., Prihar, S.S. 2002. Tillage for sustainable cropping: Food Products Press, an imprint of Haworth Press Inc.. NY.

Garside, A.L., Borrell, A.K., Ockerby, S.E., Dowling, A.J., McPhee, J.E., Braunack, M.V. 1992. An irrigated rice-based cropping system for tropical Australia. *In* Proc. 6th Aust. Agron. Conf. Armidale, NSW, Australia, 259–261.

Gebhardt, M.R., Daniel, T.C., Schweizer, E.E., Allmaras, R.R. 1985. Conservation Tillage Science. Washington DC.

Gill, M.A., Kahlown, M., Choudhary, M.A., Hobbs, P.R. 2000. Evaluation of resource conservation technologies in rice–wheat system of Pakistan. Water and Power Development Authority Report. Lahore. Pakistan.

Grace, P.R., Jain, M.C., Harrington, L., Roberston, P. 2002. The long-term sustainability of tropical and subtropical rice and wheat systems: An environmental perspective. *In* Ladha J.K., Ed. Improving the productivity and sustainability of rice–wheat systems: Issues and impact. ASA, Spec. Publ. 00. ASA. Madison, WI (in press).

Gupta, R.K., Hobbs, P.R., Ladha, J.K., Prabhakar, S.V.R.K. 2002. Resource conservation technologies. Transforming the rice–wheat system of the Indo-Gangetic Plains. Asia-Pacific Association of Agricultural Research Institutions. FAO Regional Office for Asia and the Pacific. Bangkok.

Gupta, R.K., Rickma, J. 2002. Design and improvements in exiting no-till machines for residue conditions. Rice–Wheat Consortium, Traveling Seminar Report Series No. 3. Rice–Wheat Consortium for the Indo-Gangetic Plains. New Delhi. India.

Harrington, L.W., Hobbs, P.R., Cassaday, K.A. 1992. Methods of measuring sustainability through farmer monitoring. Application to the rice–wheat pattern in South Asia. Pro. Workshop (CIMMYT/IRRI/NARC). Kathmandu. Nepal.

Hasegawa, S. 1992. Manipulation of plow pan in rice-based cropping system. *In* Murty V.V.N., Koga K., Eds. Proc. Intl. Workshop on Soil and Water Engineering for Paddy Field Management, Asian Institute of Technology. Bangkok.

Hira, G.S., Gupta, P.K., Josan, A.S. 1998. Waterlogging Causes and Remedial Measures in Southwest Punjab. Research Bulletin No. 1/98. PAU. Ludhiana. India.

Hobbs, P.R. 2001. Tillage and crop establishment in South Asian rice–wheat system: Present practices and future options. *J. Crop Production* 4:1–22.

Hobbs, P.R., Singh, Y., Giri, G.S., Lauren, J.G., Duxbury, J.M. 2002. Direct-seeding and reduced-tillage options in the rice–wheat systems of the Indo-Gangetic plains of South Asia. *In* Pandey S., Mortimer M., Wade L., Tuong T.P., Lopez K., Hardy B., Eds. Direct seeding: Research strategies and opportunities. Proc. Intl. Workshop on Direct Seeding in Asian Rice Systems: Strategic Research Issues and Opportunities. International Rice Research Institute: Los Baños. Philippines, 201–215.

Hoffman, C., Linden, S., Koch, H.J. 1991. Influence of soil tillage on net N mineralization under sugarbeet. *Z. Pfln. Bodenk* 159:79–85.

Hood, A.E.M., Jameson, H.R., Cotterall, R. 1964. Crops grown using paraquat as a substitute for ploughing. *Nature* 201:1070–1072.

Huke, R., Huke, E. 1992. Rice–Wheat Atlas of South Asia. International Rice Research Institute: Los Baños. Philippines.

Huke, R., Huke, E., Woodhead, T., Huang, J. 1994. Rice–Wheat Atlas for China. International Rice Research Institute: Los Baõs. Philippines.

Humphreys, E., Muirhead, W.A., Fawcett, B.J. 1992. The effect of puddling and compaction on deep percolation and rice yield in temperate Australia. *In* Murty V.V.N., Koga K., Eds. Proc. Intl. Workshop on Soil and Water Engineering for Paddy Field Management, Asian Institute of Technology. Bangkok, 212–219.

Jalota, S.K., Prihar, S.S., Sandhu, B.S., Khera, K.L. 1980. Yield water use and root distribution of wheat as affected by pre-sowing and post-sowing irrigation. *Agric. Water Management* 2:289–297.

Jastraw, J.D., Bearttron, T.W., Miller, R.M. 1996. Carbon dynamics of aggregate associated organic matter estimated by carbon-13 natural abundance. *Soil Sci. Soc. Am. J* 60:801–807.

Kaspar, T.C., Erbach, D.C., Cruse, R.M. 1990. Corn response to seed-row residue removal. *Soil Sci. Soc. Am. J* 54:1112–1117.

Kim, J., Kim, S., Park, S., Kang, Y., Kim, S., Lee, M., Peng, S. 2001. Wet-seeded rice cultivation technology in Korea. *In* Peng S., Hardy B., Eds. Rice research for food security and poverty alleviation, Proc. Intl. Rice Research Conf., March 31–April 3, 2000, Los Baños, Philippines, 545–560.

Kirchhof, G., So, H.B. 1990. The effect of puddling intensity and compaction on soil properties, rice and mungbean growth: A mini rice-based study. *In* Kirchhof G., So H.B., Eds. Management of clay soils for lowland rice based cropping systems. ACIAR Proc. No 70, 51–70.

Kirkby, C.A. 1999. Survey of current rice stubble management practices for identification of research needs and future policy. RIRDC Project NO. CSL-5A.

Ladha, J.K., Dawe, D., Yadav, L., Pathak, H., Padre, A.T., Hobbs, P.R., Gupta, R.K. 2002. Productivity trends in intensive rice-wheat cropping systems in Asia. *In* Ladha J.K., et al., Eds. Improving the productivity and sustainability of rice–wheat systems: Issues and impact. ASA, Spec. Publ. 65. ASA. Madison, WI.

Ladha, J.K., Fischer, K.S., Hossain, M., Hobbs, P.R., Hardy, B., (eds.) 2000. Improving the productivity and sustainability of rice–wheat systems of the Indo-Gangetic Plains: A synthesis of NARS-IRRI partnership research. IRRI Discussion Paper Series No. 40. IRRI: Los Baños. Philippines.

Lal, R. 1976. No-tillage effects on soil properties under different crops in western Nigeria. *Soil Sci. Soc. Am. J* 40:762–768.

Lal, R. 1985. Tillage in lowland rice-based cropping systems *In* Soil Physics and Rice. International Rice Res. Institute: Los Baños. Philippines, 283–307.

Lal, R. 1997a. Long-term tillage and maize monoculture effects on a tropical Alfisol in Western Nigeria. II. Soil chemical properties. *Soil Till. Res* 42:161–174.

Lal, R. 1997b. Residue management, conservation tillage and soil restoration for mitigating greenhouse effect by CO_2-environment. *Soil Till. Res* 43:81–107.

Letey, J. 1985. Relationship between soil physical properties and crop production. *Advances in Soil Science* 1:277–294.

Limon-Ortega, A., Sayre, K.D., Francis, C.A. 2000a. Wheat and maize yields in response to straw management and nitrogen under a bed planting system. *Ag. J* 92:295–302.

Limon-Ortega, A., Sayre, K.D., Francis, C.A. 2000b. Wheat nitrogen use efficiency in a bed-planting system in Northwest Mexico. *Ag. J* 92:303–308.

Lynch, J.M. 1973. Phytoxicity of acetic acid produced in the anaerobic decomposition of wheat straw. *J. Appl. Bacteriology* 42:81–87.

Mahata, K.R., Sen, H.S., Pradhan, S.K., Mandal, L.N. 1990. No-tillage and dry ploughing compared with puddling for wet-season rice on an alluvial sandy clay loam in eastern India. *J. Agric. Sci., Camb* 114:79–86.

Malik, R.K., Gill, G., Hobbs, P.R. 1998. Herbicide resistance in *Phalaris minor*—A major issue for sustainable wheat productivity in rice–wheat cropping systems in the Indo-Gangetic Plains. Rice–Wheat Consortium Rice–Wheat Paper Series 3. Rice–Wheat Consortium for the Indo-Gangetic Plains. New Delhi. India.

Maurya, P.R., Lal, R. 1979. Influence of tillage and seeding methods on flooded rice. *In* Lal R., Ed. Soil tillage and crop production, IITA Proc. Series 2, IITA. Ibadan. Nigeria, 207–290.

Maynard, M. 1991. Permanent beds—their potential role in soil management for the future. Large Area Farmer's Newsletter No. 137. Irrigation Research and Extension Committee. CSIRO Centre for Irrigation Research. Griffith, NSW, Australia, 14–18.

Mehla, R.S., Verma, J.K., Gupta, R.K., Hobbs, P.R. 2000. Stagnation in the productivity of wheat in the Indo-Gangetic Plains: No-till-seed-cum-fertilizer drill as an integrated solution. Rice–Wheat Consortium Paper Series 8. Rice Wheat Consortium for the IGP. New Delhi.

Mondal, M.K., Islam, M.J. 1991. Rice irrigation: A new water management concept for Bangladesh. *In* Proc. Regional Seminar on Water Management Practices. Federation of Engineering Institutions of South and Central Asia and Bangladesh, 901–909.

Moorman, F.R., N., van Breemen 1978. Rice: Soil, Water, Land. International Rice Res. Institute: Los Baños. Philippines.

Narang, R.S., Virmani, S.M. 2001. Rice–wheat cropping systems of the Indo-Gangetic Plains of India. Rice–Wheat Consortium Paper Series 11. Rice–Wheat Consortium for the Indo-Gangetic Plains, New Delhi, and International Crops Research Institute for the Semi-Arid Tropics. Patancheru. India.

Nayyar, V.K., Arora, C.L., Kataki, P. 2001. Management of soil micronutrient deficiencies in the rice–wheat cropping system. *J. Crop Production* 4:87–131.

Ortiz-Monasterio, J.T., Dhillon, S.S., Fischer, R.A. 1994. Date of sowing effects on grain yield and yield components of irrigated spring wheat cultivars and relationship with radiation and temperature in Ludhiana, India. *Field Crops Res* 37:169–184.

Pandey, D.S., Bath, B.S., Misra, R.D., Prakash, A., Singh, V.P. 2000. Effect of puddling and method of wheat sowing on soil health of aquic Hapludoll soil. *In* Proc. Intl. Conf. on Managing Natural Resources for Sustainable Agricultural Production in the 21st Century. *Extended Summaries* 3:1481.

Pandey, S., Velasco, L. 1999. Economics of direct seeding in Asia: Patterns of adoption and research priorities. *Int. Rice Res. Notes* 24(2):6–11.

Pathak, H., Ladha, J.K., Aggarwal, P.K., Peng, S., Das, S., B.-S., Yadvinder-Singh, Kamra, S.K., Mishra, B., A.S.R.A.S., Sastri, Aggarwal, H.P., Das, D.K., Gupta, R.K. 2002. Trends of climatic potential and on-farm yields of rice and wheat in the Indo-Gangetic Plains. *Field Crops Res* 80:23–24.

Phillips, R.E., Phillips, S.H. 1984. No-Tillage Agriculture. Principles and Practices. Van Nostrand Reinhold, NY.

Piggin, C.M., Garcia, C.O., Jamiya, J.D., Bell, M.A., Castro, E.C., Rozota, E.B., Hill, J. 2001. Establishment of irrigated rice under no-and conventional tillage systems in the Philippines. *In* Proc. Intl. Rice Research Conf. March 31–April 3, 2000. IRRI: Los Baños. Philippines, 533–543.

Pingali, P.L., Shah, M. 1999. Rice–wheat cropping systems in the Indo-Gangetic Plains: Policy re-directions for sustainable resource use. *In* Sustaining Rice–Wheat Production Systems: Socio-economic and Policy Issues. RW Consortium Paper Series 5. RWC for the IGP. New Delhi. India, 1–13.

Prihar, S.S., Ghildyal, B.P., Painuli, D.K., Sur, H.S. 1985. Physical properties of mineral soils affecting rice-based cropping systems. *In* Soil Physics and Rice. IRRI: Los Baños. Philippines, 57–70.

Radford, B.J., Kay, A.J., Robertson, L.N., Thomas, G.A. 1995. Conservation tillage increases soil water storage, soil animal populations, grain yield and response to fertilizer in the semiarid subtropics. *Aust. J. Exptl. Agri* 35:223–232.

Rasmussen, P.E., Collins, H.P. 1991. Long-term impacts of tillage, fertilizer and crop residue on soil organic matter in temperature semiarid regions. *Adv. Agron* 45: 93–134.

Regmi, A.P., Ladha, J.K., Pathak, H., Pasuquin, E., Bueno, C., Dawe, D., Hobbs, P.R., Joshy, D., Maskey, S.L., Pandey, S.P. 2002. Analyses of yield and soil fertility trends in a 20-year rice-rice–wheat experiment in Nepal. *Soil Sci. Soc. Am. J* 66: 857–867.

Rice, C.W., Smith, M.S., Blevins, R. L. 1986. Soil nitrogen availability after long-term continuous no-tillage and conventional tillage crop production. *Soil Sci. Soc. Am. J* 50:1206–1210.

Riley, H., Borrensen, T., Ekberg, E., Ryberg, T. 1994. Trends in reduced tillage research and practice in Scandinavia. *In* Carter M.R., Ed. Conservation Tillage in Temperate Agroecosystems. Boca Raton, FL: Lewis Publishers, CRC Press, 23–45.

Roy, I., Dey Sarkar, A.K. 1993. Effect of minimum tillage in wheat grown after rice. *Rachis* 12(1/2):49–51.

Sandhu, B.S., Khera, K.L., Prihar, S.S., Singh, B. 1980. Irrigation needs and yield of rice on a sandy loam soil as affected by continuous and intermittent submergence. *Indian J. Agric. Sci* 50:432–436.

Sandhu, B.S., Khera, K.L., Prihar, S.S., Singh, B. 1982. Note on the use of irrigation water and yield of transplanted rice as affected by timing of last irrigation. *Indian J. Agric. Sci* 52:871–872.

Schomberg, H.H., Ford, P.B., Hargrove, W.L. 1994. Influence of crop residue on nutrient cycling and soil chemical properties. *In* Unger P.W., Ed. Managing Agricultural Residues. Boca Raton, FL: Lewis Publishers, 99–122.

Shafiq, M., Hassan, A., Ahmad, S. 1995. Effect of crop rotation, tillage technique, and fertilization on yield of wheat (*Triticum aestivum*) and greengram (*Vigna radiatus*) under rainfed conditions. *Indian J. Agric. Sci* 65:591–593.

Sharma, B.R. 1991. Effect of different tillage practices, mulch and nitrogen on soil properties, growth and yield of fodder maize. *Soil Till. Res* 19:55–66.

Sharma, P.K., De Datta, S.K. 1986. Physical properties and processes of puddled rice soils. *Adv. Soil Sci* 5:139–178.

Sharma, R.B., Pandey, R. 1992. Tillage management for reducing water requirements of paddy in a sub-tropic sandy loam of North-Bihar, India. *In* Murty V.V.N., Koga K., Eds. Soil and Water Engineering for Paddy Field Management. Asian Institute of Technology. Bangkok:252–257.

Sharma, S.K., Tomar, S.S., Shrivastava, S.P. 2000. Improved tillage practices for increasing productivity of rice–wheat cropping sequence in Vertisols. *In* Proc. Intl. Conf. on Managing Natural Resources for Sustainable Agricultural Production in the 21st century. *Extended Summaries* 3:910–912.

Silgram, M., Shepherd, M.A. 1999. The effects of cultivation on soil nitrogen and mineralization. *Adv. Agron* 65:267–311.

Singh, S., Sharma, S.N., Prasad, R. 2001. The effect of seeding and tillage methods on productivity of rice–wheat cropping system. *Soil Tillage Res* 61:125–131.

Singh, P., Aipe, K.C., Prasad, R., Sharma, S.N., Singh, S., Singh, P. 1998. Relative effect of no and conventional tillage on growth and yield of wheat (*Triticum aestivum*) and soil fertility under rice (*Oryza sativa*)–wheat cropping system. *Indian J. Agron* 43:204–207.

Soil Science Society of America. 1987. Glossary of soil science terms. Madison, WI. *Soil Sci. Soc. Am. J.*

Sprague, M.A., Triplett, G.B. 1986. No-tillage and surface-tillage agriculture: John Wiley & Sons. NY.

Srivastava, A.P., Panwar, J.S., Garg, R.N. 2000. Influence of tillage on soil properties and wheat productivity in rice (*Oryza sativa*)–wheat (*Triticum aestivum*) cropping system. *Indian J. Agric. Sci* 70:207–210.

Staley, T.E., Edwards, W.M., Scott, C.L., Owens, L.B. 1988. Soil microbial biomass and organic component in no-tillage chronosequence. *Soil Sci. Soc. Am. J* 52:998–1005.

Sur, H.S., Prihar, S.S., Jalota, S.K. 1981. Effect of rice–wheat and maize–wheat rotations on water transmission and wheat root development in a sandy loam soil of Punjab, India. *Soil Till. Res* 1:361–371.

Sweeney, D.W., Sisson, J.B. 1988. Effect of ridge planting and N-application methods on wheat grown in somewhat poorly drained soils. *Soil Till. Res* 12:187–196.

Taylor, R. 1967. Bipyridyl herbicides and the direct establishment of crops in uncultivated soil. 1. Cereals. *In* Proc. of the 20th New Zealand Weed and Pest Control Conf: 74–79.

Tebrugge, F., During, R.A. 1999. Reducing tillage intensity—a review of results from long-term study in Germany. *Soil Till. Res* 53:15–28.

Timsina, J., Connor, D.J. 2001. Productivity and management of rice–wheat systems: Issues and challenges. *Field Crops Res* 69:93–132.

Tisdall, J.M., Hodgson, A.S. 1990. Ridge tillage in Australia: A review. *Soil Tillage Res* 18:127–144.

Tomar, S.S., Sharma, K. 1998. Soil physical conditions in relation to tillage, methods of rice cultivation and nutrient availability under rice–wheat cropping in Vertisols. In

Proc. First Intl. Agron. Congress, Extended Summaries. Indian Society of Agronomy, Indian Council of Agricultural Research. New Delhi:279.

Tracy, P., Sims, B.D., Hefner, S.G., Cairns, J.P. 1993. Guidelines for producing rice using furrow irrigation. Dept. of Agron: Univ. of Missouri–Columbia. USA.

Verma, U.N., Srivastava, V.C., Verma, U.K. 1988. Energetics of wheat (*Triticum aestivum*) production as influenced by tillage and nitrogen management. *Indian J. Agric. Sci* 58:813–816.

Verma, U.N., Srivastava, V.C., Prasad, R.B. 1989. Effect of No–tillage on wheat (*Triticum aestivum*) with rice (Oryza sativa) stubble mulching. *Indian J. Agric. Sci* 58: 669–671.

Wander, M.M., Bidart, M.G., Aref, S. 1998. Tillage impacts on depth distribution of total and particulate organic matter on three Illinois soils. *Soil Sci. Soc. Am. J* 62: 1704–1711.

Wilhelm, W.W., Doran, J.W., Power, J.F. 1986. Corn and soybean yield response to crop residue management under no-till system. *Ag. J* 78:184–189.

Willet, I.R., Higgins, M.L. 1978. Phosphate sorption by reduced and reoxidised rice soils. *Aust. J. Soil Res* 16:319–326.

Yadvinder-Singh, B.-S., Maskina, M.S., Meelu, O.P. 1988. Effect of organic manures, crop residues and green manure (*Sesbania aculeata*) on nitrogen and phosphorus transformations in a sandy loam at field capacity and waterlogged conditions. *Biol. Fertility Soils* 6:183–187.

Yadvinder-Singh, B.-S., Meelu, O.P., Khind, C.S. 2000. Long-term effects of organic manuring and crop residues on the productivity and sustainability of rice–wheat cropping system in Northwest India. *In* Abrol I.P., Bronson K.F., Duxbury J.M., Gupta R.K., Eds. Long-Term Soil Fertility Experiments in Rice–Wheat Cropping Systems. Rice–Wheat Consortium Paper Series 6. Rice–wheat Consortium for Indo-Gangetic Plains. New Delhi, India, 149–162.

Yun-Sheng, C 1983. Drainage of paddy soils in Taihu Lake region and its effects. Soil Res. Rep. 8. Inst. Soil Sci: Academia Sinica. Nanjing. China, 1–18.

Yonglu, T., Gang, H., Yao, Y., Lixum, Y. 2000. High yielding cultivation techniques for wheat under the rice–wheat cropping system in the Sichuan province of China. *In* Hobbs P.R., Gupta R.K., Eds. Soil and Crop Management Practices for Enhanced Productivity of the Rice–Wheat Cropping System in the Sichuan Province of China. Rice–Wheat Consortium Paper Series No. 9, Rice–Wheat Consortium for the Indo-Gangetic Plains. New Delhi. India, 11–23.

Zheng, J. 2000. Rice–wheat cropping system China. In Hobbs P.R., Gupta R.K., Eds. Soil and Crop Management Practices for Enhanced Productivity of the Rice–Wheat Cropping System in the Sichuan Province of China. Rice–Wheat Consortium Paper Series No. 9, Rice–Wheat Consortium for the Indo-Gangetic Plains. New Delhi. India:1–10.

11

No-Till in the Rice–Wheat System: An Experience from Nepal

D. S. Pathic
Nepal Agricultural Research Council (NARC)
Kathmandu, Nepal

R. K. Shrestha
University of Wisconsin–Madison
Madison, Wisconsin, U.S.A.

I. INTRODUCTION

The rice (*Oryza sativa L.*)–wheat (*Triticum aestivum L.*) cropping system is extremely important for food security in Nepal. In the early 1960s, the rice–wheat system occupied 0.1 million hectares. With the advent of new high-yielding, fertilizer-responsive, short-duration rice and wheat genotypes in the early 1970s, growing rice and wheat in sequence became feasible. Today, this system constitutes the major cropping system in the *terai* and midhills of Nepal, covering 0.52 million hectares (Table 1) (Pandey et al., 2001). In spite of Nepal's having adopted this system, the productivity level remains low. The major constraint to expansion of the rice–wheat area and increased yield is late planting caused by the short period available for sowing of wheat after rice and the availability of yearround irrigation (Harrington et al., 1993). The average yield of rice rarely exceeds 3 t ha^{-1} and wheat yield varies from 1 to 2 t ha^{-1}. These yields are low compared to other rice–wheat-growing countries in the region. However, rice–wheat system yields of 9 t ha^{-1} have been reported by some farmers in the midhills especially near Kathmandu and in the Kavre District. This indicates a large gap between the potential and actual yields of the rice–wheat system, and research is needed to close this gap.

Table 1 Estimated Area, Production, and Productivity of Rice–Wheat System in the Major Agroecological Regions of Nepal (Mean of 1991–1995)

Agroecological region	Area (000 ha)	Production (000 t)	Productivity (t ha^1)
Terai	342	1,306	3.82
Midhills	141	511	3.63
Mountains	37	126	3.40

Source: Pandey et al. (2001).

II. FARMERS' TILLAGE PRACTICES IN THE RICE–WHEAT SYSTEM

Rice is mostly grown under anaerobic conditions by plowing land that is saturated or has standing water and then transplanting rice seedlings and keeping the soil flooded for most of the growing period. Wheat is grown under aerobic conditions by plowing four to eight times and then seeding by broadcasting and a final plowing and planking (Harrington et al., 1993). Two to three light irrigations are given during the growth period, keeping the field dry most of the time. This conflicting soil management practice for rice creates difficulty in making a good soil tilth for wheat. Manual preparation of fields with a spade is common in some hill areas although many farmers use animal- (bullock or buffalo) drawn plows. However, in *tarai* plains, the animal-drawn plow is more common although some farmers use Chinese hand-tractors and those on larger farms use four-wheel tractors. The implements used for preparing land are the nine-tyne cultivator or disc harrow. Very few farmers use a moldboard or disc plow. Farmers who tend to apply excessive tillage are often delayed in wheat sowing.

III. LATE PLANTING OF WHEAT

About 40%–50% of wheat planting in Nepal is late (December planting), a major problem causing low yield (Giri, 1998; Harrington et al., 1993). The main reasons for late planting are excess or lack of soil moisture, late-maturing rice varieties, unavailability of labor, long turnaround time, power constraints for plowing, excessive tillage by farmers to get a good tilth, and lack of appropriate mechanization (Fig. 1). The wheat crop must be planted at the optimum time to get a good yield. A linear decline in wheat yield of 1% to 1.5% d^{-1} has been reported if wheat is planted after the end of November (Ortiz-Monasterio et al., 1994). Late planting of wheat reduces not only yield but also input efficiency. An increase

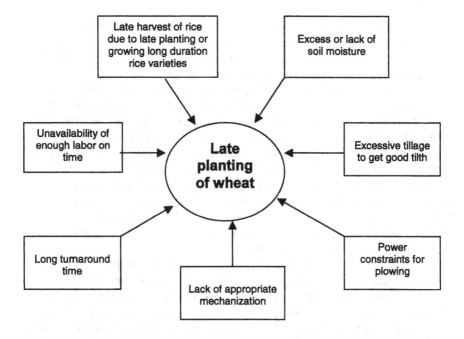

Figure 1   The most common causes of late sowing of wheat in Nepal.

in nitrogen application cannot compensate for the decline in yield due to late planting (Saunders, 1990).

IV.   ALTERNATE CROP ESTABLISHMENT OPTIONS: MINIMUM AND NO-TILL

Several options including surface seeding and minimum or no-till for wheat and direct seeding of rice have been introduced to Nepalese farmers to overcome the problem of late planting in the rice–wheat system.

A.   Surface/Relay Seeding of Wheat

Surface seeding, or placement of seed onto a soil surface without any land preparation, is gaining popularity in Nepal, especially in low-lying areas where soil moisture is in excess. Excess soil moisture after rice harvest, especially in low-lying, poorly drained fields, does not allow land preparation for wheat planting. Many farmers have either planted wheat late or left the land fallow. Surface and relay (planting wheat into standing rice) seeding of wheat is appropriate in such low-lying areas. Although this technology is best suited to low-lying heavy-textured soil, it can also be adopted in upland and light-textured soil with presowing irrigation and using soaked seed. The adoption of surface seeding has occurred in the Dhanusha, Bara, Parsa, and Rupandehi districts of Nepal, covering about 250 hectares (RARS, Parwanipur, 2000; NWRP, Bhairhawa, 2000). The surface-seeded wheat yield in farmers' field ranged from 1.06 to 3.85 t ha^{-1} in 1998–1999 and 0.51 to 4.08 t ha^{-1} in 1999–2000 in the Parwanipur command area (RARS, Parwanipur, 2000); 3.7 t ha^{-1} in 1999 and 3.4 t ha^{-1} in 2000 in Ranighat, Birgung (AIRC, 1999; 2000); and 3.4 t ha^{-1} in Bhairhawa (NWRP, Bhairhawa, 2000). The surface seeding method produced significantly higher yield compared to the farmers' practice of three to four plowings and one planking (Hobbs et al., 1997) (Table 2). This technology generated on average 42% higher net benefits compared to the farmers' practice, as cost of land preparation was negligible and timely planting of wheat was facilitated.

Wheat yield increase due to relay seeding ranged from 29% to 105% and surface seeding from 2% to 87% using different genotypes compared to conventional farmers' practice of four plowings followed by two plankings (Table 3). In relay seeding, the Bhrikuti genotype produced the highest grain yield (3.4 t ha^{-1}) followed by Achyuta (3.3 t ha^{-1}). But the percent increase due to relay seeding technology over farmers' practice was highest with BL-1135 (105%) followed by Achyuta. In surface seeding, BL-1135 (2.9 t ha^{-1}) produced the highest yield followed by Bhrikuti (2.6 t ha^{-1}) over farmers' practice. Percent increase due to surface seeding technology over farmers' practice in Bhairahawa was also highest with BL-1135 (87%) followed by NL 683 (Giri, 1998).

Surface- or relay-seeded wheat establishment generally does not have any effect on rice yield. The wheat yield from the surface seeding method is often comparable to that obtained with the Chinese seed drill, discussed later (Fig. 2).

Table 2 Effect of Reduced Tillage on Yield and Production Economics of Wheat in Rice–Wheat System, Bhairhawa, Nepal

Establishment methods	Wheat yield (t ha^1)	Plowing cost (US$ ha^1)	Net benefit (US$ ha^1)	Extra days needed to plant[a]
Surface seeding	2.78a[b]	0	425a[b]	0
Chinese seed drill	2.83a	22	448a	8
Farmer's practice	2.31b	85	299b	15

[a] Number of extra days needed for land preparation before seeding compared with the surface seeding.
[b] Numbers followed by the same letter are not significantly different at $P < 0.05$ using DMRT.
Source: Hobbs et al. (1997).

Table 3 Grain Yield (t ha[1]) of Wheat Genotypes Under Surface/Relay Seeding, Bhairhawa, Nepal

Establishment techniques	Wheat genotypes							
	BL 1022	BL1135	Bhrikuti	BL1473	NL683	UP262	Achyuta	Mean
Relay sowing	2.7	3.1	3.4	2.9	2.5	3.2	3.3	3.0 a[a]
Surface seeding	2.2	2.9	2.6	2.3	2.2	2.4	2.3	2.4 b
Farmers' practice	2.0	1.5	2.4	2.2	1.6	1.7	1.9	1.9 c
Relay sowing % age over farmers' practice	35	105	41	29	60	76	77	58
Surface seeding % age over farmers' practice	10	87	9	2	46	41	25	26

[a] Numbers followed by the same letter are not significantly different at $P < 0.05$ using DMRT.
Source: Giri (1997).

The efficiency of fertilizer application is an issue in surface seeding. The average amount of N:P:K fertilizer application by farmers is 75, 16, and 19 kg ha^{-1}, respectively. Surface-seeded wheat produced higher yield when N application was delayed and top-dressed 20 days after germination compared to a basal application or 10 d after sowing (Fig. 3). This beneficial effect of delaying N application may be due to higher efficiency by reducing denitrification losses. Due to

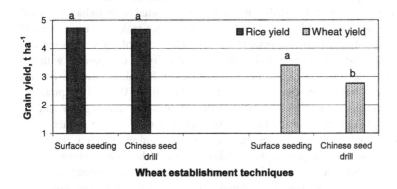

Figure 2 Effect of wheat establishment techniques on rice and wheat yield, Bhairhawa, Nepal. (Adapted from NWRP, Bhairhawa, 2000.)

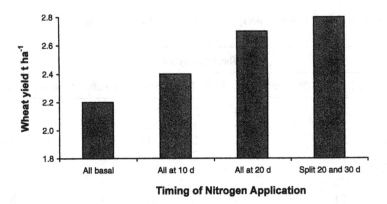

Figure 3   Effect of N application time on yield of surface seeded wheat. (From Hobbs et al., 1997.)

adoption of surface seeding, farmers were able to save about US$38 ha^{-1} in 1999 and $40 in 2000 in land preparation costs, increase production, and improve their livelihoods. Farmers were happy with the introduction of surface/relay seeding technology that was used on fallow land. It even resulted in increased land prices for these low-lying lands. There are about 450,000 ha of rice–fallow land in the country (Giri, 1998) where this technology can be used.

B.   No-Till

Direct drilling of wheat seeds into untilled fields after rice harvest is possible with a no-till seed drill. The drill used in Nepal (imported from India and called a Pantnagar no-till drill) has an inverted-T opener that makes a slit in the soil and places seed and fertilizer simultaneously. This technology was tested on-station and on farmer fields in 1999–2000 and was found comparable with the conventional method of plowing three to four times and planking one to two times when planted on the same day (NWRP, Bhairhawa, 2000). With this implement, farmers could place the seed and fertilizer directly into the standing rice stubble without any land preparation. However, the majority of Nepalese farmers does not possess tractors. Many farmers rent tractors for land preparation; therefore, one way to increase the use of this technology would be to encourage service providers to plant wheat by no-till on a contract basis. This technology saves the cost of land preparation, uses less irrigation water, reduces the turnaround time from rice to wheat, and facilitates timely planting of wheat without sacrificing yield.

C. Minimum Till

Chinese hand tractors (CHT) are becoming popular sources of power as costs of keeping animals for plowing rises. The CHT has many uses as a power source, and one of those is with a seeder attachment that rotovates, seeds, and rolls in one operation. The CHT with this seeder can be used to overcome late sowing of wheat. Seeding with the Chinese seed drill requires higher than normal soil moisture compared to conventionally tilled wheat. The demand for the CHT by farmers is increasing rapidly and its import is on the rise (NARC-CIMMYT, 2001) due to its multiple uses (Table 4). The Chinese seed drill was tested in Ranighat, Birgung, Nepal, and produced a higher yield (19%) and greater marginal return (18%) compared to the farmers' practice (Fig. 4a) (AIRC, 1999). Similar trends were also observed in farmers' fields in the Bara, Parsa, and Rautahat districts. The on-farm experiments with the Chinese seed drill produced 40% higher yield over farmers' practice (Fig. 4b). The poor yield with farmers' practice was due to late planting and uneven seed distribution at different depths since seeds were broadcast sown and then plowed for incorporation.

D. Direct Seeding of Rice

Transplanting rice is a common practice among Nepalese farmers although it is labor-intensive, a drudgery and time-consuming process. To overcome this problem, direct seeding of rice was compared with transplanted rice (Fig. 5). Direct seeding of rice using the Chinese seed drill produced comparable rice yields to those of transplanted rice. There was a heavy infestation of weeds in the direct seeded rice that required manual weeding or herbicide application. However, the direct seeded rice establishment treatment had significantly higher wheat yields

Table 4 Farmers' Perception on Chinese Seed Drill (Minimum Till) at Bara, Parsa, Nepal

Advantages	Disadvantages
Better soil pulverization	Higher investment
Higher yield	Spare parts availability
Low planting cost	Training on its use and maintenance
Weed management at sowing	
Lower irrigation water required	
Good fertilizer incorporation	
Easy for operation	
In-situ residue management	
Reduction in drudgery	

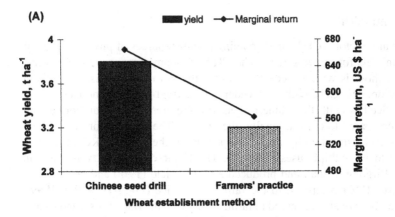

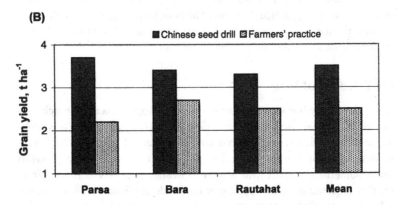

Figure 4 Effect of Chinese seed drill on (A) wheat yield and marginal return at AIRC farm and (B) yield at farmers field at three locations and mean. †Numbers followed by the same letter are not significantly different at $P < 0.05$ using DMRT. (From AIRC, 1999.)

in the next crop compared to the transplanted, puddled rice plot (Fig. 5). The wheat yield in direct seeded rice was 3.27 t ha^{-1} compared to 2.89 t ha^{-1} in the puddled rice field (NWRP, 2000; NARC-CIMMYT, 2001). The lower wheat yield in the transplanted rice field can be attributed to the effect puddling has on soil physical properties. Direct seeded rice reduced the cost of land preparation and the drudgery of the labor of transplanting and puddling soils.

V. FUTURE STRATEGIES

There is a need for policies that provide easier and prompt credit facilities to small and middle farmers, promote import of suitable and affordable farm machi-

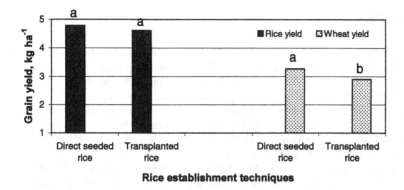

Figure 5 Effect of rice establishment techniques on yield of rice and wheat, Bhairhawa, Nepal. †Numbers followed by the same letter are not significantly different at $P < 0.05$. (From NWRP, Bhairhawa, 2000.)

neries, or encourage local manufacturers with joint ventures to make the equipment locally. Training farmers in the proper use, repair, and maintenance of equipments and machineries is also essential.

VI. CONCLUSIONS

Rice–wheat is a major cropping system of Nepal. In spite of adoption of improved varieties on a large area under rice–wheat in Nepal, the productivity of this system is low, mainly due to late planting. Timely planting of wheat can be achieved by adopting different no- or minimum-till options like surface/relay seeding, no-till, and minimal till with a Chinese seed drill. Surface and relay seeding of wheat are promising technologies for timely sowing of wheat in lower wetlands of Nepal that otherwise are left fallow or planted late. Appropriate soil moisture is a prerequisite for the success of all these technologies. If soil moisture is not enough, presowing irrigation is needed. Under surface/relay seeding, N should be applied 20 and 30 days after seeding (DAS) in equal splits to increase nutrient use efficiency (NUE) and grain yield. This technology reduces land preparation cost without sacrificing yield. Further research is needed on fertilizer management, weed management, and mechanization for the wider dissemination of these technologies. Research must also look at the total system so that these new technologies are adopted by all crops of the system for maximum benefit. Formulating policies to encourage and motivate farmers for the adoption of these new technologies and to locally manufacture equipment is essential for it to be successful.

REFERENCES

AIRC, Ranighat. 1999. Annual Report 1998/1999. Agricultural Implement Research Center, Nepal Agricultural Research Council. Ranighat, Birgung, Nepal.

AIRC, Ranighat. 2000. Annual Report 1999/2000. Agricultural Implement Research Center, Nepal Agricultural Research Council. Ranighat, Birgung, Nepal.

Giri, G.S. 1998. Surface/relay planting: An option for planting wheat on time in the lower wetlands of the Terai. Nepal. *in* Hobbs P.R., Rajbhandari N.R., Eds. Proc USAID Funded Rice–Wheat Research End-of-Project Workshop, Oct. 1–3, 1997, Kathmandu, Nepal. CIMMYT Mexico, 57–62.

Harrington, L.W., Fujisaka, S., Hobbs, P.R., Adhikary, C., Giri, G.S., Cassaday, K. 1993. Rice–Wheat Cropping Systems in Rupandehi District of Nepal Terai: Diagnostic Surveys of Farmers' Practices and Problems, and Needs for Further Research. CIMMYT. Mexico, D.F.

Hobbs, P.R., Giri, G.S., Grace, P. 1997. Reduced and Zero Tillage Options for the Establishment of Wheat After Rice in South Asia. RWC Research Paper Series No. 2. Rice–Wheat Consortium for the Indo-Gangetic Plains and CIMMYT. New Delhi, India.

NARC-CIMMYT. 2001. Three decades of NARC-CIMMYT partnership in maize and wheat research and development (1970–2000). Nepal Agricultural Research Council and International Maize and Wheat Improvement Center. Feb. 12, 2001.

NWRP, Bhairhawa. 2000. Annual Report 1999/2000. National Wheat Research Program, Nepal Agricultural Research Council. Rupandehi, Nepal.

Ortiz-Monasterio, I., Dhillon, S.S., Fischer, R.A. 1994. Date of sowing effects on grain yield and yield components of irrigated spring wheat cultivars and relationships with radiation and temperature in Ludhiana, India. *Field Crops Res.* 37:169–184.

Pandey, S.P., Pande, S., Johansen, C., Virmani, S.M. 2001. Rice–Wheat Cropping Systems of Nepal. Rice–Wheat Consortium Paper Series No. 12. Rice–Wheat Consortium for the Indo-Gangetic Plains and International Crops Research Institute for the Semi-Arid Tropics. New Delhi, India.

RARS, Parwanipur. 2000. Annual Report 1999/2000. Regional Agricultural Research Station, Nepal Agricultural Research Council. Bara, Nepal.

Saunders, D.A. 1990. Crop Management Research: Summary of Results. Wheat Research Center Monograph 5. WRC. Nashipur, Bangladesh.

12

Impact of No-Tillage Farming on Wheat Production and Resource Conservation in the Rice–Wheat Zone of Punjab, Pakistan

Muhammad Azeem Khan and N. I. Hashmi
National Agricultural Research Centre (NARC)
Islamabad, Pakistan

I. INTRODUCTION

The rice (*Oryza sativa L.*)–wheat (*Triticum aestivum L.*) system is one of the important cropping systems of Punjab, Pakistan. Both of these crops contribute significantly toward foreign exchange earnings and domestic food consumption. Diagnostic studies of this system conducted by Byerlee et al. (1984) and Sheikh et al. (2000) estimated a big yield gap between the potential and actual wheat yields realized by the farmers. Land preparation and planting practices of the rice crop conflicts with the following wheat crop and are determined to be the major yield-limiting factors.

Flooded and puddled soil (plowed in standing water) required by rice as compared to well-drained conditions deemed necessary for wheat (Hobbs, 1985) is one such example of a conflicting phenomenon. Late rice harvest, poor soil structure, and plant residues create difficulties for preparing a good seedbed for the wheat crop (Byerlee et al., 1984). This results in late planting of wheat, in which farmers usually resort to the broadcast method for wheat sowing. Rainfall at the time of land preparation further delays planting wheat by 2 to 3 weeks (Aslam et al., 1993a). Randhawa (1979) and Hobbs et al. (1988) have reported that a delay in planting wheat after mid-November caused a loss in grain yield at the rate of 1%/day/ha. Generally, "no-tillage" technology has multiple benefits, but in the rice–wheat system of Punjab, the technology was evaluated especially for timely cultivation of wheat after rice. *Basmati* rice, the major type of rice grown before wheat in Punjab, is a late-maturing crop, and conventional tillage takes a greater amount of time for seedbed preparation, whereas most of the wheat is planted late. No-tillage technology facilitates direct drilling of wheat in rice-harvested fields and also uses the residual moisture for wheat germination (Sheikh et al., 2000).

To solve the problem of stagnating wheat yield, it is essential to adopt an integrated approach that ameliorates late planting, reduces weeds, and improves fertilizer and water-use efficiency (Malik and Singh, 1995; Malik, 1996; Hobbs et al., 1997). It has been suggested that no-tillage technology reduces the *Phalaris minor* infestation and also enables timely seeding of the wheat crop (Hobbs et al., 1997). Sowing wheat with a no-tillage-cum-fertilizer drill offers an integrated solution to the problem and is appropriate in wheat fields that are hand-harvested and have anchored paddy stubble (Mehla et al., 2000).

II. NO-TILLAGE TECHNOLOGY EVOLUTION

In Pakistan, sowing wheat with the no-tillage drill has undergone various experimental and developmental phases since the mid-1980s. A collaborative study conducted by the wheat program and social sciences group of NARC in collaboration with CIMMYT Pakistan (Byerlee et al., 1984) highlighted for the first time the need for resolving the prevailing poor seedbed preparations and late planting for the wheat crop. As a result, testing an imported no-tillage wheat drill from New Zealand was recommended as a way to compare wheat tillage practices and to address the conflicts of planting wheat after rice. Results of on-farm experiments conducted on different components of no-tillage technologies from 1984 to 1989 were synthesized by Aslam et al. (1989). The results showed wheat crops established with no-tillage increased yields 10% to 40% under different land resource and no-tillage regimes. Ayub Agricultural Research Institute (AARI) agronomists under the national coordinated wheat program of NARC also conducted no-tillage wheat sowing trials in farmer fields in the Shahkot target site and further validated the promising nature of the no-tillage technologies (Nayyar et al., 1993). A comprehensive no-tillage pilot production program was then designed to expand the usage of the no-tillage drill in the rice–wheat zone of Punjab (Aslam et al., 1993a). This productivity enhancement project, launched by PARC, provided the required financial resources to launch this pilot production phase for no-tillage technologies. Social scientists at NARC conducted a socio-economic survey to highlight the problems and prospects of popularization expanded use of no-tillage wheat sowing. They found that the spread of no-tillage wheat sowing would be limited unless local manufacturers started producing operationally flexible and quality drills. The adaptation of the imported New Zealand drill by local manufacturers since 1997–1998 made it more compatible and financially affordable for farmers. The On-Farm Water Management (OFWM) department of Punjab initiated drill manufacturing (by involving local manufacturers) at a mass scale and vigorously promoted its use among farmers. Thirty thousand hectares of wheat was sown with the new, locally produced no-tillage drill during 2000–2001, and 80,000 ha in 2001–2002. The wide adoption and popularity of this technology were due to promising features like (1) timely sowing, (2) reduction in sowing cost, (3) ease in operation on hard-soil and low-lying fields, (4) increased fertilizer and water-use efficiency, and (5) considerable improvements in yields and profit.

III. YIELD COMPARISONS ON ZERO- AND CONVENTIONAL TILLAGE

Table 1 summarizes and combines the information on grain yield for no-tillage and conventional tillage over four years of experimentation conducted by the wheat program and CIMMYT, and two years by the agronomist at AARI, Faisalabad. During the initial years, farmers' tillage was equal to no-tillage. Inexperience in managing seed depth, fertilizer application, and the planting of both systems on the same day with the new drill imported from New Zealand were the main causes for no differences early on. Eventually, drill performance improved, no-tillage was given the advantage of earlier planting, and wheat yields under no-tillage were significantly higher (20%) during the years from 1986 to 1988.

Results of survey studies conducted during the pilot production phase (1991–1992), postproduction phase (1995–1996), and full extension phase (Table 2) confirmed that no-tillage technology performance continuously improved over

Table 1 On-Farm No-Tillage Trial Results

| Year | Wheat yields (ton/ha) | | | |
	No-tillage	Farmer practice	Significance	Difference
1984–1985	3.2	3.3	5%	(−) 3%
1985–1986			n.s.	
	3.6	3.5	—	(+) 3%
1986–1987	3.8	3.5	2%	(+) 9%
1987–1988	4.3	3.6	1%	(+) 19%
1988–1990	3.2	2.7	1%	(1) 20%

Sources: Aslam et al. (1989); Nayyar et al. (1993).

Table 2   Survey Results on No-Tillage Wheat Adoption

	Wheat yields (ton/ha)		
Survey year	No-tillage	Conventional practice	Difference
1991–1992	2.6	2.3	(+) 13%
1995–1996	2.8	2.4	(+) 16%
2000–2001	3.3	2.8	(+) 18%

Sources: Sheikh et al. (1994); Sheikh (1998); Khan et al. (2001).

time due to improvements in drill functioning and experience in its proper use gained by the farming community. Other important factors contributing to higher yields were the application of fertilizer and seed at the appropriate depth resulting in good germination and better wheat stand, and earlier planting of the crop.

IV.   MAJOR IMPACTS OF NO-TILLAGE TECHNOLOGY

A.   Improvements in Planting Time

Traditionally, in the rice–wheat system of Punjab, farmers plant wheat by two methods: *wad-watter* (wheat sowing under residual moisture after rice harvesting) and *rauni* (wheat sowing after applying irrigation following rice harvesting). Nearly 65% of the wheat area is planted by the "wad-watter" method where farmers disc harrow three to four times and use a cultivator another three times. In the "rauni" method, farmers cultivate with a 9-tyne cultivator seven times then cultivate two more times following irrigation. These traditional wheat-planting methods are costly and time-consuming, whereas the no-tillage drill is less time-consuming, is less expensive, and reduces irrigation. No-tillage drill owners and rental providers now plant 75% and 47% of their wheat area with no-tillage, respectively (Table 3).

Table 3   Area Allocation to Wheat Crop Under Different Planting Methods

	Drill owners	Rental users	Conventional
Methods	Percent of own wheat area		
No-tillage	74.5	46.6	—
Wad-watter	12.7	28.5	66.2
Rauni	12.8	24.9	33.8

Source: Khan et al. (2001).

Table 4 Timeliness in Wheat Planting by Conventional and No-Tillage Methods

Planting times	No-tillage adopted farms			Conventional tillage farms	
	No-tillage	Wad-watter	Rauni	Wad-watter	Rauni
Before Nov. 15	37%	35%	25%	43%	36%
During Nov. 15–30	44%	43%	55%	35%	32%
After Dec. 1	19%	22%	20%	22%	32%

Source: Khan et al. (2001).

On no-tillage adopted farms, a slight shift (3%) has been observed from late to early wheat sowing (Table 4). This shows that the additional benefits of planting wheat early with a no-tillage drill are less clear to the farmers as compared to its cost-saving advantages. Moreover, the no-tillage drill is largely (75% area) used after late-maturing fine rice varieties such as Super Basmati. However, the wad-watter method is used more (50%) after early-maturing fine rice varieties such as B-385 and B-386. The results further show earlier wheat sowing with wad-watter methods than with the no-tillage drill. In the rauni method, there was a 10% shift from December to November 30. These slight changes in timely wheat planting are encouraging; However, they need special attention during future drill promotional efforts. The potential of using no-tillage technology to help resolve rice–wheat planting conflicts still needs more demonstration. This advantage of timing will be more evident in the future when drill owners develop skills in the drill's operation.

B. Drill Performance on Different Soil Types

Variation in soil texture has major implications for rice–wheat crop stand establishment. Imperfectly drained clay and silty-clay soils are quite suitable for rice but less suitable for the production of wheat. These soils are difficult to plow and to establish a good seedbed for wheat planting. The distribution of different soil types in the rice–wheat zone is presented in Table 5. Clay and clay-loam textures are the major soil types of the area. Saline-sodic soils with high pH are also present but are less suitable for cultivation. These soil types react differently with no-tillage and conventional wheat-planting methods.

Wheat yield comparisons by tillage method on different soil types are found in Table 6. Results show that no-tillage was better than wad-watter on clay-loam, clay (rohi), and sandy-loam soils. On clay (chamb) and silt-loam soils the difference was not significant. These are preliminary results that show important

Table 5 Classification of Soil Types in the Field Surveyed

Main soil types	Local terms	% area
Clay	Rohi, Chamb	48
Clay-loam	Mera, Bhari Mera	38
Silt-loam	Halki Mera	8
Sandy-loam	Ratli	2
Kalrathi/saline	Kalar, Rohi	4

Sources: Byerlee et al. (1984); Khan et al. (2001).

relationships between planting method and soil types. More work is needed to validate such relationships and to develop specific recommendations.

C. Impact on Wheat Production

A 0.4-ton yield increase per ha (3.2 ton/ha from no-tillage compared to 2.8 ton/ha from conventional tillage) was estimated during a recent survey of no-tillage technologies in 2002 (Table 7). This increase resulted in 3,200 tons of additional wheat worth PRs. 24 million ($0.42 million) from the 80,000 ha of no-tillage wheat in 2001–2002.

D. Resource Conservation for Minimal Tillage Practices

Approximately eight different sequential land preparation operations are required in the conventional tillage farms, which cost PRs. 2525/ha ($43) compared to just

Table 6 Wheat Yield by Soil Type and Planting Methods

Soil types	Yield (ton/ha)		Difference
	No-tillage	Wad-watter	
Silt loam	3.2	3.1	(+) 3%
Clay loam	4.2	2.8	(+) 50%
Clay (chamb)	2.6	2.7	(−) 4%
Kalrathi (saline)	3.1	2.7	(+) 15%
Clay (Rohi)	3.5	2.9	(+) 21%
Sandy loam	2.3	1.6	(+) 43%
Overall	3.2	2.8	(+) 14%

Source: Khan et al. (2001).

Table 7 Increased Wheat Productions from Zero-Tillage Wheat Technology

Particulars	No-Tillage	Wad-watter
Average yield (ton/ha)	3.2 a	2.8 b
Yield increase over conventional tillage practices	0.4 ton/ha	
Yield increase for 80,000 ha no-tillage wheat	3,200 tons	
Additional income from higher no-tillage wheat yield	Rs. 24 million ($ 0.41 million)	

Source: Khan et al. (2001).

PRs. 350/ha ($6) for a single no-tillage drill operation (Table 8). This reduction in land preparation cost saved PRs. 174 million ($2.95 million) from the 80,000 ha of no-tillage wheat grown by farmers. It also saved about 3.6 million liters of diesel oil and PRs. 21 million ($0.36 million) reduction in depreciation costs for tractors. In addition to such considerable financial benefits to the farming community, there were environmental benefits in terms of reduction in air pollution (less burning of residues and diesel fuel) and improvements in the soil physical and biological properties (Table 8).

E. Irrigation Resource Conservation

In no-tillage, the reduction in frequency and duration of irrigation not only reduces irrigation cost (PRs. 700/ha or $12/ha), but also has benefits in terms of water

Table 8 Land Preparation and Energy Use Comparisons with Alternative Tillage Practices

Land preparation practices	No-tillage		Conventional tillage	
	No.	Cost (Rs/ha)	No.	Cost[a] (Rs/ha)
Cultivators/drill	1	350	4	1400
Disc harrow	0	0	2	875
Planking	0	0	2	250
Total cost	—	350	—	2525
Savings per ha				PRs. 2,175 ($36)
Savings for 80,000 ha wheat under no-tillage				PRs. 174 million ($2.95 million)
Savings in terms of diesel fuel/energy				3.6 million liters
Savings in the depreciation of tractors				PRs. 21 million ($0.36 million)

[a] 1 US PRs. 59 open market rate.

Table 9 Irrigation Resource Use Comparisons with Different Tillage Practices

Irrigation	No-tillage	Conventional tillage
Total irrigation (no)	3	4
Irrigation in hours (per ha)	19	26
Irrigation cost (Rs/ha)	1875	2625
Savings in irrigation cost per ha		PRs. 750 ($13)
Savings for 80,000 ha wheat under no-tillage		PRs. 60 million ($1.01 million)
Savings in terms of irrigation water		0.6 million ac inch
Savings in terms of diesel fuel/energy		1.4 million liters
Depreciation cost of tubewell		PRs. 3.4 million

conservation (0.6 million ac-inch), using less diesel fuel required for lifting water (1.4 million liters), and reduces the depreciation of tubewells (PRs 3.4 millions) (Table 9). The present prolonged drought condition in Pakistan and the depletion of reservoir and underground water further strengthen the case for promoting no-tillage.

F. Other Resource-Saving Advantages

Other potential benefits of no-tillage technologies are listed in Table 10. A reduction in water use and tillage practices is estimated to save as much as 0.1 million

Table 10 Other Resource-Saving Advantages of No-Tillage over Other Tillage Practices

Resource savings	No-tillage	Conventional tillage
Labor use for extra tillage irrigations (hours/ha)	20	30
Savings in labor for 80,000 ha no-tillage wheat		0.1 million man days
Savings in the length of wheat sowing time		One month to two weeks
Environmental pollution control by less land preparation		Need estimation
Improvements in soil structure and organic matter		Need estimation
Reduction in soil deterioration by less use of unfit water		Need estimation
Improved environment by reduction in the burning of rice stubbles		Need estimation
Improvements in soil fauna/flora by controlling rice stubble burning		Need estimation

man-days. There is a need to evaluate the added benefits of no-tillage in terms of reducing environmental pollution, improving soil structure and organic matter, improving soil where water of poor quality is used, and improving soil fauna and microbial diversity by better management of rice stubbles other than burning.

V. CONCLUSIONS AND RECOMMENDATIONS

No-tillage technologies have shown success in solving late wheat-planting problems, reducing cost of production, increasing productivity, reducing fuel consumption, reducing the depreciation of farm machinery, reducing water use, and improving environmental indicators. Some remaining problems in drill operation and manufacturing, if solved adequately, could further foster the adoption of this promising technology and make substantial impacts. Further improvements that need research in no-tillage wheat technologies include (1) assessing the effects of no-tillage on soil physical and biological properties, (2) developing fertility management strategies with no-tillage under different soil types, (3) identifying irrigation schedule before and after no-tillage wheat sowing, (4) further improving the operational and quality aspects of the drill, (5) selecting better wheat varieties for no-tillage wheat sown after different-duration rice varieties, (6) studying environmental and soil fertility losses due to burning of rice stubbles, (7) using participatory methodologies and training on critical aspects of no-tillage operation, (8) determining compatibility of alternative rice-planting technologies (puddling versus non-puddling) to no-tillage wheat sown in relation to different soil types, and (9) training farmers and service providers in the proper use of the machinery.

REFERENCES

Aslam, M., Majeed, A., Hashmi, N.I., Hobbs, P.R. 1993a. Improving wheat yield in the rice–wheat cropping system of the Punjab through zero tillage. *Pak. J. Agric. Res* 14(1):8–11.

Aslam, M., Hashmi, N.I., Majid, A., Hobbs, P.R. 1993b. Improving wheat yield in the rice–wheat cropping system of the Punjab through fertilizer management. *Pak. J. Agric. Res* 14(1):1–7.

Aslam, M., Majid, A., Hobbs, P.R., Hashmi, N.I., Byerlee, D. 1989. Wheat in the Rice–Wheat Cropping System of the Punjab: A Synthesis of On-Farm Research Results *1984–1988*. PARC/CIMMYT Mimeograph 89–3. Islamabad. Pakistan.

Byerlee, D., Sheikh, A., Aslam, M., Hobbs, R. 1984. Wheat in the Rice-based Farming System of the Punjab: Implications for Research and Extension. Islamabad. PakistanPARC/CIMMYT Mimeograph.

Hobbs, P.R. 1985. Agronomic practices and problems for wheat following cotton and rice. *In* Villareal R.L., Klatt A.R., Eds. Wheat for More Tropical Environments. Proc. of the Int. Sym. MexicoD.F. CIMMYT, 273–277.

Hobbs, P.R., Mann, C.E., Butler, L. 1988. A perspective on research needs for rice wheat rotation. *In* Klatt A.R., Ed. Wheat Production Constraints in Tropical Environments, Mexico, D.F., CIMMYT, 197–211.

Hobbs, P.R., Giri, G.S., Grace, P. 1997. Reduced and Zero-Tillage Options for the Establishment of Wheat After Rice in South Asia. RWC Paper No. 2. Mexico. DF. Rice–Wheat Consortium for the Indo-Gangetic Plains and CIMMYT.

Khan, M.A., Anwar, M.Z., Aslam, M., Hashmi, N.I., Hobbs, P.R. 2001. Impact of Zero-Tillage Wheat Sowing Technology in the Rice–Wheat Farming System of the Punjab. PARC & Rice–Wheat Consortium Paper No. 2001–1, Agricultural Economic Research Unit, SSI, NARC. Islamabad. Pakistan.

Malik, R.K. 1996. Herbicide resistant weed problems in developing world and methods to overcome them. Proc. 2nd Intl. Weed Sci. Cong., Copenhagen, Denmark. Rome, Italy: Food and Agricultural Organization, 665–673.

Malik, R.K., Singh, S. 1995. Littelseed canarygrass resistance to isoproturon in India. *Weed Tech* 9:419–425.

Mehla, R.S., Verma, J.K., Gupta, R.K., Hobbs, P.R. 2000. Stagnation in the Productivity of Wheat in the Indo-Gangetic Plains: Zero-Tillage Seed-Cum-Fertilizer Drill as an Integrated Solution. Rice-Wheat Consortium Paper Series 8. Rice–Wheat consortium for the Indo-Gangetic Plains, IARI campus, Pusa. New Delhi. India.

Nayyar, M.M., Arshad, M., Ali, A. 1993. Brief on no-tillage technology for wheat and cotton and direct seeding method for rice crop. Ayub Agricultural Research Institute. Faisalabad. Pakistan.

Randhawa, A.S., Jolly, R.S., Dhillon, S.S. 1979. Effect of seed rate and row spacing on the Yield of dwarf wheat under different sowing dates. *Field Crop Abst* 32(2): 87–96.

Sheikh, A.D. 1998. Economic and non-economic factors influencing the adoption of no-tillage technologies in the rice-wheat and cotton-wheat areas of the Pakistan's Punjab. Ph.D. thesis, Department of Agriculture, University of Reading. UK.

Sheikh, A.D., Zubair, M., Asif, M. 2000. Barriers in Enhancing the Productivity of the Rice–Wheat Farming System of the Punjab. AERU, NARC, Technical Report 2000.

13

Impact of No-Till Farming on the Rice–Wheat Systems in Bangladesh

N. I. Bhuiyan and M. A. Saleque
Bangladesh Rice Research Institute
Gazipur, Bangladesh

I. INTRODUCTION

[*]Rice (*Oryza sativa L.*) and wheat (*Triticum aestivum L.*) are the world's two most important cereal crops, contributing 45% of the digestible energy and 30% of total protein in the human diet (Evans, 1993). Traditionally, Bangladesh is a rice-growing country, where rice covers about 75%–77% and wheat covers about 5%–6% of the total cropped area of about 13.9 Mha (Bhuiyan et al., 2002). Rice and wheat together contribute to 100% of total cereal production and 94% of the national calorie intake (Timsina and Connor, 2001). From 1960 to 1990, the average annual growth of wheat area in Bangladesh was 8.6% (Hobbs and Morris, 1996). The area of wheat in Bangladesh increased rapidly since the mid-1980s to around 0.83 Mha in 2001 (FAO, 2002). Farmers prefer to grow *boro*[†] rice (versus wheat) if irrigation is available. Wheat and other nonrice crops are mostly

---

[*] Paper presented in the workshop on "No-till farming in South Asia's Rice–Wheat System: Rice–Wheat Consortium and USA experiences;" Columbus, Ohio, Feb. 20–21, 2002.

[†] *Boro* rice is a traditional rice crop of the eastern Gangetic Plains of South Asia that is grown in the dry winter months. The seedbeds are usually sown from mid-October through mid-December, and the seedlings transplanted into the main field in late January to early February when the average temperature rises above 12°C. The crop is harvested in May to June. It is one of the highest yielding rice crops in the region because of the cooler temperatures and high radiation receipts.

grown where irrigation water is limited and where *boro* cannot be grown (Morris et al., 1997).

The traditional technique of rice cultivation involves puddling the soil to form a saturated root zone and the following wheat is grown after dry land preparation by sufficient plowing. The salient feature of the rice–wheat system is conversion of aerobic soil to anaerobic during rice and back to aerobic during wheat. Puddling breaks capillary pores, reduces the void ratio, destroys soil aggregates, disperses the fine clay particles, and lowers soil strength in the puddle layer (Sharma and De Datta, 1986). The destruction of soil aggregates by puddling leads to the formation of surface crusts, which crack on drying and delay land preparation for the following wheat crop. If the crust is broken by tillage, large clods result, leading to poor contact with seeds, thereby restricting germination. To overcome these disadvantages of puddling in rice and tillage in wheat, alternative practices such as zero- or minimum tillage have recently been researched. The use of mechanized equipment, in particular the Chinese seed drill, is being tested in Bangladesh, which favors rapid seeding after rice. The Chinese hand tractor (CHT) seeder was found not only to increase wheat yield over conventional tillage, but also to save the cost of land preparation (Nur-E-Elahi et al., 2001). This paper reviews the impact of zero- or minimum tillage on yield, soil physical properties, and economics in Bangladesh.

II. CROP YIELD

Tillage seems to have no direct impact on the yield of a crop provided the soil is weedfree, is soft enough for young root growth, has optimum air for root respiration, and has enough moisture in the root zone for root uptake. These conditions are usually created by tillage operations, but conventional tillage (CT) involves costs and can delay the planting time under some situations. Researchers and some innovative farmers have developed no-tillage (NT) and minimum tillage (MT) practices as alternatives to CT.

There are contradictory reports on the benefits of NT and MT operations on the yield of rice and wheat. In Pakistan, Arshad et al. (1991) obtained 20% more wheat yield under NT than with CT. Aslam et al. (1993b) also reported that the wheat yield was increased under NT compared to CT. The magnitude of increase in their studies was about 10%, when both NT and CT plots were sown at the same time. In the same trials, when NT was sown earlier, soon after rice harvesting (10 to 44 days earlier than with CT), the yield increase was on average 41% higher. Aggarwal et al. (1997) reported greater wheat yield with the chisel and deep tillage system than with MT. Bajpai and Tripathi (2000) also reported that the CT of wheat produced 25% greater grain yield than that of NT. These investigators may not have practiced the appropriate method of NT or MT

Table 1 Effect of Tillage on Grain Yield of Wheat

	Grain yield (t ha^{-1})		
Tillage	1993	1994	1995
Minimum (1 furrow)	2.66	2.18	3.48 b[a]
Conventional (5–7 plow, 7–8 cm deep)	2.32	2.32	3.48 b
Deep (chiseling, 20–22 cm deep)	2.66	2.67	3.90 a
LSD$_{0.05}$	NS	NS	0.29

[a] Means followed by the same letter are not significantly different at 0.05 using LSD.
Source: Bangladesh Institute of Nuclear Agriculture (BINA), Mymensingh, 1994–1995 (1995).

and planted wheat manually with grooving tools, which allowed the soil to lose moisture. They also planted on the same day, further reducing the benefits of NT.

In Bangladesh, tillage research started in the late 1980s. Abedin and Ahmed (1986) tested tillage effects on wheat in different parts of the country. They tested four tillage practices—(1) NT (includes seeding after rice harvest without any tillage or relay cropping where wheat was broadcast into the Aman rice 15 days before harvest), (2) MT (furrow opened manually and seed placed and covered), (3) reduced tillage (only one plowing with a country plow), and (4) CT (five plowings). All four tillage systems gave similar yields under rainfed conditions at Kushtia and Thakurgaon. NT gave the highest yield in Tangail and CT the highest yield at Jamalpur.

In another study at the Bangladesh Institute of Nuclear Agriculture (BINA, 1992–1995), MT (opening a furrow), CT (7–8 cm deep), and DT (20–22 cm deep) treatments were compared for three years on Chhiata loam soils. The yield of wheat under MT, CT, and DT was similar in the first and second years, but in the third year the DT gave a better yield than the other two tillage practices (Table 1). Khan (1996) tested MT, CT, and DT for rice (main plot), and then each of the rice tillage plots were divided into three subplots to accommodate MT, CT, and DT for wheat on a silty-clay Chhiata soil and a silty-clay-loam Darsana silt-loam soil. The rice and wheat yields were lower with MT than the CT and DT at both locations (Tables 2 and 3) probably for the same reason as above—that opening a furrow manually and planting on the same day do not favor MT. The proper drill and soil moisture are very important. Weeds may have also been a problem, although not mentioned in the paper.

III. TIME OF SOWING

Adoption of NT favors timely sowing of wheat, which is very important to achieve high yield in South Asian countries like India, Pakistan, Bangladesh, and Nepal.

Table 2 Effect of Alternate Tillage Practices on the Grain Yield of Rice and Wheat on Chhiata Silty Clay Loam Soil, Gazipur

Tillage[a]		Rice yield (t ha^{-1})[b]			Wheat yield (t ha^{-1})[b]		
Main plot	Subplot	1990–1991	1991–1992	1992–1993	1990–1991	1991–1992	1992–1993
Minimum	Minimum	4.28 b	3.88 f	3.84 g	1.88 g	1.80 f	2.04 f
	Conventional		4.60 de	4.70 e	2.09 g	2.15 e	2.31 e
	Deep		5.35c	5.83 c	3.02 bc	3.02 bc	3.24 b
Conventional	Minimum		4.23 ef	4.36 f	2.22 fg	2.26 de	2.38 e
	Conventional	4.76 ab	5.19 c	5.35 d	2.56 de	2.42 d	2.58 d
	Deep		6.03 b	6.25 b	3.22 ab	3.37 a	3.50 a
Deep	Minimum		4.88 cd	4.58 ef	2.45 ef	2.38 de	2.55 d
	Conventional		6.65 a	6.08 b	2.78 cd	2.98 c	3.04 c
	Deep	5.04 a	6.63 a	6.50 a	3.35 a	3.29 ab	3.57 a
LSD$_{0.05}$		0.59	0.43	0.19	0.32	0.23	0.16

[a] Main plot for first-year rice, subplot for second and third-year rice and first-, second-, and third-year wheat. *Minimum*: The seeds were sown in the furrow made by manually pulled country plow. The soils between the rows were left undisturbed. *Conventional*: Country plow was used for land preparation followed by laddering. The plowing depths were 7.5–9.5 cm. *Deep tillage*: Plowing was done by chiseling followed by laddering, spading, and planking. The plowing depths were 20–22 cm.
[b] Means followed by the same letter are not significantly different at 0.05 using LSD.
Source: Khan (1996).

Table 3 Effect of Alternate Tillage Practices on the Grain Yield of Rice and Wheat on Darsana Silt Loam Soil, Jessore

Tillage[a]		Rice yield (t ha^{-1})[b]			Wheat yield (t ha^{-1})[b]		
Main plot	Subplot	1990–1991	1991–1992	1992–1993	1990–1991	1991–1992	1992–1993
Minimum	Minimum	4.92 b	4.16 f	4.62 f	2.14 e	2.31 e	2.52 e
	Conventional		4.58 e	5.31 e	2.53 cd	2.46 de	2.98 c
	Deep		5.24 c	6.42 c	3.02 b	2.92 bc	3.39 b
Conventional	Minimum		4.83 de	5.29 ef	2.32 d	2.50 de	2.70 d
	Conventional	5.17 ab	5.00 c	6.02 cd	2.63 c	2.68 cd	3.33 b
	Deep		6.13 b	7.42 ab	3.26 ab	3.12 ab	3.93 a
Deep	Minimum		4.98 cd	5.52 de	2.60 cd	2.61 cd	2.99 c
	Conventional		6.02 b	6.98 bc	2.98 b	3.10 ab	3.42 b
	Deep	5.75 a	6.82 a	7.73 a	3.50 a	3.39 a	4.02 a
LSD$_{0.05}$		0.58	0.33	0.67	0.29	0.35	0.12

[a] Main plot for first-year rice, subplot for second and third-year rice and first-, second-, and third-year wheat. *Minimum*: The seeds were sown in the furrow made by manually pulled country plow. The soils between the rows were left undisturbed. *Conventional*: Country plow was used for land preparation followed by laddering. The plowing depths were 7.5–9.5 cm. *Deep tillage*: Plowing was done by chiseling followed by laddering, spading, and planking. The plowing depths were 20–22 cm.
[b] Means followed by the same letter are not significantly different at 0.05 using LSD.
Source: Khan (1996).

Excess moisture after rice harvest, resulting from shallow water tables and rain in late October, often delays land opening for wheat sowing and preparing a good seedbed by CT. Farmers sometimes need to wait for a month to have congenial soil moisture content, resulting in wheat sowing after November as well as a decline in wheat yield. Aslam et al. (1993a) reported that by sowing wheat after November 15, wheat yield decreased by 30 kg ha^{-1} per day in Pakistan. In Bangladesh, an experiment was conducted with three wheat varieties, which were sown from November 15 to January 5 with 10-d intervals (BARI, 1999). The results show that the yield of wheat decreased when sown after November 25, and the magnitude of the yield decrease was 44 to 48 kg ha^{-1} d^{-1} (Fig. 1). Temperatures in South Asian countries fall rapidly after November. Because of low temperatures, wheat germination is delayed and poor crop stands can result.

Badaruddin and Razzaque (1995) showed that NT in rice saved about two weeks' time without decreasing yield (Table 4). NT and MTT practices help with timely seeding by reducing turnaround time in wheat. In Bangladesh, the Chinese seed drill, mounted on a two-wheel hand tractor, is used for MT. The machine can be used on land that is too wet to operate CT equipment. Seeds are sown in one operation: The soil is rototilled, seeds are placed in six rows and the soil is

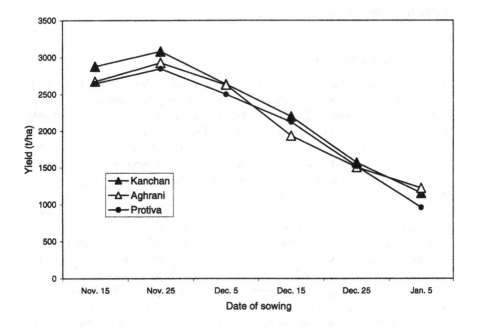

Figure 1   Effect of date of sowing on the yield of three wheat cultivars.

Table 4 Turnaround Time, Cost of Land Preparation, and Grain Yields of *T. Aus* and
T. Aman

	Aus			Aman		
Tillage	Turnaround time (d)	Cost (Tk/ha)	Yield (t/ha)	Turnaround time (d)	Cost (Tk/ha)	Yield (t/ha)
Zero	1	—	3.15	1	—	4.33
Minimum (1 plow)	2	218	3.06	2	319	3.86
High (5 plow)	3	945	3.14	4	698	4.06
Farmers' practice (7 plow)	15	1204	3.05	12	780	3.09

Source: Badarudin and Razzague (1995).

compacted by a roller. Research will study the advantages and disadvantages of this drill at the Wheat Research Center, Dinajpur, Bangladesh. However, preliminary observations showed encouraging results. BARI (1999) reported survey results showing the Chinese seeder ensured wheat sowing by the first week of December, saved seed and land preparation costs, and had a better yield compared to the traditional practice.

IV. WEED BIOMASS

Weed infestation in rice and wheat decreases yield of these crops. In general, rice yield decreases exponentially with an increase in weed biomass recorded at the harvest of the crop. Aslam et al. (1993b) claimed that there were fewer weeds under NT compared to those under CT. In their experiment, wheat was sown earlier (10–44 d) in the case of NT than in CT. The early sown wheat seedlings would have covered the soil surface before the winter weed seeds germinated and lowered the weed biomass by competition. Also, fewer weeds germinated because there was less soil disturbance.

Gajri et al. (1999) reported that the mean weed biomass at 35–40 days after transplanting under NT was 1.6 t ha^{-1} compared to 0.5 t ha^{-1} under CT in a rice crop. Weeds are a major constraint for nonpuddled rice, and solutions will be needed if this system is to be successful in the rice phase. Gajri also showed that the untilled portion of land that resulted from using strip tillage in wheat favored weed growth in the subsequent rice crop. Chisel and deep tillage systems reduced weed biomass in wheat and eventually increased yield since weed seeds were buried and did not germinate in the next crop (Aggarwal et al., 1997).

V. ROOT GROWTH

Root growth on clay-loam and silty-clay-loam soils was studied under the rice–wheat system in Bangladesh (Khan et al., 1997). They compared three tillage practices—MT, CT, and DT—under both rice and wheat crops. Root growth at 0–30-cm depth was similar for all three tillage practices. Aggarwal et al. (1997) reported lower root growth in wheat under MT than under CT, but this may be the result of improper equipment and soil moisture as mentioned above. Bajpai and Tripathi (2000) reported that puddling in rice enhanced root length density by 12% in rice but decreased root length density in wheat by 28%.

VI. SOIL PHYSICAL PROPERTIES

Tillage influences soil bulk density, soil strength, and percolation and infiltration rates. The effect of puddling on soil physical properties was more pronounced than that of dryland tillage, because puddling breaks capillary pores, reduces void ratio, destroys soil aggregates, disperses fine clay particles, and lowers soil strength in the puddle layer (Sharma and De Datta, 1986). Puddling in wetlands destroys soil aggregates and tends to form a massive structure when the soil dries. The effect of puddling on soil bulk density is pronounced. A compacted layer at a depth of 15 to 20 cm was also found to form after several seasons of continuous puddling (Aggarwal et al., 1995). Bajpai and Tripathi (2000) observed that the bulk density at 0 to 6 cm was reduced by puddling at the tillering stage of rice, but after harvest, the bulk density in the puddled plot was significantly greater than in the nonpuddled plot.

The carryover effect of tillage on soil physical properties was studied at BINA (1993–1994). The study revealed that the MT, CT, and DT treatments in wheat left no significant impact on soil bulk density, porosity, or water retention capacity after the rice crop (Table 5). However, Khan (1996) observed greater bulk density (1.49 g cm^{-3}) in the MT plot than that obtained with the CT and DT plots in case of wheat and rice. In addition, the bulk density increased with the increase in soil depth. Khan (1996) also observed that as the crop advances, the soil bulk density in wheat decreases while in rice it increases.

NT (surface seeding) tended to decrease soil strength compared to CT and DT. BARI (1998) measured soil strength after 6 days of irrigation in the wheat crop and reported that soil strength under NT was 3.43 mPa compared to 3.87 mPa with CT and 3.82 mPa with DT at 0–10-cm soil depth (Table 6). However, at 10–20-cm soil depth, where the plow pan is expected to exist, there was no difference in soil strength among the tillage practices. At the 20–40-cm and 40–60-cm soil depth, the soil strength was consistently lower under NT than CT and DT.

Table 5 Carryover Effect of Tillage on Soil Physical Properties Determined After Rice Crop (BINA 1994–1995)

Tillage	Soil depth (cm)		
	0–10	10–20	20–30
Bulk density (g cm^{-3})			
Minimum (1 furrow)	1.37	1.39	1.46
Conventional (5–7 plow, 7–8 cm deep)	1.30	1.36	1.36
Deep (chiseling, 20–22 cm deep)	1.41	1.41	1.43
LSD$_{0.05}$	ns	ns	ns
Porosity (%)			
Minimum (1 furrow)	48	48	45
Conventional (5–7 plow, 7–8 cm deep)	51	49	49
Deep (chiseling, 20–22 cm deep)	47	47	46
LSD$_{0.05}$	ns	ns	ns
Water retention capacity (cm^3 cm^{-3})			
Minimum (1 furrow)	0.48	0.49	0.49
Conventional (5–7 plow, 7–8 cm deep)	0.52	0.51	0.53
Deep (chiseling, 20–22 cm deep)	0.53	0.48	0.51
LSD$_{0.05}$	ns	ns	ns

Source: BINA (1995).

Table 6 Effect of Tillage on Soil Strength from 0–60-cm Soil Depth in Rice–Wheat Cropping System, Dinajpur, 1996–1997

Tillage	Soil strength (mPa)			
	0–10 cm	10–20 cm	20–40 cm	40–60 cm
Zero (surface seeding)	3.43 b[a]	3.74	2.83 c[a]	2.29 c[a]
Conventional (5–7 plow)	3.87 a	4.13	3.68 a	3.12 a
Deep (chiseling)	3.82 a	3.72	3.03 b	2.74 b
LSD$_{0.05}$	0.20	ns	0.10	0.10

[a] Means followed by the same letter are not significantly different at 0.05 using LSD.
Source: BARI (1998).

Table 7 Yield and Economics of Rice–Wheat Cropping Sequence with and Without Mungbean Under Conventional Tillage and Surface Seeding of Wheat[a]

Tillage[a]	Cropping pattern[b]	Yield (t ha^{-1})			TVC	Gross return	Gross margin	BCR
		Rice	Wheat	Mungbean				
Normal	R–W–M	5.00	2.28	0.31	30.41	53.25	22.83	1.74
	R–W–F	4.93	2.20	0.00	27.19	41.96	14.77	1.54
SS	R–W–M	5.11	1.91	0.36	27.78	52.79	25.68	1.89
	R–W–F	5.06	1.90	0.00	29.75	42.78	18.23	1.74

[a] Normal = 5–7 plowings, SS = surface seeding before harvesting rice (average of 10 and 20 days before harvesting).
[b] R–W–M = rice–wheat–mungbean, R–W–F = rice–wheat–fallow.
Source: Nur-E-Elahi et al. (2001).

VII. ECONOMICS

Nur-E-Elahi et al. (2001) conducted an economic analysis of wheat yield under NT compared with CT. The wheat under NT gave a comparable yield to that obtained with CT; therefore, it gave greater gross returns and benefit–cost ratio (BCR) than with CT (Table 7). The gross margin of a rice–wheat–mungbean (*Vigna radiata* L.) cropping sequence under CT was US$394 while under NT it was US$443. The gross margin in the rice–wheat–fallow pattern was lower than that of the rice–wheat–mungbean pattern. In the rice–wheat–fallow pattern, the gross margin under CT was slightly lower than that of NT.

VIII. CONCLUSIONS

The effect of tillage on the yield of rice and wheat depends very much on soil physical properties and weed infestation. If weeds can be controlled, the CT practice can be replaced with NT in loam to silty-loam soils in Bangladesh for rice and wheat. MT and NT save time, allow timely sown wheat, and give better yield and economic returns.

The changes in the soil physical properties due to different tillage practices need to be studied further. Changes in soil bulk density or soil strength are variable among several data sources, and no firm conclusion can be drawn. In the rice–wheat area, avoiding puddling in rice may be beneficial for the ensuing wheat crop. Long-term monitoring studies under different soil textures should be started to describe the impact of NT in rice and wheat on soil physical and biological properties, root growth, yield, weed, and other biotic factors and affects the livelihood and welfare of farmers.

REFERENCES

Abedin, M.Z., Ahmed, M.M. 1986. Adoption of wheat to the present cropping systems. Third National Wheat Training WorkshopWheat Research Center, Bangladesh Agricultural Research Institute, 98–123.

Aggarwal, G.C., Sidhu, A.S., Shekhon, N.K., Sandhu, K.S., Sur, H.S. 1995. Puddling and N management effects on crop response in a rice–wheat cropping system. *Soil Tillage Res.* 36:129–139.

Aggarwal, P., Parshar, D.K., Kumar, V., Gupta, R.P. 1997. Effect of kharif green manuring and rabi tillage on physical properties of puddled clay loam under rice–wheat rotation. *J. Ind. Soc. Soil Sci.* 45:434–438.

Arshad, M., Ahmad, S., Kausar, A.G. 1991. Wheat productivity through zero tillage adaptation in rice–wheat system in FSR project area Shahkot. *J. Agri. Res. Lahore* 29: 265–269.

Aslam, M, Hashimi, N.I., Majid, A., Hobbs, P.R. 1993a. Improving wheat yield in rice–wheat system of the Punjab through fertilizer management. *Pak. J. Agri. Res.* 14:1–7.

Aslam, M, Hashimi, N.I., Majid, A., Hobbs, P.R. 1993b. Improving wheat yield in rice–wheat system of the Punjab through zero tillage. *Pak. J. Agri. Res.* 14:8–11.

Badaruddin, M., Razzaque, M.A. 1995. Soil preparation, equipment, and germplasm effect on rice-wheat cropping systems. *In.* Razzaque M.A., Badaruddin M., Meisner C.A., Eds. Sustainability of Rice–Wheat Systems in BangladeshWheat Research Centre: Bangladesh Agricultural Research Institute. Nashipur, Dinajpur, Bangladesh, 37–42.

Bajpai, R.K., Tripathi, R.P. 2000. Evaluation of non-puddling under shallow water tables and alternative tillage methods on soil and crop parameters in a rice–wheat system in Uttar Pradesh. *Soil Tillage Res.* 55:99–106.

BARI (Bangladesh Agricultural Research Institute). 1998. BARI Annual report for 1997–1998.

BARI (Bangladesh Agricultural Research Institute). 1999. BARI Annual report for 1998–1999.

Bhuyian, N.I., Paul, D.N.R., Jabber, M.A. 2002. Feeding the Extra Millions by 2025: Challenges for Rice Research and Extension in Bangladesh. Keynote paper presented at the National Workshop on Rice Research and ExtensionBangladesh Rice Research Institute. Gazipur. Jan. 29–31, 2002.

BINA (Bangladesh Institute of Nuclear Agriculture). 1995. BINA Annual report for 1994–1995.

Evans, L.T. 1993. Crop Evolution, Adaptation and Yield: Cambridge University Press. Cambridge, U.K.

FAO. 2002. FAO production database, FAO, Rome.

Gajri, P.R., Gill, K.S., Singh, R., Gill, B.S., Singh, R. 1999. Effect of pre-planting tillage on crop yields and weed biomass in a rice–wheat system on a sandy loam soil in Punjab. *Soil Tillage Res.* 52:83–89.

Hobbs, P.R., Morris, M. 1996. Meeting South Asia's Future Food Requirements from Rice–Wheat Cropping Systems: Priority Issues Facing Researchers in the Post–Green Revolution EraNRG Paper No. 96–01. CIMMYT, Mexico DF.

Khan, M.S. 1996. Effect of alternate tillage practices on soil physical properties and crop production under rice–wheat cropping sequence of Bangladesh: Ph.D. thesis. Department of Soil Science, University of Dhaka.

Khan, M.S., Khan, T.H., Ullah, M.S., Shaha, R.R. 1997. Effect of alternate tillage practices on rooting characteristics of wheat under rice–wheat cropping sequence. *Annal. Bangladesh Agri.* 7:111–118.

Morris, M.L., Choudhury, N., Meisner, C. 1997. Wheat Production in Bangladesh: Technological, Economic and Policy Issues: IFPRI. Washington, D.C.

Nur-E-Elahi, M., Mollah, M.I.U., Karim, S.M.R., Khatun, A., Choudhury, N.H. 2001. New establishment methods of rice and non-rice crops in rice–wheat and rice–rice cropping systems. International Workshop on Conservation Agriculture for Food Security and Environment Protection in Rice–Wheat Cropping Systems Feb. 6–9, 2001. Lahore, Pakistan.

Sharma, P.K., De Datta, S.K. 1986. Physical properties and processes of puddled rice soils. *Adv. Soil Sci.* 5:139–178.

Timsina, J, Connor, D.J. 2001. Productivity and management of rice–wheat cropping systems: Issues and challenges. *Field Crop Res.* 69:93–132.

14

Equipment for Sustainable Cropping Systems

J. N. Tullberg, J. R. Murray, and D. L. George
University of Queensland
Gatton, Queensland, Australia

I. INTRODUCTION

Zero-tillage (ZT) seeders used for cereal crops in Australia typically involve a machine with a mass of approximately 120 kg per row pulled by a tractor that is rather heavier than the seeder, providing an engine power of around 2 kW per row. At first sight, this appears to be an excessive engineering requirement for the apparently straightforward job of inserting approximately one g/m of seed 30–60 mm below the soil surface into soft, moist soil. The idea that this task is overengineered is reinforced by the observation that many ZT seeders are based on frames similar to those of the heavier units of conventional tillage (CT) equipment.

Why has this happened? It is not because farmers love big tractors and equipment for their own sake. Most farmers are acutely aware that the tractor and its attachments, used only for a few hundred hours per year, represent a major cost in their cropping system. The reason is simply that farmers have found through hard experience that this is the only reliable way to deal with the problem of hard, uneven soil surfaces and nonuniform crop residue.

The heavy machine frame and openers are really needed only over that proportion of the machine width operating in hard soil. The expensive parallelogram depth control system is needed only over that proportion of implement width where the soil surface is uneven; large-diameter coulters are needed only where crop residue is concentrated; heavy press wheels are needed only where seeding zone tilth is inadequate. The tractor's power and weight are needed only when soil conditions are particularly difficult. Farmers buy the large tractors because it is simply bad economics to invest in a system that is unable to do the job perhaps one year in four, when the whole surface is hard. Despite the rhetoric about soil being softer in ZT—which it usually is—this is clearly not a uniform condition across the field, and a uniform state across the seasons and years.

This paper demonstrates that most of this nonuniformity is a simple consequence of wheel traffic by tractors and machines. The real significance of wheel effects is that they are largely responsible for the conditions that necessitate the use of heavier equipment and greater power in a self-perpetuating cycle. The paper summarizes some of the benefits to be achieved by controlling traffic and looks at some of the problems that occur when using equipment designed for random-traffic ZT farming in controlled-traffic and permanent-bed systems.

Most of the data quoted here have been observed during research carried out in the heavy Vertisols of subtropical Australia or the loess plateau of northern China. In most cases, there has been a striking similarity between traffic effect data from these two environments, despite their contrasting soils, climates, cropping systems, and mechanization levels.

II. TRAFFIC IMPACT—AREA AND ENERGY

A meter of seeder width is associated with a broadly predictable equipment weight, power, and tire specification. In Australian extensive agriculture, for instance, each meter of seeder width would typically weigh at least 6 kN and be pulled by a tractor of about 10kW power and 8kN weight. Tractor and seeder need tire widths of about 0.2 m and 0.15 m, respectively, per meter width of seeder.

Well-known and predictable physical relationships determine these values, which ensure that wheel traffic runs over 25%–35% of field area at seeding, when the soil is in a moist and vulnerable condition. When the field traffic is not controlled, it is difficult to plant and harvest any crop without heavy wheel impacts on greater than 50% of crop area, and this is usually greater (Kuipers and van de Zande, 1994). The impact of this wheel traffic is summarized below.

Traffic energy effects are of fundamental importance because tractor and implement wheels normally dissipate more energy per unit area of soil, over a greater depth of the soil profile, than most tillage or seeding implements. It is relatively simple to work out the energy dissipated in the soil by a tractor, based on reasonable estimates of power transmitted to the implement, power wasted in deforming soil vertically (rolling resistance), and power wasted in deforming soil horizontally (wheelslip).

With modern farm equipment, traffic energy input values are typically around 10 kJ/m². Mean tillage and seeding energy inputs to soil disturbance— which are directly proportional to the draft—are typically in the range 3–6 kJ/m².

The draft of tillage or seeder tines working soil behind tractor and implement wheels has, however, been shown to be almost double that of other tines, in both Australia and China.

These data can be used to demonstrate that the impact of preceding tractor and implement wheels can increase the energy requirements of tillage or seeding by 25% to 40%. Approximately half the total power output of a tractor can be dissipated in the process of creating and partially undoing the effects of its own wheel traffic (Tullberg, 2000), and some controlled-traffic farmers in Australia have reduced tractor size by 50%. Data presented below demonstrate the negative effects of wheeling and tillage on soil and crop.

These data also indicate the inverse relationship between tractive efficiency and soil damage. In a random traffic system, efficiency can be improved in practice only by increasing the cost of the traction system—by reducing tire pressure, replacing two-wheel drive with four-wheel drive, or replacing pneumatic tires with crawler tracks, or belts. Controlled traffic increases tractive efficiency by restricting wheels to the strong, compact soil of the permanent traffic lanes (Fig. 1).

Figure 1   Permanent wheel track. Smearing of this clay soil is quite evident in the zone beneath a permanent wheel track (vertical arrow), used twice per year for two years. The horizontal bars indicate 100-mm depth increments.

III. SOIL DEGRADATION

Traffic-induced soil degradation might be of greater economic and environmental significance than input energy considerations, but it is much more difficult to quantify. Evidence from a variety of sources indicates that the tillage process rarely removes all physical soil degradation effects (Soane and van Ouwerkerk, 1994). Effects within the disturbed layer, or seeding zone, illustrated in Fig. 2, demonstrate the negative effect of preceding wheels on aggregate size distribution and its impact on seeding timeliness. The practical implications of random harvester wheeltracks in ZT have also been commented on by Rohde (2000).

Beneath the disturbed layer, soil structural damage is often visible, and it is in this subtillage layer that "soil compaction" is commonly recognized as a "plough pan." As the depth of damage increases, mechanical rectification becomes more expensive, and natural amelioration processes occur more slowly. Interest in deep tillage is re-emerging in areas where random traffic ZT is common.

Wheel traffic impact on productivity depends on the extent to which different pore size classes are affected. Because the largest voids and pores are physically the weakest, they are the first affected by traffic, so the degradation of soil hydraulic properties is usually greater than might be expected on the basis of simple proportionality with changes in void ratio (Horton et al., 1984). Damage to larger pores might be expected to affect infiltration rates, and the literature includes ample evidence of this.

IV. WATER ENTRY AND STORAGE

Rainfall simulator results from contrasting environments (heavy clay under 100 mm/h in Australia, loess under 60 mm/h in China) confirmed that traffic effects

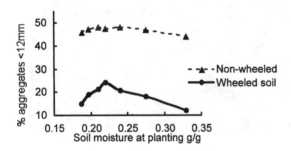

Figure 2   Wheeling effects on aggregate size in seeding zone. (From Tapaevalu, 1996, in Tullberg, 2001).

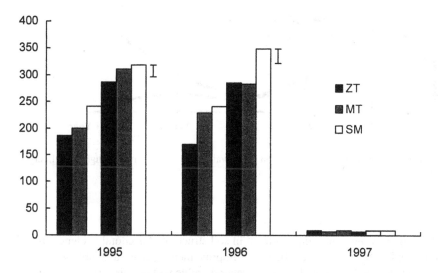

Figure 3 Traffic and tillage effects on runoff over 3 years' field experiments. Vertical bars indicate standard error of means ($P < 0.05$). ns—no significant difference, C—controlled traffic, W—wheeled, ZT—zero-till, MT—min. till, SM—stubble mulch. The 1997 rainfall pattern was unusual, with few sustained high-intensity events.

on infiltration were greater than the tillage effects (Li et al., 2001). The cumulative seasonal effect over 3 years is illustrated in Fig. 3 in terms of runoff, which is important in terms of erosion and nutrient loss. Infiltration differences (Tullberg et al., 2001) appear less dramatic but correlate well with yield.

Water in micropores of <5 microns is held at greater potential, making it unavailable to plant roots. Water in the intermediate mesopores, in the 5–50-micron diameter range, is held with sufficient potential to prevent drainage by gravity, while its accessibility to plant roots increases with pore diameter. The importance of this reserve of plant available water increases with the imbalance between rainfall and crop water use, so it is extremely important in most Australian cropping systems.

The literature (e.g., Boon and Veen, 1994) often suggests that the effects of compaction on soil water storage are small, but a substantial improvement in hydraulic conductivity, drained upper limit, wilting point, bulk density, and depth of permeable soil occurred over four years in a degraded Vertisol that was not subject to tillage or traffic (A. McHugh, personal communication). These data, illustrated in Fig. 4, indicated no difference between the effects of one and three annual wheel traffic treatments and demonstrated that a single annual tractor

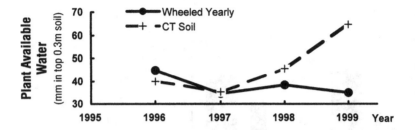

Figure 4   Annual wheeling effects on plant available water. (From McHugh, in Tullberg, 2001.)

wheeling prevented any improvement in soil structure. Anecdotal evidence from controlled-traffic farmers confirms the improvement in plant available water.

These results also underline the general problem of defining a control condition for research on soil compaction. Nonwheeled, cropped soil is usually not available, so no completely uncompacted control is available to compare with compacted treatments. In practice, soil that exhibits no apparent damage is often taken as the control condition, sometimes after deep tillage, and perhaps with reference to nearby pasture or forest areas. The control condition used in the study of heavy grain bin effects by Voorhees (2000), for instance, was soil that had not been subjected to axle loads in excess of 50 kN. This was the axle load used to impose traffic treatments in the work summarized here (Li et al., 2001; Tullberg, 2001), where the control condition was the absence of any wheel effects over a defined period.

V.   SOIL HEALTH AND CROP PERFORMANCE

Soil health is not a precisely defined parameter, but measures of soil biological activity—such as earthworm numbers—are widely regarded as the best available indicator. U. Pangnakorn (personal communication) has monitored tillage and traffic effects on several groups of soil biota over two full cropping cycles. Significant differences were found between all traffic and tillage treatments when statistical analysis was performed on incidence of occurrence data, rather than absolute numbers (to satisfy assumptions of ANOVA). Mean earthworm data are presented in Fig. 5. These data illustrate the well-known advantage of ZT versus CT. They also demonstrate a greater wheeling effect.

Mean yield data for wheeled and controlled traffic, and tilled and ZT plots over 6 years and 6 crops (4 wheat, 1 sorghum, 1 maize) of the runoff trial, are presented in Fig. 6 (Li, 2002). These data indicate that if current conventional

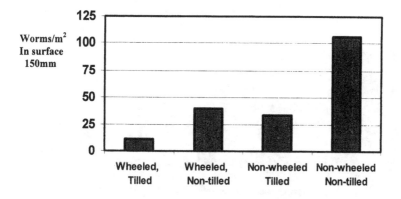

Figure 5 Tillage and traffic effects on earthworm numbers. (From U. Pangnakorn, in Tullberg, 2001.)

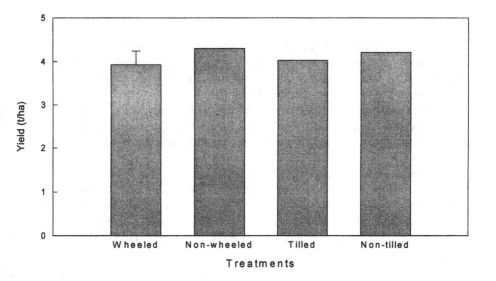

Figure 6 Tillage and traffic effects on mean yield of grain crops. (From Li, 2002.)

Figure 7 Raised-bed controlled traffic. Permanent raised beds provide all the advantages of controlled traffic, while protecting against waterlogging in the spring—important in poorly drained soils of temperate (southern) Australia.

practice is represented by wheeled tilled soil, improvements of 10% and 5%, respectively, can be achieved by controlling traffic and eliminating tillage and that these two effects appear to be additive. It is interesting to note the similarity with cumulative infiltration data for these plots.

Dryland data from West Australia and Victoria indicate a similar yield effect. In those cases where raised-bed systems (Fig. 7) overcame postseeding waterlogging problems, yield response was much greater (Hamilton and Bakker, 1998).

Permanent traffic lanes (Fig. 8) provide a straightforward guidance system for all cropping operations (Blackwell, 1998). Farmers adopting controlled traffic often report reductions in the time and material input to operations of 10% to 20%. Once permanent wheel tracks are accurately installed, the elimination of double coverage and/or gaps has a positive effect on yield (Yule, 1998).

VI. PRACTICAL IMPACT

It is easy to demonstrate that permanent wheel tracks become trafficable sooner after a rainfall event, allowing field operations to occur more rapidly. The outcome

Figure 8 Broadacre controlled traffic. This aerial photograph shows part of a large-scale controlled-traffic farming operation in subtropical Queensland. Note the "downslope" ZT system, working across the contoured banks to minimize opportunities for runoff concentration.

in northern Australia is an increased number of cropping opportunities. Farmers generally agree that controlled traffic produces greater yields than conventional farming, but they suggest that an increase in cropping frequency is the most important mechanism of improved production in an environment where seeding opportunities are usually rare and of short duration.

　　With or without a sophisticated guidance system, permanent wheel lanes improve both precision and the options available for crop management. Several farmers, for instance, now carry out ZT seeding, immediately after harvest if necessary, using a conventional seeder with tine openers seeding midway between the rows of the previous crop. This process is even easier in raised-bed systems, which usually provide a built-in guidance system. It is also much easier to use band application of fertilizer, herbicide, or insecticide and effective physical weed control.

VII.　SOIL EROSION

Regardless of other factors, controlled traffic will reduce runoff and water erosion because of the increase in cropping frequency, reduced tillage, and improved infiltration. It also, however, provides a network of channels that can be used to control overland flow. Yule *et al.* (2000) observed the destructive effects of overland flow concentration during high-intensity rainfall events. They noted the role of tine and wheel marks in exacerbating this concentration process in both tilled and ZT cropping systems, when worked on the contour. They surmised that carefully laid out "downslope" systems would overcome this problem by preventing the concentration of runoff.

Downslope systems working over, and almost at right angles to, existing broad-based contour banks (Fig. 8) have subsequently been installed by many farmers and their operation observed over several years. This approach is controversial to the extent that it appears to be the opposite of accepted practice. The evidence available so far—from preliminary modeling, from aerial survey-plus-ground inspection, and from farmer comment—all indicates a substantial reduction in erosion damage, particularly from extreme events.

VIII. CONCLUSIONS

Controlled traffic (Fig. 7) avoids the contradictions inherent in most mechanized farming systems and provides substantial, demonstrable, and consistent improvements in the economics and sustainability of cropping. The benefits demonstrated in research have been confirmed in practice, and the area of cropland managed in controlled traffic in Australia is now believed to exceed 1 Mha.

There is a growing body of farmer experience on the benefits, problems, and practical outcomes of this system. Once farmers have adopted controlled traffic, very few consider returning to conventional systems. Most controlled traffic farmers in Australia would confirm that the elimination of random wheel traffic has also eliminated their reasons for tillage. Several would claim that controlled traffic is a prerequisite of long-term ZT.

The importance of wheel traffic is not as obvious in a low-resource cropping system using smaller tractors. Nevertheless, tire pressures—and hence surface soil traffic effects—are similar across all tractor sizes. Bed-type crop production systems are a practical means of minimizing power input and structural degradation while optimizing plant available water, soil health, and crop yield. They also provide valuable opportunities to improve crop management via a consistent relationship between crop rows and management tools.

Controlled traffic has a major influence on seeder design requirements. A seeder for use in controlled traffic will not be required to work in hard soil conditions, so the whole unit can be of lighter construction and the ground and residue-engaging components smaller and "cleaner." In soft controlled-traffic soil, disc coulters tend to "hairpin" residue without a hard surface to cut against. It is more important to consider systems that avoid residue by interrow seeding, pretreatment of residue on the bed, and deflection of residue concentrations to the traffic lane. These steps to avoid the requirement for heavy disc coulters are particularly important in the design of seeders for use with smaller tractors of limited lifting capacity (Fig. 9).

Controlled-traffic/permanent-bed systems do not always provide a consistent relationship between the height of the bed and the height of the wheel lane, although the bed itself usually has a consistent shape. Seeders for these systems need to gauge depth from the bed surface, although individual opener depth control is usually not required. The ability to move the seeder laterally can also be important to allow seeding between rows of standing residue.

Both ZT and controlled traffic were demonstrated as beneficial and practicable soil/crop management systems for western agriculture at least 40 years ago. Despite the substantial benefits and sustained public and commercial sponsorship of ZT, adoption has usually been slow. Controlled traffic has overcome the problems of ZT in Australia, and early developments in China are hopeful.

Figure 9 Small-scale controlled-traffic permanent wheel tracks at 1.2-m spacing used for zero-tillage seeding in a trial carried out on the loess plateau of Shanxi Province, north China.

The challenge now is surely to change the current paradigm of increasing machinery cost, energy input, and soil degradation. Controlled-traffic ZT is a rational basis for a more economic and sustainable soil/crop management system, where a more precise, information-based approach can replace our current attempt to use brute force to achieve uniformity.

IX. SUMMARY

Zero-tillage cropping systems are relatively common in Australia, although it is rare to produce more than two or three crops without having to resort to some aggressive soil disturbance. There is, however, a stark contrast between the original vision for more environmentally sensitive cropping systems and the modern reality, where heavy ZT planters are drawn by large tractors. This paper identifies the pervasive, but nonuniform, impact of wheel traffic as the major driver of this contradiction. Controlled-traffic or permanent-bed systems—now employed on ~1 Mha cropping in Australia—are demonstrably successful in avoiding most of the problems created by random traffic and in improving the productivity, economics, and sustainability of cropping. The paper also identifies some of the issues of planting equipment design for controlled-traffic/permanent-bed cropping.

ACKNOWLEDGMENTS

Much of the work reported here was supported by the Australian Centre for International Agricultural Research under projects 9209 and 96143—"Sustainable Mechanised Dryland Grain Production"—with China Agricultural University.

REFERENCES

Blackwell, P. 1998. Customised controlled traffic farming systems. Proc. Second Natl. Controlled Traffic Conf. Gatton. Queensland. Australia, 23–26.

Boon, F.R., Veen, B.W. 1994. In Soil Compaction in Crop Production. Soane B.D., van Ouwerkerk C., Eds: Elsevier Science B.V. Amsterdam, 237–264.

Hamilton, G., Bakker, D. 1998. Approach to the development and adoption of permanent raised bed farming. Proc. Second Natl. Controlled Traffic Conf., Gatton. Queensland. Australia, 95–97.

Horton, R., Ankeny, M.D., Allmaras, R.R. 1984. In Soil Compaction in Crop Production. Soane B.D., van Ouwerkerk C., Eds: Elsevier Science B.V. Amsterdam, 141–166.

Kuipers, H., van de Zande, J.C. 1994. Quantification of traffic systems in crop production. In Soil Compaction in Crop Production. Soane B.D., van Ouwerkerk C., Eds: Elsevier Science B.V. Amsterdam, 417–446.

Li, Y., Tullberg, J.N., Freebairn, D.M. 2001. Traffic and residue cover effects on infiltration. *Aust. J. Soil Res* 39:239–247.

Li, Y. 2002. Traffic and tillage effects on dryland cropping systems in north-east Australia. Unpublished Ph.D. thesis: University of Queensland.

Rohde, K. 2000. Controlled traffic farming and fitting the header into the system. *Australian Grains* 10(5):6,8.

Soane, B.D., van Ouwerkerk, C., eds. 1994. Soil Compaction in Crop Production: Elsevier Science B.V. Amsterdam.

Tullberg, J.N. 2000. Wheel traffic effects on tillage draught. *J. Agric. Eng. Res* 75(4): 375–382.

Tullberg, J.N. 2001. Controlled traffic for sustainable cropping. Proc. 10th Australian Agronomy Conf. Hobart. Tasmania.

Tullberg, J.N., Ziebarth, P.J., Li, Y. 2001. Tillage and traffic effects on runoff. *Aust. J. of Soil Res* 39:249–257.

Voorhees, W.B. 2000. Interaction of axle load, soil water regime and soil texture on long-term subsoil compaction and crop yields in North America. Proc. Fourth Intl. Conf. on Soil Dynamics. Adelaide. Australia.

Yule, D.F. 1998. Controlled traffic farming—the future. Proc. Second Natl. Controlled Traffic Conf. University of Queensland, Gatton. Queensland, 6–12.

Yule, D.F., Cannon, R.S., Chapman, W.P. 2000. Proc. 15th ISTRO Conf. Ft. Worth, TX.

15

Mechanization Challenges in No-Till Rice–Wheat Systems for South Asia: No-Till Seeding Drills for Small Landholders

John E. Morrison, Jr.
U.S. Department of Agriculture
Temple, Texas, U.S.A.

I. INTRODUCTION

Rice is a warm-season "summer" crop. Wheat is a cool-season "winter" crop. When grown in a double-crop rotation, they are the basis for a rice–wheat system (RWS) of crop production. Such double-cropping systems increase the efficiency of production for the individual farmer and the general economy of the agricultural region, because two economically fruitful crops are being harvested each year. This efficiency is produced by maximum utilization of available soil, water, labor, input, and climatic resources.

The rice crop is traditionally grown in random plant-to-plant spacing patterns from transplanted seedlings in Asia, but it is also drilled in rows and broadcast-seeded in other regions. Field seeding conditions vary from flooded to dry. Transplanted seedling hill spacing is typically 15 to 25 cm. Seeded rows are spaced 15 to 30 cm apart. Seedling density is 100 to 400 plants/m². Established rice fields are flooded with water to control the growth of weeds as well as to water the crop. Irrigation water is removed from the fields when the crop is nearing maturity and the soil is allowed to dry for harvest. The rice grain may be stripped from the plants for harvest, but the traditional harvest method is to cut the plants 10 to 20 cm above the soil surface, leaving hills or rows of stubble. The bulk of the straw is removed from the field in the harvest operation and not returned directly to the field. If harvested by mobile combine/header harvesters, the chaff and threshed straw are dropped into windrows and subsequently burned. Because rice plants are still alive at harvest, in warmer regions, the cut plants continue to use soil water, which causes drying of the soil. Ratoon growth can produce a second crop, but not if a winter double-crop is desired; then fall growth must be prevented. In cooler regions, the rice stubble grows very little and will eventually be killed by freezing temperatures. The rice plants produce numerous tillers, which result in robust rows or hills of stubble some 10 to 16 cm wide after harvest. In traditional agriculture, the stubble is pulled or plowed loose from the soil and plowed under, used off-field, or burned, depending upon the region. Such growth termination and removal or burning require labor, reduce carbon storage in the soil, and delay the seeding of wheat. Loose stubble on the fields encumbers the use of wheat drills, resulting in the common practice of broadcast seeding of wheat.

In the RWS, wheat is traditionally seeded in prepared seedbeds after the harvest of rice and the removal of rice stubble rows. Wheat is seeded by broadcasting or by drilling into narrow rows, spaced 10 to 20 cm apart. Wheat plants form tillers to produce multiple heads and almost solid rows, similar to rice. When mature, wheat plants senesce, which eliminates the problem of termination of plant growth. At harvest, wheat grains may be stripped from the plants, but traditionally, like rice, the wheat is cut 16 to 20 cm above the soil surface and the chaff and bulk of the straw are plowed under, burned, or removed from the field for fodder. If harvested by mobile combine/header harvesters, the chaff and threshed straw are dropped into wind rows and subsequently burned. After wheat harvest, the fresh stubble is not as robust as rice stubble, but it is tightly anchored to the soil by roots. The wheat stubble row is typically no more than 5 to 10 cm wide.

Many regions have growing-season climates amenable to double-cropping, except for adequate rainfall or irrigation for each crop. With a rice culture, rainfall is supplemented with irrigation, so that water availability may not limit production. Wheat in rotation with rice may be irrigated if winter rains are inadequate, typically one or two irrigations. In South Asia, the summer and winter climates are adequate for both rice and wheat to be grown and to mature in time for the seeding of the next crop to maintain the double-crop rotation; this may be limited by available rice varieties for direct seeding. Some eastern regions can produce three or four annual crop rotations.

The RWS also encompasses other crops in order to expand total production. These crops include winter lentils, chickpeas, berseem clover for fodder, potatos, vegetables, boro rice, and mustard and sunflower oilseeds. Alternative summer crops include maize, soybeans, sorghum for fodder, mungbean, sesbania green manure, and even three years of sugar cane. Hobbs and Aadhikary (1997) noted that crop diversification is an appropriate technology for crop disease and insect control and nutrient management.

There is a lot of tradition in both rice and wheat cultural systems. Some of these traditions may be incompatible for double-cropping of the two crops. Therefore, it must be remembered that it is a "system" that is desired and not just the development and adaptation of some new technologies to fit existing practices.

A. Philosophy of the System Approach

In a successful "system," all components must be compatible. This compatibility is expressed in cropping systems by following at least two rules, namely:

1. Do not include a technical component in a system that causes a "problem" condition requiring the expenditure of additional energy, time, money, mechanization, or other input.
2. Do not include a technical component in a system that requires a considerably higher energy or mechanization input for only one operation than required for the rest of the system.

An example of rule 1 is the inclusion of a tillage operation that exposes buried weed seed, only to require additional weed control. An example of rule 2 is the inclusion of one annual high-draft-requiring deep tillage operation that requires the availability of a larger power source than any of the other system field operations.

To avoid problems of system component incompatibilities, we identify alternative technologies to accomplish desired system goals. This is done by consideration of the physical properties and physical constraints of the system, using those elements of traditional practices that are compatible, and introducing innovative technologies and approaches to complete the systems.

B. Goals for a Rice–Wheat System

The primary goal of the RWS is to produce rice and wheat grains in a double-crop sequence to maximize total annual profitability (minor crops may also be included) without unduly degrading the resource base. All other goals are secondary.

The basic process is the multiplication of planted seed in grain yields that are sufficient to provide adequate food supplies and profits to justify sustained use of the cropping system. The task is to develop an RWS, which utilizes the land, water, seed, fertilizer, labor, energy, and other inputs and produces economically viable crops by using logical, affordable, and teachable technologies in the environment of traditional agriculture.

II. MECHANIZATION IN RICE–WHEAT SYSTEMS

The following comments assume that the RWS will develop into a completely no-till system without any primary tillage or residue burial. There are four primary mechanization functions in RWS, namely:

1. Residue management
2. Weed control
3. Seeding/planting
4. Harvest

A. Residue Management

Residue management starts with the harvest of a crop and continues with each field operation, which interacts with the crop residues remaining on the field surface. Residues are typically characterized as attached and unattached, and by the length, strength, and cover provided for the soil surface. Residue strength is a particular problem with mechanization because it varies with time of exposure to climatic factors and subsequent biological deterioration (Ghidey and Alberts, 1993; Schomberg et al., 1994). Residue orientation is also a major factor, because if long pieces of residue are lying across the path of a machine, such as a direct seeder, then the residue must be pushed aside or cut to avoid machine blockages. Residues that remain as standing stubble may not impede field machines that are designed to pass between the stubble rows with little interference or blockages. Loose residues from in-field combine-harvest or those returned to the field after off-field threshing may cause machine blockages by entanglement with machine components. Thick mats of loose residues on the soil surface must be pushed aside or cut to allow the passage of field machines. If the soil is moist or soft under the residues, the path clearing, and especially the cutting, becomes problem-atic with typical direct seeder machine designs. Depending upon the climate, crop varietal differences in decomposition, etc., there may be regions where an excessively thick mat of residue develops and a portion of the residues will have to be removed from the fields to maintain a stable protective coverage without undue encumbrance to field machine operations, buildup of crop diseases, or other nuisance. This will not happen with the current rice–wheat system that includes spring plowing-under of wheat residues before rice seeding/planting. When the RWS develops into a total no-till system with year-round protection of the soil surface, technical alternatives must be chosen to continuously manage crop residues to be able to take advantage of the many proven benefits of the elimination of soil tillage.

1. Rice Residue and Stubble Management Alternatives

Rice straw and chaff residues may be managed by

1. Uniform spreading of chaff with a chaff-spreader attachment on in-field harvester combine/header
2. Uniform spreading of long loose straw with a straw-spreader attachment on in-field harvester combine/header
3. Powered chopping of loose straw into shorter lengths and uniform spreading with a chopper-spreader attachment on in-field harvester combine/header
4. Postharvest removal of loose residues from fields after harvest with in-field harvester combine/header
5. Removal of rice straw from the field with the grain for off-field threshing
6. Return of threshed rice straw residues with manures spread on the field surface
7. A portion of the residue removed to avoid buildup of an overly thick mat on the soil surface. (Burning of windrows of rice straw is not recommended.)

Assuming that the wheat is seeded after the cut-and-remove harvest of the rice crop, stubble of the harvested rice crop may be managed by one of several potential methods, including

1. Left in place with no further treatment
2. Killed with herbicide (in regions with ratoon growth) and left in place in the old rows
3. Mechanically undercut with a sweep or bar blade and left uprooted on the surface
4. Mechanically undercut with one or two sharp coulter disc blades and uprooted
5. Mechanically uprooted by a "stalk-puller" machine
6. Left on the field surface after uprooting
7. Mechanically cut close to the soil surface to minimize interference with seeders.

Northern regions will not have stubble ratoon growth problems, but it may be a consideration in southern regions. All of the mechanical uprooting technologies will require considerable tractor draft to achieve soil penetration and to sever the rice roots (our observation of mechanical pulling of sorghum and cotton stubble is that the pullers work best at relatively high travel speeds and that shallow trenches are left by the pulling operation). Uprooted stubble remaining on the field surface may cause blockage of the wheat seeders. The uprooting procedures may leave trenches of disturbed soil in the old rice rows and soil conditions, which are quite different than those between the disturbed rows. Wheat seeding will be in variable soil conditions depending on whether wheat rows fall on or between old rice rows. Close cutting of the stubble rows will not kill the stubble. Herbicidal killing of the rice stubble appears to be the least disruptive and lowest-energy approach to rice stubble control for use in southern regions with ratoon growth.

2. Wheat Residue and Stubble Management Alternatives

The following assumes that the RWS will evolve into a nonplowing no-till/direct-seeding system for both rice and wheat. Wheat residues may be

1. Cut the straw high when harvesting to leave tall 20–40-cm stubble
2. Uniform spreading of chaff with a chaff-spreader attachment on in-field harvester combine/header
3. Uniform spreading of long loose straw with a straw-spreader attachment on in-field harvester combine/header
4. Powered chopping of loose straw into shorter lengths and uniform spreading with a chopper-spreader attachment on in-field harvester combine/header
5. Postharvest removal of loose residues (for fodder) from fields after harvest with in-field harvester combine/header
6. Removal of wheat straw from the field with the grain for off-field threshing
7. Return of threshed wheat straw residues with manures spread on the field surface
8. A portion of the residue removed to avoid buildup of an overly thick mat on the soil surface.

Wheat stubble may be

1. Left in place in the field
2. Mechanically cut close to the soil surface.

B. Weed Control

Due to the heavy shading of the soil and competition for available water by the maturing rice or wheat crop, weed germination and growth are often minimal at crop harvest. Localized patches of weeds may be economically controlled with available herbicides after rice harvest. If the rice stubble is killed with herbicides, then emerged weeds will also be controlled. A review of herbicidal weed control technology is beyond the scope of this paper.

The wheat seeding operation will disturb some soil and will effectively be a shallow cultivation, which may control some freshly germinated weeds. Less soil is disturbed with direct seeding than with plowing, so that less weed seed is surfaced for germination.

Weed control during the crop-growing season will be relatively independent of the wheat seeding technology, but not independent of the tillage, residue management, and fertilizer application technologies. If crop rows are widely spaced or grouped on beds, there may be less uniform soil shading and crop competition with potential weed growth. Mechanical, animal-draft, or manual weed control, as well as fertilizer application, can be conducted in the open spaces between adjacent rows, but the desirability of using those technologies depends upon compatibility with the total system.

1. Mechanization of Weed Control

1. Fallow cultivation between crop seasons causes soil disturbance and is not compatible with the principle of minimal no-till soil disturbance. It is used effectively in stubble-mulch tillage systems for dryland regions.

2. Weed cultivation with the direct-seeder can be effective, especially with shank-type seeder furrow openers that disturb more soil than disc-type furrow openers. Excessive burial of residues by this operation may be detrimental to the total system performance, and weed seed surfaced by this operation may germinate and create more weed control problems than lower-disturbance methods.

3. In-crop cultivation between rows can damage plant roots and bury residues, with the end result being less total benefits. If crop rows are spaced more widely to allow for this in-crop cultivation, the soil will be less shaded and there will be less competition for the weeds, which may create an increased weed control problem.

4. Herbicide spraying equipment is available for all levels of mechanization, from backpack sprayers to wide self-propelled machines. These spraying machines are available from many sources, and agricultural chemical companies eagerly provide training on the calibration, mixing of chemicals, safety, and use of mechanical sprayers.

C. Seeding/Planting

1. No-Till/Direct Seeding of Rice in Wheat Stubble

For mechanization of the RWS, it is reasonable to consider the use of the same no-till/direct seeder drill for both wheat and rice seeding. The specifications for field operations are similar for both crops and should not add to the cost or complexity of the seeder, although rice must be seeded shallow so that seeder depth control must be more accurate than for wheat seeding. This should be true for operation on flat field surfaces or on raised beds, assuming that both rice and wheat are seeded on the same field surface configuration.

If wheat residues are left on the field by in-field harvesting with combine/ headers, then with double-cropping sequences the residue and stubble will not have time to be biologically reduced or deteriorated. This situation may require the use of powered residue cutters ahead of seeder row units to avoid seeder machine blockages. An economic rationalization for the use of this more costly seeder cutter attachment is that double-cropping is more profitable than annual cropping and the additional cost of specialized seeders is justified by the additional crop profits, as well as by the use of the same machine for direct-seeding both rice and wheat.

a. U.S.A. Experience. It is the author's understanding from a current survey of rice research workers that direct seeding of rice is in the initial research phase of development in the United States. Commercial seeders are being experimentally modified for use in these conditions. Until this time, the most advanced system has been the "stale seedbed" system of fall plow tillage, fallow overwintering, and spring direct seeding into the bare fallow soil.

2. No-Till/Direct Seeding of Wheat in Rice Stubble

a. U.S.A. Experience. No-till/direct seeding of wheat immediately after rice harvest is similar to the no-till seeding of soybeans immediately following wheat harvest in U.S. double-cropping systems. The urgency for seeding immediately after harvest is also similar; to seed into drying soil before it is too dry to support crop establishment. [For South Asia, Yoshida (1981) found that land preparation lost 200 mm of soil water to evaporation, and if land preparation extended more than 30 days, soil water loss is as much as 500–600 mm.] Farmers typically seed soybeans or wheat within five days after harvest. Immediately after harvest, the fresh-rice residue is tough and the stubble is strongly anchored in the soil. Stubble and loose straw residues protect the soil from rapid drying, but also present problems of machine blockage, variable soil penetration depths, and variable seed coverage during no-till seeding. Although not recommended for conservation agriculture, rice straw and stubble fields are often burned to eliminate these problems before fall seeding to winter crops. Reports are limited to anecdotal accounts of field experiences, because this system is just being developed in the United States.

3. Wheat Seeding Conditions Caused by Rice-Growing Practices

Rice may be grown with slight modifications to traditional practices to be more compatible with an RWS. Rice varieties may be selected with different maturity dates, rooting or above-ground growth characteristics, such as height, stiffness of straw, tillering propensity, leaf angle, or other characteristics for improved compatibility with the total RWS. It is possible to seed low-tillering dwarf-type varieties of rice directly into the soil for crop establishment. This may be done by "water seeding," a seed broadcasting procedure, or by drilling seed into drained soil. Such seeding practices would avoid the transplanting operation and drilling would provide evenly spaced rows, which might be advantageous for mechanized field procedures. Rice seeding would avoid the practice of soil puddling, which creates unfavorable soil structure for subsequent wheat production. If rice is seeded into closely spaced rows at sufficiently high populations, then fewer tillers may be produced. The resulting stubble rows may be narrower and less robust for stubble control after harvest and for wheat seeding. Yamada (1963) showed that rice plants at densities greater than 300 plants/m did not produce tillers, but had only one panicle per original plant; i.e., a single culm plant population. Stansel (1975) compared the propensity of many Asian rice varieties to produce 25 to 30 tillers with the tillering of U.S. varieties at only 8 to 12 tillers, and only 1 to 3 tillers in field conditions. Therefore, the selection and management of rice populations may be used to produce stubble conditions that are more compatible with direct seeding of wheat. In southern regions, after-harvest rice regrowth may be controlled by applications of an appropriate herbicide or by uprooting of the plants. If rice stubble is left undisturbed, then the roots will substantially decompose during the winter season, and the stubble will be easily cleared from the row for seeding the new rice crop the following spring after wheat harvest. Uprooting of the fresh stubble rows will place stubble clumps in the path of wheat seeders and will create open furrows in the previous stubble rows. Uprooting operations will require the availability of adequate energy sources and uprooting machines. Uprooted stubble clumps may not deteriorate as rapidly as they would have if left in the soil. These considerations do not apply to all regions, but may be constraints to wheat seeding after rice harvest.

4. Soil Type and Condition

Stansel (1975) stated that soil type is a dominant factor in the type of rice seeding. He suggested that broadcast seeding works best in heavy clayey soils where rainfall or irrigation flushing is used to water and germinate the rice seed. However, drilling is preferable in sandy loam soils, which tend to crust upon wetting. Irrigation flushing is not required and soil crusting is not a problem for direct seeding in conservation systems (M. Anders, personal communication, 2002). When too dry, clay and clay loam soils require large expenditures of energy to till and they fracture into large clods that are not appropriate for seedbeds. When too wet, clayey soils are adhesive and easily smeared and should not be disturbed with tillage tools or seeding machines. Therefore, the use of broadcast or no-till/

direct seeding is appropriate to minimize tillage energy expenditures and to take advantage of the established soil structure for seed-to-soil contact for successful crop establishment. The one-pass process of no-till/direct seeding can be used to quickly seed during a window of opportunity when the soil is at a water content, which provides a friable soil structure. Fields rutted by tires are difficult to impossible to direct-seed, so self-propelled harvesters should be equipped with tracks or flotation tires to minimize rutting.

5. Alternatives in No-Till/Direct Wheat Seeding

The no-till/direct seeding of wheat after rice harvest may be accomplished with different schemes, including

1. Seeding evenly spaced rows after the nontypical removal of rice stubble
2. Seeding of one, two, or more wheat rows between (nontypical seeded rows) rice stubble rows with individual soil furrow openers
3. Seeding evenly spaced rows with seeder equipped to cut through rice stubble
4. Seeding between killed rice stubble rows with double-shoot or triple-shoot furrow openers to seed two or three rows or wheat, respectively.

6. No-Till/Direct Seeder Functions

No-till/direct seeders for wheat and/or rice should encompass components to perform the following technical functions:

1. Cut any pieces of long, tough, loose residues that cannot be cleared to either side of the row path, to prevent blockage of the seeder or hair-pinning into the furrow.
2. Clear residues from the row path.
3. Open a furrow.
4. Deposit seed in the furrow.
5. Deposit starter fertilizer in the furrow or beside the seed line.
6. (Optional) Close the furrow to avoid making deeply corrugated field surface.
7. (Optional) Pack the soil over and around the seed to ensure water transmission to the seed.
8. Control seeding depth.
9. Apply adequate downforce on seeder row units to achieve desired seeding depth.
10. Allow independent flotation of single or grouped row units to conform to the field surface configurations.

7. Alternative Types of No-Till/Direct Seeder Furrow Openers

The two general types of seeder furrow openers are the disc type (Fig. 1) and the shank type (Fig. 2.)

In the United States, disc-type furrow openers have been the predominate furrow opener on no-till/direct seeders for row crops and for drilled crops. The discs were used to accomplish some residue cutting and to roll over uncut residues to avoid machine blockages. Seeding quality decreased when uncut residues were pressed into the bottom of the seed furrow with the disc openers, due to soil drying and toxic allelopathic effects on the seed germination and early growth (Lovett and Jessop, 1982). Disc openers were used on large seeders where additional weight could be added to improve residue cutting and to force the discs into the soil. The addition of such weight demanded stronger, heavier seeding machines and more pulling power, which were generally available and affordable on large farms.

For the smallholder farms in many global regions, mechanized seeders will be small and powered by animal draft or small tractors. Therefore, heavy disc-type seeders are not appropriate for these regions. Shank-type furrow openers tend to be self-penetrating into soils and generally can be mounted on lighter-weight machines and require lower draft than disc-type seeders. Depth control on individual shank-type furrow openers ensures uniform seeding depth and avoids excessive draft from deep seeding. Shank-type furrow openers tend to accumulate residues and generate machine blockages without positive path clearing ahead of

Figure 1 Disc-type furrow opener.

Figure 2 Shank-type furrow opener.

the openers. Therefore, for successful operation, shank-type opener seeders must be equipped with a higher level of sophistication in the design of path-clearing devices.

8. Row-Path-Clearing Devices for Shank-Type Furrow Openers

The two basic types of row-path-clearing devices for seeders are passive residue rakes (Fig. 3) and powered residue cutters/kickers (Fig. 4). Passive residue rakes can be attached ahead of individual or possibly groups of furrow openers to clear paths through low to medium amounts of crop residues that are in short pieces, loosely attached to the soil, and partially deteriorated, as generally found in annual cropping situations. This is because annual crops are usually separated by fallow periods of several months to allow enough time for biological and climatic processes to reduce the amount and strength of old crop residues to levels that can be successfully raked from the row paths.

The use of passive rolling coulter blades ahead of individual furrow openers, and in particular fluted coulter blades, has been symbolic of the practice of no-till/direct seeding in the United States since their introduction in 1966. These rolling coulters have been used to cut residue and to loosen the soil ahead of

Figure 3 Passive residue rake.

Figure 4 Powered residue cutter/kicker.

seeder furrow openers. Several technical problems are associated with the use of rolling coulters for these functions. Rolling coulters require relatively large down forces both to cut residues (Chang and Erbach, 1986) and to penetrate the soil (Tice and Hendrick, 1991). Because the amount and toughness of the residues, as well as the hardness of the soil, are spatially variable across any field, the seeder machine must be designed to apply and withstand high forces to successfully accomplish both residue cutting and soil penetration. From experience in the United States, spatial variation of applied down force is not practiced, so that uniformity of performance is not achieved. If the underlying soil is soft, the rolling coulters may push the residue into the soil without cutting and it undesirably becomes part of the seed furrow. The passive rolling coulter technology is not compatible with the specifications for successful, lightweight, and inexpensive direct seeders for the RWS.

Powered residue cutters/kickers can be attached ahead of individual or possibly groups of furrow openers to produce cleared paths through large amounts of fresh, strong residues and anchored stubble. This is appropriate for double-cropping systems, such as the RWS. It is also appropriate for use with shank-type furrow openers to avoid machine blockages. Power would be supplied by a tractor PTO drive, by a tractor hydraulic system, or by an auxiliary engine on the seeder. The essence of the situation is that if fresh, strong, thick residues and stubbles are not cut and/or removed by other field procedures, then attachments are needed on the direct seeders to positively cut row paths for the furrow openers. Additional purchase and maintenance costs associated with powered residue cutter/kickers are justified by the lowered tillage costs and the profits from successful double-cropping.

9. Other Seeder Components

Machine components to accomplish all of the identified functions of no-till/direct seeders must be mutually compatible. Details on specifications and performance of optional components are beyond the scope of this chapter.

10. New Direct Seeders Are Under Development

Numerous companies are marketing small-scale direct seeders for row crops and for drilled crops. The USDA-ARS laboratories are currently developing both row crop and drill type no-till/direct seeders for both small-scale and large-scale farming, based upon the same technologies. These machines will continue to be under test and development in 2002 for the direct seeding of maize and wheat, respectively. International sites are being identified for testing. Domestic and international companies will be encouraged to consider the commercial production of these machines.

Table 1 Rice Area in South Asian Countries[a]

Country	Upland	Deep water	Irrigated		Rainfed	
			Wet season	Dry season	Shallow (0–30 cm)	Intermediate (30–100 cm)
India	5,973	2,434	11,134	2,344	12,677	4,470
Bangladesh	868	1,117	1,710	987	4,293	2,587
Pakistan		—	1,710	—	—	—
Sri Lanka	52	—	294	182	210	22
Nepal	40	53	261	—	678	230
Bhutan	28	—	—	—	121	40
Total	6,951	3,604	13,099	3,513	17,979	7,349

[a]Total area planted (double-cropped areas counted twice) is 52,965,000 ha.
Source: Huke (1982).

Table 2 Rice Area in Southeast Asian Countries[a]

Country	Upland	Deep water	Irrigated		Rainfed	
			Wet season	Dry season	Shallow (0–30 cm)	Intermediate (30–100 cm)
Burma	793	173	780	115	2,291	1,165
Thailand	961	400	866	320	5,128	1,002
Vietnam	407	420	1,326	894	1,549	977
Kampuchea	499	435	214	—	713	170
Laos	342	—	67	9	277	—
Malaysia	91	—	266	220	147	11
Indonesia	1,134	258	3,274	1,920	1,084	534
Philippines	415	—	892	622	1,207	379
Total	4,642	1,686	7,685	4,100	12,396	4,238
% of total	13	5	22	12	36	12

[a]Total area (double-cropped areas counted twice) is 34,747,000 ha.
Source: Huke (1982).

Table 3   Estimated Rice Area by Dominant Water Regime

Location	Total rice area (thousand ha) 1978–1980	Irrigated Dry season	Irrigated Wet season	Shallow[a]	Deep water[b]	Floating[c]	Upland	% upland to total rice area
India	39,500	2,700	12,700	11,100	4,500	2,500	6,000	15.2
China	35,300[d]	e	33,600	1,800	0	0	0	0
Bangladesh	10,100	1,000	200	4,300	2,600	1,100	900	8.9
Thailand	9,300	500	1,300	4,900	1,100	400	1,000	10.8
Indonesia	9,000	2,700	4,600	600	300	200	700	7.8
Vietnam	5,200	800	1,200	1,400	900	400	400	7.7
Burma	4,800	100	700	2,000	1,000	200	700	14.6
Philippines	3,400	600	900	1,200	400	0	400	11.7
Pakistan	2,000	0	2,000	0	0	0	0	0
Nepal	1,300	0	300	700	200	50	50	3.3
Korea	1,200	e	1,200	0	0	0	0	0
Sri Lanka	700	200	300	200	0	0	50	7.1
Malaysia	700	200	200	200	0	0	100	14.3
Lao People's	700	0	50	300	0	0	300	42.8
Republic	600	0	50	200	50	100	100	16.7
Kampuchea	126,000[f]	8,800	61,400	29,000	11,200	4,950	1,070	8.5
Developing Asia	8,200[g]	0	1,200	0	900	0	0	74.4
Latin America	4,600	0	800	700	700	0	6,100	50.0
Africa	4,800	I	4,800	0	0	0	2,300	0
Other[h]	143,500	8,800	68,200	29,700	12,800	4,950	0	13.2
World								19.10
								0

[a] 0–30 cm. [b] 30–100 cm. [c] More than 100 cm. [d] Allocation by estimation rather than data. [e] All rice grown during the summer months and shown under wet season. [f] Including 1,800,000 in countries not separately identified. [g] Brazil has 6,200,000 ha. [h] Japan has 2,500,000 ha, and the United States has 1,250,000 ha.
Source: IRRI (1982).

D. Harvest

Harvest of rice and wheat for grain (and the straw for forage) may be accomplished manually or by in-field mechanized combine/header machines. The degree of mechanization of harvest will mainly affect the residue management in conservation agriculture cropping systems. Residue management, and its effects on other aspects of mechanization, has been discussed above. Other aspects of harvest mechanization are beyond the scope of this paper.

III. POTENTIAL APPLICATION FOR RICE–WHEAT PRODUCTION SYSTEM

Gupta and O'Toole (1986) inventoried the rice production areas in South Asia, Southeast Asia, and globally (Tables 1, 2, and 3). If the same mechanization concepts can be used for upland rainfed rice as for irrigated rice, the total global area for potential application is 142.5 million ha. If limited to South Asia, they estimated 12.569 million ha of wet-season rice, which might be converted to the RWS. There are an additional 25.328 million ha of upland rainfed rice in that region that may be annually cropped with the same mechanization. Hobbs (personal communication 2001) reported that 13.5 million ha is already in production under some form of the RWS.

REFERENCES

Chang, H.C., Erbach, D.C. 1986. Cornstalk residue shearing by rolling coulters. *Trans. ASAE* 29(6):1530–1535.

Ghidey, F., Alberts, E.E. 1993. Residue type and placement effects on decomposition: field study and model evaluation. *Trans. ASAE* 36(6):1611–1617.

Gupta, P.C., O'Toole, J.C. 1986. Upland Rice, a Global Perspective. Int. Rice Res. Inst., Los Baños, Philippines.

Hobbs, P.R., Aadhikary, B.R. 1997. The importance of rice–wheat systems for food security in Nepal. *In* Proc. of the Rice–Wheat Research Project End-of-Project Workshop Oct. 1–3, 1997. Kathmandu, Nepal, pp. 1–5.

Huke, R.E. 1982. Rice area by type of culture: South, Southeast and East Asia. Int. Rice Res. Inst. Los Baños, Phillipines.

IRRI (International Rice Research Institute). 1982. A Plan for IRRI's Third Decade. Int. Rice Res. Inst. Los Baños, Phillipines.

Lovett, J.V., Jessop, R.S. 1982. Effects of residues of crop plants on germination and early growth of wheat. *Aust. J. Agric. Res.* 33:909–916.

Schomberg, H.H., Steiner, J.L., Unger, P.W. 1994. Decomposition and nitrogen dynamics of crop residues: Residue quality and water effects. *Soil Sci. Soc. Am.* 58(2): 372–381.

Stansel, J.W. 1975. The rice plant—its development and yield. *In* Miller J. E., Ed. Six Decades of Rice Research in Texas. Texas Agric. Exp. Sta., Research Monograph 4. June 1975, 9–21.

Tice, E.M., Hendrick, J.G. 1991. Disc coulter forces: Evaluation and mathematical models. *Trans. ASAE* 34(6):2291–2298.

Yamada, N. 1963. Spacing. *In* Matsubayashi M., Ito R., Nomoto T., Takase T., Yamada N., Eds. Theory and Practice of Growing Rice. Ministry of Agriculture and Forestry. Tokyo, Japan, pp. 172–182.

Yoshida, S. 1981. Climatic environment and its influence. *In* Fundamentals of Rice Crop Science. Intl. Rice Res. Inst. Los Baños, Philippines, pp. 65–110.

16
No-Till Rice–Wheat Farming: The Arkansas Experience

Merle M. Anders, Jason Grantham, and Jared Holzhauer
University of Arkansas Rice Research and Extension Center
Stuttgart, Arkansas, U.S.A.

Tony E. Windham
University of Arkansas Cooperative Extension Service
Little Rock, Arkansas, U.S.A.

I. BACKGROUND

Rice cultivation first appeared in the continental United States in the mid-17th century (Dethloff, 2003, in press). The first cultivated rice variety, Carolina gold, is reported to have arrived in South Carolina from Madagascar (Littlefield, 1981). Westward expansion in the United States along with the Civil War resulted in rice production shifting from South Carolina and Georgia to the Mississippi River areas of Louisiana (Babineaux, 1967). The availability of railroads and an influx of farmers from other grain-producing areas of the United States resulted in a rapid expansion of rice production in Louisiana and Texas. Rice cultivation soon spread to Arkansas, where production increased from 6,333 kg in 1899 to 25.5m kg in 1909 (Spicer, 1964). A collection of the original varieties introduced into the United States was tested in California in the late 1800s but failed to produce grain. In 1906 William W. Mackie discovered that a short-grain Japanese variety brought from Hawaii produced sufficient grain for commercial production (Dethloff, 2003, in press). Currently, rice production in the United States is found in Arkansas, California, Louisiana, Mississippi, Texas, and Missouri (Fig. 1). Total U.S. rice production has increased steadily since 1960 (Fig. 2). Arkansas is the largest rice-producing state, with approximately 600,000 ha planted each year. In addition to this, there is approximately 20,000 ha of winter wheat grown

273

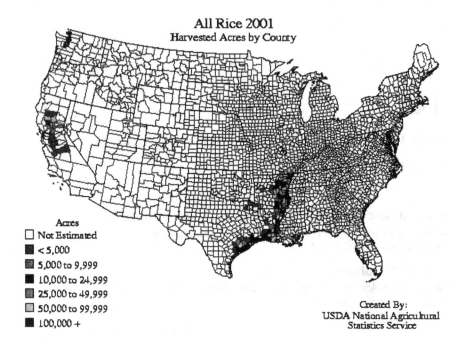

All Rice 2001
Harvested Acres by County

Acres
- ☐ Not Estimated
- ■ < 5,000
- ▨ 5,000 to 9,999
- ■ 10,000 to 24,999
- ▨ 25,000 to 49,999
- ☐ 50,000 to 99,999
- ■ 100,000 +

Created By:
USDA National Agricultural
Statistics Service

Figure 1   Area of rice harvested by county for the United States in the year 2001. (Created by USDA National Agricultural Statistics Service.)

annually in the state. Rice has been the primary crop in the silt-loam soil areas of the Arkansas Delta since it was included in the Agricultural Act of 1933 (Cramer et al., 1990). This and subsequent agriculture bills to the Federal Agriculture Improvement and Reform Act of 1996 restricted the area planted into rice, thus encouraging farmers to rotate with other crops. Soybeans emerged as the crop of choice to rotate with rice throughout much of the Mississippi Valley area. Initially rice was followed with two seasons of soybeans; today many farmers use a two-phase rice–soybean rotation. The introduction of wheat into the rice cropping systems occurred some time ago, but only recently have farmers regarded wheat as a crop that had economic potential. With approval of the Federal Agricultural Improvement and Reform Act in 1966, restrictions on the area of rice planted were removed and farmers were free to select the crop they felt would give them the best financial return. This change provided an incentive to intensify production within rice rotations through the addition of other crops such as wheat.

Rice production in Arkansas has changed considerably since its start at the turn of the century. Maturity times for the primary varieties used have decreased

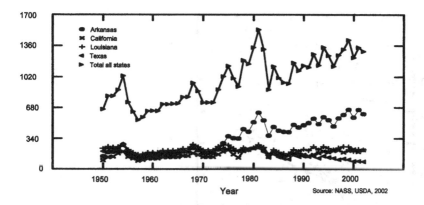

Figure 2 Area of rice harvested (ha $\times$ 1000) for four states (Arkansas, California, Louisiana, Texas) and the United States total area harvested from 1950 to 2001. (Source: NASS, USDA, 2002).

from 117 days from emergence to 50% heading in 1950 to 82 days today (Tolbert et al., 2000). Concurrent with the reduction in days to maturity have been an increase in yield potential and improved management. These two factors have resulted in an increase in Arkansas's average grain yields from 2,200 kg ha^{-1} in 1950 to 7,000 kg ha^{-1} in 2002 (Fig. 3). Today it is not uncommon for good producers to harvest as much as 10,000 kg ha^{-1} grain. Despite this shift to shorter-duration material, it was not until 1985 that the rice breeders in Arkansas initiated a specific "short-duration" breeding program (K.A.K. Moldenhauer, personal communication). It was the genetic material from this program that provided rice varieties that can be used in a rice–wheat rotation.

Of all the row crops produced in the United States, rice traditionally has the highest tillage requirements. Much of the land where rice is produced has been leveled to between 0% and 0.15% slope. Tillage generally begins immediately after harvesting a rice crop with fields disced or plowed. Often rice straw is burned before tillage begins. This practice has been banned in California, and similar actions are expected in other rice production areas. Following primary tillage, fields are lightly tilled with tined implement referred to as a "triple-K" and may be leveled with a land plane before winter rains begin. In the spring, fields will often receive multiple passes of a land plane before seeding. Most fields are dry-seeded in April and the fields flushed to promote germination. At the four- to five-leaf stage, herbicides and fertilizer are applied and the field flooded for the remainder of the season. Phosphorus, potassium, and zinc are applied to fields prior to cultivation. Fertilizer rates are determined from soil tests and will generally range between 45 and 75 kg P$_2$O$_5$ ha^{-1}, 55 and 85 kg K$_2$O

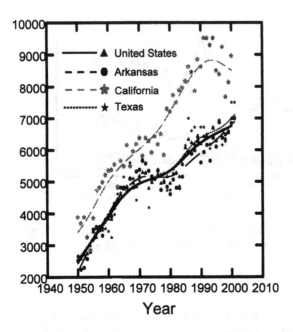

Figure 3   Mean rice grain yield (kg ha^{-1}) for the United States, Arkansas, California, and Texas from 1950 to 2001.

ha^{-1}, and 1 and 2 kg Zn ha^{-1} (Rice Production Handbook, 2001). A majority of the N fertilizer is applied at the four- to five-leaf stage and prior to flooding the field. Rates vary from 80 to 120 kg N ha^{-1}. Most producers follow this application with one or two additional applications of 20 to 40 kg N ha^{-1} as the plants develop. All fertilizer is aerially applied. When possible, farmers apply herbicides prior to flooding the field. Command 3ME (clomazone) at rates between 0.40 and 0.80 kg ai ha^{-1} tank mixed with Facet 75 DF (quinclorc) are recommended and commonly used (Recommended chemicals for weed and brush control, 2002). These chemicals are aerially applied along with any additional herbicides that might be needed for specific weed problems.

Postharvest field management has begun to shift from burning straw to rolling straw, repulling levees, and impounding rainfall on the field to provide winter waterfowl habitat. On fields that are planted into winter wheat, the rice straw is generally burned and the field tilled prior to seeding wheat.

Currently, wheat is grown as a winter crop after rice, soybeans, corn, or sorghum. Following harvest of the previous crop, fields are tilled and sown with a grain drill. Winter rain necessitates the construction of drainage furrows in the field. Sowing is generally done in October and November (Fig. 4). Winter

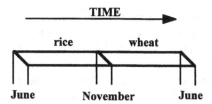

Figure 4 Cropping sequence time line for a winter wheat, summer rice rotation in Arkansas.

temperatures are such that there may be little wheat growth between November and February. Because of this it is difficult to shorten the maturity time of wheat and maintain yields. The harvesting of wheat takes place in May to June. If there is a crop after wheat it is traditionally soybeans. Soybeans are grown because it is possible to select varieties that will mature before temperatures drop in the fall. Given this temporal constraint, the first challenge in developing a successful rice–wheat rotation was to identify rice varieties that could be sown after wheat harvest and would mature before the next wheat planting in November (Fig. 4). The second challenge was to convert what is generally regarded as a tillage-intensive system to a fully no-till system.

II. RESEARCH DESIGN

To address the problem of identifying suitable short-duration rice varieties, a small number of varieties were selected from the existing breeding program and screened in 1998. By 1999 a number of potential experimental varieties had been identified. At the same time a limited number of commercial varieties that could potentially be used for a rice–wheat rotation were introduced into the market. The identification of rice varieties capable of maturing within the rotation constraints (Fig. 4) was considered essential prior to testing the system using conventional and no-till tillage approaches.

In 1999 a long-term cropping systems study was initiated at the University of Arkansas Rice Research and Extension Center in Stuttgart, Arkansas. This study contained 10 rice-based rotations, two of which addressed the issue of using rice and wheat in the same rotation. Those rotations were rice (wheat)–rice (wheat)–rice (wheat) and rice (wheat)–soybeans (wheat)–rice (wheat), where rice and soybeans are the summer crops and wheat the winter crop. A conventional tillage and no-till comparison was established (for this study no-till was defined as no-tillage at any point in the rotation). Conventional tillage consisted of burning the previous crops straw and a minimum of two disc passes and two harrow passes prior to sowing the next crop. There was no straw burning or tillage in the no-till treatments. Two fertility levels were used. They were 120 kg N ha^{-1} plus 42 kg P$_2$O$_5$ ha^{-1} plus 75 kg K$_2$O ha^{-1} for the "standard" fertility treatment and 175 kg N ha^{-1} plus 75 kg P$_2$O$_5$ plus 100 kg K$_2$O for the "enhanced" fertility treatment. Each plot contained two commonly grown varieties. Rice and wheat were sown into 170-mm rows at a seeding rate of 112 kg ha^{-1} and 90 kg ha^{-1} for rice and wheat, respectively.

III. RESULTS

Mean rice grain yield in 2000 was 6,696 kg ha^{-1} (Table 1). There was a 1,058-kg ha^{-1} reduction in grain yield in the no-till plots when compared to the conventional till plots. A large part of this difference is attributed to poor plant stands in the no-till plots. Managing the residual straw and stubble in no-till rice and wheat in a way that does not impede establishing the following crop will be a challenge. We are able to overcome some of those constraints by outfitting seeders with coulters and specialized closing wheels. Rice grain yields increased with increasing fertilizer levels—a trend not present in the full-season rice rotations (data not shown). Of the three varieties used in 2000, XL-6 yielded significantly more than the other varieties. This is a commercial variety that was recently released by a private company and had performed well in our short-duration variety screenings.

Table 1 Rice Grain Dry Weight (kg ha^{-1}) Comparisons for Short-Duration Rice Varieties Grown After Wheat Using Conventional Till or No-Till at Standard or Enhanced Fertility Levels[a]

Effect	Treatment	2000 yield kg ha^{-1}	2001 yield kg ha^{-1}	2002 yield kg ha^{-1}
All	All	6,696	6,300	7,258
Tillage	Conventional	6,451	6,703	7,510
	No-till	5,393	5,846	6,905
Rotation	Following wheat	6,696	6,300	7,258
Fertility	Standard	6,199	6,048	7,157
	Enhanced	6,300	6,552	7,358
	STG95L-28-045	6,300	5,040	Dropped
	Early LaGrue	5,544	Dropped	Dropped
Variety	XL-6	7,962[b]	7,459	Dropped
	XL-7	—	—	7,661
	RU-1093	—	—	6,854

[a]Study located at the University of Arkansas, Rice Research and Extension Center, Stuttgart, Arkansas.
[b]Includes both standard and enhanced fertility levels.

Overall rice grain yields decreased slightly in 2001 (Table 1). The difference between conventional till and no-till plots decreased to 857 kg ha^{-1} while the enhanced fertility plots outyielded the standard fertility plots by 504 kg ha^{-1}. The improved no-till yield when compared to the conventional till yield was the result of modifications made to the grain drill and subsequent improvements in plant stands in the no-till plots. Of the two varieties used in 2001, XL-6 yielded significantly better than STG95L-28-045, which was not used in 2002. We have determined that effectively dealing with wheat straw prior to sowing rice is key to this system. Chopper-spreaders on the combine combined with trash wheels and a "close till" system on the grain drill have proven effective in dealing with this problem. In South Asia, wheat straw is valued as animal feed and thus removed from the field following wheat harvest.

Mean rice grain yields in 2002 were the highest of the three years (Table 1). There was a 605-kg ha^{-1} decrease in grain yield in the no-till treatment combinations when compared to the conventional till plots. This decrease was less than in the previous two years and suggests that a goal of equal yields in the tillage comparison might be attainable. A detailed analysis of these data (not shown) indicated that differences in grain yield between tillage treatments were less in the rice (wheat)–soybean (wheat) rotation than in the rice (wheat)–rice (wheat) rotation. This same trend was present in similar rotations that did not contain wheat. The variety XL-6 was not available in 2002; thus we substituted XL-7, which was higher-yielding than RU-1093—an experimental variety identified as having potential in these systems. Grain samples of both varieties were evaluated at two local mills for milling quality. Results showed a milling yield of 37/60 for the variety RU-1093 and 60/68 for XL-7. Rice is priced at the mill on milling quality. Milling quality for RU-1093 was poor while that for XL-7 was very good. These results show the importance of selecting the correct rice variety for this system.

Net returns for the 2000 and 2001 rice crops show a net loss for all treatment combinations in 2000 (Table 2). This loss is a result not only of low yields but also of increased herbicide inputs. The same situation occurred in 2001 when additional herbicides were required for the no-till treatments. These results, along with our observations, suggest the need for specific weed control programs for no-till rice–wheat rotations. In the past we have applied no herbicides to the wheat crop, a practice that sometimes allows for a buildup of weed populations later in the season. Because there is no or little time between rotation phases, there are limited numbers of herbicides that can potentially be used. Net income for 2002 is not presented, but with higher grain yields (Table 1) and no differences in herbicide applications for the tillage treatments, it is likely there will be a significant advantage in net returns for the no-till treatments.

Establishing a no-till wheat crop after rice was difficult but resulted in grain yields similar to conventional till wheat in 2000 (Table 3). Variety comparisons

Table 2 Net Return ($U.S. ha^{-1}) Comparisons for the Tillage, Fertility, and Variety Main Effects on Rice Production from Rice Grown After Wheat[a]

Effect	Treatment	2000 net returns[b] $ha^{-1}	2001 net returns[b] $ha^{-1}
All	All	($217.41)	$190.54
Tillage	Conventional	($115.40)	$189.52
	No-till	($317.42)	$82.60
Rotation	Following wheat	($217.41)	$190.54
Fertility	Standard	($230.18)	$125.18
	Enhanced	($149.16)	$146.96
	STG95L-28-045	($224.50)	($22.25)
Variety	Early LaGrue	($227.43)	Dropped
	XL-6	($78.18)	$294.37

[a]Study Located at the University of Arkansas, Rice Research and Extension Center, Stuttgart, Arkansas.
[b]Rice was priced at $3.14 bu^{-1} and a 25% land cost included.

were the only significant differences in the 2000 wheat crop. There was no benefit from additional fertilizer nor were the rotation differences significant. We were not able to establish a wheat stand in the fall of 2000 because of early winter rains and late rice harvest. Mean grain yield over all treatment combinations for the wheat crop harvested in 2002 was similar to that in 2000 (Table 3). Unlike the 2000 wheat crop, there was a 1,213-kg ha^{-1} reduction in grain yield in the no-till system. This was further confounded by a 1,340-kg ha^{-1} increase in wheat grain yield when wheat was planted after soybeans versus planting after rice. A more detailed analysis (not presented) indicated that tillage effects were not significant in the rice (wheat)–soybean (wheat) rotation but were highly significant in the rice (wheat)–rice (wheat) rotation. These results reflect the difficulties we had in establishing an acceptable plant stand where wheat is sown after rice and all the rice straw is left on the field. There is an additional problem of water logging in the no-till plots that had a large volume of organic matter lying on the soil surface. Plant stands were less affected in plots containing the variety Pioneer 2580. This illustrates the importance of selecting appropriate rice varieties if direct seeding of both rice and wheat are used (this will be necessary if a no-till approach is used). Now that we have selected acceptable rice varieties and modified our no-till machinery, we are confident that a rice–wheat rotation in a no-till setting is possible.

Table 3 Wheat Grain Dry Weight (kg ha^{-1}) Comparisons for Short-Duration Rice Varieties Grown After Wheat Using Conventional Till or No-Till at Standard or Enhanced Fertility Levels[a]

Effect	Treatment	2000 yield kg ha^{-1}	2002 yield kg ha^{-1}
All	All	3,323	3,320
Tillage	Conventional	3,321	3,894
	No-till	3,325	2,681
Rotation	Following rice	3,340	2,873
	Following soybeans	3,288	4,213
Fertility	Standard	3,405	3,128
	Enhanced	3,241	3,447
Variety	Shiloh	2,884	3,128
	Pioneer 2580	3,761	3,511

[a]Study located at the University of Arkansas, Rice Research and Extension Center, Stuttgart, Arkansas.

IV. RECOMMENDATIONS AND COMMENTS

1. Changing from transplanted rice to no-till seeded rice in a rice–wheat rotation will require identifying short-season rice varieties that are acceptable to this system. Screening of these varieties must be done using direct seeding or the results will not apply to the end goal. Variety selection is one of the two components for success in a no-till rice–wheat rotation and should be given early attention.

2. The change from transplanting to direct seeding should be made before no-till is introduced. Once suitable varieties are identified, they should be tested in a rice–wheat rotation, using conventional tillage and no-till. Proper harvesting and sowing machinery is the second major component in a no-till rice–wheat rotation, and the development of appropriate machines needs to be started early on in the research. How the straw is dealt with is a key issue in both harvest and sowing; both processes need to be studied.

3. There are many good reasons for changing to no-till, but the bottom line is that farmers will look at labor and profit; experimental data must be designed to collect these types of information. These data need to include social issues that are impacted by these changes.

4. All research should be multidisciplinary and done at a minimum of two sites. Whenever possible, a full range of disciplines should collect

data from the same set of plots. This will necessitate larger plots but is cost-effective in the long run and avoids conflicting data sets.

5. Attention needs to be paid to possible problems in weed control. This work can only be done once a functional no-till rice–wheat rotation is established.

6. It is not advisable to sell the no-till approach as one that will greatly increase yields and profits. It will reduce input costs (including labor), improve soil quality, and perhaps improve fertilizer efficiency over time, but it may not result in significant yield increases. I might be wrong on this, but we have looked at a number of cropping scenarios and would not like to see it fail because farmers are expecting more than they will get.

7. The goal of introducing no-till to the South Asia rice–wheat area is worthwhile and should be pursued in a way that will maximize the probability of success.

REFERENCES

Babineaux, L.P. 1967. A history of the rice industry in southwestern Louisiana. M.A. thesis: University of Southwestern Louisiana. Lafeyette, LA.

Cramer, G.L., Wailes, E.J., Gardner, B., Linn, W. 1990. Regulation of the U.S. rice industry, 1965–89. *Amer. J. Agr. Econ. Nov.* pp. 1056–1065.

Dethloff, H.C. 2003. American Rice Industry: Historical Overview of Production and Marketing. pp. 67–87 in Smith C.W., Dilday R.H., Eds. Rice: Origin, History, Technology, and Production: John Wiley & Sons, Inc.. Hoboken, NJ. In press.

Littlefield, D.C. 1981. Rice and Slaves: Ethnicity and the Slave Trade in Colonial South Carolina: Louisiana State University Press. Baton Rouge, LA.

Rice Production Handbook. 2001. University of Arkansas Division of Agriculture Cooperative Extension Service Handbook MP192: University of Arkansas. Fayetteville.

Spicer, J.M. 1964. Beginnings of the Rice Industry in Arkansas. N.p.

Tolbert, A.C., Blocker, M.M., Moldenhauer, K.A.K., Slaton, N.A. 2000. Trends in grain yield and maturity of rice since 1950. Proc. 28th Rice Technical Working Group, Biloxi, Mississippi: Feb. 27–Mar. 1, 2000. Louisiana State University Agricultural Center, Louisiana Agricultural Experiment Station, Rice Research Station. Crowley, Louisiana, p. 54.

17
Impact of No-Till Farming in the United States

B. A. Stewart
West Texas A&M University
Canyon, Texas, U.S.A.

No-till (NT) farming began evolving in the United States during the Dust Bowl days of the 1930s. As a means of controlling wind erosion, stubble-mulching became a common practice. Stubble-mulching involved pulling V-shaped blades about 10 cm beneath the soil surface to cut the roots of weeds and leaving most of the crop residues on the soil surface. As herbicides became common, there was a movement to even less tillage and, in some instances, NT. Water erosion was also significantly reduced by tillage systems that left more crop residues on the soil surface. Clean tillage, either by moldboard plowing or intensive disc tillage, was the standard in the United States for many years. Today, clean tillage is discouraged and is in the minority. The Conservation Technology Information Center (2002) reported that clean tillage was practiced on 42.7% of the cropland in the United States compared to 57.3% in some form of reduced tillage including 17.6% NT.

The adoption of NT has been slow even though many advantages have been demonstrated. Intensive tillage has long been associated with good farming and has become part of the "culture" of many farming communities. Even the seal of the U.S. Department of Agriculture is embellished with a picture of a moldboard plow. A well-prepared seedbed was always pictured as one that was completely free of any vegetation or plant residues on the soil surface.

No-till farming offers many advantages, both on-site and off-site. Soil organic matter content is perhaps the most important of all soil properties, and it is well known that soil organic matter content declines rapidly with intensive tillage. Wind erosion and water erosion also generally increase with increasing tillage, and this can lead to significant off-site damages. In spite of all these advantages, however, most farmers do not adopt NT farming for a variety of reasons.

Knowler and Bradshaw (2001) reported that conservation tillage (CT), which includes NT, is practiced on about 3% of the 1.5 billion ha of arable land worldwide. Most of the land under CT is in North and South America. They found that the conventional farmer believes that tilling the soil will provide benefits to the farm and would increase tillage if economically possible. Knowler and Bradshaw (2001) studied the benefits and costs of CT. They considered on-site and off-site benefits and concluded that many of the incremental costs associated with adopting conservation tillage accrued at the farmer level, but relatively few of the benefits. In other words, society benefits more from CT than the farmer, particularly in the short term. It is also generally recognized that a higher level of management skills is required for CT farming than for conventional tillage farming.

I. RICE PRODUCTION IN THE UNITED STATES

The United States produced only 1.56% of the 586 million mt of the rice produced worldwide in 2001 (FAO, 2001). However, the average yield was 7,015 kg ha^{-1} compared to the worldwide average of 3,852 kg ha^{-1}. In contrast, India produced 22.5% of the world's rice production, but the average yield was only 2,964 kg ha^{-1}, considerably below the worldwide average. In 1961, India produced 24.84% and the United States produced 1.14% of the world's rice. The average yields in 1961 were 3,823, 1,542, and 1,867 kg ha^{-1} for the United States, India, and worldwide, respectively.

In the United States, rice is grown on some 1.3 million ha in the states of Arkansas (46%), California (14%), Louisiana (19%), Mississippi (8%), Missouri (4%), and Texas (9%). Arkansas reported almost 10% of the rice was seeded in NT conditions in 1999 (CTIC, 2002).

II. ADVANTAGES AND DISADVANTAGES OF NO-TILL IN U.S. RICE PRODUCTION

No-till systems are often promoted as a means of reducing fuel and equipment costs. These savings, however, are partially offset or even surpassed by higher herbicide costs. No-till systems also require a higher level of management skills than CT systems. Production costs, therefore, are generally not the deciding factor in the farmer's decision. Decisions by farmers are driven mostly by profit potential. In the case of rice production, the way that NT affects profit is not clear. In Texas, for example, many growers consider the ratoon crop their profit (Shepard,

2000). The ratoon crop will yield from 25% to 40% of the main crop. An early seeding of the rice crop increases the potential for a high-yielding ratoon crop. Shepard (2000) quoted Texas A&M colleague, Garry McCauley, as stating,

> An early season planting is especially where a ratoon crop is grown. If you can't plant on time in Texas, you can't get a second crop. For us, the absolute key advantage to minimum tillage is ensuring a timely planting of the main crop, which will increase the potential of the second crop. The earlier you get it in, the better your ratoon crop potential. Reduced tillage sometimes enables you to move planting back two or more weeks. That extra time increases the yield potential of your main crop and your second crop.

McCauley further explained that another benefit of reduced tillage is firm ground at harvest. The less a field is rutted, the less time and expense a grower spends on getting it back in shape. Minimizing ruts is an important factor in producing a ratoon crop.

A major concern of Texas rice producers using reduced tillage is stand establishment because the seedbed is usually not ideal for seeding. Excessive weed residues and hard soil are common problems, and a drill capable of penetrating the surface under these conditions is required.

Hill and colleagues (1998) reported that most of the rice grown in California was conventional water-seeded rice intensively managed with varying external inputs of farm equipment, fertilizers, and pesticides. No-till is an alternative system followed by a few producers because of cost, marketing, or philosophical considerations. Rice fields fallowed the previous year are most commonly used for drill-seeded rice and the land may or may not be tilled before seeding. When NT is practiced, winter annual weeds are controlled with herbicides and the seed are sown directly with an NT drill. They reported that while the principal goal of NT rice is to reduce costs by eliminating tillage operations, there were often other benefits. Aquatic and broadleaf weeds, the rice water weevil, and other pests that infest water-seeded rice during stand establishment may be reduced. However, they cautioned that NT rice requires careful water management during stand establishment. Grass weeds, normally suppressed under water-seeded culture, may grow vigorously in drill-seeded rice. Yields are usually equal to those of conventional rice-cropping systems.

Beck (1998) reported that many rice producers in Missouri have tried one or more methods of minimum tillage as part of the conventional flood irrigated production system. If rice is following another crop, however, some tillage will almost always be necessary to level the field. Weeds can be controlled by herbicides and rice can be seeded. Even when rice is following rice, the levees from the previous crop can be left standing, but some minor grading or disking is necessary to remove ruts and potholes. Beck identified the advantages of minimum tillage systems as (1) less planting season labor and rush to prepare the seedbed, (2) reduced overall seedbed preparation because of winter weathering, (3) earlier planting because of no spring tillage, and (4) cultural practices other than planting and preplant weed control can be the same as those for conventional tillage systems. He listed the disadvantages as (1) any necessary tillage and leveling needs to be done in the fall, and (2) special attention needs to be paid to chemical control of weeds.

The point that comes across clearly in all the cases discussed above is that there is a higher level of management required when NT, or greatly reduced tillage systems, are used. The success of NT systems depends on the successful use of herbicides. Although a great selection of highly effective herbicides is available, many factors such as climate, soil, and application methods can significantly affect the effectiveness of their use. This often necessitates more applications than scheduled or less weed control than anticipated.

The second point that appears as a common thread in all the studies reported is that any yield advantage is primarily due to earlier seeding. While this is very desirable in some situations, it does not apply to all. Most of the NT interest in the United States has been spurred by improved soil and water conservation, and rice production presents a totally different set of conditions when compared to other crops.

III. ADVANTAGES AND DISADVANTAGES OF NO-TILL IN U.S. WHEAT PRODUCTION

No-till wheat production in the United States has received considerable attention in recent years. Although the adoption rate has increased, it is still below the level that most scientists perceive as optimum. There are two primary reasons that U.S. farmers are interested in NT. The first is for increasing water-use efficiency in regions where water is usually the first factor that limits grain production. The second is the increased ability to plant in a more timely way. The second reason is more important under irrigated conditions and in regions of high precipitation where wheat is sometimes double-cropped with soybean, grain sorghum, corn, or other summer crops. In these situations, the window for planting is very short. Additional reasons, but often of less immediate concern to the producer, are the increased soil organic matter that occurs with NT and the reduced amounts of water and wind erosion.

Greb et al. (1979) were among the first to show the benefits of reduced tillage in dryland regions for increasing soil water storage during fallow periods and the dramatic effects that relatively small increases in plant available soil water at time of seeding would have on grain yields. They summarized more than 60 years of progress in wheat production in fallow systems in the central Great Plains, and the results are presented in Table 1. As the number of tillage

Table 1   Progress in Wheat–Fallow Systems at Akron, Colorado

Years	Tillage	Tillage operations (no.)	Fallow water storage (mm)	% of precip.	Wheat Yield (mg/ha)
1916–1930	Maximum tillage: plow harrow (dust mulch)	7–10	102	19	1.07
1931–1945	Conventional tillage; shallow disk; rodweeder	5–7	118	24	1.16
1946–1960	Improved conventional tillage; begin stubble mulch in 1957	4–6	137	27	1.73
1961–1975	Stubble mulch; begin minimum tillage with herbicides in 1969	2–3	157	33	2.16
1976–1990	Minimum tillage (projected estimate); begin no-tillage in 1983	0–1	183	40	2.69

Source: Adapted from Greb et al. (1979).

operations was decreased, there were marked increases in the amounts of water stored during the fallow periods and dramatic increases in yields. These positive effects tend to accumulate because higher yields result in more residue and increased residue results in more water storage, which translates into higher yields, creating an upward spiral. Soil physical properties are generally also improved, and soil organic matter levels tend to increase. The more than 100% increase in grain yield was due to several improved technologies in addition to increased soil water storage, but the benefit of these improved technologies largely depended on the increased water because water is generally always the first limiting factor in semi-arid regions. Greb (1979) attributed the credit of various technologies as follows: water conservation, 45%; improved cultivars, 30%; improved harvesting equipment, 12%; better seeding equipment, 8%; and fertilizer practices, 5%. The reason for the low impact of fertilizer was because of the long fallow period (16 months) between crops that resulted in sufficient mineralization of nutrients. In recent years, fertilizers, particularly nitrogen and phosphorus fertilizers, have become more important in these regions as yields have continued to increase and fallow periods have tended to become shorter. Wheat–fallow systems that produce one crop every two years have been replaced by wheat–sorghum–fallow or wheat–corn–fallow systems by many producers. These systems produce two crops in three years, and the fallow period between crops is reduced from about 16 months to about 11 months, but the amount of soil water stored during the fallow periods is changed very little. More recently, Peterson and colleagues (2000) working in the same general area have developed NT cropping systems that have more than doubled grain water-use efficiency when compared over years.

Jones and Johnson (1993), working at Bushland, Texas, the southern Great Plains, in an even harsher environment than discussed above, also found significant increases in water storage during the fallow period with NT systems as compared to tilled systems. After only 3 to 4 months of fallow, soil profile water contents averaged 36 mm greater on NT plots than on stubble-mulched tilled plots. Stubble-mulch tillage is the use of V-shaped sweeps that are pulled about 10 cm below the soil surface. Stubble-mulch tillage is itself a good water-conserving system when compared to more intensive tillage systems as shown in Table 1. No-till, however, is much more effective because more crop residue remains on the surface as a mulch and, more importantly, the soil is not disturbed. In semi-arid regions, the soil almost always becomes air-dry to the depth it is tilled.

Under irrigated conditions or in favorable precipitation areas, a primary objective of NT is to shorten the time for seeding. This is particularly important when a producer is growing two crops in the same year. An example is the harvesting of soybean or corn harvest and the seeding of wheat. Conventional land preparation requires substantially longer than simply seeding wheat into the corn or soybean residue. To be successful, however, producers must pay careful attention to details, and special considerations are required. Weisz and Heiniger (1998) discussed some of the problems, as well as the benefits, of NT wheat. They found in North Carolina, Maryland, and Kentucky that early plant growth was often reduced under NT management. They attributed this to crop residues insulating the soil surface, resulting in cooler soil temperatures. This slowed germination, reduced the rate of seedling growth, and slowed tiller development. Another common problem was that seeding sometimes was done when the soils were too wet.

Pieri and colleagues (2002) recently reviewed the potential for adoption of NT agricultural systems. They pointed out that it is difficult to induce changes in farming practices, particularly because the current practices of a given farming system have evolved over many generations. They concluded that improved land management must integrate biophysical and socioeconomic forces. Improved technology is not sufficient for adoption. Pieri and colleagues looked carefully at the rapid adoption rate in Brazil to see what lessons could be learned. From 1987 to 1997, NT practices increased 20-fold in Brazil, compared to only 4.6-fold in the United States. Brazilian farmers did not use NT until the early 1970s, but by 2000, 14 million ha were seeded using NT systems. Pieri and associates found that the initial adoption rate in Brazil was slow. The change was much more than simply changing one technical package to another. Over a period of 25 years, the adoption of NT became the backbone of an integrated approach to sustainable rural development. This included a collaborative effort on social mobilization, education and training, marketing and diversification, and environmental education. The two driving forces were the acute and highly visible land degradation issues such as erosion, and a handful of farmers who realized that radical changes in their farming systems were required to reverse the degradation trends and restore and secure their livelihoods. These innovative farmers became aware of alternative technologies that could allow them to overcome high rainfall runoff and erosion, while at the same time improving their income. At this point, however, the technologies were neither readily available nor validated. Rapid adoption came only after a complete NT development strategy created the conditions required for expansion. This required the development of appropriate rural policies and support systems including governmental and public institution support. The success of Brazil has been dramatic and is being studied by many countries as a possible model for countries throughout the world.

IV. CONCLUSION

Potential off-site damages, particularly from soil erosion, have been the driving force for the promotion and adoption of NT and minimum tillage systems. More recently, increased carbon sequestration of NT systems has become recognized and seen as a real opportunity for improving the environment. Again, however, NT rice and wheat systems present many challenges, and opportunities. The impact that NT systems can have on the environment is not clearly understood. The adoption of NT practices in Brazil has far exceeded that of most other countries in the world. The reasons for their success are being documented and studied by many countries as a means of increasing the adoption rates in their own countries.

REFERENCES

Beck, B. 1998. Missouri rice tillage systems compared: University of Missouri Exension Service. http://agebb.missouri.edu/rice/ricetill.htm.
Conservation Technology Information Center. 2002. http://www.ctic.purdue.edu/CTIC/CTIC.html.
Food and Agriculture Organization (FAO). 2001. FAOSTAT Agriculture Data: Food and Agriculture Organization. Rome. http://apps.fao.org/page/collections?subset = agriculture.
Greb, B.W. 1979. Technology and wheat yields in the Central Great Plains: Commercial advances. *J. Soil Water Conservation* 34:269–273.

Greb, B.W., Smika, D.E., Welsh, J.R. 1979. Technology and wheat yields in the Central Great Plains: Experiment station advances. *J. Soil Water Conservation* 34:264–268.

Hill, J.E., Roberts, S.R., Brandon, D.M., Scardaci, S.C., Williams, J.F., Mutters, R.G. 1998. Rice production in California: University of California at Davis. http://agronomy.ucdavis.edu/uccerice/PRODUCT/rpic12.htm.

Jones, O.R., Johnson, G.L. 1993. Cropping and tillage systems for dryland grain production. Report 93–10. Conservation and Production Research Laboratory: USDA Agricultural Research Service. Bushland, TX.

Knowler, D., Bradshaw, B. 2001. The Economics of Conservation Agriculture: Preliminary Draft for Food and Agriculture Organization. Rome.

Peterson, G.A., Westfall, D.G., Peairs, F.B., Sherrod, L., Poss, D., Gangloff, W., Larson, K., Thompson, D.L., Ahuja, L.R., Koch, M.D., Walker, C.B. 2000. Sustainable Dryland Agroecosystem Management.

Pieri, C., Evers, G., Landers, J., O'Connell, P., Terry, E. 2002. No-till farming for sustainable rural development: Agriculture & Rural Development Working Paper, The World Bank. Washington, DC.

Shepard, P. 2000. Beat the clock: Reduced tillage techniques help you plant on time. Rice Farming Magazine: Vance Publishing.

Weisz, R., Heiniger, R. 1998. Special considerations for no-till wheat. http://www.ces.ncsu.edu/resources/crops/ag580/special.htm.

18

No-Till Farming in Brazil and Its Impact on Food Security and Environmental Quality

Pedro L.O.A. Machado and Pedro Luiz Freitas
Embrapa Soils
Rio de Janeiro, Brazil

I. INTRODUCTION

No-till (NT) is also known as zero-till or direct drilling. It signifies planting grain crops or cover crops directly into the soil without primary or secondary tillage. NT is an innovative soil management system that represents a radical change in agronomic practice. It eliminates soil turnover and promotes agrobiodiversity through crop rotations and through keeping the soil surface covered by mulch from residues of cover crops. The mulch cover is essential to sustainable use of soil resources under harsh tropical and subtropical edapho-climatic conditions (Landers, 1999; Freitas et al., 1998).

The NT system has been used since ancient times by indigenous cultures, simply because they had not yet developed the power needed to till large areas manually to a significant depth (Derpsch, 1998). The modern NT system became feasible with the introduction of herbicides and the development of efficient planters that permit the preparation of a narrow seedbed in vicinity of the seed (Fig. 1). The NT system is a viable alternative for sustainable use and management of natural resources—soil, water, and biodiversity—while mitigating soil degradation, especially by water erosion (Freitas, 2002).

In Brazil, over 17 million hectares (Mha) are under the NT system. It is used to cultivate annual crops such as soybean (*Glycine max* Merr.), maize (*Zea*

(A)   **(B)**

Figure 1 NT planters pulled by tractor (A) and by animal (B) in Southern Brazil. (From M. Darolt with permission.)

mays L.), wheat (*Triticum aestivum* L.), barley (*Hordeum vulgare*), sorghum (*Sorghum bicolor* L.), sunflower (*Helianthus annuus* W.), common beans (*Phaseolus vulgaris* L.), and some cover crops (black and common oat—*Avena strigosa* and *A. sativa*; lupine—*Lupinus* spp.; hairy and common vetch—*Vicia vilosa* Roth. and *V. sativa* L.; etc.).

The history of NT in Brazil is an example of technology integration or technology adaptation to local edaphoclimatic or environmental conditions (Fig. 2) which is essential for a bio-agronomical management system. The development and adoption of NT began in southern Brazil by a handful of dedicated farmers, specialists from governmental institutions (research centers and extension service), and community leaders who recognized the potential of this integrated approach. They liaised with marketing coops, equipment dealers, and chemical providers from the private sector (Pieri et al., 2002; Freitas, 2001). Farmer-to-farmer contact and local Friends of the Land Clubs were the most effective channels for spreading NT throughout the country. No subsidies were provided to these farmers' initiatives. Adoption of NT in Brazil occurred in a rather constrained sociopolitical environment. In the 1970s and early 1980s, many politicians opposed NT farming because they assumed it would exacerbate environmental pollution and be useful to large, well-financed farms only (Pieri et al., 2002; Landers et al., 2002b).

In this chapter we review the introduction of the NT system in Brazil. It began in Southern Brazil in the 1970s, was rapidly adopted by farmers in Central Brazil (Cerrado region), and extended to the northern part of the country in the 1990s (Fig. 3). We emphasize food security and environmental benefits of the long-term use of the NT system.

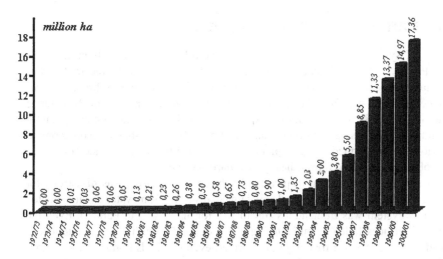

Figure 2 Expansion of NT adoption area in Brazil (1972–1973 to 2000–2001). (Adapted from FEBRAPDP, 2002a.)

Figure 3 Distribution of NT adoption in Brazil (1971: Southern region; 1990: Central "Cerrado" region).

II. SOUTHERN BRAZIL: THE PIONEERS

The NT system was first introduced in Southern Brazil, particularly in the states of Paraná and Rio Grande do Sul in 1972 (Freitas, 2001). In the mid-1960s, annual soybean (summer)–wheat (winter) crop succession was broadly introduced, replacing coffee plantations in Paraná and forest/pastures in Rio Grande do Sul. The shift to a system that required biannual soil preparation and planting required major changes in soil management practices. Plow-based seedbed preparation was performed using implements developed for the temperate conditions (moldboard and disc plows, heavy disc harrows, etc.).

Extensive tillage in tropical and subtropical Ferrasols was disastrous for the environment. Soil losses and soil damage resulting from water erosion created by this system was tremendous (up to 114 ton ha^{-1} y^{-1} in Paraná). Even the use of terraces failed to mitigate high soil losses (57 ton ha^{-1} y^{-1}; Mondardo and Biscaia, 1981). Erosion gullies formed even on fields bordered by contour dams. Soils were intensively plowed and then harrowed several times by light or heavy disc equipment. This exposed the soil surface to direct impacts of rains. In subtropical and tropical regions, storms with rainfall exceeding 150 mm hr^{-1} are not unusual. Rainfall under this condition causes soil crusting, decreases water infiltration, and thus leads to high water runoff. The presence of plant residues on the soil surface protects soils from the direct impact of raindrops and diminishes the velocity of water moving downslope as runoff (Wünsche and Denardin, 1978; Muzilli, 1979). Studies conducted by Roth et al. (1986, 1987, 1988) compared the interaction of tillage systems and crop rotation on soil cover, water infiltration, and erosion control. They showed that the water infiltration rates increased as soil cover increased. Roth and colleagues concluded that soil must be covered with at least 4–6 t ha^{-1} of residue mulch at all times. This is only possible if NT is adopted in combination with mulch-producing cover crops and appropriate crop rotations. The NT was then considered to be a promising soil management alternative to combat erosion. Initially, as soil scientists searched for strategies to combat soil erosion by water, the experiments were confined to research stations. However, spontaneous adoption by farmers in Paraná in 1972 (Mr. Herbert Bartz) and in Rio Grande do Sul in 1974 (Mr. José Veiga Mello) motivated scientists to start on-farm experiments. Also in Paraná (Campos Gerais) in 1976, several farmers (Mrs. Manoel Pereira, Franke Djikstra, and Wibe de Jagger) began to use NT due to demonstrative effects and the exchange of farmers' experiences with the support of researchers and extension agents. Consequently, the Earthworm Club was established in 1979 (later Friends of the Land Club) to provide support to farmers for this emerging but "ready-to-go" technology. These farmers' organization helped organize the Brazilian Federation for Zero Tillage into Crop Residues (FEBRAPDP) in 1992. This organization led advocacy and support for NT systems throughout the country (Landers, 1999; Freitas, 2001).

Table 1 Total Area Under Zero-Tillage in the States of Rio Grande do Sul (RS), Santa Catarina (SC), and Paraná (PR)

Year	RS	SC (1,000 ha)	PR
1980	40	—	344
1990	205	—	400
2000	3,500	950	4,000

Sources: Montoya (1984), Derpsch et al. (1991), Muzilli (2002) and, Wiethölter (2000).

Farmers adopted NT not only to control soil erosion, but also to reduce production costs related to fuel (up to 60%), labor, and machinery maintenance. Data in Table 1 show the area under NT in three states in Southern Brazil for the last three decades of the 20th century. Initially, in 1980, Paraná had the largest area under NT, comprising 65% of the total area under the NT system in Brazil. The rate of adoption was high in all states until 2000. However, as will be discussed later, the Cerrado region of Central Brazil has the largest area under NT in Brazil under continuous use.

Summer grain crops, such as soybean and maize, are mostly grown on large, intensively mechanized farms in rotation with a winter grain crop (wheat) or winter cover crops (e.g., black oat, lupine, hairy vetch, etc.). Rice (*Oryza sativa* L.) in Brazil is grown on about 3.8 M ha under different conditions including wetland irrigated or paddy, and upland rainfed conditions. More than 60% of the 11.7 million t (Table 2) is on less than 40% of the total area. The yield of paddy rice is more than three times that produced under rainfed conditions.

The NT system was first tried during the 1960s in Southern Brazil. In the state of Rio Grande do Sul, research experiments were initiated in 1980 on EMBRAPA experimental farms. They resulted in the adoption of NT by farmers such as Eurico Dorneles and in new efforts to diffuse them through the Club of No-till Irrigated Rice Farmers (Mello, 1995). By 1994, 25% of the area under

Table 2 Rice Area and Production in Brazil in 1999

System	Area (1,000 ha)	Production (1,000 t)	Yield (t ha^{-1})
Upland rainfed	2,430	4,450	1.8
Wetland irrigated	1,380	7,324	5.5
Total	3,810	11,774	3.1

Source: Adapted from Barbosa Filho and Yamada (2002).

paddy rice in the state had adopted some kind of no-till. Principal advantages included control of weeds of red and black rice (*Oryza sativa* L.), reduced nutrient and soil loss, and decreased costs of production. Important components of the NT system include annual crop rotation and succession, mulch production with cover crops such as black oat, common rye (*Secale cereale* L.), hairy and common vetch, and Chickling vetch (*Lathyrus sativus* L.), etc., and the integration with winter pastures such as annual or Italian ryegrass (*Lollium multiflorum* Lam), common rye, and barley.

The adoption of NT was slower in smallholder farming areas where no-till planters and sprayers required adaptation to animal and human traction. The first NT animal-powered planter (a prototype called "Gralha Azul") was released in 1984 by IAPAR, and many models were derived from it. To date, there are several animal-powered machines ranging from knife-rollers to sprayers (Fig. 4). Darolt (1998) reported that the area under NT in smallholder farming is approximately 20,000 ha. It comprises 7,000 rural properties in Santa Catarina and Paraná states. After two Latin American symposia on NT (1993 and 1998), the availability of improved technologies in planting, integrated weed management, crop rotation, and cover crop enhanced successful adoption of NT in smallholder farming and increased the area to 69,000 ha in 2002 (FEBRAPDP, 2002b).

III.   THE CERRADO REGION: THE WIDESPREAD ADOPTION OF THE NT SYSTEM

Once adopted in Southern Brazil, the NT system rapidly expanded northwards, especially into the subhumid tropical region (Brazilian Cerrado), which covers

Figure 4   An animal-powered sprayer in Santa Catarina, Brazil.

207 M ha of Central and part of Northern Brazil. Edaphoclimatic constraints experienced by farmers were related to chemical and physical properties of the highly weathered soils (Ferrasols, Dystric Nitosols, and Ferralic Arenosols). These soils are characterized by a low pH of 4.0 to 5.5, high Al saturation (up to 80% of the base saturation), low cation exchange capacity (CEC < 13 cmol kg^{-1}), and high adsorption capacity for phosphorus and sulphate (Spehar, 1994). This low inherent soil fertility combined with good physical conditions (e.g., soil porosity, soil depth, good draining, and workability), made these soils highly susceptible to degradation under intensive tillage with offset discs and disc plows (Blancaneaux et al., 1993; Freitas, 1994). There are two distinct seasons: the rainy season from September to May, and the dry season from June to August. Cultivation is limited to the rainy season, but dry spells in the rainy season are also common and crop growth during the reproductive phase is often seriously impaired (Assad et al., 1994). The relatively flat terrain of the Cerrados led farmers to believe that soils were free from water erosion commonly observed in Southern Brazil (Spehar, 1996). Intensive soil tillage (plowing followed by several disc harrowings) was needed to improve efficiency of preplanting incorporated herbicides, such as trifuralin (Dinitroaniline, $C_{13}H_{16}F_3N_3O_4$). A high soil erodibility (0.013 Mg ha h ha^{-1} MJ^{-1} mm^{-1}) and rainfall erosivity (8,355 MJ mm ha^{-1} h^{-1}) exacerbate risks of water erosion (Dedecek et al., 1986; Silva et al., 1997) and of decline in soil structure as soil preparation frequently occurs under unfavorable conditions (Freitas, 1994).

Farmers were motivated during the 1970s to start farming in the region because of low land prices and government infrastructural (e.g., Polocentro Program) and technical (e.g., Embrapa Cerrados) support. Soybean varieties adapted to low latitudes facilitated establishment of annual cropping systems that includes maize, upland rice, cotton (Gossypium hirsutum L.), common beans, and sunflower (Spehar, 1994). By 1990, approximately 10 M ha were under cultivation of annual crops and 35 M ha under planted pastures in the Cerrado region (Macedo, 1996).

During the mid-1980s, NT was adopted as consequences of severe economic and environmental losses and soil erosion in the region (Freitas, 2002). However, successful adoption of NT was impeded by the lack of cover crops for the dry season to produce adequate mulch. This is the main challenge to adopting NT in the tropics (Cerrado, Amazon, and semi-arid regions). The introduction of pearl millet (Pennisetum glaucum L. R. Br. and P. americanum L.), sorghum, fodder radish (Raphanus sativus L.), bengal or sunn-hemp (Crotalaria juncea L.), and more recently pasture grasses (e.g., Brachiaria spp.) and adapted crops such as quinoa (Chenopodium quinoa Willd.), amaranth (Amaranthus spp.), cajan or pigeon pea (Cajanus cajan L.), finger millet (Eleusine coracana), and teff (Eragrostis tef Zucc.), enabled production of an adequate amount of mulch essential to sustain NT systems in the tropics (Spehar and Souza, 1996). Subsequently, different methods of establishing cover crops have been tested, such as in combination with main summer crops, relay cropping with summer crops, seeding with the first spring rains, etc. These cover crops are essential because of the need for mulching to realize the full potential of crops grown in rotation under a NT system.

Upland rice, grown under rainfed conditions on well-drained soils (Ferrasols), had its heyday from the 1960s to the 1980s. It was produced on 4.5 M ha and was an affordable alternative immediately after deforestation (Embrapa Rice and Beans, 2002). Yield differences between upland and wetland systems of rice production are large (Table 2), but upland rice is better adapted to ubiquitous acid soils and demands less investments for large-scale cultivation. Due to its reputation as an excellent crop for newly deforested areas or degraded pastures in the Cerrado region and as a subsistence crop in northeast Brazil, it took awhile for it to be included in a rotation with cash crops (e.g., soybean, maize, and cotton). Upland rice production covers more than 2.4 M ha, concentrated in tropical areas in the states of Mato Grosso, Mato Grosso do Sul, Pará, and Maranhão (Barbosa Filho and Yamada, 2002). Efforts are being made to cultivate it in NT system, which can produce high yields (6,000 kg ha^{-1}), on soils with good structure involving a large volume of biopores, absence of subsurface compaction, and in regions with favorable rainfall distribution, especially during germination and sprouting (alleviating soil compaction) and flowering (avoiding water stress) stages. Achieving high yield has been made possible by advances in machinery adaptation and nitrogen fertilizer management under appropriate crop rotations under NT systems (Fageria, 2001).

The area under NT in the Cerrado region increased from 180,000 ha in 1992 to 6 M ha involving over 3,000 farmers in 2002 (FEBRAPDP, 2002c). Moreover, some areas under degraded pastures in the Cerrado region are being restored by the NT system through integration of crops with livestock (Lara-Cabezas and Freitas, 2001).

A. Crop-Pasture Rotations Using No-Till

The dominant pasture grasses in the Cerrado region are *Brachiaria* spp. and *Andropogon* spp., which cover approximately 48 M ha. About 50% of these pastures are degraded to some degree (Macedo, 1995; Spain et al., 1996). Spain et al. (1996) reported that causes of pasture degradation are complex. Despite being efficient recyclers of most nutrients, intense grazing pressure leads to N deficiency in pastures. This, combined with the lack of fertilizer application by cattle ranchers, leads to declines in productivity within three to four years. Boddey et al. (1996) reported some negative consequences of pastoral land use (mostly sown to *Brachiaria* spp.) and suggested the introduction of an N-fixing legume into the sod.

The use of a crop-pasture rotation (ley farming) has the potential of reversing pasture degradation, increasing its carrying capacity, and allowing mulch production for sustainable annual crops. The benefits to both crops and pasture include the improvement of soil fertility since the rotation strengthens nutrient recycling, increases the soil organic matter pool, enhances soil physical conditions, and increases soil biological activity (Saturnino, 2001).

Brachiaria spp. is controlled with desiccant herbicides before seeding of annual crops (soybean, maize, common beans, and others) directly into dead or dying sod. Annual crops are cultivated for two or three years, after which pasture grass is established (Spain et al., 1996; Machado et al., 1998). Improved soil fertility enables farmers to introduce more demanding grasses such as *Panicum maximum*. After crop cultivation, pastures remain green throughout the dry season, thus supplementing forage availability for the herd during the most critical period of the year from May to September (Spain et al., 1996).

The adoption of NT ley farming, however, may take a long time because of the need for more infrastructure and capital, and more complex farm administration. Spain et al. (1996) reported that farmers, who have been experimenting with these integrated systems for several years, have already overcome the greatest obstacle, namely the tradition to be only ranchers or farmers.

IV. IMPACTS OF NO-TILL FARMING ON FOOD SECURITY AND ENVIRONMENTAL QUALITY

Since the 1980s, research on soil erosion control has shown that NT without adequate mulch is not effective in controlling soil erosion by water. In Paraná, approximately 5 Mha (80% of the total area under grain crops) used to be left fallow during the winter season. Alternatively, wheat was cultivated, but its growth was often impaired by harsh weather conditions and poor rainfall distribution (Laurenti et al., 1986). Hence, IAPAR, in cooperation with the German Agency for Technical Cooperation (GTZ), research institutions such as the Embrapa Wheat Center and the ABC Foundation, the Paraná state extension service, private companies such as ICI and Semeato, and coops such as CCLPL, Batavo, and FECOTRIGO, started an intensive research program on cover crop selection. This included rotation with summer cash crops, such as maize, soybean, and cotton, and assessment of impact on soil quality. Black oat is a highly promising cover crop introduced as an alternative to fallow and wheat cultivation. It improves mulch production (2–6 Mg dry mass ha^{-1}) and enhances soil moisture storage in the root zone. In comparison, lupine and hairy vetch reduce the use of nitrogenous fertilizer by 90 kg N ha^{-1}. Growing cover crops also improves yields of maize and soybeans (Derpsch and Calegari, 1985; Monegat, 1991; Paraná, 1994). Early reports on yields of grain crops show that NT systems generally increase the productivity of soybean and wheat, particularly in years characterized by severe droughts (Embrapa, 1979; Vieira, 1981; Muzilli, 1981; Santos et al., 1995). Data in Table 3 show that adoption of NT increased the yield of wheat by a factor of 1.5 to 4 and that of soybeans by 2.5 in seasons with irregular rains.

Yield differences between NT and plowed systems contrast more during seasons with short droughts and irregular distribution of rainfall, especially in crop rotations with no cover crop. In a study conducted on the same soil type and area, Sidiras et al. (1981) reported that, compared to plowed systems, the water-holding capacity of the soil at field capacity (0.33 bar) in NT systems increased by 31%, 20%, and 5% for 3–10; 12–20; and 22–30-cm depth, respectively. In irrigated systems, there was a water economy of 10% (800 mm for two crops), with a reduction in pumping costs of US$200/ha, leading to on-farm benefits of US$20 ha^{-1} y^{-1} (Landers et al., 2001). In grain production systems, crop rotation with winter cover crops was effective in combating water erosion (Roth et al., 1986) and lowering the incidence of root diseases in wheat (Santos et al., 1990).

However, compared to plowing, the greatest benefit of NT is its higher potential to sustain adequate crop yield with minimum soil and water loss. In a comprehensive survey of literature published between 1977 and 1997 in different regions, soil types, and crop cultivation, De Maria (1999) concluded that compared to plowed systems, soil losses in NT systems decreased 75%, ranging from 0.5 to 5.0 t ha^{-1} y^{-1}. Data in Table 4 indicate the amount of nutrients lost by water erosion between 1988 and 1994 in a clay Ferrasol. The least losses of soil and water were observed in soil under a NT system. Compared to plowed systems,

Table 3   Soybean and Wheat Yield (kg ha^{-1}) in Two Tillage Systems Under Normal and Irregular Rainfall in Northern Paraná State

Rainfall	Crop	Harvest year	Management System	
			NT	CT[a]
Normal	Soybean	1977	3,280	3,280
Normal	Wheat	1979	1,281	1,854
Normal	Wheat	1980	1,789	1,867
Irregular	Soybean	1978	1,448	530
Irregular	Wheat	1977	967	609
Irregular	Wheat	1980	2,036	507

[a]CT: conventional tillage (disc plow followed by light disc harrowing).
Source: Muzilli (1984).

Table 4 Losses of Ca^{2+}, Mg^{2+}, P, K^+ (in Soil and Runoff), and SOC (in Soil Only) by Water Erosion in a Ferralsol Under Different Tillage Systems in Mato Grosso do Sul Between 1988 and 1994

Soil tillage[a]	Surface losses[b] by water erosion (kg ha^{-1}y^{-1})				
	Ca^{2+}	Mg^{2+}	P	K^+	C[c]
ChP	9.9 b	0.80 c	0.31 a	4.90 b	49.8 c
HH	15.5 a	1.10 b	0.87 a	9.10 a	93.9 b
ZT	3.1 c	0.30 d	0.15 a	1.40 c	16.8 d
DP	19.2 a	1.70 a	0.84 a	7.80 a	125.3 a
F[d]	**	**	ns	**	**
CV(%)	29.8	24.3	137.5	39.9	19.0

[a]ChP: chisel plow + light disc harrowing; HH: heavy disc harrowing + light disc harrowing; NT: no-till; DP: disc plow + light disc harrowing.
[b]Means followed by the same letter in the same column are not significantly different at $P = 0.05$ (Duncan test).
[c]SOC in soil sediments.
[d]F test: ** and ns indicate that differences between means are significantly different at $P < 0.01$ or not significant at $p < 0.05$, respectively.
Source: Hernani et al. (1999).

the adoption of NT decreased loss of total Ca^{2+}, Mg^{2+}, and K^+ loadings by 82% to 84%, and of soil organic carbon (SOC) bound to soil particles by 86%.

Reduced soil water erosion by NT systems mitigates several off-site erosion-related problems including water pollution. Data in Table 5 estimate the economic loss due to soil water erosion primarily under plowed systems to be over US$5 million y^{-1} (Santos and Câmara, 2002). Nutrients and pesticides transported with sediments and in runoff affect water quality. Uri (1999) reported a number of economic and environmental benefits associated with the NT systems in production agriculture in the United States. Similar benefits also occur in tropical and subtropical regions. For example, NT mitigates excessive nutrient runoff, which causes eutrophication, and excessive growth of algae and vegetation in natural waters. The anoxia created by decay of the biomass has an adverse impact on aquaculture. Floating algae or high siltation can clog intake pipes of hydroelectric power and filtration systems, increasing the cost of water treatment. In Paraná, a reduction of 43% in turbidity reduced water treatment costs by US$5.77/10,000 m^3 (Carroll, 1997), and it is assumed that 50% of this benefit was due to adoption of NT agriculture. Bassi (1999) observed the positive impacts of integrated management of microcatchments involving NT in the state of Santa Catarina. Data in Table 6 suggest some general environmental benefits. Muzilli

Table 5 Economic Impacts of Soil Erosion in Brazil

Negative impacts	Total value (US$ million y^{-1})
Losses of nutrients and soil organic matter	3,178.80
Land devaluation	1,824.00
Water treatment for consumption in urban areas	0.37
Maintenance of roads	268.80
Recovery of water reservoirs	65.44
Total	5,337.40

Source: Santos and Câmara (2002).

(2002) observed that due to a well-distributed network of river basins, the state of Paraná has 27 hydroelectric power stations producing approximately 46 million megawatt-hours of power, which is transferred to major industrial and urban areas in both Southern and Southeastern Brazil. Additionally, the Itaipú Hydroelectric Powerstation in the Paraná River produces 90 billion kilowatt-hours, or approximately 25% of the total Brazilian power demand. Agriculture is the dominant land use in the Paraná river basin and in the 1,350 km^2 of area surrounding the lake. Thus, adoption of NT by farmers is important to maintain adequate crop yield while also offering environmental benefits to the entire society.

The economic impact of adoption of NT was estimated by Landers et al. (2001), considering both direct and environmental benefits (Table 7). They estimated the area under NT to be 14.3 Mha during 1999–2000. The main estimated indirect benefit is the huge potential of mitigating deforestation through intensification of crop–pasture rotation and a dramatic reduction in the off-farm effects of soil erosion (e.g., road maintenance, water treatment, siltation), aquifer re-

Table 6 Impacts of the Integrated Management of Microwatersheds with No-Till in Santa Catarina, Brazil

Area under NT	From <1% in 1990 to 48% in 1996
Coliform bacteria	68% decrease between 1988 and 1998
Costs for water treatment	46% decrease between 1991 and 1996
Average crop yield	24% increase between 1990 and 1996
Area with planted forest	Increase from 8% in 1990 to 28% in 1996

Source: Bassi (1999).

Table 7 Economic Impacts of NT Adoption in Brazil

Categories	Impact (10^6 US$ y^{-1})
Direct Benefits to Farmers	**356.1**
Land value reclamation due to NT adoption (soil, water and nutrient loss reduction and crop yield increase)	332.9
Irrigation pumping economy (reduction in water use)	23.2
Reductions in public spending	**62.1**
Maintenance of rural roads	48.4
Municipal water treatment	0.5
Extended reservoir life due to reduced silting	9.2
Reduced dredging costs in ports and rivers	4.0
Environmental benefits to society	**184.1**
Increase in water availability due to greater aquifer recharge	114.4
Reduced water use due to irrigation efficiency increase	6.6
Carbon credits for the decrease in CO_2 emission (lower fuel consumption and C sequestration in soil and surface residues[a]	63.1
Benefits for mitigating deforestation through intensification crop–pasture rotation	**784.0**
Total benefits	**1,386.3**

[a]World market price of carbon at US$10.91 Mg^{-1}.
Source: Adapted from Landers et al. (2001).

charge, and perennial stream flow due to an increase in rainfall infiltration. For example, Bassi (1999) reported approximately US$90,000 in savings in rural road maintenance in a project area of Santa Catarina State from 1992 to 1999.

The use of herbicides in NT has often raised ecological concerns. In a study conducted in the state of Santa Catarina, Skora Neto (1998) observed a 97.7% decline in weed infestation after 5 years as a result of reduced weed emergence. Similar results were reported by Derpsch (2002), who observed that systematic use of cover crops and rotations combined with the proper use of chemicals reduces the amount of herbicide use in NT by 80% from that used in plowed soils. In Rio Grande do Sul, approximately 6 Mg ha^{-1} of dry matter of black oat rolled down on the soil surface with a roller knife eliminates the need of herbicide applications (Roman, 1990).

Paraguay is a neighboring country with soils similar to that of Southern Brazil. Kliewer et al. (2000) and Calegari (2001) reported that sunflower–black oat/soybean–wheat–soybean–lupine/maize rotation reduced herbicide rates to zero with a 3-year saving of US$57 ha^{-1} in Paraguay. Reduced application frequencies also reduce soil compaction and erosion. In contrast to plowed systems, herbicides used in NT systems (such as glyphosate–*N-phosphonomethyl glycine*) have low residence time in the soil and low risk of water contamination because of rapid biodegradation (50% of the original amount is decomposed within 28 days and 90% in 90 days) (Rodrigues and Almeida, 1998; Schütte, 2000). More research is needed to improve the use of integrated weed management systems, particularly those related to the persistence/solubility, bioavailability, and mammalian toxicity of both active compounds and their metabolites. Despite continued research needs to improve its performance in Brazil (Machado and Silva, 2001), environmental benefits of NT agriculture far exceed those of the plowed systems.

Compared to plowed systems, NT *per se* results in large reductions of CO_2 emissions by improving crop residue management. Residues are neither burned nor incorporated into the soil (Derpsch, 2002). Additionally, NT systems help mitigate climate change through a drastic decrease in soil erosion, which causes reduced C loss and immobilizes SOC in the long term. Despite these benefits, studies of long-term experiments on SOC sequestration through NT remain controversial in Brazil because of differences in the depth of soil sampling, correction of soil C pools on an equal soil mass basis, differences in land-use history (especially if pasture plants have been previously planted in an area under grain cropping), and crop rotation with N-fixing cover crops (Bayer et al., 2000; Freixo et al., 2002; Sisti et al., 2002). NT farming practices encompass several soil and crop management practices. Thus, results of studies to identify the most appropriate crop rotation to produce high yield and enhance SOC sequestration must also consider pest and disease control. Research information is available on both winter and summer cover crops adapted to different regions (Derpsch and Calegari, 1985; Monegat, 1991; Calegari et al., 1992; Calegari, 1995). However, farmers in southern Brazil have not changed from an oat cover to a legume cover (e.g., hairy vetch) in winter because of the high risk of *Sclerotinia* disease and the lack of availability of legume cover crop seeds (R. Molin, personal communication). Results obtained from long-term plot experiments are relevant for simulation models, but they cannot be extrapolated to a landscape or a watershed scale. In a study conducted on a temperate Chernozem soil in Canada, Bergstrom et al. (2001a, b) showed that topography, soil series alone or grouped by drainage class, provides a framework to assess the influence of management practices on SOC pool at large scales representative of farm management units. Based on a chronosequence with unequal sequence of crops, Sá et al. (2001) reported an SOC sequestration rate of 1 Mg ha^{-1} year^{-1} over a 22-year period in the top 40-cm layer of soil. Measurement of SOC pools of a toposequence under real-farm situations seems to be the most adequate procedure but represents challenges when evaluating the impact of NT systems on SOC sequestration, especially in relation to the Clean Development Mechanism (CDM) under the Kyoto Protocol.

To ensure the continuous increase in NT farming in Brazil and elsewhere in the tropics and subtropics, farmers must be given some incentives. This is justified because society benefits from NT farming. It reduces soil degradation by runoff and soil erosion and helps to conserve and preserve natural resources, including biodiversity. In addition, NT ensures food security for the growing world population by promoting high-yielding agriculture on a sustainable basis and increasing the carrying capacity of pasture through rotation with annual crops (Landers et al., 2002a; Landers and Freitas, 2001).

Expansion of NT systems to semiarid regions will also occur in the near future due to farmer-led organizations, particularly Friends of the Land Clubs. Previous findings report limited benefits of NT in the semiarid environments of West Africa (Hullugale and Maurya, 1991; Nicou et al., 1993). However, in Brazil, interest in NT is growing because of substantial social and high labor costs related to plowed systems (Mrabet, 2002). Similar ecological conditions exist in northeastern Brazil, where the NT system can be adapted with interactive participation of all stakeholders.

NT farming was initially considered to be an efficacious strategy to combat soil erosion in subtropical southern Brazil. However, it is now being rapidly adopted by farmers in tropical regions like the Cerrado, which consists of over 17 Mha. Farmers in semiarid regions in northeastern Brazil are motivated primarily by potential savings in fuel and machinery maintenance costs. Evidences exists about the positive contribution of NT farming to improved food security and about its social and environmental benefits. As stated by Pieri et al. (2002), to be successful and sustainable, the development of NT farming must be considered as a long-term process, which is highly location-specific and necessarily adapted to farmer circumstances. Benefits of NT farming extend beyond the rural property. The public, including urban communities and politicians, must be better informed about the advantages of NT systems with regard to costs of water treatment and hydroelectric system maintenance, increases in water supplies due to greater aquifer recharge, and mitigation of the global climate change by soil carbon sequestration.

V. CONCLUSIONS

NT farming represents a radical change in agronomic practices. It eliminates soil turnover, promotes agrobiodiversity through crop rotations, and keeps soil surface covered with residue mulch, essential for tropical and subtropical edaphoclimatic conditions. Over 17 million hectares are being farmed using the NT system in Brazil. Annual crops as well as cover crops are grown under this system. Brazil's adaptative effort is an example of technological integration, essential for a bio-agronomical management system.

A principal benefit of NT systems is reduced soil water erosion caused by plowed conventional management systems, which represent an annual savings of about US$5 million. NT also mitigates several off-site erosion-related problems, including water pollution. Adoption of NT in large, mechanized farms, and by smallholder farmers, not only reduces soil erosion, but also reduces the costs of production through savings in fuel, labor, and machinery maintenance. In the Cerrado subhumid tropical area, NT has helped reverse pasture degradation and mitigate deforestation by intensifying crop–pasture rotations. The adoption of the NT system is also important for maintaining adequate crop yield while extending environmental benefits to whole rural and urban society, and to the entire human-kind by mitigating global climate change.

REFERENCES

Assad, E.D., Sano, E.E., Matsumoto, R., Castro, L.H.R., Silva, F.A.M. 1994. Veranicos na região dos Cerrados brasileiros: Frequencia e probabilidade de ocorrência. In Assad E.D., Ed. Chuva Nos Cerrados: Análise e Espacialização: Embrapa Cerrados. Brasília. Brazil, 43–48.

Barbosa Filho, M.P., Yamada, T. 2002. Upland rice production in Brazil. Better Crops International, PPI. Saskatchewan. Canada. 16 (special supplement) 43–46. May 2002.

Bassi, L. 1999. Better environment, better water management, better income and better quality of live in micro-catchments assisted by the land management II project/ World Bank. Rural Week of the World Bank 24–26 March 1999. Washington, DC.

Bayer, C., Mielniczuck, J., Amado, T.J.C., Martin-Neto, L., Fernandes, S.B.V. 2000. Organic matter storage in a sandy clay loam Acrisol affected by tillage and cropping systems in southern Brazil. *Soil Till. Res* 54:101–109.

Bergstrom, D.W., Monreal, C.M., St. Jacques, E. 2001a. Spatial dependance of soil organic carbon mass and its relationship to soil series and topography. *Can. J. Soil Sci* 81: 53–62.

Bergstrom, D.W., Monreal, C.M., St. Jacques, E. 2001b. Influence of tillage practice on carbon sequestration is scale-dependent. *Can. J. Soil Sci* 81:63–70.

Blancaneaux, P., Freitas, P.L. de., Amabile, R.F., Carvalho, A.M. de. 1993. Le semis direct comme practice de conservation des sols du Brésil Central. *Cah. ORSTOM, sér. Pédol.* Paris XXVIII(2):253–275.

Boddey, R.M., Alves, B.J.R., Urquiaga, S. 1996. Nitrogen cycling and sustainability of improved pastures in the Brazilian Cerrados. In Pereira R.C., Nasser L.C.B., Eds. Proceedings of 1st International Symposium on Tropical Savannahs: Biodiversity and Sustainable Production of Food and Fibers in the Tropical Savannahs: Embrapa Cerrados. Planaltina. Brazil, 33–38.

Calegari, A., Mondardo, A., Bulisani, E.A., do Wildner, L.P., Costa, M.B.B., Alcantara, P.B., Miyasaka, S., Amado, T.J.C. 1992. Adubação verde no sul do Brasil, 2nd ed.. AS-PTA: Rio de Janeiro. Brazil.

Calegari, A. 1995. Leguminosas para adubação verde de verão no Paraná: IAPAR. Londrina. Brazil. Circular No. 80.

Calegari, A. 2001. Sistemas de rotação de culturas e seus efeitos ambientais e econômicos no Centro-Sul do Cerrado. In Hernani L.C., Fedatto E., Eds. Sustentabilidade, sim!:

Anais do 5° Encontro Nacional de Plantio Direto no Cerrado: APDC. Brasília. Brazil, 23–28.

Carroll, M. 1997. Paraná Rural—Projeto de Manejo e Conservação do Solo do Paraná. BIRD—Unidade de Gerenciamento—Brasil. Brasília. Brasil.

Darolt, M.R. 1998. Plantio direto: Pequena propriedade sustentável: Circular 101: IAPAR. Londrina. Brazil.

Dedecek, R.A., Resck, D.V.S., Freitas Júnior, E. 1986. Perdas de solo, água e nutrientes por erosão em Latossolo Vermelho Escuro dos Cerrados em diferentes cultivos sob chuva natural. *Rev. Bras. Ci. Solo* 10:265–272.

De Maria, I.C. 1999. Erosão e terraços em plantio direto. *Bol. Inf., Soc. Bras. Ci. Solo* 24(3):17–21.

Derpsch, R., Roth, C.H., Sidiras, N., Köpke, U. 1991. Controle da erosão no Paraná, Brasil: Sistemas de cobertura do solo, plantio direto e preparo conservacionista do solo. GTZ/IAPAR. Eschborn. Germany.

Derpsch, R., Calegari, A. 1985. Guia de plantas para adubação verde de inverno. Documentos, 9. IAPAR. Londrina. Brazil.

Derpsch, R. 1998. Historical review of no-tillage cultivation of crops. In Seminar No-Tillage Cultivation of Soybean and Future Research Needs in South America, 1: Foz do Iguaçu. Brazil. Proceedings [Toquio]: JIRCAS 1998(JIRCAS Working Report no. 13.), 1–18.

Derpsch, R. 2002. Sustainable agriculture. In Saturnino H.M., Landers J.N., Eds. The Environment and Zero-Tillage. Brasília. Brazil, APDC, 31–51.

Embrapa, J.N. 1979. Relatório técnico anual do Centro Nacional de Pesquisa de Trigo 1977–1978. Embrapa/DID. Brasília. Brazil, 175p.

Embrapa Rice and Beans. 2002. Sistemas de cultivo de arroz. Available at http://www.cnpaf.embrapa.br.

Fageria, N.K. 2001. Nutrient management for improving upland rice productivity and sustainability. *Commun. Soil Sci Plant Anal* 32:2603–2629.

FEBRAPDP. 2002a. Expansão da área Cultivada em Plantio Direto. Available at http://www.febrapdp.org.br/pd_area_br.htm.

FEBRAPDP. 2002b. Evolução do Plantio Direto no Paraná: Tração Animal. Available at http://www.febrapdp.org.br/pd_pr_ani_97_99.htm.

FEBRAPDP. 2002c. Expansão da área Cultivada em Plantio Direto no Brasil: áreas por Estado. Available at http://www.febrapdp.org.br/pd_area_estados.htm.

Freitas, P.L. de., Blancaneaux, P., Moreau, M. 1998. Caractérisation structurale de sols des Cerrados Brésiliens (Savanes) sous différents modes d'utilisation agricole: Etude et Gestion des Sols. Paris 5, 2, 93–105.

Freitas, P.L. de. 2002. Harmonia com a Natureza. Agroanalysis, FGV: Rio de Janeiro. Brazil Feb. 2002, 12–17.

Freitas, P.L. de. 1994. Aspectos físicos e biológicos do solo. In Landers J.N., Ed. Fascículo de experiências de Plantio Direto no Cerrado. Goiânia, APDC, 199–213.

Freitas, P.L. de. 2001. Organization of a technology platform for cooperative research and development in zero-till agriculture in Brazil. Available at http://www.embrapa.br/plantiodireto/.

Freixo, A.A., Machado, P.L.O. de A., dos Santos, H.P., Silva, C.A., Fadigas, F. de S. 2002. Soil organic carbon and fractions of a Rhodic Ferralsol under the influence of tillage and crop rotation systems in southern Brazil. *Soil Till. Res* 64:221–230.

Hernani, L.C., Kurihara, C.H., Silva, W.M. 1999. Sistemas de manejo de solo e perdas de nutrientes e matéria orgânica por erosão. *Rev. Bras. Ci. Solo, Viçosa* 23(1): 145–154.

Hullugale, N.R., Maurya, P.R. 1991. Tillage systems for the West African Semi-Arid Tropics. *Soil Till. Res* 20:187–199.

Kliewer, I., Casaccia, J., Valleros, J., Derpsch, R. 2000. Cost and herbicide reduction in the no-tillage system by using green manure cover crops in Paraguay. Proc. 15th ISTRO Conference, Fort Worth, Texas July 2–7, 2000(CD-ROM).

Landers, J.N. 1999. How and why the Brazilian Zero Tillage explosion occurred. Proc. 10th Intl. Soil Conservation Organization Conf.

Landers, J.N., Barros, G.S. de, Rocha, M.T., Manfrinato, W.A., Weiss, J. 2001. Environmental Impacts of Zero Tillage in Brazil: A first approximation. In Garcia-Torres L., Benites J., Martinez-Vilela A., Eds. Conservation agriculture: A worldwide challenge. Proc. 1st World Congress on Conservation Agriculture. Madrid, Espanha, FAO and ECAF I, Ch. 34, 317–326.

Landers, J.L., Freitas, P.L. de. 2001. Preservação da Vegetação Nativa nos Trópicos Brasileiros por Incentivos Econômicos aos Sistemas de Integração Lavoura x Pecuária com Plantio Direto. Proceeding Simpósio sobre Economia e Ecologia. Belém. Brazil.

Landers, J.N., Freitas, P.L. de, Guimarães, V., Trecenti, R. 2002a. The Social Dimensions of Sustainable Farming with Zero Tillage. Proc. 3rd Intl. Conf. on Land Degradation: Rio de Janeiro. Embrapa, IUSS, IAC, SBCS (CD-ROM).

Landers, J.N., Saturnino, H.M., Freitas, P.L. de. 2002b. Zero Tillage and Technology Transfer in the Tropics and Sub-tropics. In. Saturnino H.M., Landers J.N., Eds. The Environment and Zero-Tillage. Brasília, APDC 2002 Ch. VIII, 119–133.

Laurenti, A.C., Miranda, G.M., Alcover, M., Corrêa, A.R., Muzilli, O., Vieira, M.J., Carvalho, A.O.R., Carneiro, R.G., Nazareno, N.R.X., Bianchini, A., Cardoso, R.M.L., Biscaia, R.C.M. 1986. Culturas alternativas de inverno: Análise das potencialidades agroeconômicas: Informe da Pesquisa, 66. IAPAR. Londrina. Brazil.

Lara-Cabezas W.A.R., Freitas P.L. de, Eds. 2001Plantio Direto na Integração Lavoura-Pecuária, 2nd ed., Uberlândia, Brazil. UFU/ICIAG and APDC.

Macedo, M.C.M. 1995. Pastagens no ecossistema Cerrados: Pesquisa para o desenvolvimento sustentável. In Anais do Simpósio sobre Pastagens nos Ecossistemas Brasileiros: Pesquisa para o Desenvolvimento Sustentável. Sociedade Brasileira de Zootecnia. Brasília. Brazil, 28–62.

Macedo, J. 1996. Produção de alimentos: o potencial dos Cerrados. Documentos, 59: Embrapa-CPAC. Planaltina. Brazil.

Machado, L.A.Z., Salton, J.C., Primavesi, O., Fabricio, A.C., Kichel, A.N., Macedo, M.C.M., Zimmer, A.H., Guimarães, C.M. 1998. Integração lavoura-pecuária. In. Salton J.C., Hernani L.C., Fontes C.Z., Eds. Sistema plantio direto: O produtor pergunta, a Embrapa responde: Embrapa-CPAO. Dourados. Brazil, 217–232.

Machado, P.L.O.A., Silva, C.A. 2001. Soil management under no-tillage systems in the tropics with special reference to Brazil. *Nutr. Cycl. Agroec* 61:119–130.

Mello, I. 1995. Plantio Direto em Arroz Irrigado: Resumo Histórico. In: Revista Plantio Direto. Passo Fundo. Brazil 29 (Sep., Oct. 1995), 2–3.

Mondardo, A., Biscaia, R.M. 1981. Controle da erosão. In Fundação Instituto Agronômico do Paraná. Plantio Direto no Paraná. Circular Técnica no. 23, 33–42.

Monegat, C. 1991. Plantas de cobertura do solo: Características e manejo em pequenas propriedades: Claudino Monegat. Chapecó. Brazil.

Montoya, L. 1984. Aspectos de economicidade do manejo do solo em plantio direto. Informe da Pesquisa, 57. IAPAR. Londrina. Brazil.

Mrabet, R. 2002. No-tillage farming: Renewing harmony between soils and crops in semiarid Morocco. Proc. 3rd Intl. Conf. on Land Degradation: Rio de Janeiro. Embrapa, IUSS, IAC, SBCS (CD-ROM).

Muzilli, O. 1979. Evaluation of tillage systems and crop rotations in the state of Paraná, Brazil. Proc. 8th Intl. Conf. Soil Till. Res. Organization ISTRO 10–14 Sept. 1979: Stuttgart-Hohenheim. Germany, 1–7.

Muzilli, O. 1981. Cultura da soja. In Fundação Instituto Agronômico do Paraná. Plantio Direto no Paraná. Circular Técnica no. 23, 199–203.

Muzilli, O. 1984. Plantio direto no Brasil. In. Francelli A.L., Torrado P.V., Machado J., Eds. Atualização em plantio direto no Brasil: Fundação Cargill. Campinas. Brazil, 13–16.

Muzilli, O. 2002. Soil degradation and watershed conservation programmes in the state of Paraná, Brazil. Proc. 3rd Intl. Conf. on Land Degradation: Rio de Janeiro. Embrapa, IUSS, IAC, SBCS (CD-ROM).

Nicou, R., Charreau, C., Chopart, J.-L. 1993. Tillage and soil physical properties in semiarid West Africa. *Soil Till. Res* 27:125–147.

Paraná—Secretaria da Agricultura e do Abastecimento. 1994. Manual Técnico do Subprograma de Manejo e Conservação do Solo, 2nd ed.. Curitiba. Brazil.

Pieri, C., Evers, G., Landers, J.N., O'Connell, P., Terry, E. 2002. No-till farming for sustainable rural development. Agriculture and Rural Development Working Paper. The International Bank for Reconstruction and Development. Washington, DC.

Roman, E.S. 1990. Effect of cover crops on the development of weeds. In International Workshop on Conservation Tillage Systems. Conservation Tillage for Subtropical Areas Proceedings. Cida/Embrapa-CNPT: Passo Fundo. Brazil, 258–262.

Rodrigues, B.N., Almeida, F.S. 1998. Guia de herbicidas, 4th ed.. Londrina. Brazil.

Roth, C.H., Meyer, B., Frede, H.-G., Derpsch, R. 1986. The effect of different soybean tillage systems on infiltrability and erosion susceptibility of an Oxisol in Paraná, Brazil. *J. Agronomy & Crop Science* 157:217–226.

Roth, C.H., Vieira, M.J., Derpsch, R., Meyer, B., Frede, H.-G. 1987. Infiltrability of an Oxisol in Paraná, Brazil, as influenced by different crop rotations. *J. Agronomy and Crop Sci* 159:186–191.

Roth, C.H., Meyer, B., Frede, H.-G., Derpsch, R. 1988. Effect of mulch rates and tillage systems on infiltrability and other soil physical properties of an Oxisol in Paraná, Brazil. *Soil Till. Res* 11:81–91.

Sá, J.C. de M., Cerri, C.C., Dick, W.A., Lal, R., Filho, S.P.V., Piccolo, M.C., Feigl, B.E. 2001. Organic matter dynamics and carbon sequestration rates for a tillage chronosequence in a Brazilian Oxisol. *Soil Sci. Soc. Am. J* 65:1486–1499.

Santos, T.C.C., Câmara, J.B.D. 2002. GEO Brasil 2002—Perspectivas do Meio Ambiente no Brasil: Edições IBAMA. Brasília. Brazil.

Santos, H.P., Reis, E.M., Pereira, L.R. 1990. Rotação de culturas. XVII. Efeito no rendimento de grãos e nas doenças radiculares do trigo de 1980 a 1987. *Pesq. Agropec. Bras* 25:1627–1635.

Santos, H.P., Tomm, G.O., Lhamby, J.C.B. 1995. Plantio direto versus convencional: Efeito na fertilidade do solo e no rendimento de grãos de culturas em rotação com cevada. *Rev. Bras. Ci. Solo* 19:449–454.

Saturnino, H.M. 2001. Importancia do avanço do plantio direto nos trópicos. In Lara-Cabezas W.A.R., P.L. de Freitas, Eds. Plantio Direto na Integração Lavoura-Pecuária, 2nd ed.: APDC/UFU. Uberlândia. Brazil, 15–23.

Schütte, G. 2000. Transgenic herbicide resistant plants. Mitteilungen aus dem Institut für Allgemeine Botanik: Hamburg. Bd.28. Hamburg. Germany.

Sidiras, N., Henklain, J.C., Derpsch, R. 1981. Vergleich von drei Bodenbearbeitungsverfahren in bezug auf einige physikalische Eigenschaften, Boden- und Wasserkonservierung und Ertraege von Soja und Weizen auf einen Oxisol. *Z. Acker. und Pflanzenbau* 151:137–148.

Silva, M.L.N., Freitas, P.L. de, Blancaneaux, P., Curi, N., Lima, J.M. 1997. Relação entre parâmetros da chuva e perdas de solo e determinação da erodibilidade de um Latossolo Vermelho Escuro em Goiania (GO). *Rev. Bras. Ci. Solo* 21:131–137.

Sisti, C., dos Santos, H.P., Kohhann, R., Alves, B.J.R., Urquiaga, S., Boddey, R.M. 2002. Change in carbon and nitrogen stocks in soil under 13 years of conventional or zero tillage in southern Brazil. *Submitted to Soil Till. Res.*

Skora Neto, F. 1998. Manejo de plantas daninhas. In Darolt M.R., Ed. Plantio direto: Pequena propriedade sustentável: Circular 101. IAPAR. Londrina. Brazil, 127–158.

Spain, J.M., Ayarza, M.A., Vilela, L. 1996. Crop pasture rotations in the Brazilian Cerrados. In. Pereira R.C., Nasser L.C.B., Eds. Proc. 1st. Intl. Sym. on Tropical Savannahs: Biodiversity and Sustainable Production of Food and Fibers in the Tropical Savannahs: Embrapa Cerrados. Planaltina. Brazil, 39–45.

Spehar, C.R. 1994. Breeding soybeans to the low latitudes of Brazilian Cerrados (Savannahs). *Pesq. Agropec. Bras* 29:1167–1180.

Spehar, C.R. 1996. Prospects for sustainable grain production systems in the Cerrados (Brazilian Savannahs). In Pereira R.C., Nasser L.C.B., Eds. Proc. 1st Intl. Sym. on Tropical Savannahs: Biodiversity and Sustainable Production of Food and Fibers in the Tropical Savannahs: Embrapa Cerrados. Planaltina. Brazil, 139–151.

Spehar, C.R., Souza, P.I.M. 1996. Sustainable cropping systems in the Brazilian Cerrados. *Integrated Crop Management* 1(Technical Series, FAO), 1–25.

Uri, N.D. 1999. The use of no-till farming in U.S. Agriculture: Farmers' perceptions versus reality. *J. Sust. Agric* 15:5–17.

Vieira, L.G.E. 1981. Cultura do trigo. In Fundação Instituto Agronômico do Paraná. Plantio Direto no Paraná. Circular Técnica no. 23, 194–198.

Wiethölter, S. 2000. Manejo da Fertilidade do solo no sistema plantio direto: Experiência nos estados do Rio Grande do Sul e Santa Catarina. In Anais da XXIV Reunião Brasileira de Fertilidade do Solo e Nutrição de Plantas (FertBio 2000) (CD-ROM). Santa Maria. Brazil, 1–35.

Wünsche, W.A., Denardin, J.E. 1978. Perdas de solo e escorrimento de água sob chuva natural em Latossolo Vermelho Escuro nas culturas do trigo e soja. In Encontro Nacional de Pesquisa sobre Conservação do Solo, 2. Passo Fundo. Brazil, 289–296.

19

Potential of Tillage, Agropastoral, and Planted Fallow Systems in Low-Fertility Tropical Soils of Latin America

Bal Ram Singh
Agricultural University of Norway
Aas, Norway

Idupulapati M. Rao, Edmundo Barrios, and Edgar Amezquita
Centro Internacional de Agricultura Tropical (CIAT)
Cali, Colombia

I. INTRODUCTION

A. Latin American Savannahs

Tropical savannahs cover 45% of the land area in Latin America, or 243 million hectares (Mha), mainly in Brazil (200 Mha), Colombia (20 Mha), and Venezuela (12 Mha). The soils are mainly Oxisols and Ultisols, which are characterized by low nutrient reserves, high acidity (pH 4.0–4.8), high aluminium (Al) saturation (up to 90%), high phosphorus (P) fixing capacity (Sánchez and Logan, 1992), and a low capacity to supply P, K, Mg, and S. In addition to soil chemical constraints, these soils also exhibit high bulk density, high resistance to root penetration, low rates of water infiltration, low water-holding capacity, and low structural stability (Amézquita, 1998a, b; Phiri et al., 2001a). Because of these chemical and physical constraints, these soils are more susceptible to degradation than most soils, often degrading within 5 years of being opened up for agricultural production (Thomas and Ayarza, 1999). Furthermore, the traditional technologies in use on these soils are unsustainable. Soil organic matter (SOM) contents have gradually decreased, reducing soil fertility (da Silva et al., 1994), and excessive seedbed preparation has led to the destruction of the favorable physical soil structure, accompanied by subsoil compaction and soil sealing that eventually promoted severe erosion, even on gentle slopes (Klink et al., 1993). Soil preparation and cultivation techniques, therefore, need to be devised to reverse degradation while maintaining high yields. A highly successful strategy for intensifying agricultural production sustainability and reversing problems of degradation involves the integration of crop/livestock systems (agropastoralism) (Rao et al., 1993; Thomas et al., 1995). This strategy is based on the assumption that a beneficial synergistic effect on production and on soil occurs when annual and perennial species are combined (Lal, 1991; Spain, 1990). Available nutrients are used more efficiently and the chemical, physical, and biological properties are improved.

B. Latin American Hillsides

Agriculture in hillsides of tropical America that cover about 96 Mha (Jones, 1993) is often characterized by farming systems under which soils are degrading through erosion and loss of nutrients (Amézquita et al., 1998). Mining the environment gives farmers short-term subsistence but creates a profound discrepancy between actual systems of land use and the ecologically sound systems appropriate for fragile soils on steep slopes. Traditional agricultural systems in the Andean hillsides of Colombia are based on shifting cultivation that involves slashing and burning of the native vegetation, followed by continuous cultivation and abandonment after 3 to 5 years because of low crop yields (Knapp et al., 1996). Low-income farmers typically rely on maize and beans for household consumption, and cassava is the last crop in the cropping cycle before left to fallow. Local farmers recognize soil nutrient depletion and estimate more than six years for complete soil fertility recovery by natural fallows.

Planted fallows are an appropriate technological entry point because of their low risk for the farmer, relatively low cost, and potential to generate additional products (i.e., fuel wood) that bring immediate benefits while improving soil fertility (Barrios et al., 2003). Planting of fast-growing N-fixing trees or shrubs on degraded soils restores soil fertility through additional nitrogen (N) inputs (Giller, 2001) and improved nutrient recycling (Barrios et al., 1997) as well as nutrient uptake by trees from below the rooting depth of crops (Hartemink et al., 1996). The quantity, quality, timing, and mode of application of organic materials in different fallow systems would strongly influence decomposition processes and their contribution to SOM pools and nutrient availability (Myers et al., 1994; Palm, 1995).

C. Soil Organic Matter (SOM) Dynamics

SOM is considered a key to sustainable management of soils such as Oxisols, Ultisols, and volcanic-ash soils, and it is more important in these soils than in most soils of the temperate regions because it serves as both the important nutrient sink and nutrient source (Coleman et al., 1989). Despite the high significance of SOM to soil fertility and aggregate stability, assessing total SOM content is not sufficiently sensitive to characterizing its dynamics. It is therefore important to group SOM with similar characteristics and turnover rates in pools of different activities. Differently available fractions of SOM may explain conflicting results concerning the depletion of SOM in tropical soils (da Silva et al., 1994). However, little information exists on SOM dynamics in tropical soils, such as Oxisols and volcanic-ash soils (Neufeldt et al., 1999). The size-density fractionation of SOM as described by Barrios et al. (1996a) and Meijboom et al. (1995) could serve as an important tool to characterize SOM pools in volcanic-ash soils of the Andean hillsides.

D. Soil Phosphorus Dynamics

Low-phosphorus (P) supply is a major agronomic constraint in highly weathered tropical soils, because of strong phosphate adsorption to Fe and Al oxyhydroxides in these soils (Fontes and Weed, 1996; Mesquita Filha and Torrent, 1993). Similar to Oxisols and Ultisols, volcanic-ash soils also contain insufficient P for crop production, resulting from their allophane-rich content, which gives them a high P-sorbing capacity (Gijsman and Sanz, 1998; Rao et al., 1999). However, little is known about P cycling in tropical soils because, in the past, only readily available P was determined (Sousa and Lobato, 1988). According to Stewart and Tiessen (1987) and Beck and Sánchez (1994), this may not effectively reflect plant-available P because organic fractions are believed to contribute proportionately more with increasing P deficiency. One way of examining P transformations in soil is the use of sequential extractions, which first remove labile P, and then the more stable forms. Hedley et al. (1982) developed a sequential P extraction procedure, later modified by Tiessen and Moir (1993), that allows one to follow transformations of biologically, as well as geochemically, bound P at different levels of plant availability.

An important prerequisite to adopting a till farming system in these Oxisols of low inherent fertility is the creation of a suitable "arable layer," which can be subsequently managed with minimal soil disturbance. Thus, the objective of this chapter is to assess the impact of management practices on the creation of an arable layer for subsequent management by tillage (chisel). Management systems assessed include vertical tillage, planted fallows, and agropastoral systems. The arable layer so created is characterized for soil physical, chemical, and biological qualities.

II. TILLAGE AND AGROPASTORAL SYSTEMS AND SOIL-PLANT PROCESSES IN TROPICAL SAVANNAHS

Tillage practices by heavy machinery lead to a breakdown of macroaggregates into smaller aggregate sizes. The action of rainfall and gravity results in a repacking of these aggregates and, consequently, the total soil porosity and pore sizes are reduced. The resulting changes in macroporosity affect water flow, which in turn affects nutrient availability and thus impact negatively on the productive capacity of the soil (Preciado et al., 1998). Three important phenomena related to plant nutrition, which are negatively affected by reduction in macropores, are root growth, nutrient interception by roots, and soil drainage and aeration (Preciado et al., 1998). Reduced water infiltration encourages surface water runoff and, consequently, soil and plant nutrient losses brought about by soil erosion (Goedert, 1983). To achieve improved and sustainable crop and pasture production and to avoid degradation, key soil properties such as soil's physical constraints must be alleviated by appropriate tillage and cropping practices (Amézquita, 1998a; Phiri et al., 2001a). Research progress in determining the impacts of harrowing intensity and vertical tillage through chisel on soil properties, development of an arable layer, and P dynamics in soils of the Llanos of Colombia is summarized ahead (Phiri et al., 2001a, 2003b).

A. Soil Physical Properties

Total porosity and pore-size distribution are sensitive soil physical properties that can be used to evaluate the influence of tillage on the physical condition of the soil because they regulate the rate of water entry into the soil. They also influence soil water fluxes, which affect plant nutrient availability and growth (Amézquita, 1998a). Agriculturally productive soils have total porosity values of around 50%. Changes in macroporosity under different disc harrow passes are shown in Fig. 1. In the 0–5-cm soil layer, macroporosity ($\emptyset > 50$ μm) increased from 7.8% under native savannah (control) to around 20% with 2, 4, or 8 disc harrow passes per year for 3 years (Phiri et al., 2001a). Use of disc harrows lowered bulk densities compared to the native savannah in the 0–5-cm soil layer. Agropastoral (crop–pasture) treatments, in general, had 16% lower bulk density in the 0–10-cm soil layer and 13% lower in the 10–20-cm soil layer than those of the native savannah. There was a reduction in soil water content in the 0–5- and 5–10-cm layers as the number of disc harrow passes increased. Root growth is inhibited

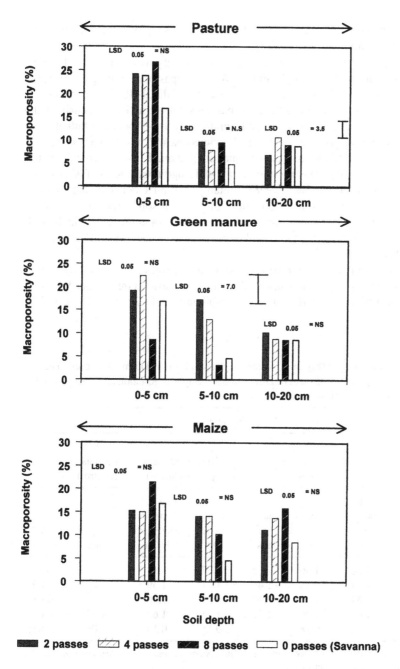

Figure 1 Effect of harrowing intensity and agropastoral systems on macroporosity at different soil depths. LSD values at $P < 0.05$. NS: not significant. The 0 number of passes represent the native savannah system. (Adapted from Phiri et al., 2001a, with permission from Elsevier Science B.V.)

when bulk density exceeds 1.4–1.6 Mg m^{-3} and is suppressed at densities near 1.8 Mg m^{-3} (Mitchell et al., 1982).

Compared with the native savannah, bulk density was reduced by the agropastoral systems and rice–soybean rotations. Agropastoral (crop–pasture) treatments, in general, had 16% lower bulk density in the 0–10-cm soil layer and 13% lower in the 10–20-cm soil layer than those of the native savannah. In the subsoil layers, all treatments presented significantly lower values of bulk density than those of native savannah (Table 1). Previous research has shown that legume-based pastures contribute to improved quantity and quality of soil organic matter with depth due to vigorous rooting ability of forage components (Fisher et al., 1994; Rao et al., 1994; Rao, 1998). Suitably low bulk densities are of great importance to sustainable management of this type of soil as they are indicative of factors that moderate root growth, infiltration, and water movement in the soil, which in turn affects nutrient availability in soil and nutrient acquisition by plants (Rao, 1998).

Results on penetration resistance for different soil layers are shown in Fig. 2. In relation to native savannah, all tillage and agropastoral treatments decreased penetration resistance, particularly in topsoil layers (0–20 cm). These results

Table 1 Bulk Density (Mg m^{-3}) Values of Soils in Profiles During the Fourth Year (June 1999) After Establishment of Different Rice–Soybean Rotation and Agropastoral Systems Compared with Native Savannah

	System						
	Rice–Soybean rotation			Rice + pastures (Agropastoral)		Native savannah (control)	
Soil depth (cm)	1 passes of chisel	2 passes of chisel	3 passes of chisel	Grass only (Ag)	Grass + legumes (Ag + Pp + Do)	Savannah	LSD (0.05)
0–5	1.36	1.36	1.33	1.37	1.38	1.61	0.11
5–10	1.49	1.42	1.46	1.44	1.39	1.64	0.09
10–20	1.54	1.57	1.50	1.55	1.56	1.73	0.08
20–40	1.60	1.60	1.57	1.56	1.62	1.73	0.06
LSD	0.16	0.18	0.16	0.15	0.19	0.09	(0.05)

LSD values are at 0.05 probability level. Ag = *Andropogon gayanus*; Pp = *Pueraria phaseoloides*; Do = *Desmodium ovalifolium*.
Source: Adapted from Phiri et al. (2003b) with the permission from The Haworth Press.

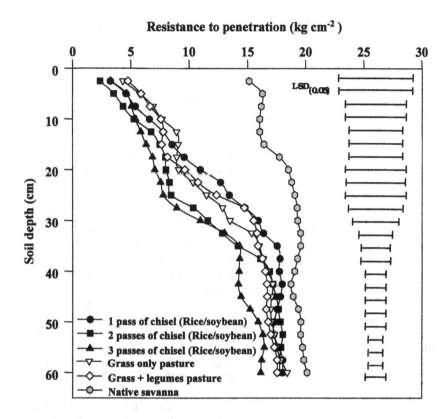

Figure 2 Changes in penetrability (measured at field capacity) with soil depth at four years after establishment of different tillage and agropastoral systems. (Adapted from Phiri et al., 2003b, with the permission from The Haworth Press.)

suggest that it is possible to improve soil physical conditions to enhance water and nutrient availability, which favor root development of the crop and forage components. Improved soil quality may also support greater crop and pasture productivity (Amézquita, 1998b). Both tillage and agropastoral treatments generally improve soil physical conditions, but whether one treatment is more beneficial than another over a longer period needs to be evaluated further.

B. Soil Chemical Properties

The intensity of disc harrowing may also affect important soil chemical properties. At the 0–5- and 5–10-cm soil layers under the green manure and the maize (*Zea*

Table 2 Soil Characteristics at Different Soil Depth Layers Under Different Harrow Passes and Cropping, Systems

Depth	Disc harrow passes per year	pH	C (%)	P (ppm)	K	Ca	Mg	Al
						cmol/kg soil		
				Native savannah				
0–5		4.8	1.1	3.2	0.1	0.2	0.2	2.6
5–10		5.0	1.3	4.6	0.1	0.4	0.3	2.0
				Grass-alone pasture				
0–5	2	4.9	2.6 a	12.0 ab	0.15 a	0.72	0.46	1.7
	4	4.9	2.0 b	15.8 a	0.13 ab	0.72	0.42	1.7
	8	4.7	2.0 b	9.5 c	0.10 c	0.53	0.36	1.9
5–10	2	4.7 ab	2.1	6.0	0.10 a	0.48	0.21	1.9
	4	4.8 a	1.9	4.6	0.08 b	0.55	0.20	2.0
	8	4.6b	1.9	4.3	0.08 b	0.57	0.20	2.0
				Green manure				
0–5	2	5.0	2.5 a	36.7 b	0.16	1.22	0.50	1.3
	4	5.0	1.8 b	38.4 b	0.15	1.00	0.48	1.4
	8	4.8	1.9 b	56.8 a	0.12	0.98	0.37	1.7
5–10	2	4.7 b	2.0 a	6.6 b	0.13	0.44	0.30	2.1
	4	4.9 a	1.6 b	9.3 ab	0.09	0.65	0.30	1.9
	8	4.7 b	2.0 a	10.7 a	0.10	0.67	0.31	2.0
				Maize				
0–5	2	4.8	2.1	20.5	0.26 a	0.96	0.45 ab	1.1 ab
	4	5.1	1.9	21.0	0.22 b	1.29	0.56 a	0.9 b
	8	4.9	1.8	27.0	0.22 b	1.01	0.42 b	1.5 a
5–10	2	4.8	1.9	18.5 b	0.15 a	0.64	0.30	1.7
	4	4.9	2.0	17.1 b	0.11 b	0.87	0.37	1.5
	8	4.9	1.8	24.2 a	0.15 a	0.79	0.35	1.8

Similar letters in the same column for each cropping system and for depth class represent no significant difference ($P < 0.05$) by Duncan multiple range test. There were only minor changes below 10 cm and hence results are not presented in this table.
Source: Adapted from Phiri et al. (2001a) with the permission from Elsevier Science B.V.

mays) cropping systems, the amount of P increased as the number of disc harrow passes increased (Table 2). The grass-alone pasture had the least amount of P and was, on average, 56% lower in the 0–5-cm soil layer than the green manure treatment, which had the largest amount. Most of K was measured in the 0–5-cm soil layer in all cropping systems, and under the maize and grass-alone pasture systems, the amount of K tended to decrease as the number of passes was increased from 2, 4, to 8 disc harrow passes per year (Table 2). The number of disc harrow passes did not significantly affect the amount of Ca or Mg in any soil layer under the grass-alone pasture and the green manure cropping systems (Phiri et al., 2001a).

The different crop rotation and agropastoral treatments improved nutrient availability and reduced Al levels. Exchangeable Al levels decreased in the first two layers, but remained similar to that of the native savannah below these depths. Differences in available P between rice–soybean rotations and agropastoral treatments were probably the result of differences in the rate of lime applied, which may have affected P adsorption in soil. Nutrient values tended to be greater in the 0–5-cm layer compared to subsoil layers. Available K was 2 to 4 times greater than that of native savannah (0.09 cmol$_c$ kg^{-1}), and that of Ca and Mg was 4 to 10 times greater than that of native savannah (Table 2).

C. Evolution of Arable Layer

To overcome soil constraints and improve soil quality for agricultural productivity, there is the potential for improved soil management through vertical tillage using a chisel plow (Amézquita, 1998a). We hypothesized that the integration of soil tillage and soil fertility together with vigorous root systems of introduced pasture species may result in the evolution/creation of an arable layer. The "arable layer" concept proposed by Amézquita (1998b) and Phiri et al. (2003b) is based on combining (1) adapted crop and forage germplasm, (2) vertical tillage to overcome soil physical constraints (high bulk density, surface sealing, low porosity and infiltration rates, poor root penetration, etc.), (3) use of chemical amendments (lime and fertilizers) to enhance soil fertility, and (4) use of agropastoral systems to increase rooting, to promote soil biological activity, and to avoid soil compaction after tillage.

Use of tillage and agropastoral treatments resulted in reducing bulk density and penetration resistance on one hand, and improving soil fertility on the other, particularly in topsoil layers (0–20 cm), and thus leading to the formation of an arable layer (Phiri et al., 2003b). Chisel treatments were moderately effective in incorporating lime and P into deeper layers. Total C and total root length development within the soil profile up to 40-cm soil depth were greater in agropastoral (rice–grass/legumes) treatment than a rice-based system with vertical tillage. Results from this study justify promoting the concept of forming an arable layer in tropical Oxisols using vertical tillage and agropastoral treatments. But to improve and sustain crop production on infertile Oxisols of the tropics, there is a need to develop better crop management strategies to overcome weed problems. We suggest that the formation of an arable layer is a necessary prerequisite to converting to a no-till system for minimizing environmental degradation in savannah soils of the Llanos of Colombia and Latin America.

D. Phosphorus Dynamics in Soils

Understanding of the P dynamics in the soil/plant system and especially of the short- and long-term fate of P fertilizer in relation to different management practices is essential to the sustainable management of tropical soils and agroecosystems. Chemical sequential extraction procedures have been and still are widely used to characterize extractable soil P into inorganic and organic fractions (Hedley et al., 1982). The underlying assumption in these approaches is that readily available soil P is removed first with mild extractants, while less available or plant-unavailable P can only be extracted with stronger acids and alkali.

In the fractionation procedure, the P fractions (in order of extraction) are interpreted as follows (Tiesen and Moir, 1993): Resin-P_i represents inorganic P (P_i) either from the soil solution or weakly adsorbed on (oxy) hydroxides or carbonates. Sodium bicarbonate 0.5 M at pH 8.5 also extracts weakly adsorbed P_i and easily hydrolysable organic P (P_o)-compounds like ribonucleic acids and glycerophosphate. Sodium hydroxide 0.5 M extracts P_i associated with amorphous and crystalline Al and Fe (oxy) hydroxides and clay minerals and P_o associated with organic compounds (fulvic and humic acids). Hydrochloric acid 1 M extracts P_i associated with apatite or octacalcium P. Hot concentrated HCl extracts P_i and P_o from more stable pools. Organic P extracted at this step may also come from particulate organic matter. Residual P, that is P that remains after extracting the soil with the already cited extractants, most likely contains very recalcitrant P_i and P_o forms.

Several long-term field studies on soil P dynamics, acquisition, and cycling in crop–pasture–fallow systems of low-fertility soils of tropical America have been conducted (Friesen et al., 1997; Guo and Yost, 1998; Oberson et al., 1999, 2001; Phiri et al., 2001a, b, 2003a, b; Lehmann et al., 2001; Bühler et al., 2002). The impact of intensive disc harrowing (2, 4, or 8 disc harrow passes per year over 3 years) and chisel plowing in combination with agropastoral systems on soil phosphorus dynamics on Oxisols has been evaluated (Phiri et al., 2001a, 2003b). The amount of biologically available P is generally concentrated in the 0–5- and 5–10-cm soil layers and differs among cropping systems. The largest amount of this fraction is obtained under the green manure followed by maize and then grass-alone pasture cropping system. Eight disc harrow passes per year resulted in the highest amount of biologically available and moderately available P under green manure and maize. Since the moderately available P is plant available in the medium term, the high amount of this fraction for 8 disc harrow passes per year may have contributed to the high P uptake of maize observed under this treatment. The largest amount of the sparingly available P was extracted for 8 disc harrow passes per year, but the treatment effects were better separated under maize than under the grass pasture cropping systems. The largest amount of organic P (P_o) was obtained for 2 disc harrow passes under green manure and 4 disc harrow passes under maize for the 0–5-cm soil layer. The Sum-P_o profile distribution showed no significant differences among treatments.

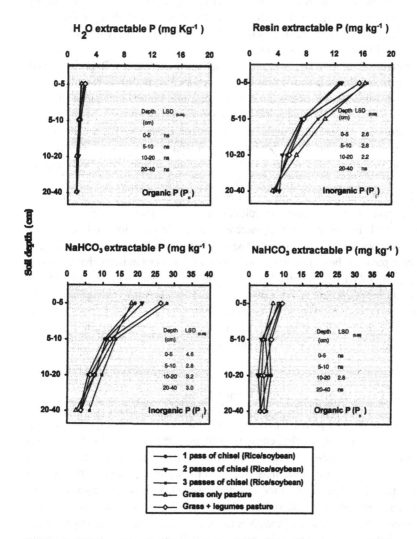

Figure 3 Soil profile distribution of the bioavailable P fractions at four years after establishment of different tillage and agropastoral systems. (Adapted from Phiri et al., 2003b, with the permission from The Haworth Press.)

In another field study the impacts of strategies including chisel tillage (1, 2, or 3 passes), crop rotations (rice–soybean), and agropastoral systems (rice–grass-alone pasture; rice–grass/legume pasture) on the buildup of an arable layer, on P dynamics, and on grain yields of upland rice and soybean were also evaluated (Phiri et al., 2003b). The biologically available resin–P_i and NaHCO$_3$–P_i each represented 5% of the total P and were significantly affected by chiseling down to 10–20-cm depth (Fig. 3). The moderately resistant NaOH–P represented, on average, 33% of total P in the 0–20-cm soil layer, and both NaOH–P_i and NaOH–P_o were significantly affected by chisel tillage.

Organic P is more important for delivering available P in improved grass–legume pastures than in continuously cropped soils. It was found that the amount and turnover of P that is held in the soil microbial biomass are increased when native savannah is replaced by improved pasture and decreased under continuous cropping (Oberson et al., 2001). Based on these studies, an alternative strategy to cropping low P Oxisols is proposed. This strategy involves strategic application of P fertilizer to crops at low rates, and planting of grass–legume pastures to enhance P cycling and efficient use of fertilizer P.

The importance of organic P in low input system has been substantiated by the isotopic exchange studies of Bühler et al. (2002). They found that resin–P_i, Bic–P_i, NaOH–P_i, and hot HCl–P_i increased with P fertilization, and the highest increase was observed for NaOH–P_i. The recovery of ^{33}P in two soils with annual fertilizer inputs and large positive input–output P balances indicate that resin–P_i, Bic–P_i, and NaOH–P_i contained most of the exchangeable P. The organic or more recalcitrant inorganic fractions contained almost no exchangeable P. In contrast, in soils with low or no P fertilization, more than 14% of added ^{33}P was recovered in NaOH–P_o and HCl–P_o fractions two weeks after labeling, showing that organic P is involved in short-term P dynamics.

E. Plant Growth and Nutrient Acquisition

Nutrient acquisition by plants depends on root system size and distribution, uptake kinetics and nutrient mobilizing capacity (Barber, 1984), and management practices, which lead to increased root proliferation and enhance nutrient acquisition. Phiri et al. (2001a) found that more than two passes of disc harrow per year in the grass pasture treatment resulted in a decreased shoot biomass production and reduced nutrient uptake (Table 3). More than two disc harrow passes may decrease the amount of SOM because of the physical breakdown of aggregates by plowing and the subsequent higher organic carbon mineralization, which may accentuate losses of N and P through leaching and fixation by soil, respectively. In Oxisols, SOM is an important component because it carries the majority of exchange sites

Table 3 Total Biomass and Nutrient Acquisition by Grass-Alone Pasture, Green Manure, and Maize as Influenced by the Number of Disc Harrow Passes per Year over Three Years

Cropping system	Disc harrow passes per year	Total shoot biomass	N	P	K	Ca	Mg
					Nutrient uptake		
					(kg/ha)		
Grass-alone pasture	2	1,756	19 a	4.7 a	34 ab	3.6	6.0 a
	4	1,613	17 a	4.8 a	43 a	2.4	4.8 a
	8	1,494	12 b	2.5 b	28 b	2.0	3.7 b
Green manure	2	6,333 b	185 b	18 ab	81	90 b	21 b
	4	7,833 a	227 a	20 a	88	135 a	35 a
	8	6,014 b	192 b	14 b	70	82 b	22 b
Maize	2	6,327 c	40 b	10 b	65 b	12 b	9 b
	4	7,466 b	54 b	8 b	94 b	18 ab	12 ab
	8	12,119 a	99 a	16 a	141 a	23 a	14 a

Grass-alone pasture = cv. Llanero (*Brachiaria dictyoneura* CIAT 6133).
Means followed by different letters within a column and within a cropping system are significantly different ($P < 0.05$) by Duncan multiple range test.
Source: Adapted from Phiri et al. (2001a) with the permission from Elsevier Science B.V.

and also participates in the formation of stable microaggregates and controls the degree of clay dispersion (Neufeldt, 1999).

The 8 passes of disc harrow per year treatment increased both aboveground production and nutrient uptake by maize (Table 3). The better performance of maize under intensive cultivation (8 disc harrow passes) may be attributed to improved rooting ability, which likely contributed to greater uptake of nutrients. The green manure cropping system had higher yields with 2 disc harrow passes per year that resulted in greater nutrient (N, P, Ca, and Mg) acquisition.

III. PLANTED FALLOWS AND SOIL-PLANT PROCESSES IN ANDEAN HILLSIDES

Expansion of agriculture on steep slopes in the Andean hillsides has made a positive impact on income and employment in the region but a negative impact on the environment, particularly on soil erosion and nutrient depletion. Soil nutrient depletion results from a negative nutrient budget, that is, when greater amounts of soil nutrients are removed from the system due to soil erosion, leaching, and crop harvest (i.e., nutrient mining) than returned to the soil in the form of biosolids and fertilizers (Smaling et al., 1997). Converting degraded soils to a natural fallow involving regeneration of the native vegetation is a traditional management practice throughout the tropics for restoration of soil fertility (Sanchez, 1999). Traditional fallows often require long periods before adequate soil fertility regeneration is achieved. Increasing population pressure on limited agricultural land has necessitated a reduction in the length of fallows. An alternative to traditional fallowing is to include plant species that replenish soil nutrient stocks faster than plants in the natural succession (Szott and Palm, 1996; Barrios et al., 1997; Buresh and Tian, 1998). Planted fallows that incorporate components that restore soil fertility are being developed for Andean hillsides to sustain agricultural production at high levels while minimizing soil degradation. The impact of the planted fallows on soil-plant processes is summarized below.

A. Soil Organic Matter Turnover

The dynamics of SOM in natural and managed ecosystems are important because SOM affects nutrient cycling, alters soil structure, and plays a significant role in the biological functions of the soil (Brown et al., 1994). Dynamics of SOM control the interaction between soil processes and plant production, and its knowledge helps understand in soil management why certain systems perform better than others (Swift and Woomer, 1993). The chemical characteristics or ''quality'' of plant materials contributing to SOM play a fundamental role in its decomposition, nutrient release (Cobo et al., 2002), and humification. Effects of SOM are often not satisfactorily explained because the turnover rate of different organic substrates in SOM is not taken into account. To overcome this problem, separation of SOM into different size and density fractions has been used for understanding SOM dynamics for managed systems (Barrios et al., 1996a). Identification of improved cropping systems for sustained production depends on our understanding how different organic fractions are affected by management systems and how they relate to changes in the soil's nutrient-supply capacity (Barrios et al., 1996b; Friesen et al., 1997; Neufeldt et al., 1999; Oberson et al., 1999; Thomas et al., 1999; Westerhof et al., 1999).

The volcanic-ash soils of the Colombian Andes usually contain a high SOM concentration. However, these soils have limited nutrient cycling through SOM because most of it is strongly bound and chemically protected by mineral particles. The rate of decomposition is thus limited and, consequently, so is the rate of nutrient cycling. In addition, these volcanic-ash soils contain insufficient P for crop production, resulting from their allophane-rich content, which is a cause of a high P-sorbing capacity (Gijsman and Sanz, 1998; Rao et al., 1999).

Planted fallows are an appropriate technological entry point because of their low risk to the farmer, relative low cost, and potential to generate additional products (i.e., fuel wood) that bring immediate benefit while improving soil fertility (Barrios et al., 2003). Therefore, the effect of contrasting planted fallows [*Indigofera* (*IND*), *Calliandra* (*CAL*), and *Tithonia* (*TTH*)], a natural unmanaged fallow (NAT), and a maize–bean rotation (ROT) on the dynamics and partitioning of SOM and P has been investigated (Phiri et al., 2001b, 2003a).

The SOM contains fractions with rapid and slower turnover rates, but the fractions with a rapid turnover (active fractions) play a dominant role in soil nutrient dynamics (Janzen et al., 1992). Phiri et al. (2001b) recovered three SOM fractions [i.e., Ludox light fraction (LL), Ludox intermediate fraction (LM), and Ludox heavy fraction (LH)] by using a size-density fractionation procedure developed by Meijboom et al. (1995). The weight of SOM fractions decreased in the order LL > LM > LH, and represented 55%, 36%, and 9% of the sum of all SOM fractions, respectively (Fig. 4). The mean amount of C, N, and P in the LL accounted for 3.8%, 1.7%, and 1.3% C; 2.3%, 1.0%, and 0.6% N; and 0.7%, 0.3%, and 0.2% P of the total soil C, N, and P in the 0–5-, 5–10-, and 10–20-cm soil layers, respectively. Significant effects of land use on this fraction were not detected in the 0–5-cm soil layer. The LM fraction accounted for 36%, 40%, and 45% of the total SOM fraction in 0–5, 5–10, and 10–20 cm, respectively. All the measured parameters for this fraction (C, N, and P contents) were significantly affected by land use and, in the 0–5-cm soil layer, they followed the trend of *Calliandra* (*CAL*) > *Tithonia* (*TTH*)> *Indigofera* (*IND*) > natural unmanaged fallow (NAT) > maize–bean rotation (ROT).The LH fraction contributed a very small percentage of the total SOM, and none of the measured parameters was affected by change in land use. This observation is consistent with other published reports (Barrios et al., 1997; Maroko et al., 1998), which indicated that changes in the land-use system might not affect the heavy-fraction SOM weight and amount of N. These results indicate that the LH fraction of SOM plays an insignificant role in nutrient recycling.

Nutrient ratios (C/N, C/P, N/P) are indicative of SOM dynamics. Although neither the soil C/P nor the soil N/P ratio was affected by the fallows and cropping system, these ratios of the SOM fractions were significantly affected by treatment. On average, the highest C/N ratios were found under NAT and the lowest ratios under *Tithonia*. The lower C/N, C/P, and N/P ratios of the SOM fractions under TTH, compared with NAT, reflect the high-quality leaf litter of the former treatment. The low C/P ratio under TTH may indicate favorable conditions for enhanced P availability, as observed by Zou et al. (1995). Nziguheba et al. (1998) found that the labile soil-P, as determined by anion exchange resin, was comparable after adding *Tithonia* green biomass and triple superphosphate at equal P rates. Results obtained for soil P_o and P_i extracted by $NaHCO_3$ and $NaOH$ were correlated with the P content in the LL and LM fractions of SOM to explore potential relationships among them. Significant correlation found between these parameters show that the P content in LL and LM fractions is an important factor

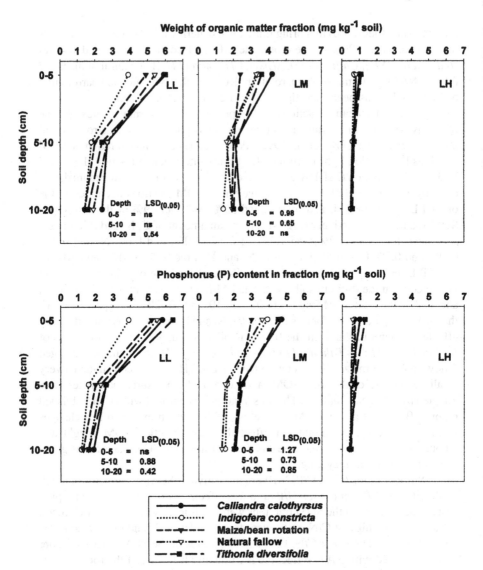

Figure 4 Profile weight distribution of light (LL), intermediate (LM), and heavy (LH) fractions of soil organic matter (SOM) and their P contents as affected by different fallows and the crop rotation system. LSD values are presented only when the differences among treatments are significant. (Adapted from Phiri et al., 2001b, with the permission from Kluwer Academic Publishers.)

in determining the amounts of $NaHCO_3$-extractable P_i and P_o. The regression equations are as follows:

$$NaHCO_3-P_i = 23.6 + 0.51 \text{ (P in LL)} + 1.93 \text{ (P in LM) R2} = 0.47$$
$$(P = 0.031)$$

$$NaHCO_3-P_o = 16.3 + 0.66 \text{ (P in LL)} + 1.73 \text{ (P in LM) R2} = 0.47$$
$$(P = 0.023)$$

These relationships suggests that systems including plant species that accumulate high amounts of P in their biomass, such as *Tithonia* (Gachengo, 1996), when used as fallow species, would increase both the "readily available" ($NaHCO_3-P_i$) and the "readily mineralizable" ($NaHCO_3-P_o$) P in the soil.

B. Phosphorus Dynamics in SOM Fractions

The amount of P in the light (LL) and medium (LM) fractions of SOM correlated well with the amount of "readily available" P in the soil (Phiri et al., 2001b). It is suggested that the amount of P in the LL and LM fractions of SOM may be sensitive indicators of "readily available" and "readily mineralizable" soil–P pools, respectively, in P-fixing volcanic-ash soils. These results also indicate that the fractionation of SOM and soil P may be more effective in detecting the impact of planted fallows on improving soil fertility than the conventional soil analysis methods.

The planted fallow and crop rotation systems had little effect on the "readily available" P fraction, which was fairly uniform across treatments and accounted for only 0.04% of the total soil–P. However, improved fallow species such as *Tithonia* can cause significant increases in the P fraction related to plant-available P. The moderately resistant P (NaOH-extractable P) accounted for the largest proportion (53% of the total soil–P) of the plant-available P. Of this, 47% was in the organic form. The large amount of P recovered from this fraction can be attributed to the high contents of exchangeable Al and Fe associated with these volcanic-ash soils. *Tithonia*, for all soil layers, had slightly but consistently higher NaOH-extractable P. The P_o contribution to the NaOH fraction by the planted fallows is desirable because the hydroxide–P_o fraction is usually more stable and may represent a relatively active pool of P in tropical soils under cultivation, especially those that are not receiving mineral P fertilizers (Tiessen et al., 1992). In general, the planted fallows and the crop rotation system had little effect on residual P. The residual P content was related more to the nature of the soil than to the treatments. These results thus indicate that short-term contributions of P from organic sources (planted fallows) are not readily incorporated into this fraction.

C. Soil Biological Activity

There has been increasing awareness of the use of *Tithonia diversifolia* as an indigenous fallow species in improving soil fertility (Niang et al., 1996), because this species has an ability to accumulate labile soil nutrients, which might be a source of plant nutrients or biofertilizers (Niang et al., 1996; Jiri and Waddington, 1998; Phiri et al., 2003b). The interest in this species is also generated by its marked association with arbuscular-mycorrhizae (AM) and its ability to mobilize and accumulate soil P.

Under field conditions, *Tithonia* is more easily established from stem cuttings than from seeds (King'ara, 1998) and hence how this establishment affects shoot and root growth characteristics, and AM association, need to be investigated (Phiri et al., 2003a). Establishment of *Tithonia* by plantlets resulted in significantly greater mycorrhizal root infection ($P = 0.05$) in both coarse and fine roots, as compared to *Tithonia* established by stakes. The higher AM infection of plants may have contributed to the greater acquisition of nutrients. Increased mycorrhizal uptake of simple forms of organic P (P_o) (Jayachandran et al., 1992) and increased net release of P from organic matter due to uptake by mycorrhizal hyphae (Joner and Jakobsen, 1994) have been demonstrated. The increased efficiency of the plantlet to associate with mycorrhizae may be related to the initial physiological competence of the plantlet compared to the vegetative stem cutting (stake). Plantlets are also likely to associate with AM faster than cuttings because they already have roots and produce photoassimilates that are an essential component for an effective plant–mycorrhizal symbiosis. This symbiosis is likely to proliferate rapidly once established.

Tithonia established by plantlets had a total shoot biomass of 16.5 t/ha, which is significantly higher ($P < 0.05$) than the 7 t/ha under vegetative stem cutting (stake) establishment (Table 4). The total root length and root biomass were not significantly affected by the method of establishment, but it appears that both thick and fine roots of *Tithonia* were colonized by mycorrhizae. *Tithonia* plant established by using plantlets had significantly higher shoot uptake and use efficiency of N, P, K, Ca, and Mg (Table 4). The higher values of these attributes in plants established using plantlets may be attributed to greater mycorrhizal (AM) colonization under this establishment method. Potassium (K) and Mg are often found in higher concentrations in mycorrhizal than nonmycorrhizal plants (Sieverding, 1991). Some experimental work suggests that in K-deficient soils the improved K uptake is related to the AM fungal species and that K may be transported by AM fungal hyphae (Sieverding and Toro, 1988). Bowen (1980) and Jehne (1980) reported that AM might play an important role as transport paths for nutrient cycling processes. AM-root external mycelia presumably can efficiently and intensively extract nutrients from a greater soil volume and thus reduce the amount of solubilized or mineralized nutrients that are chemically

Table 4 Effect of Method of Establishment (Stake or Plantlet) on Mycorrhizal (AM) Association and Nutrient Uptake Efficiency of *Tithonia diversifolia*

Plant attributes		Method of establishment		
		Stake	Plantlet	LSD$_{(P=0.05)}$
AM infection in fine roots	%	49	79	11
AM infection in coarse roots	%	48	69	12
Number of spores in 100 g of soil	%	418	509	ns
P uptake efficiency	μg/m	30	48	12
N uptake efficiency	μg/m	167	331	128
K uptake efficiency	μg/m	379	662	130
Ca uptake efficiency	μg/m	116	184	56
Mg uptake efficiency	μg/m	37	61	19

LSD values are at 0.05 probability level ($n = 3$).
ns = not significant.
Source: Adapted from Phiri et al. (2003a) with the permission from The Haworth Press.

fixed or leached. This function of AM fungi was concentrated in the 0–20-cm depth of the soil profile where most root growth occurred.

IV. SUMMARY AND FUTURE PERSPECTIVES

Conclusions drawn from some long-term experiments on Oxisols and Andisols of Latin America can be summarized as follows:

1. The effects of intensity of disc harrowing on physical and chemical properties of the soils are dependent on the cropping system used. The maize and green manure cropping systems exhibit a different response to the increased number of disc harrowing on soil physical and chemical characteristics than the grass-alone pasture cropping system.
2. Disc harrowing for 2, 4, and 8 passes per year compared to not harrowing is a better tillage practice for this soil as it improves porosity, bulk density, volumetric water content, and soil biologically and moderately available P. Tradeoffs, however, include reduction in SOM, decrease in soil macroporosity, increase in soil compaction, decrease in soil aeration, and likely negative impact on soil macro-invertebrate activity as shown by Decaens et al. (2001) for the same soil.
3. Vertical tillage with two chisel passes for rice–soybean rotation or agropastoral treatments improved soil physical and chemical charac-

teristics. However, these improved soil conditions did not translate into improved and sustained grain yields of either upland rice or soybean. This finding supports the notion that soil is a slow-changing natural resource that interacts pervasively over time with a wide range of other biological and socioeconomic constraints to sustainable agroecosystem management.

4. Buildup of an arable layer requires improvement of soil physical, chemical, and biological conditions. Introduction of tropical pasture components with legumes into the production system could provide adequate soil physical conditions, improve nutrient acquisition and recycling, and facilitate accumulation of soil organic matter below ground, leading to the buildup of an arable layer.

5. Experimental evidence to promote the concept of building up an arable layer in tropical Oxisols using vertical tillage and agropastoral treatments is provided. But to improve and sustain crop production, there is a need to develop better crop management strategies to overcome weed problems.

6. Fractionization of SOM and soil P, together with the determination of C, N, and P contents in SOM fractions, is more effective for detecting the impact of planted fallow and crop rotation systems on improving soil fertility than the conventional soil analysis methods.

7. Among the three improved fallow species tested, *Tithonia diversifolia* showed the greatest potential to improve SOM, nutrient availability, and P cycling after one year as improved fallow because of its ability to accumulate high amounts of nutrients. Nevertheless, its use may be limited in areas with seasonal drought as it is not very tolerant to extended dry periods.

8. The use of bare root seedlings (plantlets) for establishing *Tithonia* as a fallow species in volcanic-ash soil is a better method in comparison to vegetative stem cuttings (stakes). The former method resulted in increased plant growth and nutrient acquisition, which are desirable plant attributes for fallow systems because of enhanced nutrient cycling. The increased labor requirements, however, may be a limitation to the adoption of plantlet establishment.

9. The amount of P in the LL and LM fractions of SOM correlated well with the amount of "readily available" P in the soil. The amount of P_i and P_o in the LL and LM fractions of SOM could serve as sensitive indicators of "readily available" and "readily mineralizable" soil–P pools, respectively, in the volcanic-ash soils of the Andean hillsides.

10. Crop and forage cultivars differ in their ability to acquire and utilize P and other nutrients, and these differences can be exploited to improve P input use efficiency in crop–livestock systems of the tropics.

Soil degradation is a threat to the livelihood of densely populated rural communities and food security of urban populations. Degradation occurs due to improper management of soils for crop and pasture production, leading to nutrient depletion, erosion, loss of water, sealing, compaction, and loss of soil biological activity. Low soil fertility, soil erosion, seasonal drought, and overgrazing are key factors enhancing soil degradation in the tropics. Integrated soil fertility management approaches that include stress-adapted crop and forage germplasm, improved soil and crop management strategies, intensification of crop–livestock systems, and empowerment of rural communities may not only contribute to combating soil degradation but also to recuperating of degraded soils in tropical savannahs and hillsides of Latin America to profitability. Developing economically viable and environmentally acceptable crop–forage–fallow technologies to combat soil fertility degradation in Latin America remains to be a major challenge. Research and development efforts are needed to integrate multiple stress-adapted crop, forage, and fallow germplasm into production systems to intensify food and feed systems of the low-fertility soils of Latin America. Reversing soil degradation is not only a biophysical problem but also a pervasive national and global policy issue with respect to incentives, and institutional failure. Addressing soil degradation issues thus requires a long-term perspective and holistic approach.

ACKNOWLEDGMENTS

We thank N. Asakawa, G. Borrero, L.F. Chavez, J.G. Cobo, I. Corrales, D. Molina, J. Ricaurte, M. Rivera, M. Rodriguez, J.H. Galviz, C. Trujillo, and A. Alvarez for their technical assistance. A number of graduate students including T. Basamba, S. Bühler, and S. Phiri received training and contributed to the progress of this work. We gratefully acknowledge the partial financial support by a number of donors including the Ministry of Agriculture and Rural Development of the Government of Colombia; Management of Acid Soils (MAS) Consortium of Consultative Group on International Agricultural Research (CGIAR) system wide program on Soil, Water, and Nutrient Management (SWNM); Norwegian Development Cooperation (NORAD); Swiss Center for International Agriculture (ZIL) and Swiss Development Cooperation; and Australian Centre for International Agricultural Research (ACIAR).

REFERENCES

Amézquita, E. 1998a. Propiedades físicas de los suelos de los Llanos Orientales y sus requerimientos de labranza. In Romero G., Aristizábal D., Jaramillo C., Eds. Memorias ''Encuentro Nacional de Labranza de Conservación.'' 28–30 de Abril de 1998. Villavicencio-Meta, Colombia.

Amézquita, E. 1998b. Hacia la sostenibilidad de los suelos de los Llanos Orientales. *In* Memorias "Manejo de Suelos e Impacto Ambiental." IX Congreso Colombiano de la Ciencia del Suelo, Paipa, Octubre 21–24 de 1998, pp. 106–120. Sociedad Colombiana de la Ciencia del Suelo, Bogotá, Colombia.

Amézquita, E., Ashby, J., Knapp, E.K., Thomas, R., Müller-Sämann, K., Ravnborg, H., Beltran, J., Sanz, J.I., Rao, I.M., Barrios, E. 1998. CIAT's strategic research for sustainable land management on the steep hillsides of Latin America. In Penning de Vries F.W.T., Agus F., Kerr J., Eds. Soil Erosion at Multiple Scales: Principles and Methods for Assessing Causes and Impacts: CABI. Wallingford, 121–132.

Barber, S.A. 1984. Soil Nutrient Bioavailbility: John Wiley. New York.

Barrios, E., Buresh, R.J., Sprent, J.I. 1996a. Organic matter in soil particle size density fraction from maize and legume cropping systems. *Soil Biol. Biochem* 28:185–193.

Barrios, E., Buresh, R.J., Sprent, J.I. 1996b. Nitrogen mineralization in density fractions of soil organic matter from maize and legume cropping systems. *Soil Bio. Biochem* 28:1459–1465.

Barrios, E., Kwesiga, F., Buresh, R.J., Sprent, J.I. 1997. Light fraction soil organic matter and available nitrogen following trees and maize. *Soil Sci. Soc. Am. J* 61:826–831.

Barrios, E., Cobo, J.G., Rao, I.M., Thomas, R.J., Amézquita, E., Jiménez, J.J. 2003. Fallow management for soil fertility recovery in tropical Andean agroecosystems in Colombia. Agriculture, Ecosystems and Environment (in press).

Beck, M.A., Sanchez, P.A. 1994. Soil P fraction dynamics during 18 years of cultivation on a Typic Paleudult. *Soil Sci. Soc. Am. J* 58:1424–1431.

Bowen, G. D. 1980. Mycorrhizal roles in tropical plants and ecosystems. *In* Tropical Mycorrhiza Research. Mikola P., Ed: Clarendon Press. London, 165–190.

Brown, S., Anderson, J.M., Woomer, P.L., Swift, M.J., Barrios, E. 1994. Soil biological processes in tropical ecosystems. *In* The Biological Management of Tropical Soil Fertility. Woomer P.L., Swift M.J., Eds: John Wiley. Chichester, 15–46.

Bühler, S., Oberson, A., Rao, I.M., Friesen, D.K., Frossard, E. 2002. Sequential phosphorus extraction of a ^{33}P-labeled oxisol under contrasting agricultural systems. *Soil Sci. Soc. Am. J* 66:868–877.

Buresh, R.J., Tian, G. 1998. Soil improvement by trees in sub-Saharan Africa. *Agroforestry Systems* 38:51–76.

Cobo, J.G., Barrios, E., Kass, D.C.L., Thomas, R.J. 2002. Decomposition and nutrient release by green manures in a tropical hillside agroecosystem. *Plant and Soil* 240: 331–342.

Coleman, D.C., Oades, J.M., Uehera, G., Eds. 1989. Dynamics of soil organic matter in tropical ecosystems: University of Hawaii Press. Honolulu, HI.

da Silva, J.E., Lemainski, J., Resck, D.V.S. 1994. Perdas de materia organica e suas relacoes com a capacidade de troca cationica em solos de regiao de ceraados do oeste Baiano. *Rev. Bras. Cienc Solo* 18:541–547.

Decaens, T., Lavelle, P., Jiménez, J.J., Escobar, G., Rippstein, G., Schneidmadl, J., Sanz, J.I., Hoyos, P., Thomas, R.J. 2001. Impact of land management on soil macrofauna in the Eastern plains of Colombia. In Jiménez J.J., Thomas R.J., Eds. Nature's Plow: Soil Macroinvertebrate Communities in the Neotropical Savannas of Colombia: CIAT. Colombia.

Fisher, M.J., Rao, I.M., Ayarza, M.A., Lascano, C.E., Sanz, J.I., Thomas, R.J., Vera, R.R. 1994. Carbon storage by introduced deep-rooted grasses in the South American savannahs. *Nature (Lond.)* 371:236–238.

Fontes, M.P.F., Weed, S.B. 1996. Phosphate adsorption by clays from Brazilian Oxisols: Relationships with specific surface area and mineralogy. *Geoderma* 72:37–51.

Friesen, D.K., Rao, I.M., Thomas, R.J., Oberson, A., Sanz, J.I. 1997. Phosphorus acquisition and cycling in crop and pasture systems in low fertility tropical soils. *Plant and Soil* 196:289–294.

Gachengo, C.N. 1996. Phosphorus release and availability on addition of organic materials to phosphorus fixing soils. M Phil. thesis: Moi University. Eldoret. Kenya.

Gijsman, A. J., Sanz, J.I. 1998. Soil organic matter pools in a volcanic-ash soil under fallow or cultivation with applied chicken manure. *Eur. J. Soil Sci* 49:427–436.

Giller, K.E. 2001. Nitrogen Fixation in Tropical Cropping Systems, 2nd. ed.: CABI. Wallingford. UK.

Goedert, W.J. 1983. Management of the Cerrado soils of Brazil: a review. *J. Soil Sci* 34: 405–428.

Guo, F., Yost, R. 1998. Partitioning soil phosphorus into three discrete pools of differing availability. *Soil Science* 163:822–833.

Hartemink, A.E., Buresh, R.J., Jama, B., Janssen, B.H. 1996. Soil nitrate and water dynamics in sesbania fallows, weed fallows and maize. *Soil Sci. Soc. Am. J* 60:568–574.

Hedley, M.J., Stewart, J.W.B., Chauhan, B.S. 1982. Changes in inorganic and organic soil phosphorus fractions induced by cultivation practices and by laboratory incubations. *Soil Sci. Soc. Am. Proc* 46:970–976.

Janzen, H.H., Cambell, C.A., Brandt, S.A., Lafond, G.P., Townley-Smith, L. 1992. Light-fraction organic matter in soils from long-term crop rotations. *Soil Sci. Soc. Am. J* 56:1799–1806.

Jayachandran, K., Schwab, A.P., Hetrick, B.A.D. 1992. Mineralization of organic phosphorus by vesicular mycorrhizal fungi. *Soil Biol. Biochem* 24:897–903.

Jehne, W. 1980. Endomycorrhizas and the productivity of tropical pastures: The potential for improvement and its practical realization. *Trop. Grasslands* 14:202–209.

Jiri, O., Waddington, S.R. 1998. Leaf prunnings from two species of Tithonia raise maize grain yield in Zimbabwe, but take a lot of labor! Newsletter of Soil Fert Net, Harare, Zimbabwe. *Target* 16:4–5.

Joner, E.J., Jakobsen, I. 1994. Contribution by two arbuscular mycorrhizal fungi to P uptake by cucumber (*Cucumis sativus* L.) from ^{32}P-labelled organic matter during mineralization in soil. *Plant and Soil* 163:203–209.

Jones, P. 1993. Hillsides: Definition and Classification: CIAT. Cali. Colombia.

King'ara, G. 1998. Establishment method of *Tithonia diversifolia* from seeds and cuttings. Report for diploma certificate. Rift Valley Technical Training Institute, Eldoret, Kenya.

Klink, C.A., Moreira, A.G., Solbrig, O.T. 1993. Ecological impact of agricultural development in the Brazilian Cerrados. In Young M.D., Solbrig O.T., Eds. The World's Savannas: UNESCO, Parthenon Publ. Group.. London, 259–282.

Knapp, B., Ashby, J.A., Ravnborg, H.M. 1996. Natural resources management research in practice: The CIAT Hillsides Agroecosystem Program. In Preuss H-J.A., Ed.

Agricultural Research and Sustainable Management of Natural Resources: LIT Verlag. Munster-Hamburg. Germany, 161–172.

Lal, R. 1991. Tillage and agricultural sustainability. *Soil Tillage Research* 20:133–146.

Lehmann, J., Cravo, M.S., Macedo, J.L.V., Moreira, A., Schroth, G. 2001. Phosphorus management for perennial crops in central Amazonian upland soils. *Plant and Soil* 237:309–319.

Maroko, J.B., Buresh, R.J., Smithson, P.C. 1998. Soil nitrogen availability as affected by fallow-maize systems on two soils in Kenya. *Biol. Fertil. Soils* 26:229–234.

Meijboom, F.W., Hassink, J., van Noordwijk, M. 1995. Density fractionation of soil macro-organic matter using silica suspensions. *Soil Biol. Biochem* 27:1109–1111.

Mesquita Filha, M.V., Torrent, J. 1993. Phopsphate adsorption as related to mineralogy of a hydrosequence of soils from the Cerrados region (Brazil). *Geoderma* 58:107–123.

Mitchell, M.L., Hassan, A.E., Davey, C.B., Gregory, J.D. 1982. Loblolly pine growth in compacted greenhouse soils. *Trans. ASAE* 25:304–307.

Myers, R.J.K., Palm, C.A., Cuevas, E., Gunatilleke, I.U.N., Brossard, M. 1994. The synchronisation of nutrient mineralization and plant nutrient demand In. Woomer P.L., Swift M.J., Eds. The Biological Management of Tropical Soil Fertility: John Wiley and Sons. Chichester. UK, 81–116.

Niang, A., Amadalo, B., Gathumbi, S., Obonyo, C.O. 1996. Maize yield response to green manure application from selected shrubs and tree species in western Kenya: A preliminary assessment. Proc. First Kenya Agroforestry Conf. on People and Institutional Participation in Agroforestry for Sustainable Development. Muguga. Kenya, 50–358.

Neufeldt, H., da Silva, J.E., Ayarza, M.A., Zech, W. 1999. Phosphorus fractions under different land-use systems in oxisols of the Brazilian *cerrados*. In R. Thomas and M.A. Ayarza (eds.). Sustainable land management for the oxisols of the Latin American savannas: International Centre for Tropical Agriculture (CIAT). Cali. Colombia, 146–158.

Nziguheba, G., Palm, C.A., Buresh, R.J., Smithson, P.C. 1998. Soil phosphorus fractions and adsorption as affected by organic and inorganic sources. *Plant Soil* 198: 159–168.

Oberson, A., Friesen, D.K., Tiessen, H., Morel, C., Stahel, W. 1999. Phosphorus status and cycling in native savanna and improved pastures on an acid low-P Colombian soil. *Nutrient Cycling in Agroecosystems* 55:77–88.

Oberson, A., Friesen, D.K., Rao, I.M., Bühler, S., Frossard, E. 2001. Phosphorus transformations in an oxisol under contrasting land-use systems: The role of the soil microbial biomass. *Plant and Soil* 237:197–210.

Palm, C.A. 1995. Contribution of agroforestry trees to nutrient requirements of intercropped plants. *Agroforestry Systems* 30:105–124.

Phiri, S., Amézquita, E., Rao, I.M., Singh, B.R. 2001a. Disc harrowing intensity and its impact on soil properties and plant growth of agropastoral systems in the Llanos of Colombia. *Soil and Tillage Research* 62:131–143.

Phiri, S., Barrios, E., Rao, I.M., Singh, B.R. 2001b. Changes in soil organic matter and phosphorus fractions under planted fallows and a crop rotation on a Colombian volcanic-ash soil. *Plant and Soil* 231:211–223.

Phiri, S., Rao, I.M., Barrios, E., Singh, B.R. 2003a. Plant growth, mycorrhizal association, nutrient uptake and phosphorus dynamics in a volcanic-ash soil in Colombia as affected by the establishment of *Tithonia diversifolia*. *J. Sustainable Agriculture* 21:43–61.

Phiri, S., Amezquita, E., Rao, I.M., Singh, B.R. 2003b. Constructing an arable layer through vertical tillage (chisel) and agropastoral systems in tropical savanna soils of the llanos of Colombia. *J. Sustainable Agriculture* (in press).

Preciado, G., Amézquita, E., Galviz, J. 1998. Effect of time of use of soil with rice on the physical conditions. In PE 2 Staff (eds.), PE 2 Annual Report 1997, International Centre for Tropical Agriculture (CIAT), Cali, 87–88.

Rao, I.M. 1998. Root distribution and production in native and introduced pastures in the South America savannas. In Box J. E., Ed. Root Demographics and Their Efficiencies in Sustainable Agriculture, Grasslands, and Forest Ecosystems: Kluwer Academic Publishers. Dordrecht. The Netherlands, 19–42.

Rao, I.M., Zeigler, R.S., Vera, R., Sarkarung, S. 1993. Selection and breeding for acid-soil tolerance in crops: Upland rice and tropical forages as case studies. *BioScience* 43:454–465.

Rao, I.M., Ayarza, M.A., Thomas, R.J. 1994. The use of carbon isotope ratios to evaluate legume contribution to soil enhancement in tropical pastures. *Plant and Soil* 162: 177–182.

Rao, I.M., Friesen, D.K., Osaki, M. 1999. Plant adaptation to phosphorus-limited tropical soils. In Pessarakli M., Ed. Handbook of Plant and Crop Stress: Marcel Dekker, Inc.. New York, 61–96.

Sánchez, P.A. 1999. Improved fallows come of age in the tropics. *Agroforestry Systems* 47:3–12.

Sánchez, P.A., Logan, T.J. 1992. Myths and science about the chemistry and fertility of soils in the tropics. In Lal R., Sánchez P.A., Eds. Myths and Science of Soils in the Tropics: SSSA Spec. Publ. 29. SSSA. Madison, WI, 35–46.

Sieverding, E. 1991. Vesicular-Arbuscular mycorrhiza management in tropical agrosystems. Deutsche Gesellschaft für Technische Zusammenarbeit (GTZ) GmbH., Eschborn 1, Ferderal Republic of Germany.

Sieverding, E., Toro, T.S. 1988. Influence of soil water regimes on VA mycorrhiza. V. Performance of different VAM fungal species with cassava. *J. Agronomy and Crop Science* 161:322–332.

Smaling, E.M.A., Nandwa, S.M., Janssen, B.H. 1997. Soil Fertility in Africa is at stake. In Buresh R.J., Sanchez P.A., Calhoun F., Eds. Replenishing Soil Fertility in Africa: SSSA Special Publication No. 51. SSSA.. Madison, WI, 47–61.

Sousa, D.M.G., Lobato, E. 1988. Adubacao fosfatada. Proc. 6th Simposio sobre o Cerrado. Centro de Pesquisa agropecuria dos cerrados; EMBRAPA, Brasilia, 33–60.

Spain, J.M. 1990. Neotropical savannas: Prospects for economically and ecologically sustainable crop-livestock production systems. Paper presented at an International Seminar on Manejo de los Recursos Naturales en Ecosistemas Tropicales para Agricultura Sostenible Bogotá, Colombia.

Stewart, J.W.B., Tiessen, H. 1987. Dynamics of soil organic phosphorus. *Biogeochemistry (Dordr)* 4:41–60.

Swift, M.J., Woomer, P. 1993. Organic matter and the sustainability of agricultural systems: Definition and measurements. In Mulongoy K., Merck R., Eds. Soil Organic Matter Dynamics and the Sustainability of Tropical Agriculture: John Wiley. Chichester, New Jersey. UK, 3–18.

Szott, L.T., Palm, C.A. 1996. Nutrient stocks in managed and natural humid tropical fallows. *Plant and Soil* 186:293–309.

Thomas R. J., Ayarza M.A., Eds. 1999. Sustainable land management for the Oxisols of the Latin American savannas: Dynamics of soil organic matter and indicators of soil quality. International Center for Tropical Agriculture (CIAT), Cali, Colombia, 231.

Thomas, R.J., Fisher, M.J., Ayarza, M.A., Sanz, J.I. 1995. The role of forage grasses and legumes in maintaining the productivity of acid soils in Latin America. *In* Adv. Soil Sci. Series. Soil Management: Experimental Basis for Sustainability and Environmental Quality: Lewis Publishers. Boca Raton, FL, 61–83.

Thomas, R., Ayarza, M.A., Neufeldt, H., Westerhof, R., Zech, W. 1999. General conclusions. *In* Thomas R., Ayarza M.A., Eds. Sustainable Land Management for the Oxisols of the Latin American Savannas: Centro Internacional de Agricultura Tropical (CIAT). Cali. Colombia: 215–227.

Tiessen, H., Moir, O. 1993. Characterization of available P by sequential extraction. In Carter M.R., Ed. Soil Sampling and Methods of Analysis: Lewis Publishers. Boca Raton, FL, 75–86.

Tiessen, H., Salcedo, I.H., Sampaio, E.V.S.B. 1992. Nutrient and soil organic matter dynamics under shifting cultivation in semi-arid northeastern Brazil. *Agric. Ecosyst. & Environ* 38:139–151.

Westerhof, R., Vilela, L., Ayarza, M.A., Zech, W. 1999. Carbon fractions as sensitive indicators of quality of soil organic matter. In Thomas R., Ayarza M.A., Eds. Sustainable Land Management for the Oxisols of the Latin American Savannas: Centro Internacional de Agricultura Tropical (CIAT). Cali. Colombia, 123–132.

Zou, X., Binkley, D., Caldwell, B.A. 1995. Effects of nitrogen-fixing trees on phosphorus biogeochemical cycling in contrasting forests. *Soil Sci. Soc. Am. J* 59:1452–1458.

20

The Raised-Bed System of Cultivation for Irrigated Production Conditions

Kenneth D. Sayre
CIMMYT
Houston, Texas, U.S.A.

Peter R. Hobbs
Cornell University
Ithaca, New York, U.S.A.

I. INTRODUCTION

The adoption of conservation agriculture technologies, which are characterized by minimal soil disturbance (tillage) before seeding (with the ultimate aim being zero-till seeding) and by diverse strategies to increase crop residue retention on the soil surface to ensure full ground cover (leading essentially to biological tillage) over time, has dramatically increased in many countries over the past 25 years. For example, there are now over 28 million ha of zero-till seeding in Latin America with the bulk concentrated in the southern cone countries of Brazil, Argentina, and Paraguay (Derpsch, 2001). Table 1 lists the adoption of zero-till in the world up to 2001 (Derpsch, 2001). Much of this acreage is zero-till with residue retention. However, upon closer inspection, the adoption of reduced-zero-till seeding combined with surface crop residue retention in the countries mentioned above as well as other large area adopters such as the United States, Canada, and Australia, and particularly for wheat production systems, has occurred mainly by large-scale farmers and nearly universally for rainfed production systems with a few exceptions where sprinkle irrigation is used. The apparent exclusion of small-scale farmers in general and for essentially all surface-irrigated

Table 1 Global Adoption of Zero-Tillage

Country	Million hectares
United States	21.1
Brazil	17.3
Argentina	11.7
Australia	9.0
Canada	4.1
Paraguay	1.3
Rest of the world	1.4
Total	66.0

Source: Derpsch (2001).

production systems (especially where irrigated wheat is a major crop in the system) has several explanations.

There are clear-cut differences between developed and developing countries regarding how wheat and other crops are grown and who grows the crops. Less than 5% of the total wheat produced in developed countries (and also in Argentina, Brazil, and Paraguay) comes from irrigated conditions, whereas large-scale farmers are by far the most important wheat producers. In contrast, well over 50% of wheat production in developing counties (especially for large producing countries in south Asia and China) comes from irrigated conditions, and the vast majority of these farmers are very small-scale (Byerlee and Moya, 1993). These divergent circumstances have led to serious shortages of appropriate reduced-zero-till seeding implements for small- and medium-scale farmers since machinery development has largely focused on large farms. Furthermore, since there is so little irrigated wheat in the developed countries, insufficient efforts have been expended to develop appropriate crop management technologies to reduce tillage and ensure surface retention of crop residues for surface-irrigated systems, in particular, those involving wheat.

One of the main constraints for reduced tillage and crop residue retention for surface irrigation, especially for wheat, is linked to the widespread use of solid stand planting on the flat combined with flood and basin irrigation systems that are commonly used throughout the world but especially for the major, irrigated wheat producers in south Asia and China. Flood irrigation can lead to extreme difficulty in irrigation water distribution within the field when loose residues are left on the surface. In some areas (western United States, west and central Asia where irrigated wheat is grown in rotation with row crops like cotton, dry beans, and maize) it is common to find furrow-irrigated wheat, where the wheat is planted as a solid stand on the flat following tillage but then shallow furrows are made (60–100 cm apart) and used for irrigation. This system of planting results in most of the wheat plants located on top of a bed formed between the irrigation furrows, but when tillage is reduced and some crop residues are left on the surface, serious water movement problems through the furrows can occur. This chapter discusses the development of appropriate raised-bed planting systems for irrigated systems in various parts of the world.

II. EVOLUTION OF RAISED-BED PLANTING SYSTEMS IN MEXICO

A. The First Step

The Yaqui Valley is located in the state of Sonora in northwest Mexico and includes about 255,000 ha of irrigated land using primarily gravity irrigation systems to transport water through fields from either storage reservoirs (over 90%) or deep tube wells (around 10%). Over the past 25 years, more than 95% of the farmers have changed from the conventional technology of planting most of their crops on the flat with flood and basin irrigation to planting all crops including wheat, which is the most widely grown crop, on beds. Irrigation water is delivered through the fields by furrows between the beds, which range in width from 70–100 cm from furrow to furrow, depending on the distance between the tractor tires. Wheat yields for the Yaqui Valley have averaged over six tons per ha for the past several years (Aquino, 1998).

A single row is planted on top of each bed for row crops like maize, soybean, cotton, sorghum, safflower, and dry bean; 1 to 2 rows per bed are planted for crops like chickpea and canola; but 2 to 4 defined rows, spaced 15–30 cm apart depending on bed width, are used for wheat (Fig. 1). Even though most farmers still use conventional tillage, remaking the beds for each new crop, those that now grow wheat on beds obtain about 8% higher yields, use approximately 25% less irrigation water, and encounter at least 25% less operational costs compared to those still planting conventional tilled wheat on the flat, using flood irrigation (Aquino, 1998).

Based on positive interactions through visits of scientists from developing countries to the CIMMYT base agronomy program, bed planting of wheat has also been introduced in various other countries in the last five years. Table 2 lists the benefits of planting wheat on beds in irrigated systems in terms of yield and water savings from various collaborating countries. Work has also started in South Africa, Morocco, Sudan, Iran, Kyrgyzstan, Uzbekistan, and Nepal, but data were not available for this chapter. Table 2 does show that for the countries that grow spring-type wheat, there is a yield advantage for bed planting, probably a result of less lodging and a more favorable root development environment. However,

Figure 1 Bed-planted wheat with conventional tillage in the Yaqui Valley, Sonora, Mexico.

with winter wheat varieties (Turkey and Kazakhstan), this yield advantage is less apparent.

Water savings are significant for both types of wheat cultivars and range from 25%–35%, which is an extremely crucial issue in these countries, which rely on irrigation water as competition for water between urban and rural areas intensifies. Table 3 presents more detailed information for wheat biomass and grain yield for comparisons of conventional till bed planting using furrow irrigation with tilled flat planting using flood irrigation for the Shandong data, which

Table 2 The Benefits of Bed Planting for Bread Wheat in Various Parts of the World

Country	Location	Yield on beds (kg/ha)	Yield on flat (kg/ha)	Extra yield bed vs. flat (kg/ha)	Water savings using bed vs. flat (%)
Bangladesh	Dinajpur	4,710 a	3,890 b	820	25
Pakistan	Punjab	4,530 a	4,220 b	310	24
	Punjab	4,470 a	4,020 b	450	32
India	Punjab	5,750 a	5,460 b	290	33
	Haryana	5,290 a	5,010 b	280	46
	UttarPradesh	4,750 a	4,550 b	200	30
China	Shandong	7,070 a	6,110 b	960	25
	Gansu-Hex Corridor	8,770 a	7,110 b	1663	26
Turkey	Achakale	5,500 a	5,750 a	−250	20
	Diyarbakir	5,380 a	5,230 a	150	na
	Eskishehir	5,070 a	5,020 a	50	na
Kazakhstan	Almaty	5,080 a	4,900 a	180	29
Average		5,531 a	5,106 b	425	29

Means in rows followed by different letters are significantly different by LSD at the 0.05 level.
Source: Personal communications from scientists in these countries. Data to be published.

supplements the data from China presented in Table 2. These results are from trials that form part of a collaborative program with CIMMYT to extend wheat bed planting to China and other countries. It is extremely interesting to observe the marked reduction in two serious wheat diseases, sharp eyespot and powdery mildew, with bed planting, likely the consequence of modest plant canopy microclimatic differences resulting from the change in plant orientation. Similarly, plant height and crop lodging were reduced with bed planting (Table 3). Dr. Wang Fahong, the principal scientist in China for this collaborative project, has also reported a 25% reduction in irrigation water use for bed planting in these trials.

The spread of bed planting has been limited by the availability of equipment in all the countries listed above and by the need to work with and experiment with these systems with farmers in their fields. However, progress is being made to extend this technology, and Table 4 lists some data collected from collaborators in South Asia that show significant acceleration of adoption in this region for bed-planted wheat. In Shandong, China, 20,000 ha of bed-planted wheat were grown in 2001 following the initial introduction of the technology in 1997. An even larger area was planted for the 2002–2003 crop cycle in Shandong, and the

Table 3 Effect of Planting Method on Plant Height, Biomass, and Grain Yield, Sharp Eyespot and Powdery Mildew Disease Incidences, and Crop Lodging Rate for Two Wheat Varieties[a]

Wheat variety	Planting method	Plant height (cm)	Biomass yield (kg/ha)	Grain yield (kg/ha)	Incidence of sharp eyespot disease (%)	Powdery mildew disease index of (%)	Lodging rate (%)
Jimai 19	Bed planting	73	15,570**	6,195**	8	8	0
	Flat planting	77*	14,109	5,630	33**	20**	10**
Yannong 19	Bed planting	76	16,134**	6,765**	2	9	5
	Flat planting	83**	14,550	5,965	51**	23**	70**

[a]Averaged over four crop cycles (1998–2002) at Jinan, Shandong, China.
* and ** indicate variety differences within varieties at the 5% and 1% levels of probability.
Source: Data provided by Dr. Wang Fahong, agronomist/soil scientist, Shandong Academy of Agricultural Sciences, Jinan, China.

technology is spreading to several other provinces. Bed-planted areas are rising in other countries although the data were not available at the time of writing.

In South Asia, bed planting has also been used to increase the diversity of different crops grown in the Indo-Gangetic Basin (IGB). Success has been shown with vegetables, maize, legumes (chickpea and pigeonpea), oilseeds, cotton, and sugarcane. One success story in NW India involves intercropping of wheat and sugarcane, where the wheat is planted on the ridge at the same time the cane is planted in the furrow. The wheat is harvested in April and the sugarcane continues to grow. Table 5 shows some data from this new system. Yields of wheat and cane are significantly higher, as are water savings. The cane in particular benefits from the timely planting since in the conventional system it is planted late after the wheat is harvested, although there are no differences in the next ratoon cane crop. Table 6 presents further information for a variety of different crops for which comparisons between bed planting with furrow irrigation versus conventional planting on the flat with flood irrigation have been made in numerous farmer fields in NW India. In all cases, yields are higher and considerable irrigation water was saved with bed planting.

Most farmers in the Yaqui Valley and other countries listed in Table 2 have taken only the first step involving bed planting but are now working on shifting to a reduced-tillage system of seeding with appropriate management of crop residues by making the change to permanent-bed planting. They largely continue to practice conventional tillage and destroy the beds after the harvest of each crop followed by several tillage operations before new beds are formed for the succeeding crop. This tillage is accompanied by burning of crop residues although

Table 4 Adoption of Bed Planting in South Asia

Country/State	Number of districts		Area (ha) covered		Number of drills		Number of farmers	
	2001–2002	2002–2003	2001–2002	2002–2003	2001–2002	2002–2003	2001–2002	2002–2003
Uttar Pradesh (West)	11	16	1,330	2,840	23	65	200	780
Uttar Pradesh (East)	16	16	50	126	10	27	10	34
Bihar	8	8	4	125	5	21	10	125
Haryana	11	11	1,000	400	17	22	50	35
Punjab India	12	12	1,000	1700	11	17	50	73
Pakistan Punjab	9	9	1,312	1750	31	47	64	80
Nepal	3	3	5	27	5	7	8	21
Bangladesh	3	3	5	25	3	5	5	23
Total	73	78	4,706	6,993	105	211	397	1171

Source: R. Gupta, Facilitator, Rice–Wheat Consortium, 2001. Personal communication.

Table 5 Intercropping in Raised-Bed Planting System: Sugarcane + Wheat (2001–2002)

Treatment	Crop	Irrigation time (hrs/ha)	Yield (t/ha) 2000–2001	Yield (t/ha) 2001–2002
Intercropping of wheat	Wheat	39	5.82 a[a]	5.78 a
(autumn planting)	Cane	113	68.2 a (73.8)	69.4 a (74.4)[b]
Conventional (summer	Wheat	65	5.43 b	5.20 b
planted cane)	Cane	178	56.2 b (74.0)	57.3 b (73.7)

[a]Figures followed by the same letter when comparing the means of wheat or the means of cane are not significantly different at 0.05 probability using Duncan's multiple range test.
[b]Values in parentheses (x) indicate yield of sugarcane ratoon crop.
Source: Progress Report, ADB Project, Meerut, RWC–RTCC Agenda Notes. March 2003.

Table 6 Yield Comparisons for Several, Diverse Crops Planted in Beds with Furrow Irrigation and with Conventional, Flat Planting with Flood Irrigation and the Associated Irrigation Water Savings[a]

Crops	No. of farmers 2000–2002	Yield (kg/ha) Bed	Conventional
Maize	10	3,270 (35.5%)[b]	2,380
Urd bean	10	1,830 (26.9%)	1,370
Mung bean	10	1,620 (27.9%)	1,330
Green peas	15	11,910 (32.4%)	10,400
Pigeon pea	10	2,200 (30.0%)	1,500
Gram	8	1,850 (27.3%)	1,580
Wheat	22	5,120 (26.3%)	4,810
Rice	20	5,620 (42.0%)	5,290
Okra	10	3,440 (33.3%)	2,910
Carrot	15	3,630 (31.8%)	2,860

[a]For Bed Planting in Farmers' Fields Located Near Ghaziabad, U. P., India.
[b]Average percent savings in irrigation time for bed planting.
Source: Personal communication, Dr. Raj Gupta, Facilitator for the Rice–Wheat Consortium, 2002.

some maize and wheat straw are baled off for fodder and, when turnaround time permits, some crop residues are incorporated during tillage (Meisner et al., 1992). However, there has been intense farmer interest in the development of new production technologies that will allow marked tillage reductions combined with retention of crop residues, which may lead to potential reductions in production costs, improved input-use efficiency, and more sustainable soil management while allowing continued use of the gravity irrigation system. In the early 1990s, CIM-MYT wheat agronomists and their Mexican research colleagues in collaboration with farmers began to address this issue. Scientists who visited CIMMYT returned home and also started to look at permanent-bed planting with their farmers.

B. The Second Step

It became clear that the bed-planting system that is widely adopted in the Yaqui Valley and elsewhere, albeit with continued heavy tillage and crop residue burning/removal, offered some unique opportunities for tillage reductions and surface residue retention to build upon the apparent benefits mentioned below that farmers had already obtained. The bed-planting system offers simplicity for field access that can be used to improve the effectiveness of many field operations including providing placement of fertilizers, especially N fertilizers, when and where they can be used most efficiently, easier application of herbicides (tractor wheels follow the furrows), and ease in roguing. Bed planting provides an opportunity for natural, controlled traffic when tillage is reduced since all implements can be designed to track in the bottom of the furrows with only potential soil disturbance on the surface of the bed by the seeding operation, thereby concentrating compaction in the furrow bottoms and reducing compaction in the immediate area on the top of the beds where crops are seeded. The seeding of two to four defined rows of wheat on top of the bed, as opposed to the solid seeding pattern normally associated with wheat and other small grains, makes inclusion of wheat far more feasible, but it was soon established that not all wheat varieties were appropriate for bed planting; cooperation with wheat breeders helped to identify appropriate wheat plant types for bed planting (Sayre and Ramos, 1997). Seed rates could also be reduced by a third, saving the farmer the cost of this valuable input.

The next step needed to reduce tillage and manage crop residues on the surface was simply to reshape the beds as needed between each crop cycle following some degree of chopping and even distribution of the previous crop's residue (Fig. 2). By eliminating any tillage except merely reshaping the beds as needed before planting the next crop, the beds solidify so that when irrigation water is channeled in the furrows, even in the presence of the high amounts of crop residues resulting from high-yielding irrigated crops, it advances evenly without cutting across the beds. Such beds are referred to as "permanent beds" or "per-

Figure 2 Reshaping permanent beds (75 cm wide from furrow to furrow) following wheat harvest prior to maize planting in the Yaqui Valley, Sonora, Mexico.

manent raised beds''. For some soils the compaction in the furrow bottoms from machine traffic assists in enhancing lateral water infiltration and forward water advance. If residues are chopped, it is important that they are not chopped too finely in order to minimize floating residue particles as the water advances in the furrows.

Obviously there has been important implement modification and development activity associated with the evolution of the permanent bed-planting technology. The furrow makers that farmers were already using to make beds with tillage could easily be modified to reshape the permanent beds by simply attaching cutting disc coulters ahead of the furrow openers for residue management (Fig. 2). Similarly, existing zero-till planters for row crops like maize and soybean could easily be adapted for planting on permanent beds into crop residues. However, no commercial zero-till wheat drill was readily available to plant two to four rows of wheat without tillage on top of 70–100-cm-wide beds and into high levels of crop residues. Many modifications were needed. Figure 3 shows one of the wheat prototype planters that have been developed in Mexico.

Figure 3 Prototype permanent-bed wheat planter in the Yaqui Valley, Sonora, Mexico.

Once the underlying ideas for the permanent bed-planting system were developed, long-term experiments were initiated in the Yaqui Valley in 1992 to compare the current bed-planting system based on extensive tillage being used by farmers with the new permanent-bed system. The main objectives of these trials were to fine-tune the machinery and management practices and to determine the effects of the planting methods on crop productivity, emphasizing wheat, but including maize and soybean, and to monitor relevant soil parameters.

III. VALIDATION OF PERMANENT BEDS FOR SURFACE-IRRIGATED CONDITIONS

A. Research Approach

The main long-term experiments have been conducted at the CIANO/CIMMYT experiment station located near Cd. Obregon, Sonora, Mexico (lat. 27.33° N, long. 109.09° W, 38 m above sea level). The soil type is described as coarse, sandy clay, mixed montmorillonitic typic calciorthid, low in organic matter (<0.8%) and slightly alkaline (pH 7.7–8.2). Long-term weather data for the wheat-growing period (November to May) are as follows: Maximum and minimum temperatures are 26.7 and 8.7°C respectively; average growing season rainfall is 49.3 mm; and average daily pan evaporation is 4.9 mm.

The principal long-term experiment was initiated in the summer of 1992 and involves a two-crop, annual rotation with wheat planted in late November and harvested in early May and either maize or soybean planted in late May–early June and harvested in October. Since the initiation of the experiment, 10 wheat crops have been planted and harvested in the winter season, and 9 maize crops (1992, 1994, 1995, 1996, 1997, 1999, 2000, 2001, and 2002) and 2 soybean crops (1993 and 1998) have been planted and harvested in the summer season for a total of 21 crops.

The five tillage/residue management treatments included in the trial were as follows:

1. Conventional tillage with formation of new beds for each crop; all crop residues incorporated
2. Permanent beds; all crop residues burned
3. Permanent beds; crop residues baled off for fodder, leaving approximately 30% in the field
4. Permanent beds, maize residues baled off for fodder and wheat residues retained
5. Permanent beds; all crop residues retained.

All crop residues were chopped before incorporation or when retained on the surface, depending on the treatment. When residues were baled off for fodder, the maize residue was chopped before baling, but the wheat residue was not chopped. Only the wheat residue cut by the combine harvester was baled for fodder, leaving the standing stubble intact. Whenever soybean was grown, however, its residue was not removed/burned because of the comparatively low amount of residue (2–3 t ha^{-1}).

The trial also included treatments concerning rates and timing of N applications for the wheat crop, but these results are not presented here. When summer maize was grown, 150 kg N ha^{-1} were uniformly applied across the whole experiment. No N was applied when soybean was planted in the summer. All N was banded either in the center of the bed or in the bottom of the furrow, through surface residues. Similarly, 20 kg P ha^{-1} were applied uniformly over the trial area prior to planting all crops. Bed width was 75 cm and two rows of wheat, 25 cm apart, were planted on top of each bed while one row of maize or soybean was planted in the center of each bed.

Each crop was seeded into moisture following a preseeding irrigation (Figs 4 and 5). This is a common strategy used by farmers in the region as part of

Figure 4 Wheat planted on permanent beds following maize in the Yaqui Valley, Sonora, Mexico.

their weed-control strategy. At the time of planting, usually 13–17 days after the irrigation, the first flush of weeds is controlled by a shallow tillage at the time of planting with the conventional tillage system, but for the permanent-bed treatments, weeds were controlled by application of 1.5–3.0 l ha^{-1} glyphosate immediately after planting. In most years, no further weed control was needed, but appropriate postemergence, selective herbicides were used when required. Normally four to five postemerge irrigations were applied to wheat.

B. Research Results

It is common that conversion from conventional tillage to a reduced-zero-tillage system with residue retention may require several crop cycles before potential advantages/disadvantages become apparent (Blevens et al., 1984), and the results

Figure 5 Soybean planted on permanent beds following wheat in the Yaqui Valley, Sonora, Mexico.

from this long-term trial confirm this observation. Figure 6 presents the wheat grain yield trends and associated year affects for the 10 wheat crops harvested since 1993 for the 5 tillage/residue management treatments when 225 kg N/ha were applied at the first node stage. Only small albeit significant yield differences between the tillage/residue managements occurred from 1993 to 1997 (5 summer crops and 6 wheat crops), but beginning with the 1998 wheat crop, large and significant differences between the management options occurred although year-to-year yield effects were also large (Fig. 6). From 1998 onwards, the permanent-bed treatment with continuous crop residue retention has had, on average, the highest average yield closely followed by conventional tillage with residue incorporation, then the permanent-bed treatments with full removal for fodder by baling and, considerably inferior, permanent beds with residue burning.

A number of soil chemical, physical, and biological parameters have been regularly monitored throughout the experimental period. Table 7 presents a brief summary of some of these parameters measured during the 2002 wheat-growing season. Samples that were taken at the onset of the experiment in 1993 indicated uniformity for these parameters across the trial area (data not shown). The soil samples taken in 2002 indicate that, while pH was not different for the tillage/residue management treatments, Na content and electrical conductivity were significantly less for permanent beds where part or all of the crop residues had been

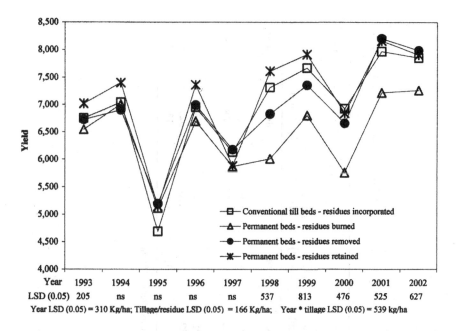

Figure 6   Effect of tillage and residue management over 10 years on wheat grain yield (kg/ha at 12% H2O) when 225 kg/ha N are applied at the first node stage at CIANO/Cd. Obregon.

left on the soil surface as compared to both tilled beds and especially permanent beds with residue burning. This is an exceedingly important result because it indicates that for soils that may tend toward the development of salinity problems, the use of permanent beds with retention of crop residues on the soil surface may help ameliorate this tendency by reducing the potential concentration of salts in the surface layers. Organic matter and total N levels were lowest for conventional tillage with residue incorporation and highest for permanent beds, increasing with levels of crop residue retention (Table 7).

    Table 7 also presents values for average soil aggregate size, which was similar for all treatments except for permanent beds with all residue removed for fodder (moderately less) and for permanent beds with residue burning, which had strikingly smaller soil aggregates. Soil strength/compaction on top of the beds (data not shown) has demonstrated a similar trend with only permanent beds with residue burning being significantly inferior to the other treatments, which had similar levels. The addition of residue mulch plays a significant role in reducing compaction at the soil surface and also improving the infiltration of water into the profile.

Table 7 Effect of Tillage and Crop Residue Management (Averaged over 4 Nitrogen Treatments) on Soil Properties for a Long-Term Bed-Planted Trial Initiated at CIANO, Cd. Obregon, Sonora in 1993[a]

Tillage/residue management	p_H (H$_2$O) 1:2	% C	% Total N	Na (ppm)	EC[b] (dS m^{-1})	Soil aggregates MWD[c]	SMB[d] C mgC (kg^{-1} soil)	SMB[e] N mg N (kg^{-1} soil)
Conventional tillage/incorporate residue	8.13	0.71	0.069	564	1.14	1.32	464	4.88
Permanent beds/burn residue	8.10	0.77	0.071	600	1.21	0.97	465	4.46
Permanent beds/remove residue for fodder	8.12	0.77	0.074	474	0.94	1.05	588	6.92
Permanent beds/retain residue	8.06	0.83	0.079	448	0.90	1.24	600	9.06
Mean	8.10	0.77	0.073	522	1.05	1.14	529	6.33
LSD (0.05)	0.02	0.01	0.001	9	0.01	0.03	13	0.16

[a]Results reported here are from 0–7-cm soil samples taken in 2002 crop cycle, 10 years after trial initiation.
[b]Electrical conductivity.
[c]Mean weight diameter by dry sieving.
[d]Soil microbial biomass–C content.
[e]Soil microbial biomass–N content.

Soil microbial biomass C and N levels in Table 7 are clearly superior for permanent beds with some or all residue retention compared to permanent beds with residue burning or conventional tilled beds with residue incorporation. This measure of potential soil health favors the permanent beds with residue retention and correlates with the observations that have been made on root disease scores and pathogenic nematode levels, which have been consistently higher for the permanent-bed treatment with residue burning (data not shown). It seems evident that the inferior grain yield performance of the permanent-bed treatment with crop residue burning as shown in Fig. 6 is strongly linked with the unremitting soil degradation that has occurred.

In 2000, a large-scale demonstration module (covering 8 ha) conducted with farmer participation was initiated to compare wheat performance with conventional tilled beds versus permanent beds with full residue retention for both planting systems. Crop performance and yield as well as variable production costs are being monitored. Figure 7 presents a comparison of the average wheat grain yields, variable production costs, and economic returns over variable costs for both planting systems for the 2001 and 2002 wheat crops. It seems abundantly clear that the permanent bed-planting system has shown both higher grain yield and a marked economic advantage. Average returns over variable costs were 75% higher for the permanent bed-planting system. Farmers are now more convinced and are confronting the issue of modifying wheat seeders to function on permanent beds with crop residue retention.

In the rice–wheat areas of South Asia, wheat is followed by rice. The challenge in this ecoregion is to find ways to grow rice on beds if permanent-bed systems are to be practical. Work has started in this area, but it soon became apparent that integrated weed control, selection of appropriate varieties, and careful water management were needed to ensure success. Table 6 presents some of the extremely promising initial results for bed-planted rice in NW India. Furthermore, information from the Rice–Wheat Consortium for the 2002 crop cycle in northern India, based on an average of 17 farmer field trials, shows average yields for bed-planted rice was 5.25 t/ha compared to 5.87 t/ha for normal, puddled, and transplanted rice. The main reason for the lower average yield was poor weed control in the bed-planted rice, partly attributed to a severe drought and poor water management in this year. However, some farmers were able to obtain higher yields with bed-planted rice with 40%–60% savings in water and thus providing incentive for future emphasis on this important component of permanent bed-planted rice–wheat. Also in South Asia, the use of permanent beds can lead to a whole array of other crops that can be grown, as indicated in Table 6.

IV. CONCLUSIONS

The permanent raised-bed irrigated planting system for wheat and other crops being developed in Mexico and elsewhere may finally provide a coherent technol-

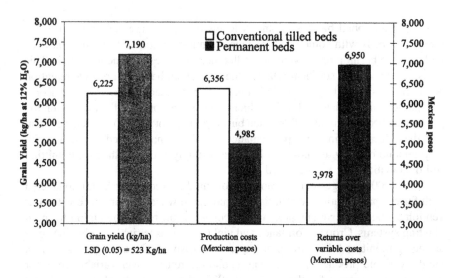

Figure 7 Comparison of average wheat grain yields, variable production costs, and returns over variable costs for wheat produced on conventional tilled beds versus pérmanent beds at CIANO, Cd. Obregon for the 2000–2001 and 2001–2002 crop cycles.

ogy to extend marked tillage reductions with appropriate management of crop residues for surface-irrigated production systems including those where wheat is a major crop. It is not a zero-till system per se since some soil disturbance occurs by the occasional reshaping of the bed in the bottom of the furrows between the beds, but it can be seen as a zero-till system on the top of the bed where the crops are seeded. The results presented here indicate that tillage reduction for surface-irrigated production systems resonates in the same positive way in terms of production profitability and sustainability as found for rainfed production systems. They also indicate that retention of crop residues will be essential to ensure that the required enrichment of critical levels of the chemical, physical, and biological soil parameters, which are crucial to ensure long-term production sustainability, can be achieved. Irrigation farmers must realize that some residue retention will be essential if they attempt to adopt permanent bed-planting systems even though their primary goal may be simply to realize lower production costs.

REFERENCES

Aquino, P. 1998. The adoption of bed planting of wheat in the Yaqui Valley, Sonora, Mexico. Wheat Special Report No. 17a. Mexico, DF: CIMMYT.

Byerlee, D., Moya, P. 1993. Impacts of International Wheat Breeding Research in the Developing World. Mexico, DF: CIMMYT.

Blevins, R.L., Smith, M.S., Thomas, G.W. 1984. Changes in soil properties under no-tillage. *In No–Tillage Agriculture—Principles and Prac*tices Phillips R.E., Phillips S.H., Eds: Van Nostrand Rheinhold Company. New York, 190–230.

Derpsch, R. 2001. Conservation tillage, no-tillage and related technologies. *In* Conservation Agriculture: A Worldwide Challenge. García-Torres L., Benites J., Martínez-Vilela A., Eds: ECAF/FAO. Córdoba. Spain, 161–170.

Meisner, C.A., Acevedo, E., Flores, D., Sayre, K., Ortiz-Monasterio, I., Byerlee, D. 1992. Wheat production and grower practices in the Yaqui Valley, Sonora, Mexico. Wheat Special Report No. 6. Mexico, DF: CIMMYT.

Sayre, K.D., Moreno Ramos, O.H. 1997. Applications of raised-bed planting systems to wheat. Wheat Special Report No. 31. Mexico, DF: CIMMYT.

21

Control of Rice Water Weevil in a Stale or Tilled Seedbed

M. O. Way, R. G. Wallace, M. S. Nunez, and G. N. McCauley
Texas Agricultural Experiment Station, Texas A&M University, Beaumont, Texas, U.S.A.

I. INTRODUCTION

Conservation tillage—stale seedbed—is becoming more widespread in Texas rice. Advantages include earlier/more timely planting; earlier harvest, which reduces the likelihood of inclement weather during combining operations; fuel and labor savings; less soil compaction; reduced "wear and tear" on equipment; decreased water usage for the main crop due to earlier harvest and fewer flushings if seed is drilled to moisture; less soil erosion; reduced red rice populations; and increased opportunity to ratoon crop. Disadvantages include purchase of expensive planting equipment, possible rapid development of weed resistance to nonselective herbicides, potentially poor stand establishment relative to conventional tillage, and buildup of organic matter in soil, which can lead to production of toxic byproducts of decay under anaerobic conditions. However, very little data are available on response of insect-damaged rice and rice insect pests to conservation tillage. Thus, the Entomology Project at Texas A&M University, Beaumont, has begun investigations to better understand the relation between stale-seedbed tillage and rice water weevil (RWW), *Lissorhoptrus oryzophilus*, populations and damage to rice.

II. MATERIALS AND METHODS

Experiments were conducted at the Texas A&M University Agricultural Research and Extension Center at Beaumont. Plot size was 18 ft by 4 ft with metal barriers surrounding plots to prevent interplot movement of fertilizer and pesticides. Each experiment consisted of four replications. Soil was Labelle formerly known as Beaumont clay. Seed treatments (Icon 6.2FS) were applied using the "Le Sak" method in which rice seed and a slurry of the pesticide were mixed by hand-shaking (10 minutes) in a plastic bag followed by air drying. Pesticidal sprays (herbicides and insecticides) were applied by hand using either a one- or two-person, CO_2-powered spray rig (4-ft or 22-ft spray width). For herbicides, final spray volume was about 12 gpa; for insecticides, final spray volume was about 28 gpa. Nitrogen fertilizer (urea) was applied by hand to all plots at rates and timings recommended by the Texas Cooperative Extension as described in the 2000–2001 Rice Production Guidelines. RWW immatures were sampled twice during the growing season by removing five, 4-in.-diameter by 4-in.-deep mud cores (each core contained at least one rice plant) from each plot. Cores were washed and immature RWW recovered from the roots of rice plants using 40 mesh screen buckets. Rice was harvested with a small plot combine and yields adjusted to 12% moisture. Immature RWW counts were transformed using $\sqrt{X + 0.5}$ and all data analyzed by ANOVA and LSD.

A. Experiment 1

The experiment was designed as a split plot with tillage (conventional or stale seedbed) as main plots and treatments (Icon 6.2FS, Karate Z or untreated) as subplots. During the fall of 2000, soil to be in conventionally tilled plots was disced as needed and as weather permitted to control weeds. No land preparation was performed on soil to be in stale-seedbed plots. Just before planting on April 19, 2001, conventionally tilled plots were disced and "roterraed" while stale-seedbed plots were treated with Roundup at 2 qt/acre 10 days prior to planting. At the time of planting, good weed control was achieved in both conventional and stale-seedbed plots. All plots were drill-planted at 90 lb Cocodrie/acre with selected plots receiving Icon 6.2FS -treated seed at 0.0375 lb (AI)/acre. Karate Z was applied to selected plots at 0.03 lb (AI)/acre about 4 hours before the permanent flood, which was applied about 3 weeks after emergence of rice through soil. Rice in stale-seedbed plots emerged more uniformly and quicker (about 3 days) than rice in conventionally tilled plots. Rice in all plots was flush irrigated as needed from emergence to the permanent flood. Rice stand counts were taken 2 days before the permanent flood. Immature RWW were sampled 3 and 4 weeks after the permanent flood.

B. Experiment 2

The experiment was designed as a split plot with tillage (conventional or stale seedbed) as main plots and treatments (described in Table 2) as subplots. Prior to the first planting, soil in all plots was disced and harrowed. Plots were drill-planted with Cocodrie at 90 lb/acre on April 6, 2001. Selected plots were planted with Icon 6.2FS-treated seed at 0.0375 lb (AI)/acre. After planting, plots were flushed. About 2 weeks after rice emergence through soil, all plots were sprayed with Roundup at 2 qt/acre (this spray resulted in the death of rice and weeds in all plots). About 10 days after Roundup was sprayed, selected plots were tilled with a garden rototiller (depth of tillage was about 6 in.) while remaining plots were not tilled. Immediately after tillage, plots were drill-planted with Cocodrie at 90 lb/acre. Selected plots received Icon 6.2FS-treated seed. Thus, some plots were replanted into a stale seedbed while some were replanted into a tilled seedbed. After planting, plots were flushed. Replanted rice emerged May 17 (as well as first planted rice planted at the same time as replanted rice). About 3 weeks later, the permanent flood was applied. Between rice emergence through soil and the permanent flood, plots were flushed as needed. About 3 and 4 weeks after the permanent flood, immature RWW were sampled.

III. RESULTS

A. Experiment 1

As mentioned previously, rice emerged quicker and more uniformly in the stale seedbed, which was probably due to a shallower depth of planting. The drill seeder tended to plant deeper in the tilled seedbed because soil was less compacted and more friable. However, across treatments, rice plant stands were similar for both tillages (Table 1). RWW populations were similar for both tillages on the first sample date but higher in conventional tillage on the second sample date. In addition, stale seedbed outyielded conventional seedbed by more than 1,000 lb/acre. Perhaps better RWW control in the stale seedbed (on the last sample date) contributed to this yield difference. Across tillages, Icon 6.2FS outperformed Karate Z and outyielded the untreated by about 950 lb/acre. Although not significant, Karate Z outyielded the untreated by about 420 lb/acre. Plant stands in Icon 6.2FS-treated plots were significantly higher than in untreated plots.

B. Experiment 2

Across treatments, RWW populations on the first sample date and yields were similar for both tillages (Table 2). However, as in Experiment 1, RWW were more abundant on the second sample date in tilled plots. Regardless of tillage, first planted Icon-treated seed protected untreated, replanted seed and produced almost a 1200 lb/acre yield increase.

Table 1 Control of Rice Water Weevil (RWW) in Stale Seedbed and Conventional Tillage, Beaumont, TX, 2001

Tillage	Trt.[a]	Rate [lb (AI)/acre]	Timing	Rice plants/ 3-ft row	RWW/5 cores June 15	RWW/5 cores June 28	Yield (lb/acre)
Conventional	Untreated	—	—	34.3	46	15.8	5,938
	Icon 6.2FS	0.0375	st[b]	43.5	4.3	7.0	7,191
	Karate Z	0.03	bpf[c]	36.3	15.3	20.5	6,561
Stale seedbed	Untreated	—	—	41.0	45.5	5.5	7,361
	Icon 6.2FS	0.0375	st	44.8	2.0	2.0	7,993
	Karate Z	0.03	bpf	41.3	12.5	4.8	7,580

[a]90-lb/acre seeding rate for all treatments.
[b]st = seed treatment.
[c]bpf = immediately before permanent flood.

Main plot effects[a]:

Tillage	Plants/3-ft row	RWW/5 cores June 15	RWW/5 cores June 28	Yield (lb/acre)
Conventional	38.0	21.9	14.4a	6,563b
Stale seedbed	42.3	20.0	4.1b	7,644 a

Subplot effects[a]:

Treatment	Plants/3-ft row	RWW/5 cores June 15	RWW/5 cores June 28	Yield (lb/acre)
untreated	37.7 b	45.8 a	10.7ab	6,650b
Icon 6.2FS	44.2 a	3.2 c	4.5 b	7,592a
Karate Z	38.8 ab	13.9 b	12.7a	7,071b

[a]Means in a column followed by the same or no letter are not significant at the 5% level (ANOVA, LSD).

	Interactions		
Plants/3-ft row:	*RWW/5 cores—June 15:*	*RWW/5 cores—June 28:*	*Yield:*
tillage × trt. = ns	tillage × trt. = ns	tillage × trt. = ns	tillage × trt. = ns

Table 2 Control of Rice Water Weevil (RWW) with Icon 6.2FS Applied to First or Replanted Rice in a Stale or Tilled Seedbed, Beaumont, TX, 2001

| Replant tillage | First planting | Replant | RWW/5 cores | | Yield (lb/acre) |
			June 27	July 6	
Till	Untreated	Untreated	41.8	63.8	6,149
	Icon[a]	Untreated	3.5	3.8	7,628
	Icon	Icon	0.5	2.5	7,370
	Not planted	Icon	3.3	6.8	6,921
Stale seedbed	Untreated	Untreated	38.8	37.5	6,426
	Icon	Untreated	2	5.5	7,302
	Icon	Icon	2.5	1.5	6,857
	Not planted	Icon	3.3	2.8	7,383

[a]Icon 6.2FS @ 0.0375 lb (AI)/acre

Main plot effects[a]:

| Tillage | RWW/5 cores | | Yield (lb/acre) |
	June 27	July 6	
Till	12.3	19.2 a	7,017
Stale seedbed	11.7	11.8 b	6,992

Subplot effects[a]:

| First planting | Replant | RWW/5 cores | | Yield (lb/acre) |
		June 27	July 6	
Untreated	Untreated	40.3 a	50.7 a	6,287 b
Icon	Untreated	2.8 b	4.7 bc	7,465 a
Icon	Icon	1.5 b	2.0 c	7,114 a
Not planted	Icon	3.3 b	4.8 bc	7,152 a

[a]Means in a column followed by the same or no letter are not significant at the 5% level (ANOVA, LSD).

Interactions

RWW/5 cores—June 27:	RWW/5 cores—July 6:	Yield:
tillage × trt. = ns	tillage × trt. = sig.	tillage × trt. = ns

22

Social and Economic Issues in Agricultural Intensification

G. Edward Schuh
HHH Institute of Public Affairs
Minneapolis, Minnesota, U.S.A.

This chapter addresses three issues that tend to be neglected when promoting agricultural intensification in the developing countries: (1) the influence of the resource endowment on the development and adoption of production technology; (2) the role of externalities as these affect the merits of subsidies to promote the use of fertilizer and other modern inputs; and (3) institutional design and development as a component of policy reform. The main focus of this discussion is on Sub-Saharan Africa, but these issues are important in other parts of the world as well. On the third issue, the emphasis is on institutional design for operating in an open international economy.

The modernization of agriculture on the African subcontinent has for the most part been disappointing, and for a long period of time. The hypothesis advanced here is that the failure of agricultural modernization in Sub-Saharan Africa is rooted in (1) an inappropriate bias toward resource-saving effects in the technology promoted as the means to modernize the agriculture in the region, (2) a rather misguided and excessively orthodox bias when assessing the benefits and costs of subsidies used to promote the use of fertilizer and other modern inputs in the region, and (3) a failure to develop appropriate institutional arrangements that are essential to the modernization of the sector, especially in an open economy. These issues are discussed in that order, with some concluding comments at the end.

I. RESOURCE ENDOWMENTS AFFECT ADOPTION OF NEW PRODUCTION TECHNOLOGY

A prominent feature of the discussion of agricultural development for Sub-Saharan Africa is the perceived need to promote a Green Revolution in the region similar to the one that occurred in Asia. The emphasis is on producing and adopting yield-enhancing crop varieties and the technological package that would support them. Other elements of the package include the use of modern fertilizers, pesticides, and irrigation.

The World Bank and other international development agencies such as the Sasakawa Global 2000 program have also sought to develop extension services that could promote the use of that technological package. Some attention has also been given to changes in economic policy that would provide stronger incentives for farmers to adopt the new technology and would integrate the domestic economy more effectively into the international economy.

However, the promotion of this technological package and the other instruments of agricultural development have, with a few exceptions, been notably unsuccessful in the region. Food production in sub-Saharan Africa has not kept up with the growth in population, with the result that food production per capita has, overall, declined. There is limited progress in alleviating poverty in the region, and the economies for the most part have stagnated in aggregate terms.

It is hypothesized here that the failure of agricultural modernization in the region is due in part to a failure to promote appropriate kinds of technology at the farm level. This comes from a failure to appreciate the significance of the underlying resource endowment as a factor influencing the adoption of new technology.

There are two ways of thinking about the decisions pertinent to the introduction of new production technology in agriculture. [For the importance of the modernization of agriculture in promoting economic development, see Schuh (2002).] One way is to recognize, rather simplisticly, the high social rates of return to investments in agricultural research in the past (Alston et al., 2000) and to concentrate on making such investments as the basis of economic growth and development. Most of those studies have been of an *ex post* nature and have been largely oblivious to underlying resource endowments as they affect the adoption of any new technology. The social rates of return to the creation and introduction of new agricultural technology tend to be high, ranging from 25% to 35% on the low end to over 100% on the high end (Alston et al., 2000). Since most developing countries can borrow money for investing in agricultural research at quite low rates of interest (5% and less in recent years) and on very soft terms, this seems like an obviously good investment.

A second, more operational or instrumental way of thinking about the process is to conceive of the new production technology as alleviating the constraints that impede increases in agricultural output, with an emphasis on which resources in the endowment are particularly important for affecting the adoption of the new technology. Hayami and Ruttan (1970, 1985) pioneered in this way of thinking about the process. Their analysis showed that if labor scarcity is the main constraint to increasing agricultural output, the need is to alleviate that constraint by concentrating on labor-productivity-enhancing technology, such as mechanization. If land is the relatively scarce resource, the new production technology should be land-productivity-enhancing, such as fertilizers. This perspective raises important issues of science and technology policy. In particular, it raises the issue of evaluating the resource-saving effects of any new production technology, together with a more acute *ex ante* perspective in the allocation of resources to agricultural research.

Hayami and Ruttan also posit a positive (in contrast to the above normative) perspective to this issue. They hypothesize that the process of innovation is responsive to relative resource endowments, and thus that these endowments can be used to make predictions about the pace and resource-saving dimensions of the process of technological change. These propositions are formalized in a model of "induced technological innovation" [see Hayami and Ruttan (1985)]. They go further, however, and focus on the importance of having the right institutional arrangements if the resource-saving effects of the new technology are to be focused on an efficient growth path in the resource-resource dimension. Ruttan and Hayami (1984) have extended the induced-innovation perspective by advancing the hypothesis that the institutional arrangements themselves are also responsive to the underlying resource endowments.

The perspectives developed by Hayami and Ruttan can be used for both positive and normative analysis. First, they can be used to explain the resource-saving direction that a new production technology takes as it is adopted in the economy, and to explain the direction that institutional innovations evolve in the economy as economic development proceeds. Alternatively, the model can be used in a normative manner, as a guide to policy in allocating resources to research and institutional development. From the second perspective, both the research process and the process of institutional change need to be guided by an understanding of relative resource scarcity if development policy is to be efficient and succeed.

In this section, we focus on the resource-saving dimension of new production technology. Hayami and Ruttan showed in their 1985 book that because early in Japan's agricultural modernization process the economy was land-scarce and labor-abundant, what was needed for agricultural development was a set of biological innovations consisting of improved varieties and a package of modern inputs, including fertilizers, that would raise land productivity. The United States, on the other hand, early in its development process had a resource endowment that was labor-scarce and land-abundant. This induced a process of mechanization in U.S. agriculture that raised labor productivity.

Both processes proved to be efficient and successful, with the result that these two countries led the process of agricultural modernization around the world, even though the respective resource-saving effects were quite different in the two countries. Even more impressive was the fact that in a later period of their respective histories, the relative resource scarcities of the two countries reversed themselves, and the respective innovative processes also reversed themselves. In the case of the United States, the frontier was eventually closed and land began to become increasingly scarce in a relative sense. This induced a process of biological innovation in agriculture, and the productivity of land became an objective. In Japan, a similar reversal occurred. As industrialization proceeded, labor became increasingly scarce. The result was a process of mechanization in Japanese agriculture.

This leads us to the situation in sub-Saharan Africa where, unfortunately, all too often the thrust of policy-making in the region, especially in science and technology policy, has been to try to replicate there the successful "Green Revolution" of Asia. The emphasis has been on introducing improved cereal varieties and the use of commercial fertilizers. The result has been for the most part a failure, and for an obvious reason—including the one described in the next section. The situation is more like that in the United States and Latin America in their earlier days when there was relative labor scarcity in the economy.

Uma Lele (1975) made this point some years ago. Unfortunately, the data to test this hypothesis are limited. Wage data and data on the price of land, which give evidence on relative resource scarcity, although accessible for Japan and the United States are not available for Africa. Given this limitation, Lele made a somewhat different argument with more confidence. She called attention to the low productivity of labor in sub-Saharan Africa, due in large part to poor health and nutrition. One can infer from this that productivity-enhancing innovations should be focused on raising the productivity of labor rather than on land, as has been the case.

One of the interesting things about the region is that it has not benefited from the animal-driven mechanization that was such an important phase in other now-modernized agricultures (Pingali et al., 1987) More recently, however, herbicides such as Round-Up have been introduced. Such herbicides make it possible for a farmer to cultivate as many as five hectares of land instead of just one. This represents a significant increase in labor productivity and a significant increase in agricultural output for the region should it be possible to spread the use of this labor-saving innovation throughout the region.

If this labor-scarcity hypothesis is correct, and certainly more empirical research is needed on this issue, the dominant innovation process in sub-Saharan Africa should be either some of the many variants of mechanization, or the use of herbicides to raise productivity of labor. It is not that improved varieties and the use of fertilizers and associated modern inputs cannot raise the productivity of labor. The underlying production functions are obviously not completely separable. However, when the productivity of labor is so low, it seems clear that a shift in emphasis is needed to reduce or eliminate that constraint as a barrier to increasing agricultural output.

It should be noted that the adoption of the fertilizer/high-yielding variety package in practice increases the demand for labor. If there is a constraint on the supply of labor, this can inhibit the adoption of the new package. In a study of such a technical package in Plan Puebla in Mexico some years ago, it was found that the availability of off-farm employment was a determining factor in the adoption of the new technology. Families in which the head of the family was working off-farm did not adopt the technology; those in which the head of the family did not work off the farm did adopt it. Interestingly, the average family incomes of the two groups of families, both adopters and nonadopters, were approximately the same (Villa Issa, 1976).

The rapid spread of AIDS in sub-Saharan Africa is giving added impetus to this problem. Projections are that in some countries in the region, the population may actually decline in the future. The impact on the labor force will be even greater, given AIDS' pattern of mortality. It is especially morbid and in turn debilitating in its effects on the productivity of labor. Unfortunately, these issues of science and technology policy are sorely neglected in current decision-making, especially consideration of relative resource endowments and the relative scarcity of the main inputs in agriculture.

II. SHOULD THE USE OF COMMERCIAL FERTILIZERS BE SUBSIDIZED?

The problems created by the failure to recognize the importance of adapting the design of technological packages to the endowment (or relative scarcity) of resources in sub-Saharan Africa has been exacerbated by strong pressure from the World Bank and other international development agencies against subsidies for the use of commercial fertilizers. These proscriptions have become part of the general policy of international development agencies to eliminate the use of subsidies and other government interventions. This perspective came to the fore during the economic crisis of the 1980s, when debt burdens became excessive and finding the means to service those debts became important. The international development community turned against import-substituting development policies and shifted toward an export-promoting policy stance. Efficiency in resource use became a policy imperative, and pressures against any policy distortions grew in importance.

Promoting resource efficiency in the domestic economies of sub-Saharan Africa and integrating them into the international economy were obviously important for an efficient growth policy. However, the prevailing policy proscriptions have ignored some important reasons for intervening in the economy and became too restrictive against the use of fertilizer subsidies in particular. The rigid approach taken by the international development agencies has ignored that there can be externalities that make fertilizer subsidies a rational way of promoting the use of such inputs.

The analytical justification for a bias against subsidizing fertilizers is that resources in the domestic economy need to be used most efficiently if maximum sustainable economic growth is to occur. The argument is that tradable products or commodities should be priced at their border price equivalents (that is, they should have domestic prices equivalent to the prices in the international markets as converted through the value of the domestic currency in foreign exchange markets). Providing subsidies for the use of fertilizers would cause their domestic prices to be less than their border price equivalents and thus would lead to "excessive" use of fertilizer. This in turn would place excessive demands on the market for foreign exchange and thus in most cases would bid resources away from alternative uses that presumably could contribute more to economic growth. Such subsidies also tend to make it difficult to balance budgets in the domestic economy and thus make it more difficult to have an effective macroeconomic policy.

This neoclassic economic argument against the use of subsidies is correct in principle, but it ignores certain externalities that are important elements in the economic environment that producers face when making their production decisions. In particular, the logic of the above case ignores important conditions in the economic environment of most agricultural producers, and especially those in sub-Saharan Africa. These economic conditions constitute important externalities that must be taken into account when deciding fertilizer and subsidy policy. In particular, a case can be made for subsidizing the use of fertilizer, given the following conditions: (1) efficient financial intermediaries for the agricultural sector are absent, so farmers are unable to obtain the optimal amount of credit for their operations; (2) investments in roads and other transportation infrastructure have been inadequate (less than optimal); (3) there is high complementarity between fertilizers and improved seeds, and eventually between fertilizers and irrigation water and other modern inputs; and (4) external pecuniary economies are associated with agricultural modernization, which means that the benefits of such modernization would be widely distributed in the economy, and in favor of the poor.

Fertilizer subsidies were fairly high when recent efforts to modernize agriculture began in sub-Sahara Africa over a decade ago. These subsidies were rooted in at least two of the traditional reasons given for providing such subsidies. The first was a learning-by-doing argument, based on the belief that subsidies provided over a relatively short period of time would induce farmers to try the new input. In the process, they would learn how to use it and become persuaded of its benefits. This suggests that the subsidies could eventually be phased out.

The second traditional reason for providing subsidies for the use of fertilizers was an attempt to offset the risk and uncertainty in adopting a new innovation such as improved varieties and fertilizers. Poor farmers are justifiably risk-averse, so subsidies lower the financial stakes for farmers trying new varieties and fertilizer. This is related to the learning-by-doing argument but different in that subsidies would not only help to offset the risk associated with the new input but would make its use more profitable. It could be argued in this case also that subsidies could eventually be phased out.

Eventually, the World Bank and other international development agencies turned against the use of such subsidies, however. The attempt to phase out the use of such subsidies was part of a general attempt to phase out the use of import-substituting industrialization policies and to move developing economies toward more efficient use of their domestic resources. A related issue was reducing the recurring budget deficits associated with these subsidies, and the need for institutional reforms to collect more domestic taxes to pay for the subsidies. The deficits created recurring problems in managing domestic inflation. Finally, there was a rather narrow promotion of orthodox neoclassical economic policies as the basis of economic development policy—a perspective that this chapter challenges. In the following sections we address each of the above issues.

A. Imperfect Capital Markets and Capital Rationing for Agriculture

Pedro Sanchez, 2002 World Food Prize laureate, has noted that plant nutrients have been drained from the soils in sub-Saharan agriculture due to a long process of weather-induced leaching. When the demands for agricultural output were minimal, replacement of the nutrients could be done by fallowing and other organic means. As population growth rates in the region have increased and as greater demands have been placed on the soils, however, organic means have proved to be inadequate to replace the recurring losses.

If farmers in the region were able to borrow money at reasonable rates, it would be rational for them to borrow funds to purchase and apply the fertilizers needed. Some of that borrowing could be justified by increases in the current crop. With the right terms and conditions, however, farmers might also borrow to replace the depleted stock of nutrients for future production. These would be optimal and efficient decisions on their part.

Unfortunately, the agricultural sector in sub-Saharan Africa does not presently have an effective and efficient set of credit and financial institutions. They neither make sufficient credit available on appropriate terms nor mobilize savings and reinvest them in the economy in efficient ways. Most government credit institutions, intended to make up for deficiencies in the private sector, are inadequate and inefficient. The inability of farmers to borrow funds at reasonable rates is a serious impediment to the use of fertilizers. Subsidies for the use of fertilizer would be one way of offsetting this institutional deficiency.

One alternative would be to invest in the development of viable credit and financial institutions, and especially in the development of true financial intermediaries. The development of these institutional arrangements should be pursued as a matter of high priority in most developing countries, irrespective of their implications for efficient fertilizer use. However, the development of these intermediaries cannot be expected to occur overnight, nor can farmers be expected to learn how to use them overnight.

B.   Deficiencies in the Transportation Infrastructure

With a few exceptions, the transportation infrastructure is sorely inadequate in most sub-Saharan African countries—a point that international development agencies seem to ignore. Agriculture is a highly dispersed sector of the economy, with the product typically being produced long distances from the consumer. At the same time, farmers live and work long distances from the sources of the modern inputs such as fertilizers needed to produce those products.

The result is all too familiar. The prices offered for the commodities produced are low at the farm gate, and the prices of the modern inputs such as fertilizers are relatively high. This makes the use of fertilizers unprofitable. This problem is especially important in the case of fertilizers because they tend to be a bulky commodity.

Again, an obvious solution would be to strengthen the transportation infrastructure. This would have many benefits and externalities. But this solution also is not likely to occur overnight. In the interim, a good case can be made for providing subsidies for the use of the fertilizer—subsidies that would offset the high transportation costs.

C.   The Complementarity Between Fertilizers and Improved Varities

Still another factor that seems not to have been recognized by analysts in the international development agencies and by policy makers alike is the high degree of complementarity between the adoption of improved varieties and the use of fertilizers. An important characteristic of the improved varieties of Green Revolution fame is that they have a greater and more prolonged response to the use of fertilizers. Moreover, they produce that response without lodging or collapsing of their own weight.

These improved varieties and the associated modern inputs such as fertilizers and irrigation are the source of the huge increases in food output in Asia that have helped feed a rapidly growing global population, with a consequent decline in the price of most food commodities. Unfortunately, if it is not profitable to use fertilizers, it will not often be profitable to adopt the improved varieties. This explains in part the slow rate of adoption of the improved varieties in sub-Saharan agriculture.

We know that if improved varieties are adopted, the social rate of return for investing in the research that produces these improved varieties is very high—probably from 35% to over 100%. The evidence on this is overwhelming. Thus, improved varieties are a powerful source of economic growth. In fact, there are few investments to promote economic growth and development that have such a high social rate of return.

Unfortunately, the failure to have a proper fertilizer policy means that improved varieties are simply not adopted, so this powerful source of economic growth is sacrificed. Subsidies for the use of fertilizers could help make this source of economic growth realizable.

D. External Pecuniary Economies Associated with Agricultural Modernization

There are substantial external pecuniary economies associated with the modernization of agriculture—a point I have examined in an article in the University of Chicago's *Policy Review* (Schuh, 1999). These externalities occur both in the case of domestic staples and in the case of tradable commodities. In the case of domestic staples, the result of adopting the new production technology is typically to lower the price of the food commodity. That is equivalent to an increase in the real income of all consumers of food. That is the true significance of the modernization of agriculture and of the importance of the sector in the development process. There are few other goods or services in the economy that everybody in society consumes. This is the main reason why the social rate of return from investing in agricultural research is so high. Although the increase in per capita incomes for each individual may be small, the fact that everybody benefits means that when the benefits are summed across the society, the total is large.

But the adoption of the new production technology has other advantages as well. Low-income groups and the disadvantaged benefit more in relative terms because they spend a larger share of their income on food than do middle- and upper-income groups. There are few goods or services in the modern economy for which this is the case. The fact that agricultural modernization has in recent years been sorely ignored by most international development agencies means that this powerful and pervasive source of economic growth has been sacrificed in favor of lower-payoff investments with more inequitable sources of economic growth.

Although the form in which the benefits are realized is different, there are similar externalities associated with the modernization of tradable commodities. In that case, the benefits of the modernization are realized in the form of savings or earnings of foreign exchange. That benefits the economy as a whole because the foreign exchange earnings help to service the foreign debt and/or to finance higher rates of economic growth. It also causes the real value of the nation's currency to rise, other things being equal, and those benefits are again widely distributed in the economy.

Finally, the immediate adoption of improved technology for tradable commodities means that those sectors become more competitive in international markets. Export commodities in particular tend to be labor-intensive, and thus this development tends to create new employment for the low-income, disadvantaged worker.

To conclude this section, we emphasize the fact that two negative externalities and two positive externalities are associated with the use of fertilizer subsidies. All four externalities help to make a case for providing subsidies for the use of fertilizers. The failure to capitalize on these externalities has contributed to the slow pace of modernization of sub-Saharan agriculture and to the slow pace of reducing poverty in that part of the world.

III. SUGGESTIONS FOR POLICY AND INSTITUTIONAL DESIGN

These begin from the micro level, with an emphasis on the design of ways to provide subsidies for the use of fertilizer. This is where a more immediate impact can be felt. The discussion will then move to more aggregative levels of analysis.

A number of goals come to mind when thinking about fertilizer subsidies. The first is to find a way to offset the negative externalities that now play such an important role in impeding the use of fertilizers—the imperfect credit and capital markets, and high transportation costs. A second is to provide subsidies in such a way that they indeed get passed along to the farmers and are not captured by others in the fertilizer and food chains. The third is to provide the subsidies in such a way as to bring about the most institutional innovation. Realization of these objectives will be determined in part by how competitive the respective sectors of the economy are.

My colleague, Tom Foster, has suggested that subsidies might best be provided for the cost of transportation. As long as the transportation sector is reasonably competitive, that can be an effective alternative. Moreover, it has the added advantage of creating a pressure group that would promote investments in roads and other improvements in the physical infrastructure.

In the past, some of the subsidies for fertilizer consumption have been provided in the form of subsidized credit. The difficulty with this approach is that it is not likely to reach the very small producer. Past experience suggests also that this does not bring about much institutional innovation. It is also fairly easy, in the absence of a competitive credit market, for bankers to capture the subsidy without passing it along to the producer.

Finally, there ought to be a way of providing the subsidy directly on the use of the fertilizer. From this perspective, greater use might be made of the voucher system now in use in many sub-Saharan African countries. These vouchers are paid for with PL 480 food-aid funds and can be used by farmers to purchase any of the modern inputs. Such a system can help offset the negative incentives currently created by the monetization of food aid—a veritable crime against poor farmers in many developing countries.

If extensive use were to be made of such vouchers, it would provide a stimulus to further develop the fertilizer distribution system. Stockers would learn what fertilizers to acquire and would develop skills for promoting their use. They would learn where to acquire fertilizers and how to purchase them. Middle-persons further up in the distribution system would also learn how to import fertilizer and about the international distribution system.

In broadening the range of policy reforms, the next step should be to develop a system of true financial intermediaries. Microcredit schemes are one means of increasing the availability of credit to small borrowers. However, other means are also possible and should be pursued. The development of such institutional innovations is an imperative for the development of the economies in most developing countries. Such countries have in the past relied too heavily on foreign aid and on foreign investment in their economies. In point of fact, most of the capital for domestic economic development must come from domestic sources. The failure to develop effective domestic financial intermediaries to mobilize savings effectively and channel them to efficient uses is a serious constraint on the development of the developing countries.

A next set of reforms is to give more attention to shifting the resource-saving bias in the new technological innovations, especially in sub-Saharan Africa but not limited to that region. Staff development efforts in extension services need to give more attention to transmitting knowledge on how to raise labor productivity. More research is needed to design labor-enhancing production technology, including for the household. More research is also needed to better understand the resource-saving effects of alternative technological packages.

Much work still needs to be done on economic policy issues in the developing countries. The extent to which trade and exchange-rate policies discriminate against the agricultural and rural sectors is a critical issue. Without efficient relative prices, farmers are not likely to have the incentive to adopt the new technology and to modernize the agricultural sector in their respective countries.

A related issue is the importance of preparing to take advantage of any improved trade opportunities resulting from the Doha round of multilateral trade negotiations. With the leadership of Brazil and India, the agenda has been set for the developing countries to have a greater role in the next round of negotiations. To be effective, participants in these negotiations from the developing countries need to know how much they will benefit from particular results of the negotiations, and how much opening their markets will benefit other countries.

A related issue is the huge costs that farm subsidies of the European Union and the United States impose on developing countries. These countries need to develop estimates of the costs of these subsidies to their respective economies, and to publicize these costs. In addition, developing countries need institutional arrangements that will evaluate these policies on a continuous basis and develop and publicize estimates of their costs.

Finally, the institutional arrangements that will make it possible for the developing countries to engage in international trade on a more extensive basis are an imperative. International trade can be an important source of economic growth and development. These contributions should not be sacrificed.

IV. CONCLUDING REMARKS

This paper concludes with two points. First, science and technology policy to promote economic and agricultural development has proceeded without a sound understanding of past economic development. For too long, development policy-makers in sub-Saharan Africa and other parts of the world have drawn too strong and simple a parallel with Asia. The basic resource endowments in various regions of the world are quite different. Asia was traditionally a land-scarce, labor-surplus region, and thus the goal of science and technology policy was to raise the productivity of land. In Africa and some other parts of the world, the resource endowment is just the reverse. Labor has been scarce and land abundant. There the goal of policy-makers should be to raise the productivity of labor, not of land. It is the low productivity of labor that is the constraint on economic growth and development, not of land. It is also critical to raising per capita incomes.

The failure to recognize this point is somewhat surprising. It was clearly made by Hayami and Ruttan, who showed its importance with their "induced innovation" and "induced institutional change" models. Their contribution to the agricultural development literature and to agricultural development policy is widely recognized, but this important point of theirs was lost.

Finally, there continues to be an imbalance between the biological and natural sciences and economics and the social sciences in the design of agricultural development policies. The search for "silver bullets" in the form of dramatic new technological breakthroughs still tends to command attention, based more on hope than scientific analysis, while the contributions of the social sciences are neglected. In the final analysis, complementarity between the biological and natural sciences, on the one hand, and the social sciences, on the other, is essential. Achieving such complementarity in crafting solutions to development problems should have a powerful effect on agricultural development and modernization all around the world. For this reason, I want to commend my Ohio State colleagues for organizing multidisciplinary seminars and workshops that lead to volumes such as this one.

ACKNOWLEDGEMENTS

In preparing this chapter, the author benefited from the research assistance of Rebeccah Hooper, and received helpful comments on an earlier draft from Vernon W. Ruttan.

REFERENCES

Alston, J.N., Chan-Kang, C., Marra, M.C., Pardey, P.G., Wyatt, T.J. 2000. A Meta-Analysis of Rates of Return to Agricultural R&D: Ex Pede Herculem?. Washington, DC: International Food Policy Research Institute.

Hayami, Y., Ruttan, V.W. 1970. Factor prices and technical change in agricultural development: The United States and Japan, 1885–1960. *J. Political Economy* 78(1):5–14.

Hayami, Y., Ruttan, V.W. 1985. Agricultural Development: An International Perspective. Baltimore, MD: Johns Hopkins University Press.

Pingali, P., Bigot, Y., Binswanger, H.P. 1987. Agricultural Mechanization and the Evolution of Farming Systems in Sub-Saharan Africa. Baltimore, MD: Johns Hopkins University Press.

Ruttan, V.W., Hayami, Y. 1984. Towards a Theory of Induced Institutional Innovation. *J. Development Studies* 20:203–223.

Schuh, G.E. 1999. Agriculture and economic development. Contribution to Special Issue of. *Chicago Policy Issues* 3(1):57–65.

Villa Issa, M. 1976. The effect of the labor market on the adoption of new production technology in a rural development project: The case of Plan Puebla. Mexico. Unpublished Ph.D. thesis.

23
Impacts of Conservation Tillage in a Temperate Environment*

D. Lynn Forster
The Ohio State University
Columbus,, Ohio, U.S.A.

I. ECONOMIC EFFECTS OF CONSERVATION PRACTICES USING FARMERS' ACCOUNTING DATA

A. Objectives

First, farm-level data are needed to investigate factors affecting two measures of performance: relative overall efficiency, and return on assets (ROA). Factors hypothesized to be associated with differences in relative overall efficiency and ROA include CT and other farming practices, capital structure, and farm operator characteristics. Of particular interest are the effects of CT and crop rotations, which are capable of reducing agricultural pollution in Lake Erie and its tributaries.

Numerous studies [as exemplified by Williams (1988) and Klemme (1985)] have shown alternative tillage practices (CT, NT, and ridge till) to be more profitable than conventional systems (moldboard plow and disc twice) under a wide range of operating environments. However, Featherstone et al. (1991) reported no statistically economic differences among tillage systems (conventional, ridge-till, and NT) used on a sample of farms. One of the concerns of Featherstone et al. (1991) was that in spite of reported positive attributes of alternative tillage systems, there may be some short-term penalties associated with their adoption.

---

* This chapter is a summary of results reported in Forster (2002) and Forster et al. (2000).

B. Model and Data

Return on assets (ROA) and relative overall efficiency are the two measures of performance used to assess the economic effects of alternative farming practices, capital structure, and farm operator characteristics. Return on assets is an accounting-based measure that reflects an economic return to assets deployed in the production process (Blue and Forster, 1997).

The measure of overall efficiency is derived from an expenditure-constrained profit optimization model proposed by Färe et al. (1990). Overall efficiency is defined as the ratio of actual profit to the short-run unconstrained profit derived from the profit optimization model.[*] It measures by how much a firm's actual profit is falling short of a theoretically derived maximum profit because of production choices and expenditure constraints. The set of Lake Erie Basin farms participating in the Ohio Farm Household Longitudinal Study (Stout et al., 1992) for 1987, 1988, 1990, and 1992 constitutes our sample. This sample is restricted to farms having gross farm income larger than $40,000 in order to represent commercial farming and exclude rural residents with a peripheral interest in agriculture. Demographic, off-farm employment, financial, production, and marketing data were collected each year of the survey. Numbers of farms in the samples totaled 98, 112, 127, and 113 in each of the respective years.

C. Regression Analysis of Return on Assets and Efficiency Scores

Overall efficiency and return on assets are used as dependent variables in regression models, which included operator and farm characteristics as independent variables. Similar types of analyses have been performed to assess the factors associated with nursing home efficiency (Fizel and Nunnikhoven, 1993), educational efficiency (Lovell et al., 1989), New York dairy farm efficiency (Tauer, 1993), and West Bengal farm efficiency (Ray, 1985).

In this analysis, gross sales and its squared term are used as a measure of farm size. Overall efficiency and ROA are expected to increase as farm size gets larger because of technological and pecuniary economies of scale. The squared term allows results to capture the expected curvilinear relationships caused by economies of scale.

On one hand, personal characteristics such as motivation and willingness to accept risk change over the operator's lifetime may contribute to a life cycle of growth and decline of the farm business (Nalson, 1968). On the other hand,

---

[*] For details on how overall efficiency is generated from the expenditure-constrained profit optimization model, see Färe et al. (1990).

older farm operators may have acquired skills to allocate resources more efficiently to end uses. Thus, in this analysis, age of the farm operator is used as an independent variable in overall efficiency and ROA equations.

The number of years of education the farm operator possesses may positively influence ROA and overall efficiency because more highly educated producers may be better at evaluating new information, quicker to adopt innovations, and more technically efficient (Asplund et al., 1989; Rogers et al., 1988; Tauer, 1993; Kalirajan and Shand, 1986). Years of education are identified by dummy variables indicating the highest level of education: less than 12 years, 12 years, 12–16 years, or more than 16 years.

Larger farm size and improved profits are often accompanied by financial leverage, that is, increased debt in financing farm assets. This may be due to technical change (Shepard and Collins, 1982) or personal characteristics such as motivation, ambition, and willingness to accept risk (Upton and Haworth, 1987). Bravo-Ureta and Pinheiro (1993) demonstrated that credit use has a positive impact on technical efficiency. The use of debt may ameliorate expenditure constraints some firms face and thus lead to increased overall efficiency and ROA. In this study, use of debt is measured as the debt-to-asset ratio.

Information collection and its use are important managerial activities. As a farm becomes larger, more management expertise is required. Often this information comes from outside sources via consultants or through the use of computers. Farmers who seek greater amounts of information from numerous sources are more likely to adopt innovations (Feder and Slade, 1984; Asplund et al., 1989). Bravo-Ureta and Pinheiro's (1993) review of efficiency in developing country agriculture verified that information use affects efficiency positively. Tauer (1993) showed that the use of a more elaborate accounting system improves dairy farm efficiency. In this study, a 0–1 dummy variable measures information use, with 1 indicating that the farm operator used computers, consultants, or extension agents, and 0 indicating that these information sources were not used.

Rotations and tillage practices were used as explanatory variables to examine the effects of various farming systems on actual and financial efficiency. The array of rotations and tillage practices used on farms in the sample were categorized into four rotations and four tillage systems; they were represented by three rotation and three tillage system dummy variables.

D. Regression Results

1. Return on Asset Regression

Gross sales and age are the only variables that consistently influenced ROA year to year (Table 1). In three out of the four years, larger farms have statistically higher return on assets, as expected. In addition, the squared term of gross sales

Table 1 Factors Associated with Farm Economic Performance, Analysis of Lake Erie Basin Farmers' Accounting Data (1987–1992)

Hypothesis tested:	Returns on assets and efficiency depend on
	Farm size
	Capital structure (debt/assets)
	Operator characteristics (age, education)
	Information use
	Tillage
	Rotations

Results	Return on assets	Efficiency
Dependent variables (Means)	5.2%	54%
Independent variables		
Farm size	++	++
Debt/assets		+
Age	−	
Education	No	No
Information use	No	No
Tillage	No	No
Rotations	No	No

++, + = positive relationship between dependent and independent variable at .05 or less, and .05–.10 level of significance.

− = negative relationship between dependent and independent variable at .05 or less level of significance.

No = relationship between dependent and independent variable at less than .10 level of significance or more.

Source: Forster (2002).

also is significant for all years. In 1987 and 1990, the squared term was negative, indicating that ROA increased at a decreasing rate; in 1990, the squared term was positive, implying that ROA increased at an increasing rate.

Older age had a negative impact on ROA in three of the four years (1987, 1988, and 1990). This result suggests that, in general, younger farmers achieve a higher ROA, possibly because of their lower degree of risk aversion. Education, information use, rotations, and tillage systems showed little effect on ROA.

2. Overall Efficiency Regression

Gross sales and capital structure are the only variables that have consistent effects on overall efficiency. Other variables such as age, education, rotation, and tillage are statistically significant in a particular year, but not in all four years.

Larger farms had a higher overall efficiency in 1987, 1990, and 1992. The 1990 regression results indicated that as farm size increased, overall efficiency decreased slightly, reaching a minimum for medium-sized farms, and then increased for larger farms.

The strong positive relationship of debt use (DEBT/ASSET) with overall efficiency implies that farms incurring more debt have higher overall efficiency. Again as suggested earlier, this result suggests that the use of debt ameliorates the effect of expenditure constraints and enhances overall efficiency.

II. ECONOMIC PERFORMANCE OF CONSERVATION PRACTICES USING SIMULATION MODELING

A. Objectives

The above results imply that tillage systems have no statistical effect on ROA or overall efficiency. However, the conventional wisdom during the past decade seemed to be that CT was more profitable than conventional tillage. This reflected the fact that the proportion of farmers using CT has tended to increase in the past two decades.

To analyze farmers' decision to adopt CT, a bioeconomic simulation model was developed that linked alternative farming practices with production, profits, and pollutants. Two geographic sites are modeled to simulate representative farms on two types of farmland typical of the Western Lake Erie region of Ohio: flat, i.e., Hoytville soils, and sloping, i.e., Blount-Glynwood-Pewamo soil associations. On each of these geographic sites, the performance of representative farms is assessed under alternative tillage technologies. Then various scenarios of CT adoption in a region are compared by extrapolating results of the analysis of these representative farms (Smith, 1997).

B. Model

The bioeconomic model consisted of two components. First, the Erosion Productivity Impact Calculator (EPIC model) is used to simulate crop yields and pollution parameters (Sharpley and Williams, 1990). Second, using production results from the EPIC model, a farm-level integer programming model is used to select a set of profit-maximizing farming practices (i.e., crop rotations, level of fertilizer and pesticide inputs, and farm size) on representative farms. Decision variables in the model are type of farming practice and amount of each practice, for example, acres of corn, soybeans, and wheat using conventional tillage, mulch tillage, or NT.

Once optimal rotations, crop acreages, and input levels for each representative farm have been determined by the integer programming model, the results are used in a regional aggregation of representative farms. These analyses allow comparison of the structural, economic, and environmental impacts brought about by alternative levels of CT adoption. Structural impacts are measured by changes in the number and average size of farms making up the region as well as by changes in crop production. Economic impacts are measured by changes in the region's revenues and costs. Finally, environmental impacts are measured by changes in individual pollution parameters, namely sediment, organic nitrogen, phosphorus, nitrates, and pesticides.

There are four regional tillage scenarios on hypothetical units of 100,000 cropland acres. The four scenarios, shown in Table 2, depict alternative rates of CT adoption. In each of these four scenarios, the three tillage systems were used in varying proportions ranging from predominantly conventional tillage (Scenario 1) to predominantly NT (Scenario 4).

C. Results

The conventional-tillage-dominated scenario (50% conventional, 35% mulch, and 15% NT) exemplified tillage practices in the western Lake Erie Basin during the early 1990s. Table 3 illustrates the three alternative tillage scenarios' impacts for farms on Hoytville soils. Average annual returns, costs, and farm size are compared to those of the conventional-dominated scenario. Percentage values represent the increase or decrease in a parameter relative to its value in the conventional-tillage-dominated scenario.

As mulch-till and NT acreage increased, profits improved for a variety of reasons. First, optimal wheat acreage decreased and corn acreage increased in the three alternative conservation-dominated tillage scenarios. The cause of this substitution was the reduced spring field time requirements for corn and soybeans when the conventional tillage system is replaced by CT. In the three alternative scenarios, wheat acreage was reduced by 13%–41% compared to the conventional tillage scenario.

Table 2 Regional Tillage Scenarios

	Conventional Scenario 1	Uniform Scenario 2	Mulch-till Scenario 3	No-till Scenario 4
Conventional	50%	33.3%	25%	15%
Mulch-till	35%	33.3%	50%	35%
No-till	15%	33.3%	25%	50%

Source: Forster et al. (2000).

Table 3   Economic Comparisons of Scenarios, Hoytville Region

		Conventional scenario average/hectare	Change from conventional scenario		
			Even scenario	Mulch scenario	No-till scenario
Total revenue		$664	2%	0%	5%
Variable costs	Nitrogen	$49	2%	0%	3%
	Phosphorus	$25	−4%	−4%	−8%
	Herbicide	$35	9%	3%	18%
	Labor	$17	−2%	−5%	−4%
	Machinery	$25	−6%	−9%	−12%
	Other costs	$143	2%	0%	4%
	Land cost	$185	0%	0%	0%
Fixed costs	Machinery	$99	−8%	−7%	−15%
Total costs		$578	0%	−2%	−1%
Returns above	total cost	$84	22%	16%	43%
Average farm size (hectares)		613	5%	6%	10%

Source: Forster et al. (2000).

Second, in the mulch-till and NT-dominated scenarios, labor and operating machinery costs decreased. In part this was because labor and machinery costs were linked essentially to the management system's efficiency in field time utilization. A 2%–5% decrease in labor costs was realized on a regional scale. The same was true for operating machinery costs, which were 6%–12% lower for the region in the alternative scenarios.

A large decrease in aggregate fixed machinery cost occurred for the region as tillage shifted from conventional to the three alternative scenarios, especially NT. First, at the individual farm level, NT fixed costs were less than conventional and conservation fixed costs. Second, the number of farms in a region decreased by about 10% in the NT-dominated scenario. In short, fixed costs decreased in the three alternative scenarios because the implements used were less expensive implements and they could farm more acres at the same cost.

Several input costs remained the same or increased slightly from the base scenario to the alternative scenarios. Nitrogen fertilizer use increased in the three alternative scenarios because of increased corn acreage. Transportation, seed, insurance, and interest (''other costs'') also increased slightly in the alternative scenarios because of increased corn acreage; however, the magnitude of these changes was not large.

The only cost that substantially increased when moving toward CT and NT systems was herbicide. In the fourth scenario, where NT acreage comprised 50%

of the total, herbicide costs increased by 18% compared to the base scenario. For the sake of brevity, results for simulations on Blount-Glynwood-Pewamo soils are not discussed here; however, results for the two soil regions were generally similar.

In summary, CT systems improved regional net farm income because of three factors: reduced labor and machinery inputs, shifts in the enterprises' crop mix, and improved economies of scale. On representative farms, the time available for fieldwork played a major role in the selection of NT or mulch tillage as preferred systems. For example, with NT cultivation, field time in spring was used almost exclusively to plant corn and soybeans, without having to perform any of the seedbed preparation required by conventional tillage. This allowed more acreage to be planted under the NT system. Because more acreage was planted with NT methods, fixed machinery costs were spread over a larger volume of production, and per unit costs of production declined accordingly. At the same time with CT, labor costs per acre were less. Higher herbicide costs offset these lower machinery and labor costs to some degree.

A second difference for farms using more CT was that the enterprise crop mix changed with the adoption of mulch tillage and NT systems. With conventional tillage, there are greater labor requirements for corn and soybeans in the spring and fall. This makes wheat, which does not require much labor during the critical spring and fall months, more attractive for growing on some of the farm's cropland, even if it is not the most profitable crop in absolute terms. Wheat acreage sharply declined with the use of NT and mulch till systems. Thus, the technological transformation toward CT has meant that there are larger and fewer farms, which are more specialized.

The bioeconomic model results indicated that one of the consequences of adopting CT practices was that erosion lessened and future soil productivity was

Table 4 Environmental Comparison of Scenarios, Hoytville Region

		Conventional scenario average/hectare	Change from conventional scenario		
			Even scenario	Mulch scenario	No-till scenario
USLE	tons	1.73	−8%	−17%	−17%
YNO3	kg	6.38	2%	5%	5%
YON	kg	31.06	−21%	−33%	−44%
YP	kg	5.48	−21%	−33%	−45%
PSRO	g	24.81	1%	0%	2%
Average			−9%	−15%	−20%

Source: Forster et al. (2000).

enhanced. Potential pollutants that result from soil erosion, such as sediment, organic nitrogen, and total phosphorus loadings, also decreased (Table 4). However, some pollution parameters increased, such as those for nitrates and herbicides.

III. CONCLUSIONS AND IMPLICATIONS

On average, Ohio farms in the Lake Erie Basin exhibited 54% in overall efficiency and a 5.25% return on assets for these four years. Given their resource base, most northwest Ohio farms were capable of improving profits, starting from a position of relatively low overall efficiency.

Larger farm size generally improved ROA and overall efficiency. However, in years when low output prices or drought resulted in financial losses for many farmers, we saw that large farms had little, if any, competitive advantage.

Farms having higher debt/asset ratios had higher overall efficiency, which implies that farmers successfully used debt to alleviate financial constraints. However, the use of debt had no significant positive effect on return on assets.

Statistical analysis of farm-level data did not ascertain significant differences in profitability among tillage systems or crop rotations. These findings imply that Lake Erie Basin farmers who adopted crop rotations and CT practices did so without sacrificing profits or efficiency. However, there were substantial differences in chemical, labor, and machinery expenses among tillage systems. Farmers in the sample using CT systems reduced their labor, fuel and machinery costs per acre and their machinery investment per acre, but increased their herbicide costs per acre.

If this analysis is correct and CT systems offer little if any financial advantages to farmers, what accounts for the increased adoption of these tillage systems? Results from modeling farm-level decision-making using the bioeconomic simulation model are that tillage system, farm size, and crop selection are determined jointly. Conservation tillage enables farms to be larger and more specialized and, as a result, farm profitability improves. Multivariate statistical analysis in the previous section was unable to show the effect of tillage on profitability because it was unable to account for endogeneity of variables (or joint effects of tillage, size, and crop selection) in production decisions.

Results from the bioeconomic simulation model indicate that adoption of CT technologies can be expected to have far-reaching consequences for a region's agricultural economy and environment (Table 5). First, soil erosion is reduced with NT agriculture and future soil productivity is enhanced. Second, potential pollutants that are directly related to soil erosion, such as sediment, organic nitrogen, and total phosphorus loadings, also decrease. Third, farm profitability im-

Table 5   Impacts of Conservation Tillage on Region, Summary of Simulation Results

Adoption of conservation tillage has both positive and negative impacts on region's
economy and environment.

Positive:	Negative:
Some pollutants decrease	*Some pollutants increase*
Sediment	Herbicides
Organic nitrogen	Nitrates
Total phosphorus loadings	*Crop mix changes* (less wheat)
Improved farm profits	*Regional economic impact*
Rate of return	Less purchased inputs
Net farm income	Less labor
	Larger farm size

Source: Forster et al. (2000).

proves. These effects have positive implications for a region's farmers, its environment, and the long-term sustainability of its agriculture.

However, these positive effects are somewhat offset by some unintended
consequences of CT adoption. First, certain pollution parameters may increase,
including nitrates and herbicides. Second, a region's crop mix and input usage
may change; in this case, wheat production decreased and row crops increased.
Third, there is a decline in labor use and input purchases (e.g., machinery and
fuel), which implies that agriculture's contribution to a region's economy also
decreases. Finally, average farm size increases, resulting in fewer farms in a
region. It is not certain that any or all of these changes would occur with CT
practices in a different environment. The results reported here are probably more
a consequence of the institutional setting than the difference between temperate
and tropical climates.

REFERENCES

Asplund, N.M., Forster, D.L., Stout, T.T. 1989. Farmers' use of forward contracting and
    hedging. *The Review of Futures Markets* 8(1):24–37.
Blue, E.N., Forster, D.L. 1997. Factors affecting performance measures of northwest Ohio
    farms. ESO 2394, Department of Agricultural Economics, The Ohio State Univer-
    sity.
Bravo-Ureta, B.E., Pinheiro, A.E. 1993. Efficiency analysis of developing country agricul-
    ture: A review of the frontier function literature. *Agric. and Resource Econ. Rev*
    22(1):88–101.
Färe, R., Grosskopf, S., Lee, H. 1990. A non-parametric approach to expenditure con-
    strained profit maximization. *Amer. J. Agric. Econ* 72(3):574–581.

Featherstone, A.M., Fletcher, J.J., Dale, R.F., Sinclair, H.R. 1991. Comparison of net returns under alternative tillage systems considering spatial weather variability. *J. Production Agric* 4(2):166–173.

Feder, G., Slade, R. 1984. The acquisition of information and the adoption of new technology. *Amer. J. Agric. Economics* 66(3):312–320.

Fizel, J.L., Nunnikhoven, T.S. 1993. The efficiency of nursing home chains. *Applied Econ* 25(1):49–55.

Forster, D.L. 2002. Effects of conservation tillage on the performance of Lake Erie Basin farms. *J. Environ. Quality* 31(1):32–37.

Forster, D.L., Smith, E.C., Hite, D. 2000. A bioeconomic simulation model of farm management practices and environmental effluents in the western Lake Erie Basin. *J. Soil and Water Conserv* 55(2):177–182.

Kalirajan, K., Shand, R.T. 1986. Estimating location-specific and firm-specific technical efficiency: An analysis of Malaysian agriculture. *J. Econ. Devel* 11(1):147–160.

Klemme, R.M. 1985. A stochastic dominance comparison of reduced tillage systems in corn and soybean production under risk. *Amer. J. Agric. Econ* 67(3):550–557.

Lovell, C., Walters, L., Wood, L. 1989, Exploring the distribution of DEA scores. IC² Conference on new uses of DEA in management. IC² Institute, University of Texas at Austin.

Nalson, J.S. 1968. The Mobility of Farm Families. Manchester: Manchester University Press.

Ray, S.C. 1985. Measurement and test of efficiency of farms in linear programming models: a study of West Bengal farms. *Oxford Bull. Econ. and Stat* 47(2):371–386.

Richards, R.P, Baker, D.B., Eckert, D.J. 2002. Trends in agriculture in the LEASEQ watersheds, 1975–1995. *J. Environ. Quality* 31(1):17–24.

Rogers, E.M., Burdge, R.J., Korsching, P.F., Donnermeyer, J.F. 1988. Social Change in Rural Societies: An Introduction to Rural Sociology, 3rd ed.: Englewood Cliffs, NJ. Prentice Hall.

Sharpley A.N., Williams J.R., Eds. 1990EPIC—Erosion/Productivity Impact Calculator 1. Model Documentation. USDA Technical Bulletin No. 1768.

Shepard, L.E., Collins, R. 1982. Why do farmers fail? Farm bankruptcies 1910–1978. *Amer. J. Agric. Econ* 64(3):610–615.

Smith, E.C. 1997. Farm management practices and environmental effluents in the Western Lake Erie Basin of Ohio: An economic optimization of far systems. Unpublished M.S. thesis, The Ohio State University.

Stout, T.T., Forster, D.L., Edgington, G. 1992. Organization and Performance of Ohio Farm Farm Operations in 1990. Research Bulletin No. 1189. Wooster, OH, Agricultural Research and Development Center.

Tauer, L.W. 1992. Short-run and long-run efficiencies of New York dairy farms. *Agric. and Resource Econ. Rev* 22(1):1–9.

Upton, M., Haworth, S. 1987. The growth of farms. *Eur. Rev. of Agric. Econ* 14(3): 351–366.

Williams, J.R. 1988. A stochastic dominance analysis of tillage and crop insurance practices in a semiarid region. *Amer. J. Agric. Econ* 70(1):112–120.

24

Policy Challenges for the Rice–Wheat Farming Systems of the Indo-Gangetic Plains

Peter Hazell and Mark Rosegrant
International Food Policy Research Institute
Washington, D.C., U.S.A.

The vast irrigated lands of South Asia, and the Indo-Gangetic Plains in particular, played a key role in the Green Revolution and helped move the South Asian subcontinent from national food crises and starvation to a position of food security and relative plenty. But these areas are now struggling with second-generation problems, including low prices and declining profitability of rice and wheat within a larger context of national and global cereal surpluses, unsustainable water use, and farming practices that threaten the viability of continued productivity growth. Just what should the future role of these irrigated regions be, and what kinds of policies are relevant to solving their second-generation problems? We address these issues in turn, with a focus on India, the largest practitioner of the rice–wheat system.

I. FUTURE CHALLENGES FACING THE INDO-GANGETIC PLAINS

A. Declining Profitability

India now has large surpluses of rice and wheat, and to support declining farm incomes, the government has to procure and store grains to shore up farmgate prices. Growth in demand for rice and wheat has slowed considerably in recent

389

years and is now expected to grow not much faster than the rate of population growth, which is also slowing. Any significant growth in cereal demand is likely to arise from the need for more livestock feed, which means growing more maize and coarse grains rather than more rice and wheat (Bhalla et al., 1999).

Unfortunately, rice and wheat production in the Indo-Gangetic Plains is not particularly competitive in world markets (Gulati and Kelley, 2000) and is handicapped by the continuing distortions in the agricultural markets of OECD countries that both limit the potential for Indian exports and promote unfair competition in India's domestic market. For many of the smaller farmers who predominate in the region, the amount of net income generated per acre from rice and wheat is no longer sufficient to support their families.

At the same time, the cost to the government of continued price supports and the subsidies on power, water, fertilizers, and credit are becoming a major financial burden. Subsidies currently consume more than half of total government spending on agriculture (World Bank, 1999). Many of these subsidies played a useful role in launching the Green Revolution in the late 1960s and helped ensure that small farmers gained as well as large farmers. But today they are largely unproductive and simply detract from the public resources that are available for investment in future agricultural growth.

Gulati and Narayanan (2000) have estimated that the subsidies for power and fertilizer alone now cost the government about $6 billion per year, equivalent to about 2% of national GDP. There are also large subsidies for water, credit, and the food distribution system. If these are considered as well, it seems likely that at least $10 billion is spent each year on subsidies. These are an enormous financial burden on the country and represent public resources that could be invested much more productively in agriculture and the rural sector to address remaining high levels of rural poverty in the country.

In the future, greater emphasis needs to be placed on diversification into higher-value crops and products rather than growing just more wheat and rice. This strategy could help increase farm incomes and relieve some of the environmental pressures that arise from monoculture farming in water-constrained irrigated areas. Fortunately, diversification opportunities are growing rapidly in India. These are emerging from a number of sources.

1. The national diet is changing as a result of the accelerated national economic growth achieved in recent years, and the rising affluence of the middle classes is increasing domestic demand for livestock products, especially milk and milk products, as well as for fruits, vegetables, flowers, and vegetable oils. This creates opportunities for farmers to diversify, even specialize, in higher-value products. This benefits especially those farmers who have ready access to markets, information, inputs, and so forth.

2. Ongoing policy reforms are also slowly opening up export markets for Indian farmers. This, together with the removal of restrictions on interstate trade within India, should enable more farmers to specialize in certain crops in which they have a comparative advantage and can best compete in the market. These opportunities should also improve if the next round of world trade negotiations succeeds in further opening of more agricultural markets around the world.

3. There are also opportunities for generating greater value added in agro-processing, particularly if agroindustry were no longer covered by current protective policies and became more competitive with imports (World Bank, 1999; Gulati and Kelley, 2000). Oil seeds processing, for example, is highly protected at present, making domestic vegetable oils noncompetitive with imports. While producers can compete as growers of vegetable oil seeds, they are penalized when trying to compete in the vegetables oils market because their products have to be processed by a highly inefficient domestic industry (Gulati and Kelley, 2000).

Agricultural growth of these types could make important contributions to income increases in the Indo-Gangetic Plains. But these markets are highly competitive, placing a high premium on quality and cost. Achieving competitiveness will require investments in rural infrastructure and technology (roads, transport, electricity, improved varieties, disease control, etc.) and improvements in marketing, grading, and distribution systems for higher value, perishable foods (refrigeration, communications, food processing and storage, food safety regulations, etc.).

A key challenge will be ensuring that small farmers can get and hold a share of these higher-value markets. This will require some direct government assistance rather than leaving everything to market forces alone. Helping small-scale producers capture part of these growing markets will require that agricultural research systems give adequate attention to the problems and endowments of smaller farms and not just large ones. The private sector seems likely to play a greater role in conducting the research needed for many higher-value products. But private research firms are attracted more to the needs of larger farms than small ones, and to regions with good infrastructure and market access. Public research institutions will need to play a key role in ensuring that small farmers and more remote regions do not get left out of these new market opportunities.

Smallholder farmers will also need to be organized more effectively for efficient marketing and input supply. While smallholders are typically more efficient producers of many labor-intensive livestock and horticultural products, they are at a major disadvantage in the market place because of (1) poor information and marketing contacts, and (2) their smaller volumes of trade (both inputs and outputs) that lead to less favorable prices than available to larger-scale farmers.

Contracting arrangements with wholesalers and retailers has proved useful in some contexts, but for the mass of smallholder farmers in India, cooperative marketing institutions probably offer the more realistic option. Operation Flood is a good example of what can be done. Under this project, dairy cooperatives collect, process, and market milk collected from millions of small-scale producers, including landless laborers, women, and smallholder farms, many of whom produce only 1 or 2 liters per day. In 1996, Operation Flood reached 9.3 million farmers yet still accounted for only 22% of all marketed milk in India (Candler and Kumar, 1998). The government assists the program through technical support (e.g., research and extension, veterinary services, and the regulation of milk quality). Otherwise, the program is run by the cooperatives themselves with no direct financial support from government.

B. Growing Water Scarcity

Irrigation in India today has increasing difficulty competing with other sectors for financial resources as well as water. While irrigated area and water withdrawals continue to grow, the pace of irrigation development has slowed. Projections of water and food supply and demand using IFPRI's IMPACT-WATER model show that irrigation in India will also increasingly have difficulty in competing with other sectors for water.

1. IMPACT-WATER Model

IMPACT-WATER is an integrated global water and food modelling framework that is applied to simulate the complex relationships among water availability and demand, food supply and demand, international food prices, and trade at basin, country, and global levels over a 30-year time horizon. IMPACT-WATER combines an extension of International Model for Policy Analysis of Agricultural Commodities and Trade (IMPACT) with a Water Simulation Model (WSM). Crop water demand and water supply for irrigation are simulated, taking into account the hydrologic fluctuations, irrigation development, growth of industrial and domestic water uses, environmental and other flow requirements, water prices, and water supply and use infrastructure. Crop production is determined for irrigated and rainfed crops as a function of crop and input prices and consumptive use of water [see Rosegrant et al. (2002) for a detailed description of model methodology and global results].

Even under a business-as-usual scenario that presumes continuation of existing policies and trends, growing water shortages will significantly reduce rice and wheat crop yield growth in India, particularly in the Indus and Ganges basins. If India were to move toward the sustainable management of groundwater, reducing overdrafts, rice and wheat production will become even less competitive than now.

Irrigation currently accounts for over 90% of water withdrawals in India, but with rapid urbanization and industrialization, domestic and industrial demands for water are increasing at a much higher pace than demands for irrigation. Under the business-as-usual scenario, consumption of water for nonirrigation uses will more than double in India by 2025.

With the limitations on development of new water (especially around most major cities) and the high economic, environmental, and political costs of developing new systems, these rapidly growing demands in many cases compete with irrigation for the water presently used in agriculture. Therefore, irrigation water consumption in India will be increasingly supply-constrained, with a declining fraction of potential demand being met over time. Irrigated agriculture in India will be constrained by competing demands for water. While potential irrigation demand in India (the amount of irrigation water needed to meet full crop water-use needs) could increase by 16.6% (Rosegrant, Cai, and Cline, 2002), the actual increase in irrigation water consumption is likely to be only 3.1%.

This tightening irrigation water constraint is shown by the irrigation water supply reliability index (IWSR), defined as the ratio of water supply available for irrigation over potential demand for irrigation water. Table 1 shows the IWSR for India compared to other South Asian countries and the rest of the world. In India as a whole, the IWSR is projected to decline from 0.80 today to 0.69 by 2025. In the Indus Basin within India, the IWSR will decline from 0.83 to 0.75, while in the Ganges Basin, which faces more rapid growth in domestic and industrial water demand, the IWSR is likely to decline very sharply, from 0.83 to 0.65 by 2025.

The increasing water scarcity for irrigation will be a significant constraint on growth of cereal yields in India, as shown by the projected relative crop yields

Table 1 Irrigation Water Supply Reliability Index, 1995, 2010, and 2025

Basin/countries/regions	Water available/water requirement		
	1995	2010	2025
India	0.80	0.72	0.69
Indus	0.83	0.78	0.75
Ganges	0.83	0.70	0.65
Pakistan	0.78	0.75	0.72
Bangladesh	0.79	0.76	0.73
Other South Asia	0.88	0.86	0.85
Developed countries	0.86	0.84	0.89
Developing countries	0.80	0.73	0.71

Source: IMPACT-WATER simulations.

for irrigated cereals. Relative crop yield is the ratio of the projected crop yield likely to be attained with available water compared to the economically attainable yield at given crop and input prices under conditions of full water adequacy. The relative crop yield for cereals in irrigated areas is projected to decline from 0.84 in 1995 to 0.72 in 2025 for all India, from 0.84 to 0.73 in the Indus, and from 0.83 to 0.69 in the Ganges (Table 2).

Thus, even under a business-as-usual scenario where irrigation and other water users continue to mine groundwater beyond recharge rates, crop yield growth will be significantly constrained in India and elsewhere in South Asia. Northern and western India have experienced significant groundwater depletion due to pumping in excess of groundwater discharge, so irrigated production there is clearly not sustainable in the long term, and eventually policy measures will have to be introduced to reverse this imbalance.

What would be the impact on water availability and food production in India and elsewhere if unsustainable groundwater pumping were eliminated? Using the IMPACT-WATER model, a low groundwater pumping (LGW) scenario was simulated assuming that groundwater overdraft in India will be phased out over the next 25 years through a reduction in the ratio of annual groundwater pumping to recharge at the basin or country level to below 0.55. Compared with levels in 1995, under the LGW scenario, groundwater pumping will decline by 69 km^3 in India, equivalent to 10% of total water withdrawals in India in 1995.

Phasing out of groundwater mining would reduce irrigated cereal production in India by 14.4 million mt (8%) compared to baseline projections for 2025. The cereal production losses are concentrated in the Ganges region, where production would fall by 8.3 million mt (11%) compared to the 2025 baseline values, and by 4.4 million mt (11%) in the Indus region compared to the baseline production.

Table 2   Relative Cereal Yield Index, 1995, 2010, and 2025

Basin/countries/regions	Realized yield/potential yield		
	1995	2010	2025
India	0.84	0.76	0.72
Indus	0.84	0.75	0.73
Ganges	0.83	0.70	0.69
Pakistan	0.84	0.74	0.68
Bangladesh	0.88	0.85	0.82
Other South Asia	0.85	0.81	0.79
Developed countries	0.89	0.86	0.86
Developing countries	0.86	0.80	0.74

Source: IMPACT-WATER simulations.

It would be very difficult to compensate for the losses in production from reduced groundwater pumping through resources generated only in irrigated agriculture. Although improvements in irrigation efficiency in the specific over-drafting basins could in theory compensate for these declines in groundwater use, the required efficiency improvements would have to be huge and likely unattainable. In the Indus region, an improvement in basin efficiency from 0.59 today to 0.76 by 2025 would be required in order to generate enough cereal production to compensate for the reduction in groundwater overdraft; in the Ganges region, the increase would have to go from 0.62 to 0.85.

Improvement of basin water-use efficiency depends both on technological improvements in irrigation systems, domestic and industrial water use, and recycling systems and on institutional settings related to water allocation, water rights, and water quality. In the irrigation sector, improvements can include adoption of advanced technology such as drip irrigation and sprinklers; better conjunctive use of surface and groundwater; improvements in water management through adoption of demand-based irrigation scheduling systems and improved maintenance of equipment; and institutional improvements such as establishment of more effective water user associations and tradable water rights, introduction of water pricing, and improvements in the legal environment for water allocation. In the industrial sector in developing countries, the amount of water used to produce a given amount of output is far higher than in developed countries. A major technology backlog exists that can be tapped by developing country industry to save water. The domestic water sector also has potential to improve water-use efficiency through leak detection and repair in municipal systems, to install low-flow showerheads and toilets and ecological sanitation, and to adjust lifestyles to consume less water.

However, as shown above, there is relatively little room for improving water-use efficiency in the most severely water-scarce basins, i.e., the Indus and Ganges, and food production and farm incomes could fall significantly if policy reforms are put into place to increase allocations for environmental purposes or to eliminate groundwater overdraft. In these basins, alternative interventions will be required in order to compensate farmers for the negative effects of the environmental water diversions including more rapid crop yield growth from agricultural investments, diversification into less water-intensive crops, or broader economic diversification to reduce the relative role of agriculture over time.

C. Other Environmental Problems

The Green Revolution, which transformed productivity levels in the Indo-Gangetic Plains, also caused environment problems, particularly in combination with inappropriate pricing and subsidy policies. In addition to the groundwater mining problems discussed above, it was accompanied by water management practices that led to the salinization of some of the best irrigated lands, to fertilizer and pesticide contamination of waterways, and to pesticide poisoning of humans and beneficial insects. The problems began in the 1970s and seem to be getting worse. There is mounting evidence showing that yield growth in many of the intensively farmed areas has now peaked and in some cases is even declining (Pingali and Rosegrant, 2000). There are growing voices arguing that Indian farmers should revert back to the low external-input farming technologies of pre–Green Revolution days (e.g., Shiva, 1991). This would have a disastrous impact on yields and food supplies and would destroy the environment on an even larger scale because of the need to rapidly expand the planted area in many rainfed areas.

There are realistic prospects for making modern technologies more environmentally benign and reversing resource degradation problems on a national scale (Pingali and Rosegrant, 2000). But it will take significant and determined action by the government. Needed actions include

1. Development and dissemination of technologies and natural resource management practices that are more environmentally sound than those currently used in many farmers' fields. Some of these technologies already exist and include precision farming, crop diversification, ecological approaches to pest management, pest-resistant varieties, no-till farming, improved water management practices, and the system of rice intensification (SRI) discussed in Chapter 5. The challenge is to get these technologies adopted more widely in farmers' fields. Managed properly, some of these technologies can even increase yields while they reduce environmental damage. Further agricultural research is needed to create additional technology options for farmers, and this should include interdisciplinary work on pest control, soil management, and crop diversification, but also use of modern biology to develop improved crop varieties that are even better suited to the stresses of intensive farming but with reduced dependence on chemicals (e.g., varieties that are more resistant to pest, disease, drought, and saline stresses). SRI reduces the need for water application and the use of agrochemicals while raising yield. Agricultural research and extension systems will need to give much higher priority to sustainability problems than they have in the past.

2. Reform of policies that create inappropriate incentives for farmers in the choice of technology and natural resource management practices. The large subsidies for water, power, fertilizer, and pesticides mentioned earlier make these inputs too cheap and encourage their excessive and wasteful use with dire environmental consequences. Pricing these inputs at their true cost would save the government much money

while also improving their efficient use. This would reduce environmental degradation while having beneficial fiscal effects.

3. Reform of public institutions that manage water to improve the timing
and amounts of water delivered relative to farmers' needs, and the
maintenance of irrigation and drainage structures. When farmers have
little control over the flow of water through their fields, they are less
able to use water more efficiently and prevent waterlogging or salinization of their land. Forestry departments also need to work more closely
with local communities, devolving responsibilities where possible, to
improve incentives for the sustainable management of public forest
and grazing areas, which improves watersheds and the recharge of
groundwater by slowing monsoon runoff.

4. Assist farmers in diversifying their cropping patterns to relieve the
stress of intensive monocultures. Investments in marketing and information infrastructure, trade liberalization, more flexible irrigation systems, and so forth can increase opportunities for farmers to diversify.

5. Resolve widespread "externality" problems that arise when all or part
of the consequences of environmental degradation is borne by people
other than the ones who cause the problem, e.g., pollution of waterways
and siltation of dams because of soil erosion in watershed protection
areas. Possible solutions include taxes on polluters and degraders, regulation, empowerment of local organizations, and appropriate changes
in property rights. Effective enforcement of rules and regulations is
much more difficult to achieve than the writing of the new laws that
create them.

II. IMPLICATIONS FOR NATIONAL FOOD SECURITY

What would happen if some of the policy changes recommended in this chapter for
the Indo-Gangetic Plains were actually implemented? In particular, what would
happen to India's national food security? This is a complex issue that is not easy
to resolve, but we offer an approximate answer using the IMPACT-WATER
model (See Sec. I.B.).

    As already discussed, if India wanted to seriously address its water-mining
problem by phasing out the groundwater overdraft by 2025, this would require
significant improvements in water management and efficiency and agricultural
diversification into higher-value crops. These changes, in turn, would address
many of the other income and environmental problems that have emerged in the
Indo-Gangetic Plains. The downside is that such a policy could lead to significant
reductions in rice and wheat production unless productivity is significantly raised.

    The IMPACT-WATER simulation presented earlier suggests a reduction in
irrigated cereal production of 14.4 million metric tons (8%) compared to baseline

projection for 2025. The cereal production losses are concentrated in the Ganges and Indus regions, where production would fall by 11% compared to the 2025 baseline projected levels. These reductions in staple food production could be unacceptable to policy-makers from the perspective of national food security.

One solution to this problem is to give greater emphasis to increasing cereal production in rainfed farming areas. Such a strategy would offer at least two major advantages. First, because the lion's share of India's remaining rural poor now live in rainfed agricultural areas, and most of them depend on farming and related activities for their livelihoods, increasing cereal production in rainfed rather than irrigated areas would not only relieve the pressure on water and the environment in irrigated areas, but it would put more food in the hands of the people who really need it. By simultaneously operating on demand and supply, this strategy could reduce hunger and poverty while avoiding additional production destined for public storage or subsidized exports. Second, recent work at IFPRI shows that for many public rural investments (especially agricultural research, roads, and education) the marginal productivity returns are now higher in many rainfed areas compared to irrigated areas, with also a more favorable impact on poverty reduction (Fan et al., 2000).

But could increased rainfed cereal production be sufficient to compensate for the irrigated production decline due to reduced groundwater pumping? This question is addressed by an IMPACT-WATER model scenario combining elimination of groundwater overdraft with higher rainfed agriculture development. The reduction of irrigated cereal production due to reduced groundwater pumping can be offset by an increase in rainfed area and yield within the same regions, but the required increase in rainfed cereal yields would be very large. In 2025, average rainfed cereal yields would need to be 20% or 0.3 metric tons per hectare higher in India than in baseline projections. In addition, rainfed cereal area would need to increase by 0.8 million hectares in India. Such an area expansion is itself environmentally damaging, requiring encroachment on fragile lands. Moreover, the yield increase would require substantial additional investments in agricultural research and management for rainfed areas, water harvesting, and market access, and it is not clear that these yield increases are achievable in rainfed areas, even with increased investments.

These results show that an appropriate strategy would need to target both improved water-use efficiency in irrigated areas, and policy and investments to boost rainfed crop yields, and enhance agricultural diversification in water-scarce areas.

The global food production impact of the elimination of groundwater overdraft in India is relatively small, but the impacts for specific countries and basins are quite large. While the seriousness of Indian shortfalls in demand and increases in imports should not be minimized, they may be a worthwhile tradeoff for restoring sustainability of groundwater supplies. It may be necessary for these countries to rely more on imports to meet the decline in irrigated production compared to the baseline. Agricultural research investments should be increased; and particularly in the hardest-hit river basins such as the Ganges, investments and policy reforms (including elimination of power subsidies for pumping) should be implemented to increase basin efficiency and encourage diversification out of irrigated cereals into crops that give more value per unit of water.

Combined with emerging insights from the literature on rainfed agriculture, the results here point to a strategy for investments and policy reforms to enhance the contribution of rainfed agriculture.* Water harvesting has the potential to improve rainfed crop yields in some regions and could provide farmers with improved water availability and increased soil fertility in some local and regional ecosystems, as well as environmental benefits through reduced soil erosion. Nevertheless, despite localized successes, broader farmer acceptance of water-harvesting techniques has been limited because of high implementation costs and greater short-term risk from additional inputs and cash and labor requirements (Rosegrant et al., 2002; Tabor, 1995). Water-harvesting initiatives frequently suffer from lack of hydrological data, insufficient planning regarding important social and economic considerations, and the absence of a long-term government strategy to ensure the sustainability of interventions. Greater farmer involvement at the planning stages for maintenance and data collection, and provision of appropriate educational and extension support could help realize the potential contributions of water harvesting (Oweis et al., 1999).

The rate of investment in crop breeding targeted to rainfed environments is crucial to future cereal yield growth. Strong progress has been made in breeding for enhanced crop yields in rainfed areas, even in the more marginal rainfed environments (Byerlee et al., 1999; Lantican and Pingali, 2001). The continued application of conventional breeding and the recent developments in nonconventional breeding offer considerable potential for improving cereal yield growth in rainfed environments. Cereal yield growth in rainfed areas could be further improved by extending research both downstream to farmers (Ceccarelli et al., 1996) and upstream to the use of tools derived from biotechnology to assist conventional breeding and, if concerns over risks can be solved, from the use of transgenic breeding. There are also possibilities for yield improvement through changes in soil and water management.

Crop research targeted to rainfed areas should be accompanied by increased investment in rural infrastructure and policies to close the gap between potential yields in rainfed areas and the actual yields farmers achieve. Important policies

---

* Pilot projects extending SRI to rainfed rice in the Philippines have shown similar relative yield increases for unirrigated rice as for irrigated production, using mulch for water conservation and weed suppression.

include higher priority for rainfed areas in agricultural extension services and access to markets, credit, and input supplies (Rosegrant and Hazell, 2000). Successful development of rainfed areas is likely to be more complex than in high-potential irrigated areas because of their relative lack of access to infrastructure and markets and their more difficult and variable agroclimatic environments.

Investment, policy reform, and transfer of technology to rainfed areas, such as water harvesting, will therefore require stronger partnerships among agricultural researchers, local organizations, farmers, community leaders, NGOs, national policy-makers, and donors (Rosegrant and Hazell, 2000). Progress in rainfed agriculture may also be slower than in the early Green Revolution because new approaches will need to be developed for specific environments and tested on a small scale before wide dissemination, but enhanced rainfed crop production growth would be an important source of water savings.

III. CONCLUSIONS

Rapid growth in the productivity of irrigated agriculture in the Indo-Gangetic Plains has helped move South Asia from food deficits in the 1960s to a degree of surplus today, but there are still problems of maldistribution, so hunger has not yet been banished from the subcontinent. This region is now struggling with different problems, ones arising in the wake of the Green Revolution. These include most notably low prices and declining profitability of rice and wheat together with diminishing relative quantity and quality of land and water resources available to support production. Unsustainable water use and farming practices threaten the future viability of continued productivity growth.

To overcome these problems, policy reforms are needed to encourage (1) diversification into higher-value crops and activities and a reduced reliance on rice and wheat production in irrigated areas, (2) improved efficiency in the use of irrigation water and a substantial reduction in the mining of groundwater resources, and (3) improved management of agricultural chemicals and reductions in water pollution and land salinization.

A combination of policy and institutional reforms is required, including the phasing out of many input subsidies; market reforms; trade liberalization (including removing trade protection for agroindustries); reform of public sector water institutions; introduction of tradable water rights; and increases in productive investment in agriculture and the rural sector. Appropriate reforms of farming systems in the irrigated areas would lead to a significant reduction in cereal production in countries like India, and this would need to be offset by complementary investments in cereal production in rainfed areas. The latter would not only help fill the national food gap, but would also directly address the food security needs of a large share of India's remaining rural poor. Analysis shows that this would also give favorable economic returns.

REFERENCES

Bhalla, G. S., Hazell, P., Kerr, J. 1999. *Prospects for India's Cereal Supply and Demand to 2020.* 2020 Vision Discussion Paper 29. Washington, DC: International Food Policy Research Institute.

Byerlee, D., Heisey, P., Pingali, P. 1999. Realizing yield gains for foods staples in developing countries in the early 21st century: Prospects and challenges. Paper presented to the study week on Food Needs of the Developing World in the Early 21st Century, The Vatican, Jan. 27–30.

Candler, W., Kumar, N. 1998. *India: The Dairy Revolution.* Operations Evaluation Department, World Bank. Washington, DC.

Ceccarelli, S., Grando, S., Booth, R.H. 1996. International breeding programmes and resource-poor farmers: Crop improvement in difficult environments. In *Participatory Plant Breeding.* Eyzaguire P., Iwanaga M., Eds. Proc. of a Workshop on Participatory Plant Breeding, July 26–29, 1995, Wageningen, The Netherlands. Rome: International Plant Genetics Research Institute.

Fan, S., Hazell, P., Haque, T. 2000. Targeting public investments by agro-ecological zone to achieve growth and poverty alleviation goals. *Food Policy* 25(4):411–428.

Gulati, A., Narayanan, S. 2000. Demystifying fertilizer and power subsidies in India. *Economic and Political Weekly,* March 4:784–794.

Gulati, A., Kelley, T. 2000. Trade Liberalization and Indian Agriculture. New Delhi: Oxford University Press.

Lantican, M.A., Pingali, P.L. 2001. Growth in wheat yield potential in marginal environments. In the Proceedings of the Warren E. Kronstad Memorial Symposium. March 15–17, 2001. Mexico City, CIMMYT.

Oweis, T., Hachum, A., Kijne, J. 1999. *Water Harvesting and Supplementary Irrigation for Improved Water Use Efficiency in Dry Areas.* SWIM Paper 7. Colombo, Sri Lanka, International Water Management Institute.

Pingali, P., Rosegrant, M.W. 2000. Intensive food systems in Asia: Can the degradation be reversed? In *Tradeoffs or Synergies? Agricultural Intensification, Economic Development and the Environment.* Lee D.R., Barrett C.B., Eds. Wallingford. UK: CABI Publishing.

Rosegrant, M.W., Cai, X., Cline, S. 2002. World Water and Food to 2025: Dealing with Scarcity. Washington, DC: International Food Policy Research Institute.

Rosegrant, M.W., Cai, X., Cline, S., Nakagawa, N. 2002. *The Role of Rainfed Agriculture in the Future of Global Food Production.* Environment and Production Technology Division Discussion Paper No. 90. Washington, DC: International Food Policy Research Institute.

Rosegrant, M., Hazell, P. 2000. *Transforming the Rural Asian Economy: The Unfinished Revolution.* Hong Kong: Oxford University Press for the Asian Development Bank.

Shiva, V. 1991. The Green Revolution in the Punjab. *The Ecologist* March/April.

Tabor, J.A. 1995. Improving crop yields in the Sahel by means of water-harvesting. *Journal of Arid Environments* 30(1):83–106.

World Bank. 1999. Towards rural development and poverty reduction. Paper presented at the NCAER-IEG-World Bank Conference on *Reforms in the Agrilcultural Sector for Growth Efficiency, Equity and Sustainability*, India Habitat Centre, New Delhi, April 15–16.

25

Soil and Water Sustainability Issues Related to Australian Grain Cropping Systems

Frank M. Vanclay
University of Tasmania
Hobart, Tasmania, Australia

I. INTRODUCTION

Agriculture is an important industry for Australia. In 2000, some 23 million hectares were under cultivation (ABS, 2002). Beef and wheat are the most important agricultural commodities, each producing over US$2.5 billion per annum. Wheat utilizes the largest area of land of any crop (over 12 million hectares) (ABS, 2002). Various other grain crops are grown, including barley and canola, often in rotations with wheat. Rice is also grown, sometimes in conjunction with wheat.

Australia's agriculture has come at a price. Severe environmental impacts have resulted from agricultural activities. Agricultural sustainability is therefore a major discussion in Australia, especially in 2002 when Australia experienced a severe drought. Debates about sustainability are complicated in that they involve a range of considerations. They are especially complex when the social aspects of sustainability are considered. This chapter provides a brief overview of grain agriculture in Australia and a discussion of sustainability issues focusing especially on socioeconomic concerns.

II. OVERVIEW OF AGRICULTURE AND THE AUSTRALIAN ENVIRONMENT

There were 146,371 establishments undertaking agricultural activity in Australia in 2000. Almost all had agriculture as their primary activity. Some 16,463 pro-

Table 1 Statistics for Selected Commodities

Crop/commodity	Area cropped (thousands of hectares)	Total yield (thousands of tonnes)	Annual gross value (U.S. $million)
Wheat	12,168	24,757	2,657
Barley	2,596	5,032	476
Canola	1,911	2,460	418
Rice	131	1,084	159
Cotton	435	698	779
Grapes	111	1,311	615
Sugarcane	428	38,165	485
Vegetables	—	—	1,024
Fruit and nuts	—	—	969
Dairy products	—	—	1,565
Beef/veal	8.6 million head slaughtered	1,988	2,778
Mutton/lamb	33.5 million head slaughtered	680	580
Poultry	405 million head slaughtered	646	567
Pork/pig meat	5.0 million head slaughtered	363	436
Wool	143 million head shorn	695	1,182

Source: Table constructed for this chapter from 2000 data modified from ABS (2002)

duced grain alone, and another 18,232 were mixed grain and livestock enterprises (ABS, 2002).

In a normal year, the gross value of Australia's agricultural production is about US$17 billion per annum (ABS, 2002). Wheat is Australia's most valuable crop and contributes about US$2.66 billion per annum. However, beef has a higher gross value of production at about US$2.78 billion.[*] Cotton and grapes also each return more than US$500 million per annum (ABS, 2002). Other significant commodities include barley, canola, rice, wool, dairy products, vegetables, and fruit (Table 1).

Most grains are produced in rotation with wheat. Therefore, the broadacre grain-growing area of Australia is generally known as the wheat-belt (Fig. 1).

---

[*] Based on 2000 figures in Australian dollars produced by the Australian Bureau of Statistics converted at an indicative (relatively low) 0.55 to the U.S. dollar.

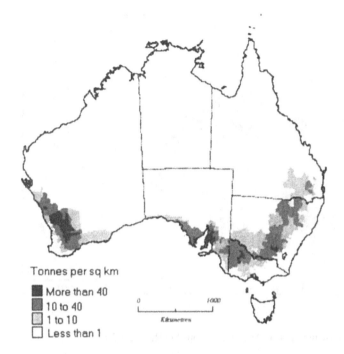

Figure 1   The wheat-belt of Australia (shading depicts wheat production in tonnes per square kilometer and was generated using small area Agricultural Census data for 1996–1997, Agstats 7117.0). (From the Australian Bureau of Statistics, 2002.)

Wheat is grown as a winter cereal because of the winter-dominant rainfall and the warm temperate Mediterranean climate of the wheat-belt. It is planted in mid- to late autumn following the first rains and it is harvested from late spring to early summer. Farm size varies across the wheat-belt, but 1,000 hectares represents a typical wheat farm. Farms are as large as 5,000 hectares, but those with 2,000 hectares or more are generally considered to be large farms. Most farmers run mixed enterprises with a primary focus on wheat and sheep production. They also produce a variety of other crops, such as canola, oats, and lupins in rotation with wheat.

Nearly all of Australia's rice is grown in the Riverina District in the state of New South Wales (NSW) in the Murrumbidgee, Coleambally, and Murray Valley Irrigation Areas (Fig. 2). This is an area with an average annual rainfall of around 400 mm. There are 2,200 family farms of 300 to 500 hectares in size (SunRice, 2002). Industry practice favors five-year rotations comprising: rice,

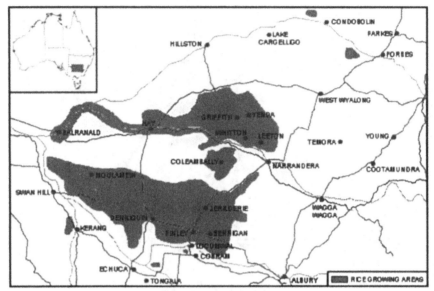

Map (C). Australian Surveying and Land Information Group (AUSLIG) www.auslig.gov.au

Figure 2   The rice-growing area of Australia. (From SunRice, 2002.)

rice, a grain crop such as wheat or barley, and two years of pasture with livestock. Typically, the farm is broken up into three subunits with rice being grown only on one third of the total area at any time (SunRice, 2002). The growing season includes the Australian summer, with planting in October and harvesting from March to May. The crop is grown in flooded bays using an average of 16 megaliters of water per hectare (ML/ha). On sandy soils, which do not readily pond water, crops can use more than 20 ML/ha.

Australia's environment is influenced by its flatness and by the Great Dividing Range, a mountain barrier that extends the full length of the east coast. This barrier results in a marked reduction of rainfall for areas inland of the range, making Australia the world's driest continent after Antarctica. Low rainfall is exacerbated by high variability in annual rainfall caused by the El Niño Southern Oscillation phenomenon (Bureau of Meteorology, 1995).

Irrigation schemes have been developed including the famous Snowy Mountain Scheme, which has diverted water from snowmelt that would have flowed south to the ocean into the westward flowing rivers of the Murray Darling Basin (MDB) (Fig. 3). This scheme has enabled the Murray Darling Basin to produce, in normal years, about US$7 billion, or 41% of Australia's gross value of agricultural production. The MDB comprises 1,061,469 square kilometers, or

Figure 3 The Murray–Darling Basin of Australia. (From Murray–Darling Basin Commission, 1999.)

about 14% of the Australian landmass. It includes Australia's three longest rivers, the Darling (2,740 km), the Murray (2,530 km), and the Murrumbidgee (1,690 km). About 2 million people reside within the Basin (MDBC, 2002a). They represent 10% of the Australian population.

The geological age of the Australian continent means that few good soils remain. Most are deficient in phosphorous and nitrogen. This deficiency was addressed by adding superphosphate.* The soils in many parts of Australia are

---

* Much of the superphosphate was sourced until recently from Nauru, leading to the almost total denudation of this small Pacific island.

highly susceptible to erosion by wind and water, and to structure decline and waterlogging. However, large-scale adoption of no-till agriculture over the last 20 years has substantially reduced water and wind erosion losses, except in periods of drought. High amounts of salt in the soil deposited over time make salinization of the productive surface layer a major risk (McTainsh and Boughton, 1993).

A complex situation exists. Australia has developed an agriculture that is unsustainable and that creates very real and immediate threats to the environment in the form of salination and acidification. These conditions threaten biodiversity, farmers' livelihoods, the viability of rural communities, and the quality and availability of drinking water for many Australian people, especially residents of Adelaide at the bottom of the Murray Darling Basin. Many farmers, however, feel that they are locked into a production system that makes it difficult to consider alternatives to their current production methods (Lawrence and Vanclay, 1992, 1994; Vanclay and Lawrence, 1993, 1995).

III. SALINITY

Dryland salinity is arguably the most serious environmental issue facing Australia. Currently touted figures indicate that 2.5 million hectares are affected and that dryland salinity costs Australia $850 million per year. Over 15 million hectares could become affected within the next 50 to 100 years (Australian Academy of Science, 2002).

Induced salinity was initially associated with irrigation. However, it is now prevalent in dryland agriculture and in inland cities and towns. Modern agroecosystems use less water than the pre-existing natural ecosystems. The excess water results in accessions to the water table, and over time the water table rises to become close to the surface. The soil profile in many parts of Australia contains trace quantities of salt, for a variety of geological reasons. Rising water tables dissolve the salt and bring it to the surface. Through evaporation and "capillary action", salt is concentrated in the topsoil whenever the water table is less than two meters below ground surface (MDBC, 2002d; Walker et al., 1999).

Areas that are affected by salt are known as discharge areas. All areas that contribute to rises in water-table levels are known as recharge areas. However, accessions to the water table are not uniform over a catchment area. The underlying geology of the terrain can profoundly affect the potential for the water table to rise. Recharge and discharge areas can be close together or far apart connected by aquifers. Usually recharge occurs in upland areas, and discharge occurs in valley floor areas. Salinity is problematic because it is spatially distorted by separation of recharge and discharge. It is also temporally distorted since the rise in water-table levels took many decades. Vegetation change over the last century is largely responsible for salinity today. Thus salinity must be seen as a community problem, which is not a result of current management practices of farmers. In fact, salinity is a direct result of farmers' adopting the agricultural practices promoted in the past. Once salinized, it is nearly impossible to return the land to a presalted state, although some restoration can be undertaken (Vanclay and Cary, 1989; Vanclay and Lawrence, 1995).

The final stage of soil salting is bare land covered with a salt crust, but there is a considerable lead-up to this situation. The first signs of potential soil salting are reduced yields in crops and/or reduced biomass in pasture species. However, these symptoms may also be due to other factors. The composition of pasture species also changes. Salt-sensitive species, such as subterranean clover (*Trifolium subterraneum*), give way to the more salt-tolerant species, such as strawberry clover (*Trifolium fragiferum*). As the concentration of salt increases, other less palatable salt indicator species, such as sea barley grass (*Critesion marinum*), become prominent. Only very salt-tolerant species, such as ruby salt bush (*Enchylaena tomentosa*) or Samphire (*Halosarcia pergranulata*), grow in soils at very high concentrations of salt (DNRE, 2002; Walker et al., 1999).

Seepage of groundwater and resultant waterlogging can occur because salting is associated with high water tables. Rushes, such as spiny rush (*Juncus acutus*), may indicate the presence of waterlogging by saline seepage. Seepage of groundwater and increased surface runoff due to reduced ground cover can lead to water erosion and gullying, particularly in areas where ground cover has already declined as a consequence of salting (Vanclay and Cary, 1989).

Agronomists are aware of practices that can ameliorate problems caused by salting. In discharge areas, salt-affected land should be fenced off to keep stock out; salt-tolerant species, such as tall wheat grass (*Lophyrum elongatum*), should be sown. Trees can be planted around the perimeter of salt-affected areas to provide a windbreak in order to shelter the newly sown plants and to lower the water table in the immediate vicinity. However, the best form of control minimizes further accessions to the water table by preventing the percolation of water through the soil. This can be achieved by reforesting the recharge areas, or by growing deep-rooted pasture species such as lucerne (*Medicago sativa*) or phalaris (*Phalaris aquatica*). Deep-rooted pasture species and trees, especially Australian native species such as *Eucalyptus*, are both high in water use. They have high evapotranspiration potential. Because of their deeper root systems, they are able to utilize more of the available water. Rain that falls onto paddocks of deep-rooted species has less chance of entering the water table than rain falling onto other pasture species or crops (especially annuals) (Walker et al., 1999).

The effects of salting are widespread. The more obvious are the loss of productivity, reduced profit levels, and reduced land value experienced by individual farmers. Every section of a rural community experiences the consequences of widespread salting. Farm and town water supplies are affected, household and farm equipment can be damaged by salt water, the biological integrity of creeks and rivers is threatened, wildlife habitats are threatened, and road foundations are undermined with inevitable deterioration of road surfaces.

In urban areas such as Wagga Wagga in inland NSW, rising water tables have been caused by the excessive watering of lawns and by leaking water reticulation systems. The common building practice in the early 20th century of channelling roof stormwater runoff to backyard rubble pits rather than to the stormwater system, which normally feed into local watercourses, has also contributed. Salinity has made it difficult to grow grass on playing fields and showgrounds. It has killed trees in suburbs and has brought about premature rusting of underground water reticulation systems, thus worsening the problem. The salt levels in rising damp in houses and other buildings has dissolved the mortar holding brickwork together. Damage to the bitumen surfaces of streets is also widespread (Quinn, 1998).

A major concern is downstream areas, especially in the Murray Darling Basin. Salt enters watercourses along the whole length of a river. The volume of salt (measured as tonnes) at the bottom end of catchments can be very large. Of greater concern is the concentration of salt in the water, which is exacerbated by the use of river water to irrigate high water using crops, such as cotton and rice.

IV. WATER MANAGEMENT ISSUES

Water management is the second most complex natural resource management issue in Australia, especially in the Murray Darling Basin. In some water diversion schemes, such as the Snowy Mountains Scheme, water has been diverted into westward flowing rivers to supply irrigation and other uses in the MDB, instead of flowing south through coastal plains to the ocean. This diverted water is also used to produce hydroelectricity.

Approximately 71% of Australia's irrigated land is in the MDB. It consists of just under 1.5 million hectares. About 14,743 farms are irrigated. They represent about 28% of farms in the basin. Over 95% of water in the basin is used for irrigation—and over 70% of all water used in Australia is for irrigation in the MDB.

Development of irrigation areas and associated dams and weirs has had a profound effect on river flow regimes. Instead of high flows in early spring, rivers now have high flows in summer. Instead of low flows in late summer, the rivers now have low flows in winter. This altered flow regime has greatly affected the ecology of the river, upsetting native fish breeding cycles. Native fish populations have also been affected by invasion of European carp (*Cyprinus carpio*), and considerable attention is given to reducing carp numbers in rivers. Issues such as salinity concentration in the river and blue-green algae blooms, or outbreaks of toxic cyanobacteria, are also of considerable concern (Bowmer, 1998; Smith, 1998).

Flood management, and the capturing of winter rains for irrigation purposes, has meant a major change to the flooding regimes. The Murray River now has a continuous flow although not always at the mouth of the river. Without regulation, flows in the Murray River would have fluctuated considerably. The river would have ceased to flow completely during five drought periods in the last century (MDBC, 2002b). This changed flooding regime has had profound ecological effects on Australian floodplains. For example, many billabongs (oxbow lakes) are only replenished through flooding. Many native vegetation species, such as River Red Gums (*Eucalyptus camaldulensis*), are adapted to periodic flooding, and the loss of flooding is impacting Australia's biodiversity. Forests such as Barmah and Millewah have been considerably affected (MDBC, 2002e).

Over the past few decades, the southward flowing Snowy River (which is not in the Murray Darling Basin) has only had around 1% of its prediversion flows. A massive public campaign had been waged by environmental groups to restore at least some of the flow to the Snowy. For many years, the campaign had little success. However, in late 1999, there was a delicate balance in the political situation in the state of Victoria. Three rural Independent politicians who were strongly in support of restoring flows to the Snowy were able to strike a deal with the minority State Labor Party, which allowed it to form the government. A plan was forged to restore river flows in the Snowy River to 15% in 7 years, 21% in 10 years, and eventually to 28%. The scheme would cost about US$165 million over 10 years. On August 28, 2002, the tap was turned to allow the first of these environmental flows to be released. Because of much concern about this plan by irrigators and from South Australia (who are dependent on Murray River water for domestic consumption), the scheme to increase environmental flows had to be achieved without compromising the security of irrigators' water rights and without affecting supply to Adelaide. The water therefore had to be achieved through water savings made elsewhere in the system (Premier of Victoria, 2002).

Adelaide is not on the Murray River. However, in a normal year it draws about 40% of its drinking water from the Murray. In dry years when other water sources are depleted, this can increase to 90%. Water quality in the Murray therefore has a significant effect on the quality of Adelaide's drinking water. Without water management and diversions, it is doubtful that South Australia could sustain the level of population it has. Even with the addition of diverted snowmelt water into the westward flowing rivers, the extraction of water for irrigation and other purposes means that the Murray River discharges to the sea on average only 21% of the water it would if there had been no development (IAG, 1995).

With considerable extractions from the Murray River, in low-rainfall years there is a danger that the mouth of the Murray silts up completely, an event that happens periodically*. Long-term closure of the mouth of the Murray would be detrimental, leading to the stagnation of the important Coorong Wetlands near the mouth of the Murray. Consequently, there has been considerable discussion of the need to "flush" the Murray River, particularly in those years when the silting over of the mouth is a reality. With intense and contested demand for water already, a proposal to expend millions of liters on flushing is highly debated. Counterresponses to the recommendation for flushing include calls for examination of the sources of salt and other nutrients in the river system, consideration of the amount of water lost to evaporation, and examination of the environmental effects of the many weirs on the river. In 2002, there was commitment to flushing the river, as much to prevent outbreaks of toxic blue-green algae. However, with the Murray Mouth already closed, a short-term solution in the form of dredging was utilized. On November 22, 2002, after nearly 12 months choked closed, the Murray River once again flowed to the sea (*The Weekend Australian* November 23–24, 2002).

The expansion of cotton growing in northern NSW and southern Queensland increases the demand for water. Cotton growing requires considerable volumes of water, and cotton growers have diverted water from river systems into holding ponds for later use. These holding ponds tend to have large surface area and are relatively shallow. As they are located in areas with high evaporation potential, the loss is considerable. The net effect has been a considerable reduction in the water flows in the MDB.

The Snowy Mountains Scheme was specifically developed to provide water for inland irrigation. Originally, the intention was to provide free water to irrigators, with the costs recouped through the sale of hydroelectricity. Farmers were granted water allocations. In demarcated irrigation areas, these were based on a volumetric allowance. In other areas, farmers had an entitlement to irrigate a certain area of land. In the late 1990s, all water licenses have been converted to volumetric allocations.

The history of settlement of many rural areas is the history of irrigation. Irrigation schemes began in several places in the 1880s, such as Kerang, Pyramid Hill, Swan Hill, and Shepparton-Tongala. Renmark and Mildura have their history tied to the irrigation schemes of the Canadian Chaffey brothers. The Murrumbidgee Irrigation Area (MIA), which is centered around the township of Griffith NSW, was developed between 1906 and 1913 with water supplied by Burrinjuck Reservoir (completed in 1929). Soldier Resettlement schemes after World War

---

* The Web site of the Murray Darling Basin Commission provides some excellent photos of the mouth of the Murray River. See http:/www.mdbc.gov.au/river_murray/mouthpic/rivermouth.htm. In addition, there is an animated display at http://www.mdbc.gov.au/river_murray/ river_murray_system/murray_mouth/mouth_animation.htm.

I and World War II led to the expansion of many smaller irrigation areas and to the development of the Coleambally Irrigation Area (CIA) with water supplied by Blowering Dam (completed in 1968) (MDBC, 2002c). In irrigation areas, the structure of the society is based around the abundant supply of free or cheap irrigation water.

Flood and furrow systems are relatively inefficient, while other irrigation methods are less wasteful. Inefficient systems are problematic, not only because of wastage and the cost to farmers, but because they contribute to salinity, and because they consume more water. Part of the problem has been the capital cost associated with more efficient systems, and the fact that the economics did not encourage a change in irrigation practices when farmers do not pay high charges for the water. The price of water has not reflected the value of the water when all externalities are included, and water generally has been subsidized by public authorities. In recent times, water management authorities have been trying to bring about a change in the way the system operates.

Different crops use different amounts of water, and the economic return per unit of water also varies (Table 2). The price per megaliter varies considerably from place to place, but in late 1999, water was being sold to rice growers for AUD$31.50 (about US$17) a megaliter. This was a price they thought was too high. Social and economic factors mean that water-inefficient crops may be desirable crops for farmers to grow. This is substantiated by the increase in rice and cotton production. In the end, the cost of water is only a relatively small proportion of the total cost of all inputs. Ironically, increasing the price of water may bring about an increase in water consumption because farmers may switch to higher profit, but greater water-consuming, crops. The creation of a free market in water through tradable rights has also meant an increase in water usage, as farmers who previously did not utilize their water allocations now sell them.

Table 2 Water Required to Make AUD$100 Profit

Agricultural commodities	Water required to return AUD$100 profit (thousands of liters)
Fruit	200
Vegetables	460
Dairy products	500
Cotton	760
Rice	1,850
Pasture	2,780

Source: http://www.mdbc.gov.au/education/encyclopedia/Irrigation/
Irrigation.htm.

A comprehensive audit of water use was completed in 1995 as part of the water management process. One major finding of this audit was that on average only 63% of the water that is permitted to be used is actually used (IAG, 1995). This means that water consumption could increase significantly even without a change to entitlements. If this were to happen, it would place further strain on the environment. In response to the audit, the MDBC Ministerial Commission agreed at its June 1995 meeting to introduce a cap on extractions of water from MDB rivers.

In the past, farmers were given water allocations free on application, but they were tied to the ownership of land. Some farmers never used their entitlements. They are called "sleepers." Others used their entitlements only periodically and are called "dozers." Most irrigators had licenses for water to be taken from surface water resources. Other farmers had licenses for water extraction from groundwater sources. In some cases, farmers had entitlements to extract water from groundwater supplies where there was no (longer) groundwater. They are called "ghosts." The concern with the sleeper and dozer licenses is that if these license holders begin to use or trade their water allocation, it would mean lower allocations for those who had previously been using their entitlements.

The Murray Darling Basin comprises sections of four Australian states and one territory. Therefore, water management in the basin needs a collaborative approach. However, in Australia, water management is a state (not federal) responsibility, and different systems apply in each state. These different systems add to the complexity of management in the basin and to the difficulty of introducing a uniform and fair system.

In NSW, for example, different forms of water entitlement exist. A water entitlement was not a guaranteed right to water. A limited number of high security entitlements exist on regulated streams. Regulated streams were those supplied by water storage. High security entitlements would normally be filled except in the very worst of droughts. Most water entitlements were normal security entitlements. A normal entitlement was a right to water provided that it can be supplied given the available water in the system. Each year, farmers were told at several times throughout the year, the percentage of water that they were guaranteed to receive. For example, in 1998–1999, farmers in the Murrumbidgee were given an initial allocation of only 40% of their entitlement because of water shortages. Farmers pay annual charges and volumetric fees, which vary considerably among locations.

Unregulated streams are those not supplied by storages. In this situation, most entitlements were based of the area of the farm or land to be irrigated. There were often restrictions specifying the conditions under which a farmer could irrigate, such as the minimum rate of flow of the stream, the maximum rate of pumping, and the number of hours or times of day. Because the stream is unregulated, water may not be available to an individual user when the farmer needs it.

In the past, water entitlements were only about managing a system of supply of water. There was little concern about the environmental impacts or the availability of water for consumption in Adelaide. With the new system and the implementation of a cap on diversions, the focus of attention has changed to managing water better for all purposes. In the old system, the use of water in unregulated areas was unimportant since it did not affect the availability of water in the regulated system. However, the changed focus of water management in the new system means that the unregulated rivers have become important because they also contribute to environmental flows and to Adelaide's water supply. Consequently, entitlements in the unregulated streams have been changed from area-based licenses to volumetric licenses.

Various licenses for water harvesting also exist. Initially, water harvesting meant the extraction of water from river systems into storage dams for later use. Water harvesting was permitted in very high-flow regimes. During these times, no limit was imposed on the amount of water extracted. This type of entitlement was logically consistent with a water supply conceptualization but was also responsible for reducing the amount of water available for environmental flows. Because of this, the concept of water harvesting now also extends to dam water that is collected from rain falling on a piece of land. The water management authorities are now laying claim to the rainfall on farms and are developing regulations relating to farm dams. Farmers very much resent this.

Transfers of water entitlements have only been possible recently. Under the old model of management, the ability to transfer was not necessary. In the philosophy of the new model, the ability to transfer water is important. As a result, permanent water transfers have only been available in NSW since 1989, and temporary transfers of water since 1983. Now, when water is short, a farmer can buy the water entitlement of another farmer. Some farmers may make more money selling their water entitlements than growing low-value crops using their water allocation.

There is no provision within the cap to reduce water consumption, or to increase environmental flows. The cap is simply a process for ensuring that water consumption does not continue to grow, and for ensuring equity—at least that all the water is not used by water users at the top end of the catchment leaving no water for Adelaide. Independent of the cap (which is a transboundary, multistate initiative), each state is attempting to address issues of water reduction. This will also be on the agenda for the MDBC.

Somewhat separate from the Irrigation Management Strategy of the MDBC and the cap, the Council of Australian Governments (COAG), a joint forum of the Commonwealth and state governments, developed a "strategic framework for reform of the Australian water industry" in 1994. Consistent with economic rationalist trends in Australia and the requirements of the National Competition Policy (NCP), a process of full cost pricing, institutional reform, water trading, and improved public consultation was to be implemented. The COAG water reform process, as it is called, is taking place. Responsibility for implementation is at the state level. NCP payments totaling $2.4 billion are being offered to the states from the Commonwealth government to encourage compliance and effective implementation of the reforms.

The Industry Commission's (1998) report identified three pillars for the "ecologically sustainable management of natural resources in agriculture" to be (1) to impose on natural resource owners, managers, and users a "duty of care" for the environment, (2) to improve the markets for natural resources, and (3) to encourage conservation on private land. The strategy being invoked is one of privatization. It makes individuals responsible and it commodifies the environment. The report promotes the use of voluntary standards and codes instead of mandated standards as much as possible. Embedded in the strategy is an attempt to encourage water use that will achieve highest value. It is assumed that this can best be done through water trading. Ideally, "highest value" should take into account consumptive and nonconsumptive uses. Consequently, the value of environmental flows is being measured. One problem with this approach is that the rising price of water being traded brings about a shift in production, not to the most environmentally sound or water-efficient crops, but to those crops that return most income, which, in the case of cotton, use more water. The system has also encouraged structural adjustment in agriculture, decreasing the number of small farmers. This is having social and economic implications for communities.

Water management in the MDB includes issues related to managing the volume of water and water use. However, it also relates directly to water quality. Dryland salinity management and water management are intertwined. Dryland salinity is, of course, a primary cause of salt in Australia's waterways.

In Australia, water salinity is measured by electrical conductivity (EC). The World Health Organization threshold for desirable drinking water is 800 EC, and the threshold for irrigation is 1,500 EC. At 5,000 EC, an ecological transition occurs, creating the division between fresh water and salt water. The average salinity of the sea is around 45,000 EC (MDBMC, 1999).

Within the MDB, a number of rivers already exceed 800 EC on average. Other rivers in the MDB will exceed this level within 20 years (MDBMC, 1999). The worst situation in the MDB is Barr Creek in northern Victoria, which drains some of the Kerang Irrigation Area. It has recorded salinity at over 60,000 EC!

Morgan in South Australia is the monitoring station just below where the water for Adelaide is extracted from the Murray River. Thus, it is a key locality indicator for the health of the MDB. In 1988, average salinity at Morgan was 583 EC, and between 1993 and 1999, salinity levels at Morgan were less than 800 EC 92% of the time. A major report, the Salinity Audit (MDBMC, 1999),

Table 3 Predicted Average Salinity Levels for Selected Major Rivers in the MDB

River	1998	2020	2050	2100
SOUTH AUSTRALIA				
Murray River at Morgan	570	670	790	900
VICTORIA				
Murray River at Merbein	360	400	450	510
Avoca River at Third Marsh	1,440	1,470	2,200	2,990
Campaspe River	600	600	610	610
Loddon River	870	880	900	970
NEW SOUTH WALES				
Darling River at Menindee	360	430	490	530
Murrumbidgee	250	320	350	400
Murrumbidgee at Wagga	140	170	190	220
Lachlan	530	780	1,150	1,460
Macquarie	620	1,290	1,730	2,110
Namoi	680	1,050	1,280	1,550
Gwydir	560	600	700	740
Macintyre	450	450	450	450
QUEENSLAND				
Warrego	210	1,270	1,270	1,270
Condamine-Balonne	210	1,040	1,040	1,040
Border Rivers	310	1,010	1,010	1,010

Source: Adapted from Murray Darling Basin Ministerial Council, 1999, pp. 14–15.

suggests that by 2050, average salinity would have risen to 790 EC, and to 900 EC by 2100 (Table 3). The contribution of various rivers to salinity in the MDB has been estimated, with large increases in salinity levels expected, especially for the Queensland rivers. While Morgan is used as the key measuring point for the MDB, and the water quality in Adelaide is a prime consideration, many localities within the MDB have poor-quality water. For example, Boorowa and Yass in NSW have recorded salinity levels of over 1,400 EC in their town water supply (MDBC, 2002d).

V. CONCLUSION: THE FUTURE OF AUSTRALIAN AGRICULTURE

Salinity and water management are presently Australia's two leading sustainability issues. Acidity, soil erosion, soil structure decline, and agronomic practices that lead to the development of resistance in weeds are also major agro-environmental issues (Pratley and Robertson, 1998). Government commitment to biodiversity translates into the need for protection of on-farm remnant vegetation and to wetland protection and restoration. This has caused tension in agricultural communities. However, in 2002, the intensity of the drought across much of Australia transcended and exacerbated many other issues.

The 2002 drought sparked new discussions about "drought-proofing" Australia. One group, the Farmhand Foundation, backed by many large corporations, espoused more schemes to turn rivers inland. A counter group of leading natural resource scientists, called the Wentworth Group, opposed this idea by arguing that it is impossible to drought-proof Australia and that Australians need to learn to live with the environment.

Land management issues are social issues as much as they are technical issues (Vanclay and Lawrence, 1995). They are social issues partly because of the social impacts (negative consequences) that rural people experience as a result of these environmental problems. More cogently, how farmers perceive these issues and their solutions will need to be understood, appreciated, and acted upon if solutions are to be adopted. The nonpoint source of many of these environmental issues means that contributors are often undetectable, regulation is ineffective, and voluntary cooperation is necessary.

The future for Australian agriculture remains bleak. Environmental issues, particularly salinity, will inevitably worsen. Australian agriculture will have to adapt to a changing environment. Living with salinity rather than fighting salinity will become a focus for research. New salt-tolerant crops will need to be grown, and salinity will need to become an accepted feature of large areas of the Murray Darling Basin. Increasing pressure will be exerted on water management, with increasing expected returns from irrigation. This will lead to an increase in the production of high-value crops, which, despite efforts to make irrigation water use more efficient, may actually increase water consumption.

The return of large areas of farmland to forestry for environmental benefit will have an inevitable effect on the viability and vitality of small rural communities. Together with the general cost-price squeeze from declining terms of trade in agriculture, it will accentuate the downward spiral of rural communities. Information technology will help overcome some of the negatives associated with rural living. Nevertheless, the vastness of Australia will still be a major barrier to deal with, especially in the more remote areas (Vanclay and Lawrence, 1995).

REFERENCES

Australian Academy of Science. 2002. Feeding the future—sustainable agriculture: http://www.science.org.au/nova/071.htm.

Australian Bureau of Statistics. 2002. Australia now: A statistical profile: http://www.abs.gov.au/Ausstats/abs@.nsf/2.6?OpenView.

Bowmer, K 1998. Water: quantity and quality. In. Pratley J., Robertson A., Eds. Agriculture and the Environmental Imperative. CSIRO Publishing. Collingwood, Vic., 35–69.

Bureau of Meteorology. 1995. Climate and weather. In. Douglas F., Ed. Australian Agriculture: The Complete Reference on Rural Industry, 5th ed. National Farmers Federation: Morescope Publishing. Canberra, 43–54.

Department of Natural Resources and the Environment (DNRE). 2002. Salinity Indicator Plants: A Guide to Spotting Soil Salting. Department of Natural Resources and the Environment. Melbourne.

Independent Audit Group (IAG). 1995. An Audit of Water Use in the Murray–Darling Basin: Murray Darling Basin Ministerial Council. Canberra.

Industry Commission. 1998. A Full Repairing Lease: Inquiry into Ecologically Sustainable Land Management: Report No. 60. Industry Commission. Canberra.

Lawrence, G., Vanclay, F. 1992. Agricultural production and environmental degradation in the Murray–Darling Basin. In. Lawrence G., Vanclay F., Furze B., Eds: Agriculture, Environment and Society. Macmillan, Melbourne, 33–59.

Lawrence, G., F., Vanclay 1994. Agricultural change in the semi-periphery: The Murray–Darling Basin, Australia. In. McMichael P., Ed. The Global Restructuring of Agro-Food Systems: Cornell University Press. Ithaca, NY, 76–103.

McTainsh, G., Boughton, W., Eds. 1993. Land Degradation Processes in Australia: Longman Cheshire. Melbourne.

Murray Darling Basin Commission. 2002a. Basin statistics: http://www.mdbc.gov.au/naturalresources/basin_stats/statistics.htm.

Murray Darling Basin Commission. 2002b. Backgrounder 2: Regulation and distribution of River Murray waters: http://www.mdbc.gov.au/river_murray/river_murray_system/hume/ops_review/regulation.htm.

Murray Darling Basin Commission. 2002c. Irrigation: http://www.mdbc.gov.au/education/encyclopedia/Irrigation/Irrigation.htm.

Murray Darling Basin Commission. 2002d. Water and land salinity: http://www.mdbc.gov.au/education/encyclopedia/water_and_land_salinity.htm.

Murray Darling Basin Commission. 2002e. The Barmah-Millewa forests and Barmah-Millewa forum: http://www.mdbc.gov.au/naturalresources/policies_strategies/projectscreens/barmah_forum.htm.

Murray Darling Basin Ministerial Council. 1999. The salinity audit of the Murray—Darling Basin: Murray Darling Basin Commission. Canberra.

Pratley, J., Robertson, A., Eds. 1998. Agriculture and the Environmental Imperative: CSIRO Publishing. Collingwood. Australia.

Premier of Victoria. 2002. Carr and Bracks open tap on $300 million–10-year Snowy River plan. Aug. 28, 2002 (press release): http://www.premier.vic.gov.au/newsroom/news_item_archive.asp?id = 124.

Quinn, P 1998. Urban salinity: The shift to collaborative partnerships. *Rural Society* 8(1): 29–38.

Smith, D.I 1998. Water in Australia: Resources and Management: Oxford University Press. Oxford.

SunRice. 2002. Rice farming: http://www.sunrice.com.au/rice_farming.htm.

Vanclay, F., J., Cary 1989. Farmers' Perceptions of Dryland Soil Salinity. School of Agriculture and Forestry: University of Melbourne.

Vanclay, F., G., Lawrence 1993. Social and environmental impacts of economic restructuring in Australian agriculture. *Int. J. Sociol. of Agric. and Food* 3:97–118.

Vanclay, F., G., Lawrence 1995. The Environmental Imperative: Ecosocial Concerns for Australian Agriculture: Central Queensland University Press. Rockhampton. Australia.

Walker, G., Gilfedder, M., Williams, J. 1999. Effectiveness of current farming systems in the control of dryland salinity: CSIRO Division of Land and Water. Canberra.

26

Technology for Small Farms: The Challenge of Diversity

Peter E. Hildebrand
University of Florida
Gainesville, Florida, U.S.A.

I. ARE LARGE FARMS DIFFERENT FROM SMALL FARMS?

All farms are not created equal. No one would argue with that statement. Obviously, irrigated farms are different from rainfed farms, and temperate farms are different from those in the tropics. Livestock "ranches" are different from crop "farms." But are large farms different from small farms? Are family farms different from industrial farms? Many would argue that large and small or family and industrial are not characteristics that differentiate farms, particularly when new or "modern" technology is concerned. I once heard the research dean of a U.S. land-grant university declare, "Certainly our technology works on small farms as well as large ones. We test it on small plots, don't we?"

A. Average Farms Do Not Exist

An "average" Punjabi rice–wheat farm household might have 1.2 adult males, 1.4 adult females, 1.7 adolescent males, 0.6 adolescent females, and 2.4 children. Half would have more land, and half would have less. It could have from 0.4 to 1.6 cattle or buffalos. Other averages could be mentioned, but few such farm households could be found. Yet we tend to use averages when discussing yields, farm size, available resources, or capabilities to adopt new technologies. But we ignore the fact that most rice–wheat farms do not achieve the average yield, nor are they average size, nor do they have the resources or capabilities appropriate for adopting new technologies such as minimum tillage that suit the average situation.

B. Land Is the Most Limiting Resource on Small Farms

Increased production usually means increasing the amount of product produced per unit of land area, because arable land is most often the limited resource. Thus we tend to look for technologies that increase yield per unit of land area even if they require other resources such as more labor or more cash than many smallholders have available, such as mechanization, inorganic fertilizer, or pesticides. Yet if we measured the productivity (yield) of any of these inputs when used with the new technology in terms of its product per unit of cash expenditure, this might well be lower than what the farmers are already getting. Alley cropping agroforestry, as an example, may produce less per unit of labor than farmers are already producing.

Because labor, cash, or seed is often more limiting than land on small-scale, family farms, we first need to consider increasing the productivity of these critical resources. Increased productivity of land may follow. In crops like potatos or beans, which are staple foods and can readily be sold, yet somewhat difficult to store under rustic conditions, the amount of seed can be more of a limiting factor than land.

> In Nariño, Colombia, a well-known minifundio area, farmers planted potatos in low densities to maximize the productivity of each potato planted. The size of their potato field was fixed by when they ran out of seed. Even on such small farms, seed was the limiting resource for potatos, not land (Andrew, 1970).
> In the Pakistani Punjab in the 1960s, farmers with limited water spread it out on as much land as possible to maximize the productivity of the water. This, of course, gave less yield per unit land area. But if they had increased the per hectare application of water, this would have reduced the productivity of their water, resulting in less product (Andrew and Hildebrand, 1982).

C. "Our" Crops Are Not Necessarily the Farmers' Priority Crops

We also tend to emphasize the crop or crops in which we are interested, forgetting that the farmer must allocate resources among all the activities and needs of the household. Intercropping can reduce the yield of one particular crop but increase the overall productivity of the land and labor involved in production. Also, "late" planting that reduces yield of a particular crop is often a compromise, enabling the farmer to get some production from another complementary crop, giving higher total yield between the two crops.

The farmer may put a higher priority on another crop rather than the ones in which we are interested. Planting the other one first and ours later than we think is optimal can be a very rational management decision. Seasonal needs of farmers can lead to what we think of as inappropriate technologies. For instance, early (short-season) crops usually have lower yields than later (longer-season) crops. Yet farmers grow them to provide food or cash at a time when these are needed even if the yield is less. For them, food harvested earlier has more value than higher yield later on when food is abundant.

D. We Are in a Hurry to See Results

We tend to define "results" in different ways. Many of us define results as statistically significant differences among treatments. If our trials do not result in statistically significant differences, we do not see "results." This leads to tight controls on nontest variables in experiments to reduce unexplained variance, thus increasing the probability of achieving statistically significant differences among test treatments. When this practice is followed in on-farm trials, it creates an artificially superior environment that farmers are unable to provide over time on a field basis (Hildebrand and Russell, 1996).

Another method of helping to ensure statistically significant results is to have higher yields that achieve a lower coefficient of variance, thereby increasing the probability of finding significant differences among treatments. In on-farm trials, this leads to selecting the best farms or the best fields or even the best spots in fields on which to conduct trials. The director-general of one national agricultural research organization for which I worked said, "Pete, if you work in those hills [where the small farms were located], you won't get any response." By "response," he meant results indicated by statistically significant differences.

But results can also be measured by breadth of adoption. Whether it is a cultivar, another input, or a cultivation practice, broader adoption is always better than limited adoption. This is appreciated by industry. A limited number of broadly adoptable products is better than a larger number of narrowly applicable items. This biases our research toward practices and materials that require the modification of the field environment so that they can be productive, similar to what is seen in experiments, in varied locations. This practice has been very effective in areas such as the "Corn Belt" of the United States. It can also be effective in flooded rice paddies, which homogenize agronomic conditions. But small farmers with limited resources cannot modify the environments of their fields to suit the requirements of new technologies. Therefore, technologies must be developed to match the constraints of the farmers' environments.

II. THE CASE FOR ATTENDING TO SMALL FARMS

A. They Are Not Reducing in Number

In 1970, 70% of the population of Pakistan was involved in agriculture (FAO Database Collections). By 2000, only half of the population was involved in agriculture. During this time in India, the agricultural population dropped from 67% in 1970 to 54% in 2000. In Bangladesh, the decline was from 84% to 56%. This interpretation makes it sound like these countries are urbanizing. Could it be that they are following the path of Western countries, where farms are becoming larger and larger and farmers fewer and fewer? No, this is not the case. In these three countries, between 1970 and 2000, the agricultural population increased from 475 million people to 620 million people. The 145 million added people obviously are not operating large farms. If they all are on small farms, this would mean about 20 million more small farms in the region over the last 30 years.

Over this same 30-year period, even after realizing that Green Revolution technology has had little effect on limited-resource farmers on marginal lands, we have avoided the challenge of working with the great biophysical and socioeconomic diversity in which the world's small farmers struggle to survive. Perhaps we thought small farms would go away: "Get bigger or get out." Perhaps we thought that our technology was scale-neutral and that it would just take time for innovations to trickle down to the late adopters and laggards, the small farmers who had not adopted them. Perhaps we thought that average farms represented all farms. Perhaps we felt we have to work where we can easily measure results as we are accustomed to doing. Perhaps we are convinced that land is the most limiting resource on small farms and that yield should be measured only in terms of unit land area. Perhaps we have forgotten, or did not know, that farmers have many priorities, and theirs may not coincide with ours.

B. Not All Farms Adopt New Technology

For whatever reasons, over this period of time, only a small fraction of our research and technology development efforts have been oriented directly and adequately at intensification of the still increasing numbers of small-scale, family farms with limited or marginal resources. For 40 years we have been convinced by Everett Rogers (1962) that many farmers are slow adopters or even laggards or nonadopters of new technological innovations. Because we have faith in our technology, we accept uncritically what Rogers tells us. We like to work with the "innovators" or the "early adopters," those inclined to adopt our technology.

The theory tells us that it is not our responsibility when some farmers do not adopt what we consider to be good technology. It is not our fault that many farmers do not have access to credit, or the cash resources to acquire our good technology, or that this is not available in local markets. We are not surprised if it takes time for the benefits of our technology to be understood and for farmers to be "motivated" to adopt it (Hildebrand, 1980).

As a result, the research affects a relatively small number of larger farmers, or those who work in better environments or with more resources (Fig. 1). It

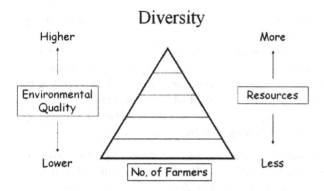

Figure 1 Farm resource diversity and relative numbers of farmers. (Adapted from Hildebrand, 1993.)

seems natural that the innovation does little or nothing for the very large number of limited-resource farmers at the bottom of the pyramid. We have taken the easy road of asking those farmers with sufficient resources to change their agricultural environment to suit the technology. Now it is time for us to accept the challenge of creating technologies that suit the diverse biophysical and socioeconomic environments of most of the world's small farmers.

C. Agricultural Intensification on Small Farms is Possible

To be effective in helping intensify highly diverse small-farm agriculture, it is necessary to comprehend the livelihood systems of these small farmers. A livelihood system is comprised of all the on- and off-farm activities available to farmers in an area from which they select their strategies to survive and thrive. This includes not only all the crops and livestock they raise, but different ways or times of raising them. Besides production activities, it is also important to understand reproduction and community activities as well because they also use scarce farm resources.

Production activities are those that result in the production of goods such as food (for consumption or sale) or cash. Farming, fishing, carpentry, cottage industry, migrant work, paid labor, government jobs, etc. can be considered production activities whether on or off the farm.

Reproduction activities are those like maintenance and care of the family unit that result in the survival and succession of the family or household. Meal preparation, hauling water or fuel, childcare, laundry, house clean-

ing, reroofing, house-building, or caring for elderly or disabled are among
reproduction activities.

Community activities are more vague and more difficult to quantify in terms
of inputs and outputs; however, they play a key role in understanding
how households and communities function. These include attending or
organizing meetings, forming or participating in women's, men's, children's, or producer groups, acting as part of a village or community
council, household food sharing, and the like.

Seasonal activities and periods of cash or labor scarcities are important to understand as well as which of the household members is involved in each activity.

All households do not adopt the same strategies. Livelihood strategies are
a function of the characteristics of each household such as wealth, gender of the
household head, relative age of each household, and household composition (sex,
age, and relationship of household members). Even though all households in a
livelihood system have access, at least in principle, to the same set of activities,
the constraints and resources reflected in these characteristics cause members to
choose different subsets of activities as strategies. To help diverse households
intensify production, it is critical to assess the capabilities of each type of household to help them mold technologies to the needs and constraints of each type.

This sounds like anthropology, and it is feared that anthropologists need
years to do ethnographic studies in remote villages and then do not really want
anyone or anything to change "their" village. This is an unfortunate stereotype.
Anthropologists with solid agricultural backgrounds can be productive members
of multidisciplinary teams. Also, an increasing number of agronomists have anthropological training. Incorporating these kinds of scientists in teams working
to intensify diverse small-farm agriculture is highly productive. Economists with
agronomic and/or anthropological training can also be useful members of such
teams.

Modeling small-scale, limited-resource family farm livelihood systems,
such as by ethnographic linear programming (Bastidas, 2001; Breuer, 2000; Cabrera, 1999; Grier, 2002; Gough, 2002; Kaya, 2000; Kaya et al., 2000; Litow,
2000; Mudhara, 2002; Pomeroy, 2000; Sullivan, 2000; Thangata, 2002) is one
effective way to integrate crop, animal, anthropologic, and economic knowledge
gained through farmer participation to help predict which households may be able
to adopt different kinds of new technologies even prior to their being developed or
offered to farmers in the community.* These models help us understand the kinds
of technologies that are needed by the different types of households in an area

---

* See Hazell and Norton (1986) for details on linear programming; see Hildebrand (2002) for details
on ethnographic linear programming. For examples of ethnographic linear programming, see the
authors listed above.

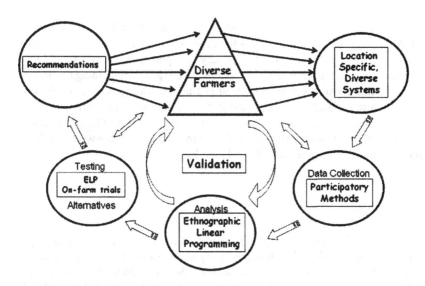

Figure 2 Schematic representation of the methodology for ex-ante evaluation of potential technological, infrastructure, or policy changes. (Based on Bastidas, 2001.)

and thus can guide innovative thought into unique approaches to agricultural intensification that can help even the poorest kinds of small-farm households survive and thrive.

Figure 2 represents a highly efficient and very effective methodology for incorporating these different perspectives in a participatory process with farmers that incorporates diversity, in problem or constraint assessment, in technology, infrastructure or policy development, and in recommendations. With the availability of laptop computers, it is now feasible for modelers to work in the field with farmers in the process of creating, validating, and using their models. When these models are validated (adequately simulate the existing livelihood system), alternatives can be pretested in the models, even while on-farm trials are being conducted, both to help researchers better understand how the alternatives would fit into the strategies of the different kinds of households, and to help characterize the recommendation domains for which the technologies are appropriate. The results of the on-farm trials and knowledge gained from continuous contact with the farmers can be used to improve the ethnographic linear programming models, which should constantly be modified to make them even more useful.

D. It Can Be Done

Miniaturization of computer hardware and advances in software have generated the potential to create and use sophisticated models in the field while working with farmers in their diverse environments in a participatory, ethnographic mode. We know these methods work. Now is the time to put them all to work together. This will require concerted, multidisciplinary and participatory efforts and the will to shed many approaches to which we tend to cling. The remaining challenge is for research and extension personnel, infrastructure managers, and politicians to become more innovative in their search for technologies, infrastructure, and policies specifically aiming to assist the still increasing number of highly diverse, small-scale, limited-resource family farms in many countries of the world, including those in the rice–wheat area of the Indo-Gangetic Plains.

REFERENCES

Andrew, C.O. 1970. Problemas en la modernización del proceso de producción de papa en Colombia. Departamento de Economía Agrícola, Instituto Colombiano Agropecuario, Boletín Departamental No. 6. Bogotá. (Cited in Andrew and Hildebrand.).

Andrew, C.O., Hildebrand, P.E. 1982. Planning and Conducting Applied Agricultural Research: Westview Press. Boulder, CO.

Bastidas, E.P. 2001. Assessing potential response to changes in the livelihood system of diverse, limited-resource farm households in Carchi, Ecuador: Modeling livelihood strategies using participatory methods and linear programming. Ph.D. dissertation, Food and Resource Economics Department: University of Florida.

Breuer, N.E. 2000. The role of medicinal plants in rural Paraguayan livelihoods. M.A. thesis, Latin American Studies Program: University of Florida.

Cabrera, V.E. 1999. Farm problems, solutions, and extension programs for small farmers in Cañete, Lima, Peru. M.S. thesis, Agricultural Education and Communication Department: University of Florida.

FAOSTAT Agricultural Data. http:dRapps.fao.org/page/collections?subset = agriculture.

Grier, C.E. 2002. Potential impact of improved fallows on small farm livelihoods, Eastern Province, Zambia. M.S. thesis, Food and Resource Economics Department: University of Florida.

Gough, A.E. 2002. The Starter Pack Program in Malawi: Implications for household food security. M.S. thesis, Agricultural Education and Communication Department: University of Florida.

Hazell, P.B.R., Norton, R.D. 1986. Mathematical Programming for Economic Analysis: Macmillan. New York.

Hildebrand, P.E. 1980. Motivating small farmers, scientists and technicians to accept change. *Agric. Admin* 8:375–383.

Hildebrand, P.E. 1993. Targeting technology diffusion through coordinated on-farm research. Presented at the Association for Farming Systems Research-Extension North American Symposium on Systems Approaches in North American Agriculture and Natural Resources: Broadening the Scope of FSRE: University of Florida.

Hildebrand, P.E. 2002. Ethnographic linear programming of limited-resource family-farm livelihood systems. Class handout, AEB 5167: Economic analysis in small farm livelihood systems. Food and Resource Economics Department: University of Florida.

Hildebrand, P.E., Russell, J.T. 1996. Adaptability analysis: A method for the design, analysis and interpretation of on-farm research-extension. Iowa State University Press, Ames.

Kaya, B. 2000. Soil fertility regeneration through improved fallow systems in southern Mali. Ph.D. dissertation, Forest Resources and Conservation Department: University of Florida.

Kaya, B., Hildebrand, P.E., Nair, P.K. 2000. Modeling changes in farming systems with adoption of improved fallows in southern Mali. *Agric. Syst* 66:51–68.

Litow, P.A. 2000. Food security and household livelihood strategies in the Maya Biosphere Reserve: The importance of milpa in the community of Uaxactún, Petén, Guatemala. M.S. thesis, Agricultural Education and Communication, Farming Systems Department: University of Florida.

Mudhara, M. 2002. Assessing the livelihood system of diverse smallholder farm households: Potential adoption of improved fallows in Zimbabwe. Ph.D. dissertation, Food and Resource Economics Department: University of Florida.

Pomeroy, C. 2000. An evaluation of a crop diversification project for low-resource hillside farmers in the Dominican Republic. M.S. thesis, Agricultural Education and Communication Department: University of Florida.

Rogers, E.M. 1962. Diffusion of Innovations: The Free Press. New York.

Sullivan, A.J. 2000. Decoding diversity: Mitigating household stress. M.S. thesis, Agricultural Education and Communication Department: University of Florida.

Thangata, P. 2002. The potential for agroforestry adoption and carbon sequestration in smallholder agroecosystems of Malawi: An ethnographic linear programming approach. Ph.D. dissertation, Ecology, Natural Resources and Environment Program: University of Florida.

27
Adapt or Adopt?
Prospects for Conservation Tillage
in the Indian Farming System

Paul Robbins
The Ohio State University
Columbus, Ohio, U.S.A.

The promise of conservation tillage and other innovations for increasing stagnant yields in food grain production in South Asia is evident from successes observed throughout Latin America and the global North (Unger, 1984). The impoverishment of Indian soils in the wake of the Green Revolution seems an obvious invitation for transition to techniques that stabilize erosion, conserve soil nutrients and moisture, and create a better biological environment for plant growth (Lal, 2000). The rapid adoption of these techniques in the contexts of North America and southern Latin America is notable, since it demonstrates a lack of typical constraints on diffusion, including infrastructure and pricing effect. One might therefore expect that once the benefits of no-till production are known and understood by local producers, adoption should occur rapidly in the Indian context.

As will be demonstrated here, however, the Indian farming system remains one in which (1) all forms of standing biomass are farm assets of significant value, (2) pastoral integration with intensified production is high, and (3) the gendered division of labor introduces differential costs to alterations of tillage practice. As a result, no-till practices may not be universally welcome nor immediately amenable to incorporation on Indian farms. Treating the production logics of north and northwest India as a farming system (Brush and Turner, 1987), this chapter points to certain characteristics of integrated land use that present challenges for conservation tillage. Using a case study from an intensively cropped and irrigated district of Rajasthan, this chapter demonstrates the practical problems of rapid transition to no-till techniques, with specific attention to processes of nutrient cycling and the role of livestock. It concludes with a call for research into ways in which resource-conserving techniques (RCTs) might be adapted to suit local production economies and ecologies, pointing to an approach where conservation is advanced by learning from existing tillage practices.

I. THE INDO-GANGETIC AGROPASTORAL PRODUCTION SYSTEM

> From the farmers' viewpoint, there was never one typical farming system or even a reasonably limited number of them. There were, rather, the general information systems and the uses that could be made of them. Household farming systems were like individual games, and general information systems were the rules. (Leaf, 1987, p. 274)

To speak of an "Indian farming system" is to invoke an analytical fiction. The variation in levels and types of production, soil and moisture conditions, and socioeconomic circumstances are vast. As we see from Leaf's description above of Punjabi production systems, elements of production vary widely as producers attempt to solve a complex production equation.

Even so, some common characteristics can be identified, specifically those system and information components that differ greatly from conditions in Latin America and the United States where no-till was first introduced and developed, and which raise questions of applicability and adaptation of the innovation. As characterized elsewhere in this volume, the Indo-Gangetic rice–wheat system (RWS) is one of relatively high intensity, with an increasingly industrialized orientation, and recently faced with stagnant or declining yields resulting from soil degradation and other factors. Factor productivity has been evidently declining. Despite these similarities with other global farming systems that have made the transition to conservation tillage, however, due to its unique ecological history, the Indo-Gangetic farming system differs in several respects.

A. The Historical Development of the Agropastoral Equation

The antiquity of double cropping systems in North India is well established. Complex rotations have been practiced at least since the time of the Mughals, and the fertilizing benefits of legumes were known well before that time. The range and complexity of cropping patterns in the precolonial period is notable and included carefully managed rice–wheat double cropping as well as triple cropping of rice–tobacco–cotton (Habib, 1999).

Developing in parallel to a semi-intensive system of production were allied systems of pastoral production that emerged during the Mughal and colonial eras, systems that maximized the efficiencies of natural resources and labor. Though not originally banned from slaughter [ancient Vedic texts actually prescribed meat eating; see Zimmerman (1987) and Robbins (1999)], the rising aversion to cattle consumption in the medieval period left a surplus of animal traction and dung as key resources in the agrarian system. Manure came to provide the primary fertilizer input into farming. Direct grazing of animals on fallow fields was practiced throughout all but the most densely populated regions (Habib, 1999). Pastoral breeding specialists emerged during the period, whose periodic migrations brought large herds into agricultural regions (Royal Commission on Agriculture in India, 1928).

At the same time, formalized controls emerged in land law for the enclosure of forest and grazing land and the establishment of *padat* (waste) lands available for cultivation but assigned to long-term fallow. This land management system together with large animal herds set into motion a nutrient cycling regime where forest and waste-fodder nutrients were deposited onto agricultural land while fallow and field stubble were increasingly made available to both local and migrating herds (Agrawal, 1999).

As increased land was brought under the plow and greater intensity followed multiple cropping, environmental services (nutrient supplies in the form of fodder) increasingly came from grazing lands and fallow within and outside the region (Cincotta and Pangare, 1994, following Boserup, 1965) and from field stubble and fodder crops, especially *jowar* (Sorghum bicolor). Some labor and nutrient resources were invested in crop production, while some was reserved for animal production. Some nutrient inputs came from external supplies, while some came from local grazing animals or from contracting with migrant herds.

The resulting pattern is one of a complex agropastoral equation (following George, 1990). Balances were struck between fodder versus food production, raising animals versus crops, and using local versus external nutrients by producers seeking production stability and subsistence, while avoiding unnecessary labor inputs and other sources of ''drudgery'' (Chayanov, 1986).

B. The Current Configuration

The radical intensification of production that followed the Green Revolution throughout northwest India altered many of the specific solutions to this equation and changed the overall structural configuration of the system without fundamentally altering the key decisions in household production. In Punjab, for example, increased land consolidation and a rise in wage labor followed high-input farming systems, but farmers still faced the same kinds of decisions regarding the balance of inputs, the value of fallow, and the sources of nutrients for agropastoral production (Leaf, 1987).

Overall, as discussed elsewhere in this volume, the adoption of high-yielding varieties (HYVs) of major cultivars, especially rice and what, has been a widespread success, bringing with it quickly increasing production indices and an increasingly capitalized economy facing new production demands. The primary changes in production have been prompted by the need to meet the increasing demand for inputs to sustain yields. Table 1 shows Indian nitrogen fertilizer consumption, production, and shortfalls between 1961 and 1997.

It is important to note that levels of production and demand vary highly year to year. Some years show nitrogen surpluses while others have large shortfalls. The general trend, however, is toward increased fertilizer consumption, increased state subsidies for fertilizer production and consumption, and insufficient availability. The shortfalls in industrial fertilizer availability vary highly within and between regions. Even in states where fertilizer consumption is relatively low (Madhya Pradesh and Maharashtra both fall below the 74 kg/ha average fertilizer input), there are many areas where low availability of industrial inputs and low capital availability result in higher relative shortfalls. Even within localities, varying access to capital and uneven soil characteristics result in differences in use and availability of appropriate nutrient inputs.

In even the most highly intensive farming operations, however, shortfalls in purchased chemical nutrient inputs are met with inputs in the form of animal waste. Cattle dung (*gober*) and sheep/goat dung (*mingni*) are nutrient sources throughout the world, but have an especially well-evolved role in north and northwest Indian farming systems. Experimental analysis of sheep urine and dung has demonstrated positive effects of dung input on millet production, in terms of threshed panicle, leaf, stem, and yields; many of the effects of annual manure application last two or three cropping seasons after application (Powell et al., 1998).

The direct grazing of animals on fallow fields also enhances production through the application of urine (Semmartin and Oesterheld, 2001). Local produc-

Table 1 National Nitrogen Fertilizer Production and Consumption

	1961	1981	1997
Consumption[a]	210	3,678	10,302
Production[a]	98	2,164	8,599
Import[a]	399	1,510	1,155
Shortfall[a]	−287	4	548
State subsidy[b]	0.00	10.74	158.00

[a] 1,000 tonnes.
[b] 10 million $US.
Source: Government of India (1998).

ers actually prefer these inputs to industrial urea, which they insist "burns" the soil and tends to reduce crop yields in seasons subsequent to application. This effect is likely the result of poor organic carbon availability in north Indian soils, which can be offset by dung inputs; residual organic carbon is evident in soils even several years after application (Williams and Haynes, 1995). Thus animals remain sources of key agricultural inputs, and dung inputs are frequently more expensive than their industrial alternative because of farmers' assessment of the superior results from organic sources of nutrients.

So, too, while tractorization has swept through the region, displacing animal draught power, the highly unstable price of fuel and the maintenance demands for machines make cattle, buffalo, and even camel traction still a viable option for many households. Indeed, the demand for camel traction appears to be quickly outpacing the dwindling breeding stock (Kohler-Rollefson, 1999). Clearly, livestock are still an important component in farming systems in terms of consumption, providing key protein sources especially for the rural poor. But the demand for livestock for production purposes has been sustained and indeed has seen overall increases in recent years.

Table 2 shows the growth in livestock populations since independence and demonstrates the relative growth of national herd of small stock in particular. Small livestock, especially goats, are preferred by the rural poor not only for the easy availability of milk and meat proteins they provide, but also for the assurance they provide to quick cash from the growing national and international market for meat (Robbins, 1994, 1999). Goats earn 400–600 Rps each on local markets and represent a $125 million export market for India, comparable to that of chemicals ($147M) or tobacco and tobacco products ($81MUS). Equally important from the producers' point of view, small stock dung provides especially important inputs into agricultural production. This is supported by experimental analysis that shows sheep dung to be more effective than that of cattle in increasing yields (Powell et al., 1998).

The fodder requirements of this large and growing livestock population are met through means quite different from the Western agricultural model. While fodder crops like *jowar* remain important, their overall production has fallen relative to food crops, even while livestock populations have increased (consider

Table 2   Change in Indian Livestock Populations, 1951–1992

	1951	1992	% change
Cattle	155,295	204,584	31.7
Small stock	86,207	166,062	92.6

Sources: Government of India, 1972; Government of India, 2000

changes during the 1980s shown in Table 3). Experiments in the transition to
intensified stall-feeding operations, moreover, especially those based on hybrid
animal varieties, have resulted in consistent failure (Kohler-Rollefson and Rathore, 1998). Thus, the large and growing livestock population is increasingly dependent on fallow crop wastes, forests, and long-fallow grass and scrubland.

In some cases, the planting of high-yielding varieties (HYV) of important
cultivars has been constrained by demands for crop residues and chaff since dwarf
hybrid varieties often provide too little in the way of biomass to meet forage
needs. Many agricultural producers in the semi-arid, cash crop-producing areas
of eastern Rajasthan and Madhya Pradesh, for example, explain that it is necessary
to balance HYV of wheat with more traditional seed bases, in order to provide
adequate fodder. Moreover, even in the high-intensity zones of Punjab and Haryana, migrant herds of goat and sheep are welcomed during fallow periods for the
grazing of stubble and their deposition of dung. In almost all cases, the animals
are allowed to browse for no fee, and in many cases, the herder is compensated,
in grain or cash, for the presence of the animals for even a few days.

The emergent picture of agricultural production in India is one of a specific
"regional modernity" (following Agrawal and Sivaramakrishnan, 2002)—a
unique regionalized hybrid system that neither eschews globalizing influences,
nor mimics them in any simple way. Green Revolution inputs have been adapted
to the specific and peculiar conditions of local producers, giving rise to "hybrid"
technologies. These embrace some facets of high-input/high-yield production,
while retaining many traditional elements, especially those that help to spread
risk and hedge against downturns in rainfall, prices, or input availability.

Farmers have adapted high-intensity systems in order to adopt them, maintaining elements of the extensive system, especially the role of livestock and
crop waste, to make the more "modern" system productive under the prevailing
ecological and economic conditions. The synthesized system functions by (1)
maintaining standing crop waste as a key farm input, (2) providing nutrient inputs
and grazing for a large and growing livestock population, and (3) closely integrating these into agricultural production through the deposition of supplementary
nutrients.

Table 3 Crop Production in Area (thousands of hectares)

	1981	1990	% change
Jowar	15,809.4	14,948.0	−5.4
Wheat	22,278.8	23,456.9	+5.3
Rice	40,151.5	42,176.6	+5.0

Source: Government of India, 1991.

C. Crop Waste, Herding, and Intensification: A Case from Rajasthan's Irrigated Belt

A case example from contemporary North India is demonstrative. The case region is the southern Pali district of Rajasthan, which, unlike many other parts of the state, has undergone a dramatic transition to intensified production in the last decade and a half. Table 4 shows the rapid intensification of production, as demonstrated through an increasing use of HYVs, increasing crop yields, increasing double and triple cropping, and a decline in culturable wasteland as irrigation development has extended cropping to previously long-fallow lands.

The production system of the region, however, continues to rely upon elements of extensive production. These not only persist in the wake of intensification; they account for much of its success. This is because increased application of water is not the only prerequisite for higher yields; the soils of the region are poor in the availability of N and P, and increased cropping mines available nutrients (Singhania and Somani, 1992; Mongia and Bandyopadhyay, 1993). Extension packages that attempt to offset nutrient limits of crop production by external inputs (Pandey et al., 1987) are supplemented by nutrient subsidies in the form of livestock dung, especially that of goats and sheep. Dung is valuable not only in chemical contributions to soil but also in its contribution to more active and diverse soil microbial communities, which perform many services for increasing soil–plant interactions (biological N fixation, P solubilization, etc.). The preference for traditional sources of nutrient inputs is in fact as great or even greater in households engaged in capitalized, intensified, and cash-crop farming than in more traditional extensive-production units. The system of support for nutrient-supplying animals is, moreover, dependent on crop waste from agricultural fields.

Herds with between 40 and 500 animals move between pasture, wastelands, and fallow fields throughout the region, depositing there half of the nutrients harvested through daily browsing. The other half is deposited in household pens where animals are stalled nightly. These materials are collected and stored in piles visible in most villages and sold by the cartload to agricultural producers,

Table 4 A Green Revolution Transition in Pali, Rajasthan, 1980–1999

	1980	1999
HYV wheat (%)	27.0	62.2
Average wheat yield (kg/ha)	1,464	2,487
Culturable waste (ha)	54,817	41,277
Area sown more than once (ha)	42,484	157,132

Source: Government of India, 1999.

commonly growing HYV wheat and maize. The nutrient application is usually sufficient for an entire year, through multiple harvests.

The nutrient-exchange system in the southern part of the district can be calculated from detailed estimates of annual quantities based on interviews with farmers and herders as shown in Fig. 1.

Dung production from night stalling: 84 kg/animal/year.

Fertilizer demand in twice-cropped land per bigha (0.16 ha): 2–4 cattle carts, or 500 to 1,000 kilos, or 6 to 12 animal/years, strewn of fields prior to planting.

Cost of dung cattle cart: 200–225 Indian Rupees (US$4.88–$5.49) paid directly to the animal-holder in cash or value of wheat flour.

With a 1993 regional estimated population of 566,682 small stock, therefore, approximately 47,000 metric tons were harvested in one year. This is equivalent to 190,000 cattle carts applied on between 48,000 and 95,000 bighas (7,700–15,000 hectares). With 28,665 hectares under double cropping in 1992, dung contributes between 27% and 52% of total fertilizer inputs into the cash-crop agricultural system. Many of these nutrients are recycled through supporting the animals on crop waste in the subsequent short-fallow period. The total value of the dung fertilizer economy in these two tehsils alone (though the entire product does not reach market) is in the range of tens of millions of rupees (Robbins, 2002). The contribution of dung deposition from herds directly browsing on

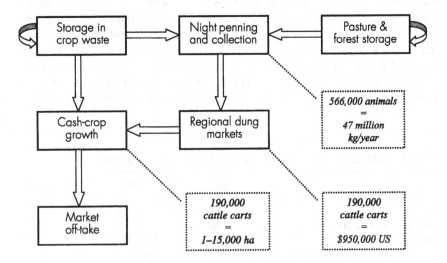

Figure 1   Nutrient cycling in Southern Pali, Rajasthan. (From Robbins, 2002.)

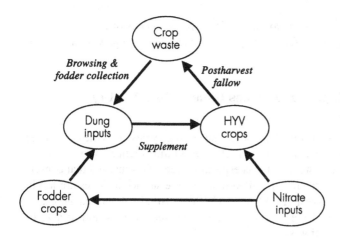

Figure 2 The contemporary production system.

fallow crop waste is not included in this calculation, so it represents a very conservative estimate. The case, moreover, points to the emergence of an intensified system of fallow storage, crop waste, and dung inputs in production that represents an adapted hybrid of Green Revolution production techniques with appropriate and available local natural resources (Fig. 2).

D. Potential Hidden Costs of the No-Till Transition

This is not to argue that local producers do not need more and better means to sustain yields and increase productivity or that the current system is in any way "optimal." Stagnant and falling yields remain a problem for producers, especially as increasing demands for inputs of all kinds, whether locally produced or purchased, narrows their profit margins and makes production more difficult to sustain.

A transition to nonconventional tillage, therefore, could provide many advantages for rural producers. Increases in soil moisture will help to stabilize yields across highly variable climate conditions, especially in the arid and semi-arid parts of the country (Sharma and Acharya, 2000). The number of man-days required for planting can be significantly reduced, and the opportunity costs, therefore, of other kinds of production will be increased (Fowler and Rockstrom, 2001). The degree to which demand for herbicides might rise is beyond the scope of this assessment, but the reduction in tractor demands may well offset other input costs. Animal traction can be maintained in planting, utilizing available resources (Kaumbutho and Simalenga, 1999). In this way, the incorporation of conservation tillage practices offer solutions to many of the specific needs of local producers, even while presenting a risk for the sustainable functioning of many of the current system components.

E. Loss of Key Inputs and Cycles Under a Zero-Till, Dry Mulch Regime

The removal of crop wastes from resource-harvesting cycles, as is the case under zero-till systems, could become a serious change in farm practice, even in intensified Indian production systems. This is especially true if it is insisted that conservation benefits only accrue in zero-till systems where surface residues are left in place following the harvest (Moldenhauer, 1985). Briefly described here is a qualitative assessment of the costs associated with such a transition to a zero-till, dry mulch conservation system.

First, such a transition involves a loss of fodder to support of small and large stock. Small-scale livestock production continues to sustain household economies in the region and represents one of the few avenues for access to cash and animal proteins. Even a small number of goats in a household (<10) can yield several thousand rupees annually with little or no labor input. The efficiency of this part of the farming system depends upon the cycled nutrients of the sort described previously, foraged from fallow fields, which are accessed either through open grazing or crop waste harvest. Costs of purchased fodder inputs are prohibitively expensive and have risen more quickly than that of food grains. In 1997, the wholesale price index for *jowar* (1952–1953 = 100) was 2007, relative to 661 for rice and 1,519 for wheat. The 1999 harvest price for *jowar*, moreover, was 490 rupees/quintal, surpassing traditional subsistence crops like *bajra* (4.13 Rs/Kg), though less than primary cash crops (wheat @ 5.96 Rs/Kg) (Government of India, 1999). Thus, the crucial benefits of livestock production are linked to field biomass availability.

Limits on fodder availability have serious implications for dung production, moreover, and for nitrogen input availability for agricultural production. As the state moves away from subsidized fertilizer markets under the recommendation of a recent state pricing policy review, increases in fertilizer prices are predicted. While this policy will be accompanied by increasing availability of farm credit, the need for cheap and effective nutrient inputs will rise rather than fall as a result (Government of India, 1998). While zero-till will increase yields, many of the benefits accruing will therefore be offset by the combination of rising input costs and loss of dung resources.

Nor will these potential costs to the farm household be unknown or unpredictable to local producers. Previous experience in the African context has shown that adoption of conservation tillage remained slow as long as researchers advocated zero-tillage techniques and retention of crop waste on the field (Fowler and

Rockstrom, 2001). Producers' failure to adopt in this case was clearly linked to rational concerns about hidden costs in production.

F. Gendered Division of Labor and No-Till Practice

A further hidden cost of the transition comes in the area of women's labor. In most north/northwest Indian farm households, although there is wide variation by *jati* and class status, plowing the soil is men's work while field maintenance and weeding are commonly managed by women. Moreover, management of field fodder and dung collection is also commonly women's work, as is small stock management in nonpastoral specialist households. Thus, many of the above-described efficiencies and nutrient cycles are maintained through women's labor.

Reducing immediately available fodder for animals might result in an increased labor burden for women, especially if collection of fodder and dung increasingly become off-farm tasks. Reductions, on the other hand, in hand weeding under conservation tillage may decrease women's labor burden. Overall, however, the potential costs of transition for women will be high if gains are measured only in terms of decreased man-hours for tillage.

G. Adapt or Adopt? A Research Agenda for Incorporating Local Tillage Practice

We see, therefore, that the current agropastoral system of production in North India depends heavily upon the integration crop residues, animal wastes, grazing-land nutrients, and intensive crop production. For this system to be changed, there will need to be adaptations of no-till principles to local socioecological conditions. Some form of conservation tillage system, with shallower cultivation, reduced fallow time, and weed-seed control combined to decrease soil moisture loss and erosion, has its attractions. A radical transition to an idealized zero-till system, with dry and green mulching, however, introduces economic and social costs that most producers will find prohibitive. A new research agenda is required, therefore, to learn how to adapt conservation tillage to the regional agropastoral system.

H. Research Mandate I: Conservation Tillage with Crop Residue Harvesting Variations

To the degree that increased yields depend heavily on the retention of crop waste on the field as a form of dry mulch, as is argued in much of the literature (Moitra et al., 1996; Singh et al., 1998), current configurations of conservation tillage are incompatible with the farming systems of the region's producers. Crop residues are a crucial input and throughput in agrarian nutrient cycling and production.

Experimental crop research has already demonstrated, however, that in many cases minimum tillage, with or without straw, can enhance soil moisture conservation and moisture availability during crop growth. For many crops, including chickpea grain, yield is statistically similar for zero-tillage and minimum tillage. While the absence of mulch appears to reduce soil moisture somewhat, it is not consistently a requirement for improved yields through conservation tillage (Rathore et al., 1998).

This research should be extended to *in situ* production systems, examining methods that encourage minimal tillage while retaining the possibility of residue harvesting and the removal of some or all green and dry mulch from the field. Participatory field appraisals that view all standing biomass as a local resource must come prior to any extension or "education" work in conservation tillage.

I. Research Mandate II: Conservation Tillage Compatibility with Stubble Grazing

Almost no research has focused thus far on the potential compatibility of on-field grazing and browsing with conservation tillage. This seems a serious oversight, given the input values of dung and urine in agricultural production, the preference for on-field grazing by most producers, and the compatibility of this system with patterns of animal movement and rotation through the area resulting from growth in the regional herd. What research has been conducted appears to confirm that intercropping of fodder species and on-field extraction can be incorporated into some forms of conservation practice (Morris et al., 1998; Rao and Phillips, 1999).

Again, however, *in situ* research is lacking, especially in the Indian context, where varying the grazing/browsing behaviors of cattle, sheep, and goats leads to varying effects on standing biomass and rootstock. Field-based examinations of stocking rates and rotations are imperative to gain a better understanding of nutrient and energy exchanges between on-field and off-field grazing/browsing.

J. Research Mandate III: Integrated Fodder Production and Silvicultural Techniques

As conservation removes pastoral resources from the regional farming system, especially for small stock whose dung will remain an important soil nutrient supplement, these costs must be offset with on-farm production systems for fodder. Since intensified and capitalized animal management schemes have shown no significant success, moreover, the prospect for stall feeding with purchased fodder seems limited.

As a result, integration of tree fodder crops in silviculture and agroforestry into conservation tillage is imperative (Devendra, 1999). No-till research should examine combinations of tree culture, fodder harvesting, and *in situ* browsing/grazing. To determine relative productivity of varying arrangements, field tests should evaluate the incorporation of different local tree species, different densities of tree planting on conservation plots, and different systems of seasonal harvesting and coppicing of both tree fodder and on-field mulch, including indigenous shrubs and herbaceous plants. By recreating the elements of nutrient cycling displaced in the implementation of no-till cultivation, the high costs of transition might be avoided.

K. Research Mandate IV: Analysis of Gender, Labor, and Control

While some conservation innovations are arguably "gender-neutral" and more dependent on matters of household wealth, resource endowments, and labor availability (Bonnard and Scherr, 1994), the gender differences that exist in control over decision-making and labor in agricultural production suggest that conservation tillage holds implications for household social relations.

Research is required that examines the time and labor constraints under current tillage regimes and the areas where decreases or increases in labor demand are created through conservation practices. Specifically, will weeding tasks decrease or increase? Will off-field fodder collection demands increase or decrease? By assessing the hidden costs of transition for differing members of households, extension efforts to expand no-till adoption will be better positioned to adapt and offset inequitable solutions. This will likely require the provision of alternative resources to decrease labor, the alteration of elements of the conservation plan to suit household labor constraints, and the involvement of women in decision-making at the outset.

L. Learning from Local Practice

In sum, the introduction and incorporation of conservation tillage in South Asia must be seen as a process that incorporates, and is not hostile to, tillage practices currently in place in the region. Field-based research will need to consult resident expert farmers, use simplified trial designs, respect farmer knowledge, and measure the very real and immediate costs of adoption against the somewhat more distant and unrealized potentials (Fowler and Rockstrom, 2001). The traditional prevailing extension approach—where a farmer's selection of some elements of conservation and not others is viewed as a failure in adoption (see, for example, Nowak, 1985)—must therefore be abandoned in favor of a more plural and less restrictive adoption policy.

Accordingly, approaches to no-till extension should be formulated from bottom to top. The experiment station and laboratory science components need

to be fitted to field-based questions based on farmers' needs and regional impera-
tives. Extension officers should be better trained to listen for location-specific
issues and priorities, adapting themselves to the realities of production, rather
than vice versa.

Only if no-till science and extension come to resemble farmer practice—where no single farming system prevails but instead a vast information
system is geared to immediate and changing needs—can it hope to succeed on
a large scale, reaching and benefiting most of the agricultural community. The
alternative is a potentially expensive and impractical failure in an area of development where there is little room for error.

REFERENCES

Agrawal, A. 1999. Greener Pastures: Politics, Markets, and Community among a Migrant
    Pastoral People: Duke University Press. Durham, NC.
Agrawal, A., Sivaramakrishnan, K., Eds. 2002. Regional Modernities: Yale University
    Press. New Haven, CT.
Bonnard, P., Scherr, S. 1994. Within-gender differences in tree management: Is gender
    distinction a reliable concept? Agroforestry Systems 25(2):71–93.
Boserup, E. 1965. Conditions of Agricultural Growth: The Economics of Agrarian Change
    Under Population Pressure: Aldine. Chicago.
Brush, S.B., Turner, B.L.I. 1987. The nature of farming systems and views of their change.
    In Comparative Farming Systems. Turner B.L.I., Brush S.B., Eds: The Guilford
    Press. New York.
Chayanov, A.V. 1986. The Theory of Peasant Economy: University of Wisconsin Press.
    Madison, WI.
Cincotta, R., Pangare, G. 1994. Population growth, agricultural change, and natural re-
    source transition: Pastoralism amidst the agricultural economy of Gujarat: Overseas
    Development Institute, Pastoral Development Network, 36a.
Devendra, C. 1999. The relevance and implications of livestock-tree interactions in agro-
    forestry systems in developing countries. Annals of the Arid Zone 38(3–4):399–414.
Fowler, R., Rockstrom, J. 2001. Conservation tillage for sustainable agriculture: An agrar-
    ian revolution gathers momentum in Africa. Soil and Tillage Research 61, 93–107.
George, S. 1990. Agropastoral equations in India: Intensification and change of mixed
    farming systems. In The World of Pastoralism. Galaty J.G., Johnson D.L., Eds:
    Guilford Press. New York, pp. 119–144.
Government of India. 1972. All-India Statistical Abstract: Government of India. Delhi.
Government of India. 1991. Area and Production of Principal Crops in India: Directorate
    of Economics and Statistics. Delhi.
Government of India. 2000. All-India Statistical Abstract: Government of India. Delhi.
Government of India Department of Economics and Statistics. 1999. Statistical Abstracts
    Rajasthan: Directorate of Economics and Statistics. Jaipur.
Government of India, Ministry of Finance. 1998. Economic Survey: 1997–98: Government
    of India. New Delhi.

Habib, I. 1999. The Agrarian System of Mughal India: Oxford University Press. Oxford.

Kaumbutho, P.G., Simalenga, T.E., Eds. 1999. Conservation Tillage with Animal Traction: Animal Traction Network for Eastern and Southern Africa. Harare. Zimbabwe.

Kohler-Rollefson, I. 1999. Personal communication. Discussion concerning Pushkar camel fair 1999.

Kohler-Rollefson, I., Rathore, H.S. 1998. NGO Strategies for Livestock Development in Western Rajasthan (India): An Overview and Analysis: League for Pastoral Peoples. Ober-Ramstadt. Germany.

Lal, R. 2000. Soil management in the developing countries. *Soil Science* 165(1):57–72.

Leaf, M. 1987. Intensification of peasant farming: Punjab in the Green Revolution. In *Comparative Farming Systems*. Turner B.L.I., Brush S., Eds: Guilford Press. New York, 248–275.

Moitra, R., Ghosh, D.C., Sarkar, S. 1996. Water use pattern and productivity of rainfed yellow sarson (*Brassica rapa L. var glauca*) in relation to tillage and mulching. *Soil and Tillage Research* 38(1–2):153–160.

Moldenhauer, W.C. 1985. A comparison of conservation tillage systems for reducing soil erosion. In *A Systems Approach to Conservation Tillage*. D'Itri F.M., Ed: Lewis Publishers. Chelsea, MI, 111–122.

Mongia, A.D., Bandyopadhyay, A.K. 1993. Soils of the Tropics: Vikas Publishing House. New Delhi.

Morris, J.L., Allen, V.G., Vaughan, D.H., Luna, J.M., Cochran, M.A. 1998. Establishment of corn in rotation with alfalfa and rye: Influence of grazing, tillage, and herbicides. *Agronomy J.* 90(6):837–844.

Nowak, P.J. 1985. Farmers' attitudes and behaviors in implementing conservation tillage decisions. In *A Systems Approach to Conservation Tillage*. D'Itri F.M., Ed: Lewis Publishers. Chelsea, MI, 327–340.

Pandey, P., Prasad, M., Oraon, A. 1987. The impact of new technology on dryland agriculture In *Dryland Agriculture in India*. Shafi M., Raza M., Eds: Rawat Publications. New Delhi, 79–98.

Powell, J.M., Ikpe, F.N., Somda, Z.C., S., Fernandez-Rivera 1998. Urine effects on soil chemical properties and the impact of urine and dung on pearl millet yield. *Experimental Agric.* 34(3):259–276.

Rao, S.C., Phillips, W.A. 1999. Forage production and nutritive value of three lespedeza cultivars intercropped into continuous no-till winter wheat. *J. Production Agric.* 12(2):235–238.

Rathore, A.L., Pal, A.R., Sahu, K.K. 1998. Tillage and mulching effects on water use, root growth and yield of rainfed mustard and chickpea grown after lowland rice. *J. the Science of Food and Agric.* 78(2):149–161.

Robbins, P. 1994. Goats and grasses in Western Rajasthan: Interpreting change: Overseas Development Institute Pastoral Development Network 36a, 6–12.

Robbins, P. 1999. Meat matters: Cultural politics along the commodity chain in India. *Ecumene* 6(4):399–423.

Robbins, P. 2002. Pastoralism inside-out: The contradictory conceptual geography of Rajasthan's Raika, *Nomadic Peoples*, forthcoming.

Royal Commission on Agriculture in India. 1928: Report of the Royal Commission. London.

Semmartin, M., Oesterheld, M. 2001. Effects of grazing pattern and nitrogen availability on primary productivity. *Oecologia* 126(2):225–230.

Sharma, P.K., Acharya, C.L. 2000. Carry-over of residual soil moisture with mulching and conservation tillage practices for sowing of rainfed wheat (*Triticum aestivum* L.) in northwest India. *Soil and Tillage Res* 57(1–2):43–52.

Singh, B., Chanasyk, D.S., McGill, W.B. 1998. Soil water regime under barley with long-term tillage-residue systems. *Soil and Tillage Res* 45(1–2):59–74.

Singhania, R.A., Somani, L.L. 1992. Soils of Rajasthan. In *Geographical Facets of Rajasthan*. Sharma H.S., Sharma M.L., Eds: Kuldeep Publications. Ajmer, pp. 55–67.

Unger, P.W. 1984. Tillage Systems for Soil and Water Conservation: Food and Agriculture Organization of the U.N., Rome.

Williams, P.H., Haynes, R.J. 1995. Effect of sheep, deer and cattle dung on herbage production and soil nutrient content. *Grass and Forage Science* 50(3):263–271.

Zimmerman, F. 1987. The Jungle and the Aroma of Meats: An Ecological Theme in Hindu Medicine: University of California Press. Berkeley.

28

Innovation, Degradation, and Conservation in South Asia's Rice–Wheat System: Farmers' Choices and Lessons from U.S. Experience

Keith D. Wiebe and Paul Heisey
USDA Economic Research Service
Washington, D.C., U.S.A.

Recent analyses indicate that resource degradation is adversely affecting agricultural productivity in South Asia, particularly in the rice–wheat system. Despite current food grain surpluses driven by price policies, a number of factors—including poverty, population growth, and environmental concerns—mean that it remains important to sustain productivity over the long term. Many factors have contributed to past productivity growth in South Asia's rice–wheat system, including advances in agricultural research and development, expansion of irrigation infrastructure, and increased use of inorganic fertilizer. Likewise, many factors will be necessary to sustain productivity in the future.

We focus here on the social and economic factors that influence farmers' choices regarding resource-conserving management practices. Critical among these are the timing and magnitude of costs and returns to the adoption of alternative practices, and the duration and security of farmers' property rights in land (as recognized in a long South Asian tradition of economic analysis on such

The views expressed in this chapter are those of the authors, and may not be attributed to the Economic Research Service or the U.S. Department of Agriculture.

Table 1   Area in Rice–Wheat Rotations,
South Asia, 1988

Country	Rice–wheat area (million ha)
Bangladesh	0.5
India	9.1
Nepal	0.5
Pakistan	1.6
South Asia	11.7

Sources: Hobbs and Morris (1996), based on
Huke et al. (1994a, 1994b); Woodhead, Huke, and
Huke (1994); and Woodhead et al. (1994).

issues). Data requirements for careful analysis of technology adoption patterns
are demanding and remain scarce for South Asia. Evidence from the United States
suggests that an increasing number of farmers are finding it profitable to adopt
conservation tillage (CT), but that their choices depend significantly on characteristics of farmers, land, technology, markets, and policy.

I.   THE RICE–WHEAT SYSTEM IN SOUTH ASIA

A.   The Biophysical Environment

The rice–wheat cropping system of South Asia is the dominant system across
the Indo-Gangetic Plains of northern India and Pakistan. The system is also important in the Terai region of Nepal and much of northwestern Bangladesh. It is
characterized by a *kharif*, or "summer" (monsoon), wetland rice crop followed
by a *rabi*, or cool, dry season wheat crop. In some areas, additional crops may
enter the rotation, for example, a second crop of rice, an oilseed crop between
rice harvest and wheat planting, or a legume crop between wheat harvest and
rice transplanting (Hobbs and Morris, 1996). Analysis of district-by-district crop
data indicated that by about 1990 about 12 million hectares were planted annually
to the rice–wheat rotation, over three quarters of which were in India (Table 1).*

South Asia's rice–wheat area is generally subtropical, ranging from humid
in the east to arid in the west. Despite this general pattern, winter rainfall tends
to follow the reverse direction, somewhat higher in the northwest than in the
southeast. The availability of winter rainfall, as well as residual moisture from

* Recent Government of India data on wheat and rice production in India suggest that the rice–wheat
cropping system may now cover a larger area.

the preceding rice crop (which is partially dependent on soil type), largely determines the feasibility of producing a wheat crop without supplemental irrigation. Temperatures are also higher to the east and south. This limits the growing season for wheat (Wood et al., 2000; Huke and Huke, 1992; Hobbs and Morris, 1996).

A substantial proportion of the agricultural area is estimated to be relatively free of inherent soil constraints, and most soils in the region are highly productive, although the soils are quite variable. The soils of the Indo-Gangetic floodplain tend to be calcareous and characterized by high pH levels, compared with the floodplain soils in Bangladesh, which are noncalcareous and more acidic. Soils in more elevated areas are coarser, while lower-lying soils are finer (Hobbs and Morris, 1996; Wood et al., 2000). Despite the good quality of many soils, agricultural intensification has often been associated with soil and water degradation in the rice–wheat system. Soil degradation results from erosion, chemical degradation (including salinization, fertility decline, and reduced organic matter), and physical degradation (including compaction, crusting and sealing, and waterlogging). Soil degradation may be more problematic in the western part of the rice–wheat system (Wood et al., 2000). Water puddling for rice production tends to influence soils in ways that are "favorable for rice but unfavorable for wheat" (Hobbs and Morris, 1996, p. 30). Water problems include both declining availability with lowered water tables, again particularly noticeable in the northwest portion of the rice–wheat zone, and declining quality caused by salts and minerals, particularly related to the use of tubewell irrigation (Hobbs and Morris, 1996).

B. The Evolution of Production Technology

Since about 1950, an interrelated series of technological changes has profoundly influenced the size and performance of the rice–wheat system in South Asia. Starting around 1950, irrigated rice area began to expand. The area devoted to irrigated wheat also began to expand slowly at that time and accelerated in the late 1960s. Whereas in 1950, only about one third of India's total wheat area (at that time about 10 million hectares) had some form of irrigation, by the early 1990s about 83% of a wheat area of almost 25 million hectares was irrigated. This expansion was fueled both by a conversion of irrigated area to wheat and rice and, perhaps more importantly, by conversion of previously unirrigated land to irrigation. Over all of South Asia, at least half of the total rice area is now irrigated, and this fraction is undoubtedly higher in the rice–wheat system (Hobbs and Morris, 1996; Rejesus et al., 1999). Following the Green Revolution diffusion of high-yielding varieties (HYVs) of rice and wheat that began in the late 1960s, an increasing proportion of the investment in new irrigation was in tubewells rather than in new large-scale irrigation systems.

This acceleration of irrigated area from the late 1960s was intimately tied to the best-known new technology to diffuse through the rice–wheat system, the rapid spread of high-yielding varieties of rice and wheat. These new semidwarf varieties in many cases performed at least as well as the tall varieties they replaced at low levels of fertilizer application and more erratic water availability. They diffused particularly rapidly because they were also more responsive to fertilizer application and better water control. Over a wide range of input levels, the marginal response of the HYVs to fertilizer and to water was greater than the marginal response of the varieties they displaced.

In India, diffusion of HYV rice began in the late 1960s and adoption reached more than three quarters of the total area by the early 1990s. Despite the rapid diffusion of HYV rice in the rice–wheat system, it did not quite match the speed with which HYV wheat diffused. The slightly slower diffusion of rice resulted particularly because under some market conditions it was more profitable to grow traditional high-quality *basmati* aromatic rice varieties, produced primarily for export (Hobbs and Morris, 1996). Beginning in Pakistan during the late 1980s, improved high-value rice varieties became available to farmers and spread rapidly (Sharif et al., 1992). Diffusion of HYV wheat also began in the late 1960s. By the late 1990s, over 90% of the total wheat area in South Asia was planted to HYVs. One hundred percent of the irrigated wheat area was planted to these high-yielding, semidwarf varieties (Heisey et al., 2002).

The third major component of technical change in the rice–wheat system was the rapid increase in the use of inorganic fertilizer. In India, Pakistan, and Bangladesh, the average amount of inorganic nutrients applied to rice was well under 10 kg ha^{-1} in the early 1960s. By the late 1980s, the average application rate of fertilizer to rice was nearly 70 kg ha^{-1} in Bangladesh and almost 90 kg ha^{-1} in India and Pakistan (Hobbs and Morris, 1996). Fertilizer use on wheat in India and Pakistan was also very low in the 1960s but stood at nearly 130 kg nutrients ha^{-1} for all of India, and about the same level for Pakistan's Punjab, by the mid-1990s.[*] In some of the most intensively cultivated areas, where fertilizer and HYV wheat diffused earliest and fastest, fertilizer application rates were even higher. In the Indian Punjab, total nutrients applied to wheat reached about 200 kg ha^{-1} by the late 1980s and early 1990s and appeared to be leveling off at about that amount (Rejesus et al., 1999).

II. EVIDENCE OF RESOURCE DEGRADATION IN THE RICE–WHEAT SYSTEM

The diffusion of HYV rice and wheat, the rapid increase in fertilizer use, and the expansion of irrigated area had many important consequences in South Asia. Total area planted to these crops increased steadily but relatively slowly over the 1960s, 1970s, and early 1980s. Area expansion for these crops has shown many

---

[*] In Pakistan's Punjab, irrigated wheat area is about 90% of the total, and fertilizer application rates are higher on irrigated than on rainfed wheat. (This estimate is for the entire Punjab of Pakistan.)

signs of leveling off since the 1980s. The area in the rice–wheat system appears to have grown somewhat faster than the rate of growth in aggregate areas planted to these two crops. Despite moderate and decreasing growth in crop areas, total production of rice and wheat in South Asia grew dramatically, fueled by high rates of growth in yield following the introduction of HYVs (Hobbs and Morris, 1996).

A. Yields

For a variety of reasons, there were significant differences in the patterns of yield growth among different countries and between different regions within countries. In general, the earliest adoption of Green Revolution technologies was in the more favorable production environments of northwest India and Pakistan, and rapid yield growth began in those areas.[*] For example, both rice and wheat yields grew much faster in the Punjab of India than in the state of Bihar in eastern India. At the beginning of the Green Revolution, rice yields were roughly equal in both states, between 1 and 1.3 mt ha^{-1}, but by 1990 rice yields in Bihar were still less than 2 mt ha^{-1}, while they were fluctuating around 3 mt ha^{-1} in Punjab. Wheat yields were already higher in Punjab (1.2 mt ha^{-1}) than Bihar (0.8 mt ha^{-1}) before the Green Revolution. By 1990, the gap had increased, with yields over 3.5 mt ha^{-1} in Punjab, and between 1.5 and 2 mt ha^{-1} in Bihar.

In recent years, however, concerns have been raised about the sustainability of agricultural intensification following the Green Revolution, both because of unease about the high use of external inputs and due to fears about degradation of the resource base (Faeth, 1993). One cause for concern is that in the most intensively cropped Green Revolution areas the "economically exploitable gap" has narrowed between yields on experiment stations and the best yields in farmers' fields (Pingali and Heisey, 2001). This certainly appears to be the case in parts of the rice–wheat system, particularly in areas like Northwest India (Byerlee, 1992; Hobbs and Morris, 1996).

The first real concerns about the effects of possible soil and water degradation were raised by long-term yield trials, in which input levels were kept constant over time. In many cases, these trials showed declining yields at the same input levels (Hobbs and Morris, 1996). The first attempts to confirm declining yields in farmers' fields often calculated partial factor productivity measures, for example, by dividing crop output over fertilizer application rates and viewing this ratio over time. Quite naturally, this measure of average fertilizer productivity fell over time, as would be expected if fertilizer response had the standard concave shape and rates of fertilizer application increased over time. A more relevant partial

---

[*] One qualification was that there was somewhat slower growth of rice yields in areas where high-quality but low-yielding basmati rice was important.

factor productivity measure would be the marginal fertilizer response—the change in crop output in response to an incremental increase in fertilizer—holding all other inputs constant. If, over time, the marginal response to fertilizer fell when measured at the same level of fertilizer application, this could indicate resource degradation (Byerlee, 1992). This kind of calculation has shown more ambiguous results in the rice–wheat system. In Haryana, India, for example, the marginal fertilizer response of rice might have declined between the late 1970s and the late 1980s, but the marginal fertilizer response of wheat may have increased (Chaudhary and Harrington, 1993).

B. Total Factor Productivity (TFP)

To avoid some of the ambiguities caused by analysis of partial factor productivity measures, such as yield or output per unit of fertilizer, economists have increasingly turned to total factor productivity indices to assess the sustainability of particular agricultural systems (Lynam and Herdt, 1989).* These indices are not without their own problems. For example, Byerlee and Murgai (2002) argue that total factor productivity can be a conceptually flawed measure of sustainability since inclusion of nonmarket inputs and outputs and purported social prices for these inputs and outputs often violate the theoretical basis of TFP estimates. To better analyze agricultural sustainability, they recommend excluding nonmarket inputs and outputs from the estimation of TFP, and then relating trends in the resulting measure of TFP to underlying trends in technology, human capital, physical infrastructure, and indicators of resource quality. Furthermore, Murgai (2001) contends that in periods of rapid factor-biased technological change, standard methods of TFP calculation will underestimate the rate of TFP growth because it is not possible to separate the contributions of technical change from those of greater factor use without knowing the parameters of the underlying production technology.

Most studies of agricultural TFP focus on the aggregate agricultural sector, or on crops or livestock production in the aggregate. However, a few TFP analyses have been conducted that are relevant either to the rice–wheat system of South Asia or to the two major crops, rice and wheat. Ali and Velasco (1993) found that TFP in rice–wheat systems in Pakistan declined between 1970 and 1989, with the rate of decline increasing in the 1980s. Cassman and Pingali (1995) found that in Ludhiana District, Punjab, India, TFP increased steadily until the mid-1970s, then leveled off for about 10 years before beginning a slow decline. Mohan Dey and Evenson (1991) argued that in Bangladesh, despite slowdowns

---

* Some economists prefer the term "multi-factor productivity," since such indices cannot generally incorporate *all* inputs (or outputs). The studies cited in this section use the term TFP, however, so we follow this convention.

in yield growth toward the end of the period, TFP in both rice and wheat grew for much of the period from 1952 to 1989. In the Punjab of India, Sidhu and Byerlee (1992) found that TFP in wheat fluctuated around the same level during the initial Green Revolution, but grew at 2% annually immediately after the Green Revolution, in the so-called input-intensification period. Kumar and Mruthyunjaya (1992) calculated rates of TFP growth for wheat in five states in Northwest India. Though results varied, their analysis showed patterns roughly similar to those calculated by Sidhu and Byerlee. Kumar et al. (1995) concluded that for India as a whole, the rate of TFP growth in wheat fell slightly between the 1970s and the 1980s, but was still over 1% per year in the latter period.

Several recent studies have attempted to refine TFP estimates and to make them more relevant to the questions of resource degradation and sustainability. Fortunately, two of these studies have information pertinent to the rice–wheat system, in particular to the most input-intensive part of that system in northwestern India and Pakistan. Murgai (2001) contrasted traditional TFP calculations for the Indian Punjab with estimates that took into account the factor-biased (land-saving) nature of Green Revolution technical change. The traditional estimates showed higher TFP growth in the input-intensification phase immediately following the Green Revolution, and then lower TFP growth in the so-called Post-Green Revolution period that followed the input-intensification period. Corrected estimates showed considerably higher TFP growth initially, during the Green Revolution, making the subsequent decline in TFP growth appear much more pronounced.

Although Murgai (2001) hypothesized that some of the decline in the rate of TFP growth might be attributed to declining quality of soil and water, Ali and Byerlee (2001) calculated TFP estimates for various cropping systems in Pakistan's Punjab that explicitly took into account measures of soil and water quality over time. All these systems included wheat as one of the crops evaluated; they differed in the crop with which wheat was rotated. The rice–wheat system was one of those analyzed. The data presented by Ali and Byerlee showed that at least three measures of soil quality and two measures of tubewell water quality worsened over time (Figs. 1a and 1b). Their estimates showed that between 1966 and 1994—a period covering the Green Revolution, input-intensification, and Post-Green Revolution phases—TFP growth was positive in all systems except for the rice–wheat system. Although TFP growth finally turned positive in the rice–wheat system in the Post-Green Revolution period, this was not enough to counteract the negative trends of the preceding years. In all systems, but particularly in the rice–wheat system, the negative trends in soil and water quality were strong factors counteracting the positive effects of technological change and public investment in roads and literacy on TFP growth. Such trends pose clear challenges for policymakers concerned with shaping farmers' choices to sustain productivity growth in South Asia.

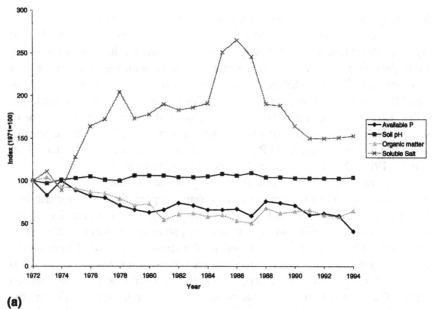

(a)

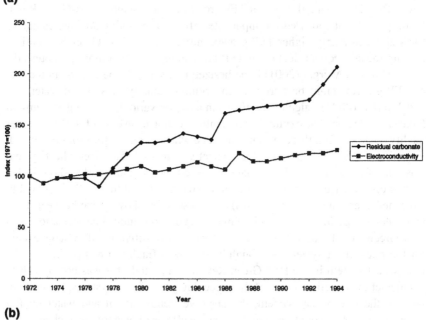

(b)

Figure 1   Indices of trends in (a) soil quality and (b) tubewell water quality in Punjab, Pakistan. (From Ali and Byerlee, 2001.)

III. CONSERVATION TECHNOLOGIES FOR SOUTH ASIA'S RICE—WHEAT SYSTEM

Current conservation technology options for the rice–wheat system in South Asia are analyzed in detail in other chapters of this book. For the most part, they focus on one key agronomic element of the system: tillage and related seeding practices. The puddled anaerobic soils that have traditionally been thought to be optimal for rice production have traditionally been converted to aerobic soils for wheat production through plowing.* This has led both to delays in wheat planting and to reduced seed germination in the wheat crop. Technologies aimed at addressing these problems fall into two general categories: (1) no-tillage (NT) or reduced tillage with various options for wheat seeding; and (2) more significant modification of water and soil management through raised beds. Though these technologies lead to further short-run management tradeoffs, preliminary evaluation of some of the minimum tillage options suggests favorable short-run cost and yield effects. Interactions between machinery investment costs, farm size, land tenure, and equipment rental markets are likely to be important in the intermediate run and have yet to be fully analyzed. Longer-term effects on resource quality indicators and general equilibrium economic effects also deserve further consideration.

IV. SOCIAL AND ECONOMIC FACTORS AFFECTING FARMERS' CHOICES

Characteristics of the biophysical environment and of the state of technology clearly affect farmers' choices regarding crops and cropping practices. Social and economic factors are also important. In any particular production environment, for example, different technologies involve different costs and benefits. Farmers consider not only the magnitude but also the timing of these costs and benefits when comparing alternative technologies and practices. Such considerations drove the dramatic increases in irrigation, use of HYVs, and inorganic fertilizer noted above in South Asia's rice–wheat region.

In making such comparisons, farmers generally have less incentive to consider costs and benefits that they do not feel directly than those that they do. For example, resource degradation may result if farmers disregard adverse impacts on land or water that they believe will occur only far in the future or beyond the boundaries of their land. Careful understanding of farmers' incentives is thus critical if we are to better understand the likelihood that resource degradation will occur, the likely economic and environmental consequences if degradation

---

* As discussed in Chapter 8, there is now disagreement on the desirability of keeping irrigated rice soils continuously flooded, i.e., anaerobic.

does occur, and the various ways in which these consequences can be mitigated or avoided.

Pagiola (1999) provides a simple but useful graphical framework for comparing net returns to alternative practices/technologies over time (Fig. 2). Note that practices generating higher net returns today may not do so indefinitely and (conversely) that practices offering higher net returns over time may require initial investments and thus generate lower returns in the short term. This is likely to be the case for many conservation practices. (Note also that returns received in the future are generally worth less than the equivalent nominal amount received today and must thus be discounted at an appropriate rate.)

Such a tradeoff between short-run and long-run net returns introduces a number of critical social and economic factors into the farmer's decision-making framework. Perhaps most basically, in order to benefit from a conservation practice that requires an initial investment, the farmer must have some expectation that he or she will continue farming a particular plot of land long enough to realize the benefit. This introduces the factor of tenure security. (Tenure-related influences on farmer behavior have long been the subject of theoretical analysis by South Asian scholars, including Pranab Bardhan, Kaushik Basu, Mukesh Eswaran, and Ashok Kotwal.) A farmer with a lease that expires at time t_1, for example, receives only a fraction of the benefit that would be realized by a farmer with a lease that runs through t_2, and both of them receive less benefit than would

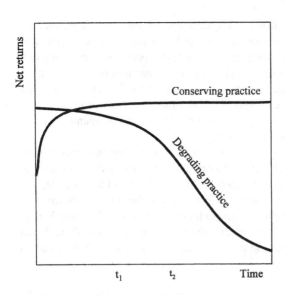

Figure 2   Private net returns to alternative practices. (From Pagiola, 1999.)

a farmer who owns his or her land (in perpetuity). Similarly, a farmer whose ownership of a parcel of land is secure would be more confident of realizing long-term benefits from adoption of a conservation practice than would a farmer whose ownership is being contested (whether through formal or informal channels).

Assuming secure tenure and long-term gains in net returns to be realized from adoption of a conservation practice, a farmer might still be prevented from adopting the practice if he or she is unable to afford the initial investment required. This might be the case because of poverty, for example, or because of constraints on access to credit. Even with sufficient cash reserves or access to credit, a farmer might lack the information necessary to carefully compare long-run costs and benefits to alternative practices, particularly in situations characterized by a high degree of market or environmental uncertainty.

All these factors illustrate the point that technologies or practices are not uniformly optimal or inferior for all farmers, even within a given production environment. Rather, individual farmers make their own decisions regarding alternative technologies based on biophysical characteristics such as soil quality and access to water, as well as social and economic characteristics that include land tenure, income and wealth, and access to credit and information. Careful empirical analysis of these decisions is difficult (and thus scarce), both because data requirements are demanding and because many of these factors are causally interrelated. Data on the adoption of reduced tillage practices in South Asia's rice–wheat system are scarce, so we turn to the United States for a closer look at factors that affect farmers' choices.

V. ADOPTION OF CONSERVATION PRACTICES IN THE UNITED STATES

According to the FAO (2001), "conservation agriculture," of which NT or minimum tillage is an important element, is currently practiced on about 57 million hectares (about 3% of cultivated area worldwide). Most of this area is located in North America and Latin America. Support from public, private, and farmer NGOs is cited as important, but demonstrated improvements in long-run net returns are probably the deciding factor in most farmers' decisions. Evidence suggests that adopters have—on average—experienced less soil erosion, improved soil quality, increased cropping intensities and yields, reduced input costs, and higher and more stable incomes relative to farmers using conventional practices.

No-till agriculture currently accounts for 21 million hectares in the United States, about 18% of total planted area and nearly half of the 44 million hectares under conservation tillage (CT) (Magleby et al., 2002). No-till area has expanded almost four-fold since 1989, accounting for virtually all of the increase in CT. Use of NT (and CT overall) is highest for corn, soybeans, and small grains (Fig. 3), and thus it is most extensive in the Corn Belt, the Northern Plains, and the Appalachian region of the east-central United States.

Among the direct benefits cited in relation to adoption of NT or other forms of CT are improved moisture retention (especially under dry conditions), lower costs for machinery, fuel, and labor (which may be partially offset in some cases by higher chemical costs), increased opportunity for double-cropping and timelier planting as a result of reduced field preparation time, and thus economic returns that are generally higher and more stable. Magleby et al. (2002) note, however, that actual results depend critically on farmers' choices and other underlying factors.

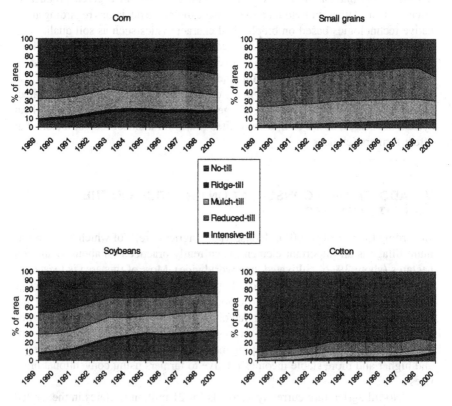

Figure 3 Tillage trends on major crops in the United States. (From Magleby et al., 2002.)

To analyze farmers' choices about conservation practices, it is necessary to take a long-run perspective that captures biophysical and economic interactions. One such approach is to formalize our earlier discussion of farmers' choices using a dynamic economic analysis. For example, Hopkins et al. (2001) analyze optimal levels of two practices—fertilizer application and residue management—to address irreversible soil erosion and reversible nutrient depletion on nine different soils in the north-central United States.

In general, they find that differences in optimal choices across soils and in soil nutrient levels exceed the differences across erosion phases (topsoil depth) within a given soil. Optimal residue levels generally increase (at a decreasing rate) with soil nutrient levels but vary only slightly with soil depth, indicating that the benefits of residue management derive primarily from protecting soil nutrient stocks rather than from slowing the rate of soil loss. Likewise, it is generally optimal to apply fertilizer at relatively high rates to build up nutrient stocks when they are low, and to apply no fertilizer and draw nutrient levels down when they are high, but optimal application rates vary little with soil depth. Optimal strategies also vary with the length of the farmer's planning horizon. For a farmer with a horizon of only a few years, it would be optimal to deplete soil nutrient levels and disregard residue management (corresponding to the "degrading practice" in Fig. 2), but for a farmer with a longer planning horizon, it would be optimal to sacrifice some short-term returns by choosing fertilizer and residue management levels that sustain yields over the longer term (corresponding to the "conserving practice" in Fig. 2).

Soule and Tegene (2003) analyze land tenure and other factors affecting the adoption of conservation practices among 941 U.S. corn producers in 1996 and 1,417 U.S. soybean producers in 1997. Patterns of adoption vary by crop, tenure, conservation practice, and other factors (Table 2). Adoption of CT is significantly and positively associated with farm size, with the proportion of the farm in corn or soybeans, and with designation as highly erodible land (HEL) for both corn and soybeans. For corn, younger and more educated farmers are more likely to adopt CT, while limited-resource or part-time (LRPT) farmers and improved drainage are negatively associated with adoption. In addition, corn farmers in the South are more likely to adopt CT than farmers in other regions. For soybeans, adoption of CT is positively associated with government program participation, and farmers in the north-central region are less likely to adopt than farmers in the Corn Belt. Renters as a group do not behave significantly differently than owner-operators, contrary to conventional wisdom, but subdividing this group reveals that cash-renters are less likely than owner-operators to adopt conservation practices for corn production, while share-renters are less likely than owner-operators to adopt conservation tillage when it comes to producing soybeans.

Table 2 Results of Logit Regression Models for Adoption of Conservation Practices, U.S. Corn Producers (1996) and Soybean Producers (1997)

Variable	Conservation tillage		Medium-term practices	
	Corn	Soybeans	Corn	Soybeans
Intercept	4.97[a]	−0.43	7.35[a]	1.13
Farm size	0.02[a]	0.03[a]	−0.02[b]	−0.01
Operator's age	−0.02[a]	−0.01	−0.03[a]	0.00
College education	0.63[b]	0.22	−0.02	0.17
Program participation	−0.54	0.50[a]	−0.04	0.41
LRPT farmer	−0.09[a]	−0.22	−0.59[a]	−0.10
Corn-soybeans area	0.75[a]	1.30[a]	−0.56[b]	−0.36
HEL designation	0.77[a]	0.99[a]	2.48[a]	1.65[a]
Improved drainage	−1.14[a]	0.17	−0.06	0.21
Urban proximity	0.08	0.06	0.13	−0.08
Precipitation	−1.01	0.33	4.53[a]	1.69[b]
Temperature	−0.05	−0.03	−0.22[a]	−0.08[b]
Plains	−1.59[a]	0.40	0.92	0.43
North Central	−2.04[a]	−0.46[b]	0.51	−0.32
Corn Belt	−1.50[a]	—	0.76	—
South	—	−0.12	—	−1.05[a]
Renter	−0.35	−1.11	−0.67[a]	−0.12
Cash-renter	−0.54[b]	0.16	−0.66[a]	−0.29
Share-renter	−0.12	−0.33[b]	−0.68[a]	0.06
% correctly predicted	68.5%	62.2%	73.4%	76.4%

[a] Indicates significance at the 5% level.
[b] Indicates significance at the 10% level.
Source: Soule and Tegene (2003).

For corn producers, adoption of medium-term practices follows similar patterns to adoption of CT. Operator's age and LRPT farmers are significant and negatively associated with adoption, while HEL designation has a positive and significant impact. Farm size, the proportion of the farm in corn or soybeans, and temperature are negative and significant, while precipitation is positive and significant, indicating that such practices occur more frequently in cooler areas where water or wind erosion may be a concern. For soybeans, HEL, precipitation, and temperature were the only significant variables explaining the adoption of medium-term conservation practices. This may be due at least in part to the government policy of conservation compliance, which requires farmers to have a conservation plan for HEL in order to receive certain government payments. The coefficient on the renter dummy variable is negative and significant for corn but not significant for soybeans. Both cash-renters and share-renters are less likely than owner-operators to adopt such practices in corn production, but do not behave significantly differently from owner-operators in soybean production.

VI. POLICY CONSIDERATIONS

Analysis of total factor productivity in the rice–wheat region of northwestern India and Pakistan indicates worrisome trends in resource quality that threaten to undermine productivity growth. A variety of promising technologies is being developed to address this threat, but their ultimate success in stemming resource degradation and enhancing productivity growth will depend on the extent to which they are adopted by farmers. This will depend in turn on the magnitude and timing of costs and benefits that alternative technologies offer to farmers, whose incentives and opportunities will vary with differences in soils, terrain, climate, and hydrology.

South Asia and the United States obviously differ in significant ways in terms of biophysical, social, institutional, and economic characteristics. Nevertheless, farmers in the two regions share a common interest in sustaining and increasing productivity over the long term. Evidence from empirical analyses of data available in the United States thus provides broadly applicable insights into the social and economic factors that shape farmers' choices about conservation technologies and practices. Critical among those factors that are policy-sensitive are education, identification of lands that are especially vulnerable to degradation, and the nature, security, and duration of property rights in land and water. Requiring approved conservation practices on environmentally sensitive land as a condition for participation in government programs (and for receipt of government payments) can also be influential in encouraging adoption, although this involves potentially significant monitoring and enforcement costs. Additional factors not included in the U.S. analyses that might be expected to influence adoption decisions in South Asia's rice–wheat system include levels of income, wealth, and access to credit.

Sustaining agricultural productivity growth in South Asia—together with its effects on income and food security—offers a compelling motivation for policy efforts to encourage adoption of appropriate technologies and conservation practices. The challenge for policy-makers is to find the mix of investments in research and development, infrastructure, education, and institutions that best achieves a suitable balance between private and public concerns about productivity and efficiency as well as broader public goals in terms of equity and the environment.

REFERENCES

Ali, M., Byerlee, D. 2001. Productivity growth and resource degradation in Pakistan's Punjab. In. Bridges E.M., Hannam I.D., Oldeman L.R., Penning de Vries F.W.T.,

Scherr S.J., Sombatpanit S.J., Eds. *Response to Land Degradation*: Science Publishers. Enfield, NH, 186–199.

Ali, M., Velasco, L.E. 1993. *Intensification-Induced Resource Degradation: The Crop Production Sector in Pakistan*: IRRI. Los Baños. Philippines.

Byerlee, D. 1992. Technical change, productivity, and sustainability in irrigated cropping systems of South Asia: Emerging issues in the post-Green Revolution era. *J. Int. Develop* 4(5):477–496.

Byerlee, D., Murgai, R. 2002. Sense and sustainability revisited: The limits of total factor productivity measurements of sustainable agricultural systems. *Agric. Econ* 26(3): 227–236.

Cassman, K.G., Pingali, P.L. 1995. Extrapolating trends from long-term experiments to farmers' fields: The case of irrigated rice systems in Asia. In. Barnett V., Payne R., Steiner R., Eds. *Agricultural Sustainability in Economic, Environmental and Statistical Terms*: John Wiley and Sons, Ltd.. London.

Chaudhary, M.K., Harrington, L.W. 1993. *The Rice–Wheat System in Haryana: Input–Output Trends and Sources of Future Productivity Growth*. C.C.S. Haryana Agricultural University Regional Research Station (Karnal) and CIMMYT. Mexico, D.F..

Faeth, P., Ed. 1993. *Agricultural Policy and Sustainability: Case Studies from India, Chile, the Philippines and the United States*: World Resources Institute. Washington, DC.

FAO. 2001. *The Economics of Conservation Agriculture*. Food and Agriculture Organization of the United Nations. Rome.

Heisey, P.W., Lantican, M.A., Dubin, H.J. 2002. *Impacts of International Wheat Breeding Research in Developing Countries, 1966–97*: CIMMYT. Mexico, D.F..

Hobbs, P., Morris, M. 1996. *Meeting South Asia's Future Food Requirements from Rice–Wheat Cropping Systems: Priority Issues Facing Researchers in the Post-Green Revolution Era*: NRG Paper 96-01. CIMMYT. Mexico, D.F..

Huke, R., Huke, E. 1992. *Rice–Wheat Atlas of South Asia*: IRRI and CIMMYT. Los Baños, Philippines, and Mexico, D.F.

Huke, E., Huke, R., Woodhead, T. 1994a. *Rice–Wheat Atlas of Bangladesh*: IRRI, CIMMYT, and BARC. Los Baños, Philippines. Mexico, D.F., and Dacca, Bangladesh.

Huke, R., Huke, E., Woodhead, T. 1994b. *Rice–Wheat Atlas of Nepal*: IRRI, CIMMYT, and NARC. Los Baños, Philippines. Mexico, D.F., and Kathmandu, Nepal.

Hopkins, J.W., Lal, R., Wiebe, K.D., Tweeten, L.G. 2001. Dynamic economic management of soil erosion, nutrient depletion, and productivity in the North Central USA. *J. Land Degradation and Develop* 12:305–318.

Kumar, P., Mruthyunjaya. 1992. Measurement analysis of total factor productivity growth in wheat. *Indian J. Agric. Econ* 47(3):451–458.

Kumar, P., Rosegrant, M.W., Hazell, P. 1995. *Cereal Prospects in India to 2020: Implications for Policy*: IFPRI 2020 Policy Brief No. 23. IFPRI. Washington, DC.

Lynam, J.K., Herdt, R.W. 1989. Sense and sustainability: Sustainability as an objective in international agricultural research. *Agric. Econ* 3:381–398.

Magleby, R. et al. 2002. Soil management and conservation. Chapter 4.2 in *Agricultural Resources and Environmental Indicators, 2000*: Economic Research Service, U.S. Department of Agriculture. Washington, DC: http://www.ers.usda.gov/Emphases/Harmony/issues/arei2000/.

Mohan Dey, M., Evenson, R.E. 1991. *The Economic Impact of Rice Research in Bangladesh.* Bangladesh Rice Research Institute, International Rice Research Institute, and Bangladesh, Agricultural Research Council. Gazipur, Bangladesh. Los Baños, Laguna, Philippines, and Dhaka, Bangladesh.

Murgai, R. 2001. The Green Revolution and the productivity paradox: Evidence from the Indian Punjab. *Agric. Econ* 25:199–209.

Pagiola, S. 1999. Economic analysis of incentives for soil conservation. In. Sanders D.W., Huszar P.C., Sombatpanit S., Enters T., Eds. *Incentives in Soil Conservation: From Theory to Practice*: World Association of Soil and Water Conservation. Ankeny, IA.

Pingali, P.L., Heisey, P.W. 2001. Cereal crop productivity in developing countries: Past trends and future prospects. In. Alston J.M., Pardey P.G., Taylor M., Eds. *Agricultural Science Policy: Changing Global Agendas*: IFPRI/The Johns Hopkins University Press. Baltimore and London, 56–82.

Rejesus, R.M., Heisey, P.W., Smale, M. 1999. Sources of productivity growth in Wheat: A review of recent performance and medium- to long-term prospects. CIMMYT Economics Working Paper 99/05. CIMMYT. Mexico, D.F.

Sharif, M., Longmire, J., Shafique, M., Ahmad, Z. 1992. Adoption of Basmati 385: Implications for time conflicts in rice–wheat system of Pakistan's Punjab. In. Byerlee D., Hussain T., Eds. *Farming Systems of Pakistan*: Vanguard Books. Lahore, 89–104.

Sidhu, D.S., Byerlee, D. 1992. *Technical Change and Wheat Productivity in the Indian Punjab in the Post-Green Revolution Period*: CIMMYT Economics Program Working Paper 92/02. CIMMYT. Mexico, D.F..

Soule, M.J., Tegene, A. 2003. Land tenure and the adoption of conservation practices in the United States. In Wiebe K., Ed. Land Quality, Agricultural Producitivity, and Food Security: Biophysical Processes and Economic Choices at Local, Regional, and Global Levels. Edward Elgar. Cheltenham, UK, and Northampton, MA, USA, 319–336.

van Lynden, G.W.J., Oldeman, L.R. 1997. *The Assessment of the Status of Human-Induced Soil Degradation in South and South-East Asia.* United Nations Environment Programme (UNEP), Food and Agriculture Organization of the United Nations (FAO), and International Soil Reference Information Centre (ISRIC). Wageningen, The Netherlands.

Wood, S., Sebastian, K., Scherr, S.J. 2000. *Pilot Analysis of Global Ecosystems: Agroecosystems*: International Food Policy Research Institute and the World Resources Institute. Washington, DC.

Woodhead, T., Huke, R., Huke, E. 1994. *Rice–Wheat Atlas of Pakistan*: IRRI, CIMMYT, and PARC. Los Baños, Philippines>. Mexico, D.F., and Islamabad, Pakistan.

Woodhead, T., Huke, R., Huke, E., Balababa, L. 1994. *Rice–Wheat Atlas of India*: IRRI, CIMMYT, and ICAR. Los Baños, Philippines, Mexico, D.F., and New Delhi.

29

Toward Promoting Conservation Agriculture for the Rice–Wheat System

José R. Benites
Food and Agriculture Organization of the United Nations (FAO)
Rome, Italy

I. INTRODUCTION

South Asia's Indo-Gangetic Plains is a 120,000-sq. km area stretching from Pakistan, through Nepal and India to Bangladesh. This area was the cradle of the 1960's Green Revolution. Using improved wheat and rice varieties, irrigation, and higher doses of fertilizers, farmers doubled rice production and boosted wheat output by almost five times in just three decades. This resulted in a steady rise in rural incomes and employment and vastly improved food security in the region.

Today, however, these farmers, and South Asia's expanding population in general, face uncertainty. The area under rice and wheat has stabilized, and further expansion seems unlikely. At the same time, evidence suggests that the growth in cereal yields has begun to slow in many high-potential agricultural areas, because of soil nutrient mining, declining levels of soil organic matter content, increasing salinity, falling water tables, and buildup of weeds, pathogen, and pest populations. The challenge facing the region today is to further increase productivity while making agriculture more efficient economically, ecologically sound, and sustainable for future generations.

The answer does not lie in more irrigation and higher doses of chemical fertilizers. Recent research indicates that farmers can produce more, while conserving natural resources, by abandoning current land plowing and harrowing practices in favor of conservation tillage (CT). Recommended practices are based on effective soil cover, crop rotation and "no-till (NT)." The latter refers to the simple technique of drilling seed into the soil with little or no prior land preparation and leaving as much residue on the soil surface as possible.

Many examples exist from around the world of how to achieve sustainable production systems, simply by applying basic principles of good farming. The terminology being adopted by FAO and other organizations to describe these systems is "conservation agriculture" (CA), which is consistent with the following general principles:

1. No mechanical soil disturbance as achieved through direct seeding or planting
2. Maintaining a permanent soil cover by making use of crop residues and cover crops
3. Judicious choice of crop rotations

This chapter includes a brief overview of past and recent efforts by FAO to support CA, including NT rice–wheat systems in South Asia. Lessons learned from past successes and failures are presented, as are avenues to promote CA for rice–wheat systems.

II. THE NEED FOR CONSERVATION AGRICULTURE

The rice–wheat (RW) belt is home to more than 600 million people of the region. It is estimated that approximately 240 million people consume rice and/or wheat produced in RW cropping systems. Population growth for all of South Asia currently exceeds 2% per year. Even if this rate slows as predicted to around 1.8% per year, the population in this belt is likely to exceed 850 million by the year 2010. Regional food production must increase by 2.5% annually to meet the needs of growing populations, raise their incomes, and generate employment opportunities. Income levels in the region are very low, and 40% of the population lives in poverty.

Widespread concerns exist about environmental impacts of the RW cropping system. Under this system, farmers grow rice in the monsoon summer season followed by wheat in the dry winter season. Farmers use this system on nearly 12 million ha in South Asia, especially in Pakistan, northern India, Nepal, and Bangladesh. China has an additional 10 million ha of land under cultivation using the RW system (Gomez-McPherson, 2000).

In conventional agriculture, soil tillage is considered to be one of the most important ways to create a favorable soil structure, as well as to prepare the seedbed and to control weeds. But machinery, particularly that drawn and/or driven by tractors, destroys the soil structure by reducing the size of soil particles. Today, conventional tillage methods are a major cause of severe soil loss and desertification in many developing countries.

Conservation agriculture is the most promising concept for sustainable agriculture. It conserves, improves, and makes more efficient use of natural resources through integrated management of available soil, water, and biological resources combined with external inputs. It contributes to environmental conservation as well as to enhanced and sustained agricultural production. It can also be referred to as resource-efficient/resource-effective agriculture (Bot and Benites, 2001). It keeps soils in good condition through continuous crop residue coverage. This minimizes erosion, increases rainfall infiltration, and increases the nutrient-exchange capacity of the soil.

Important benefits can result for farmers, national economies, and the environment, from adopting CA principles in the Indo-Gangetic Plains. They are evident in wheat fields planted using NT, including 8,000 ha in Haryana and 5,000 ha in Pakistan Punjab. Many tangible benefits accrue to CA, including reduced costs of production, generally higher yields, fewer weed problems (mainly *Phalaris minor*), water conservation, reduced lodging, no yellowing of wheat plants after first irrigation, fuel conservation, less use of tractors, better germination in salt-affected soils, reduced herbicide needs, savings in time, and reduced labor requirements.

A transition to CA generally increases resource-use efficiency. This is especially evident in areas of low and uncertain rainfall. The combination of CA with improved soil moisture management and innovative supplementary irrigation, such as low-pressure drip irrigation systems, minimizes water and plant nutrient inefficiencies. It matches crop needs, which vary over the season, with water and nutrient availability. Conservation agriculture permits better infiltration and in-soil storage of rainwater and automatically also reduces soil and water movement and transport in the arid and semi-arid tropics.

It is important to consider land management practices, which affect rainfall catchment, before considering those that aim to control runoff in order to enhance water availability and retain soil productivity. When considered sequentially, they are complementary rather than competing alternatives. Conserving water and soil requires careful management of soil structure and vegetative cover in order to enhance infiltration and maintain above- and below-ground drainage. This is especially true for areas that experience high rainfall and that are prone to water-logging.

A. On-Farm Benefits

Adoption of CA practices by farmers often results in substantial yield increases. Other benefits at the farm level are reduced labor requirements and reduced energy needs for sowing, because soils become softer and easier to work if CA practices are used on them. Plowing is by far the most energy-intensive and time-consuming operation for farmers around the world because they must use rented equipment, which is not always available on a timely basis. Major on-farm benefits include

> Labor savings
> Improved soil structure, organic matter, and biological processes
> Enhanced nutrient availability through better responses to fertilizers
> Reduced fertilizer requirements
> Reduced erosion through runoff, resulting in increased moisture retention
> Greater opportunities for crop diversification
> Increased income
> Improved quality of life

In Pakistan, Gill et al. (2000) found that NT is the most feasible and economically attractive option for the RW farming system. It is more attractive than laser leveling and bed and furrow techniques. Wheat cultivation by conventional methods proved economically less feasible.

Several general conclusions can be made regarding the on-farm benefits of CA, namely,

> The benefits of CA can be increased and sustained through better irrigation agronomy and water management extension services.
> Governments should encourage the private sector to help promote CA.
> Governments should ensure the quality control of implements used for CA.
> The capacity of extension services to disseminate CA techniques to farming communities should be strengthened.
> The availability of NT drills and laser levelers should be ensured in local markets.
> Equipment improvements can help reduce costs and increase production efficiencies.
> Rice crop yields can be increased by about 25% through modified transplanter designs that eliminate gaps between plants.

B. Off-Farm Benefits

Experience has shown that CA systems achieve yield levels as high as comparable conventional agricultural systems with less yearly fluctuations due to natural disasters such as droughts, storms, floods, and landslides. Therefore, CA contributes to food security. It improves health status, living conditions, and water. It also lowers government subsidies to maintain roads, waterways, and other infrastructure. Several major off-farm benefits are

> Lower flood peaks and reduced water runoff
> Reduced water treatment costs
> Reduced herbicide and pesticide applications

Improved recycling of animal waste
Enhanced biodiversity
Reduced carbon emissions
Increased carbon sequestration.

C. FAO Conservation Agriculture Working Group

FAO has promoted the CA concept for more than 10 years, particularly in Latin America. Now that CA is becoming a success story in that region, FAO is expanding the program to others such as Africa, Eurasia, and Central and South Asia, particularly in the Indo-Gangetic Plains. Table 1 shows a comparison of crop yields and economic return to labor under conservation agriculture and conventional agriculture in Parana, Brazil.

The formation of groups of individuals who have a common interest in the concepts and practices of CA has resulted in the establishment of an interdisciplinary Conservation Agriculture Working Group (CAWG) within the Agriculture Department of the FAO. Participants come from the divisions of Land and Water, Agricultural Engineering, and Crop and Animal Production. The CAWG has actively promoted field projects, created awareness, fomented networking, and advised on related policy. It has promoted dissemination of information through workshops and international meetings. The 2001 Madrid Congress is a high-profile example. The CAWG has also fomented technical publications, bulletins, guidelines, and training manuals on CA. Detailed information and numerous publications relating to CA are available on the conservation agriculture website now prepared by FAO, which may be accessed at: http://www.fao.org/ag/ags/

Table 1 Crop Yields and Economics Return—Maize (Parana, Brazil, 1997–1998 to 1998–1999)

Index	Conservation agriculture	Conventional agriculture
Total production costs (R$/ha)	256.00	274.00
Yield (kg/ha)	5,929.00	5,723.00
Market price (R$/kg)	0.10	0.10
Total output (R$/ha)	604.76	583.75
Net profit (R$/ha)	**348.76**	**309.75**
Labor (hs/man/ha)	92.00	122.00
Return to labor (R$/ha)	3.79	2.54

1 US$ = 1.20R$ (average 1997–1998–1999).
Source: Ribeiro personal communication, 1999.

AGSE/Main.htm. The CAWG also has numerous publications in the pipeline about the principles of CA and has helped with preparation of a CD-ROM on CA.

D. Conservation Agriculture Linkages with International Initiatives

A concerted approach to the CA issue is urgently needed in countries that have many donor agencies involved in supporting the agricultural sector. This will ensure coherence with other components of the agricultural programs of these agencies. FAO provides such mechanisms for intercountry networking, as do a range of other research networks. However, international level activities should not take priority over those at a country or local level.

The full range of stakeholders—from local to global levels—must be considered. They include the Global Environment Facility (GEF) and related environmental pacts, such as

The conventions on desertification (CCD; http://www.unccd.int/main.php), biological diversity (CBD; http://www.biodiv.org/), and climate change (FCCC; http://unfccc.int/)

The World Bank group, IFAD, the international NGOs, and other entities operating at a global scale

The CGIAR, especially CIMMYT and IRRI

National governments and entities that are committed to implementing the conventions, reporting on progress, and facilitating sustainable land use and management, land improvement, and the restoration of degraded lands

National and subnational research and development entities that are engaged in land-related activities by governments, NGOs, and the private sector

Local authorities.

E. Networks

The transfer of concepts, principles, and technologies of CA is facilitated by network interchanges between countries. They help scientists and others share known solutions and avoid some of the problems identified during the continual learning process. Such networks can accelerate the rates of advancement of knowledge and techniques being steadily accumulated by both national institutions and farmer-led community groups in their efforts to reverse land degradation on a global scale during the 21st century.

Eleven participating countries founded the South Asia Conservation Agriculture Network (SACAN) at the International Workshop on Conservation Agriculture held in Lahore, Pakistan, in February 2001. This workshop was co-sponsored by FAO, CIMMYT, and various national partners. SACAN focuses on the RW ecosystem in the region. It is already generating momentum for adoption of CA technologies in the region. These technologies are relevant to the diverse socioeconomic and agro-ecological conditions. The exchange of CA expertise and experiences among the stakeholders and member countries is encouraging. SACAN also promotes active integration of CA and IPM approaches in order to achieve a more holistic and systemic response to sustainability and production needs of farmers in the region, particularly small farmers. SACAN is coordinated by the On-Farm Management Research Institute at Lahore, Pakistan.

F. Technical Cooperation Among Developing Countries (TCDC)

FAO's TCDC program can facilitate efficient utilization of regional and national expertise, thus speeding the adoption of CA principles and practices by farmers in the developing world. The TCDC program is part of a more comprehensive partnership program. It encourages visits by professionals to other developing countries participating in the scheme. Cost for visits are fully underwritten by FAO. Visits facilitate the exchange of knowledge and experiences from one country to another, thus benefiting each country. The TCDC program has helped FAO promote CA concepts, principles, and technologies and strengthen the capacity of national organizations to develop sustainable land-use systems.

Experience interchanges resulting from TCDC missions have helped to create a critical mass of knowledge and expertise that has the potential to lead beneficiary nations into a new era of environmentally friendly prosperity.

G. FAO's Regular Program and Field Projects

FAO has been implementing field projects that place greater emphasis on helping farmers to improve the care of their land through practices of CA over the last decade. This contrasts with past efforts, which were focused on combating erosion and other land degradation processes.

Better land husbandry practices were introduced in Indore, India, from 1976 to 1980. They produced results that were superior to those resulting from traditional soil conservation practices. The practices affected soil conditions, soil moisture, and other factors in the ways that is now described as a conservation agriculture approach (T.F. Shaxson, personal communication, 2001). Although this work was not specifically undertaken by FAO, the following section on networking highlights current support for SACAN activities.

The major conservation tillage challenge in Pakistan over the last decade has been to change rice puddling. It should be abandoned in favor of CA practices and water-saving approaches. Cover crops need to be introduced into the crop rotations. Furthermore, ample scope exists in the region to integrate livestock into agriculture through the judicious introduction of forage crops in the rotations. More adoption research is needed on nonmechanized CA options for marginal landholders and tenant farmers and for areas in which mechanization is rarely available. Greater emphasis needs to be given to addressing adoption of CA by smallholders, including national institutional requirements for appropriate extension services to support farmer-led approaches.

FAO, together with CIMMYT and other partners, is currently promoting direct drilling of wheat in rotation with paddy rice. Feedback from farm fields in the South Asian RW area indicates that 50% net benefit increases over those attained after conventional tillage operations are possible when wheat seed is drilled directly into the rice crop or stubble. About half of these benefits result from reductions in production costs and the remaining half from increased yields due to water saving and improved water-use efficiency (Hobbs et al., 1997).

FAO is in the process of approving a pilot CA test program for various crop rotations in Pakistan, which include RW, wheat–maize, and wheat–cotton rotations. The purpose of the project will be to enhance productivity, profitability, and sustainability of major food crops in Pakistan. Further objectives of the pilot test are to develop improved techniques and disseminate them into the farming system for further large-scale replication.

III. KEY ISSUES FOR SUSTAINABLE RICE–WHEAT PRODUCTIVITY

FAO (2002) recently indicated that

1. For land management, NT and reduced-till systems may expect to be increasingly adopted for postrice nonrice crops during 2002 to 2015.
2. For postrice irrigated wheat production in the Indian and Pakistan's Punjab, NT and decreased-till procedures—including raised beds—that use four-wheel, tractor-drawn seeders, are already finding favor with farmers (Fig. 1).
3. Water-conserving procedures that involve preparatory laser-guided land leveling have been used for *wheat* production in the RW area of the Punjab region of Pakistan (Fig. 2). Subsequent minimum-tillage and postrice seeding of wheat into raised beds in farmers' fields has reduced field-level water requirements by 25% to 40%. This system also results in a 17% increase in wheat yield and in fertilizer-use efficiency.
4. *Integrated (insect-) pest management (IPM)* procedures have been adapted to several varied rice-system ecozones. This cultivar and infor-

Figure 1   Four-wheel, tractor-drawn seeder.

mation-and-training-based management has helped lessen insecticide-induced pest outbreaks, fomented development of insecticide resistance, and helped decrease production costs and operator-health dangers.

5. Appropriate *integrated nutrient management* (*INM*) in rice-based sequences and systems can replenish nutrients removed by the crops and can increase the soil biomass. This will result in higher "soil health," and it will also facilitate adoption of high-yielding crop cultivars. The $N + P + K$-aggregate (mineral) application rate exceeds 200-kg nutrients/ha *per annum* (not per crop) in the People's Republic of China, the Republic of Korea, and Vietnam. Interest rates charged are sufficiently high that there is risk of adverse environmental impacts.

6. For many smallholder rice-system farmers in India, the lack of traction power (whether gasoline- or animal-derived) delays field operations and reduces productivity. The cost in Nepal of a People's Republic of China-manufactured two-wheel tractor with basic cultivation attachments is US$1,300. This compares to the average Nepal income of US$200/person/year. In Pakistan, the cost of a locally manufactured wheat-seed drill that requires four-wheel tractor power is US$600.

Figure 2 Laser-guided land leveling.

7. Integration of livestock into RW systems and competing alternative uses of crop residues are important issues (Mueller et al., 2001). Livestock contributes 20% to 30% of household income for farm households that practice RW and RW–mustard rotations in unfavorable areas of Eastern Uttar Pradesh, India. Nonfarm activities contribute an additional 50% of total farm income. Throughout the rice-based systems, ruminants can constitute sources of savings for smallholder households.

IV. NEEDED ACTION

Farmer mentalities must be changed in order to overcome the fears and gaps and to abandon common and traditional practices that endanger natural resources, including soil, water, and plant nutrients. Science offers scope to review, reinterpret, and recombine past results and observations. There is also considerable scope to introduce nonconventional agronomic practices that have been briefly discussed in this chapter to trainers. They can teach them to students, new recruits, and in-service staff of agricultural advisory services in the public and private sectors, thereby strengthening farmer support services.

Experience has shown that single-pointed and isolated interventions are unlikely to address the biological and socioeconomic realities farmers face. This highlights the importance of dialogue among partners to decide the best ways to improve particular ecosystems and livelihoods.

Programs that encourage implementation of CA practices on a wide scale will need public sector financial and advisory support for farmers. It will offset monetary and nonmonetary costs associated with the needed transitions. Positive and negative externalities should be considered when calculating real benefits and costs to individuals, to the agricultural sector, and to the state or nation as a whole. This will compensate in part for public benefits received by farmers. Examples of important externalities, which justify this type of support, include reductions in pollution levels, carbon sequestration, and minimizing downstream damages from flooding and erosion. Also needed are essential research and advisory functions, which commercial firms will not provide because they are not profitable (Fresco, 2001).

Many challenges remain, and successful application techniques for CA principles still need to be identified, particularly for very dry areas and for resource-poor farmers.

V. CONCLUSIONS AND RECOMMENDATIONS

A. CA Technologies

The availability of mechanization may determine the extent to which conservation agriculture, based on NT procedures, will be adopted for other postrice crops, such as wheat.

For *rice*, NT management may to some extent be adopted in Asian *upland-rice* systems—within conservation agriculture procedures—and perhaps in combination with perennial rice by 2020.

For *submerged-soil rice*, NT systems might not be adopted widely. However, dry-soil rather than wet-soil pre- and early-monsoon tillage is likely to be preferred, because it facilitates *dry seeding* of rice seeds. Water economy may be a major motivating factor. Necessary cultivars and land-leveling and drainage techniques are likely to become more widely available, thus allowing for cost-effective control of associated weed, insect, and fungal-disease pests.

B. Linkages Among Levels

A strategy and methods to facilitate rapid adoption of NT RW systems should be developed to ensure effective linkages among various stakeholders. They will also facilitate the cultural learning process, the communications process, and a common language. Strategies and methods should be developed for use at global, regional-subregion, or river basin, national, landscape/lifescape, and local catchment and community levels.

C. Capacity

A Virtual Research and Development Center (VRDC) should be developed for RW systems. The center should provide decision support tools, electronic conferences, online discussion forums, expert consultations, digital library facilities, and electronic newsletters. It should be jointly launched by CGIAR/FAO/U.S. universities/AID and other partners.

The ability to identify lessons learned from successes and failures in preventing or reversing RW yields decline should be nurtured. The same should occur for land, water, and biological resource degradation and to identify what needs to be done at each level.

Above all, the need to empower local land users, communities, and personnel exists at all levels. This can be accomplished by providing them with better knowledge and information and by building their capacity to assess their situation, to identify priorities and options, and to test their improved RW systems and management practices, while accounting for individual and common resources and basic principles. Institutional building and ecology disciplines are essential parts of capacity building.

In South Asia, female literacy rates are inferior to those found in sub-Saharan Africa. Thus, rice-system technical assistance to member governments by FAO and its partners must consider the vital need for women's education and development. Attention to rural women is crucial for overall rural development and for alleviation of rural hunger and poverty. Education of women is probably the single most effective investment in development that any country can make.

D. Policy and Planning

FAO has the mandate and the expertise to help member governments devise and implement policies that can facilitate smallholder enterprise, rural livelihood, and food security.

Water-related policies shall be prominent for the rice systems during the next three decades. It will be impossible to ignore the legitimate demands of nonrice users for diverted water irrespective of whether such policies shall be determined by central government or by decentralized government. Thus, farmers who irrigate rice will need to adhere to policies that promote and enforce productive water use. Such policies are likely to be based on equitable, but contentious, systems of water markets, water and electricity pricing for tubewells, and of water rights and water obligations.

Pro-rural and pro-smallholder policies that reward sustainable natural-resource management will need to be ensured. Such pro-smallholder policies should ensure equitable access to land, water, and common property resources. The latter is especially important to poor families since they are an important source of fuel and fodder.

Appropriate policies with attendant regulations and enforcement systems are needed. Subsidies that encourage and reward environmental degradation and resource misuse need to be identified and phased out in many rice-growing nations. Subsidies that may cause great environmental damage are fertilizer, livestock-feed concentrates, water, electricity, and credit.

Environmentally friendly and cost-effective procedures exist for water, fertilizer, and pest management in systems that include rice production. No-till establishment of post-RW can save 1,000 m^3 water/ha/crop for irrigated postrice crops, reduces tractor-fuel use, and eliminates crop-residue burning. The latter can also decrease the release of carbon dioxide (CO_2) to the atmosphere by 13 t/ha/year.

The RW Consortium should help develop locally driven processes to identify and facilitate incorporation of local inputs into national action plans and commitments. Thus, these inputs will help meet local and national objectives and priorities established for programs by governments. This will be facilitated if CA influences local planning processes, thereby supporting sustainable agriculture and environmental goals.

VI. SUGGESTED FOLLOW-UP

A need exists for a follow-up electronic forum to further develop methodologies and different technical aspects of the NT RW systems.

First-stage participating countries should be identified. They should have good experiences and partnerships on which to build. India and Pakistan are ongoing examples because of their wide range of ecozones, their positive and negative experiences, and their capacity to use remote sensing at the field level.

A multidisciplinary group of experienced persons should be identified at national and regional levels to guide the adoption process. This group should use innovative mechanisms, such as study tours and induction training programs, to develop new ways of thinking, including a sustainable human livelihoods' perspective and locally driven approaches.

National RW activities need to build on existing operational programs and capacities. For example, they might build on experiences and methodologies developed by SACAN.

The preparatory process and the Earth Summit itself should be used to stimulate buy-in to an NT RW system by different countries and partners. Efforts should be made to ensure a coordinated strategy among FAO, UNEP, GEF, and convention secretariats.

The RW Consortium should involve bodies and processes that have already conducted reviews of different options to prevent RW yield declines and to rehabilitate degraded lands. Options could include water harvesting, conservation agriculture, livestock and environment, agroforestry, and participatory planning processes. The RW Consortium should liaison with existing technical advisory bodies and technical groups under CCD, NAPs, the CBD, other CGIAR centers, technical groups of FAO, and other national, regional, and international partners.

REFERENCES

Bot, A., Benites, J. 2001. Conservation Agriculture. Cases studies from Latin America and Africa. FAO Soil Bulletin 78: FAO. Rome.

FAO. 2002. Asia's Rice-Based Livelihood-Support Systems: Strengthening Their Role in Lessening Hunger and Rural Poverty Through Sustainable Growth in Agricultural Enterprises (forthcoming).

Fresco, L. 2001. Agriculture and conservation: Key for food security and poverty alleviation. Paper presented at the 1st World Congress for Conservation Farming. Madrid, 1–5, Oct. 2001.

Gill, M., Akram Kahlown, M., Hobbs, P. 2000. Evaluation of resource conservation technologies in rice–wheat system of Pakistan. Mona Reclamation Experimental Project. Wapda Bhalwal, Lahore.

Gomez-McPherson, H. 2000. Back to the office report on India and Pakistan: FAO.

Hobbs, P., Giri, G., Grace, P. 1997. *Reduced- and zero-tillage options for the establishment of wheat after rice in South Asia.* RCW Paper No. 2. Rice–Wheat Consortium for the Indo-Gangetic Plains and CIMMYT. Mexico, D.F.

Mueller, J.P., Pezo, D.A., Benites, J.R., Schlaepfer, N.P. 2001, Conflicts between conservation agriculture and livestock over the utilisation of crop residues and cover crops. Paper presented at the 1st World Congress for Conservation Farming. Madrid, Oct. 1–5, 2001.

30

The Role of the South Asian Conservation Agriculture Network (SACAN) in No-Till Farming in Pakistan

Mushtaq A. Gill and Maqsood Ahmed
On-Farm Water Management
Lahore, Punjab, Pakistan

I. BACKGROUND

Asia covers about one fourth of the globe's land mass and accommodates over 60% of the world's population. It constitutes one quarter of the agricultural area and 36% of arable land of the world. The continent produces 43% of the total wheat of the world and shares over 90% of total rice production (Gill, 1999). Nature has blessed the region with abundant agricultural and manpower resources and a favorable climate. The diversity in natural resources, climate, people, and socioeconomic conditions has given rise to diverse farming systems in the region. The Green Revolution technologies introduced in this region during the mid-1960s and 1970s led to a dramatic increase in yields and production of wheat and rice, the two most important staple food grains in this region. These crops occupy 54% of the total world's area planted to cereals and provide 41% of the calories consumed by over six billion people around the globe. In the first decade of the Green Revolution, grain production grew rapidly, mainly due to adoption of high-yielding varieties (HYV), increased use of fertilizers, irrigation, and cropping intensity (two crops instead of one), and increased cropped area. Since then the land devoted to rice and wheat has stabilized and there is very little scope for further expansion in area for these crops.

The total land area of South Asia is 437 million hectares, with an agricultural area of about 210 million hectares. A little expansion in agricultural area took place in Nepal and Pakistan during the last 40 years, but total area under agriculture has been almost the same as some decrease has occurred in agricultural area of other countries. On the other hand, the population of the region has risen to 1.323 billion from 0.55 billion during the last 40 years and is not expected to stabilize during the next 20 years. Although water resources and irrigation have expanded due to construction of dams and canals, and sinking of tubewells during the last four decades, development of water resources has not kept pace with population growth.

South Asia supports 35% of the total population of Asia and produces 35% and 31% of the total wheat and rice production, respectively, produced in the Asian continent. Crop productivity is lower than potential in South Asia, with the majority of the farmers still practicing traditional production techniques. Moreover, the cost of production has increased many times due to rising prices of fuel and other agricultural inputs, making farming less profitable. Existing, traditional production technologies also do not lead to efficient utilization of natural resources, especially water. Extremely low efficiency of input use has led to wastage and depletion of natural resources and environmental degradation.

Evidence is accumulating that growth in rice and wheat yields has started slowing down in the high-potential agricultural areas of South Asia. This may be the result of degradation of the resource base devoted to rice–wheat systems, including mining of soil nutrients, declining organic matter, increasing salinity, fluctuating water tables, and buildup of weeds and pests. Another important factor in the stagnation of productivity could be the traditional way of cultivation that uses multiple tillage operations. To the contrary, the population of the region is increasing at an annual rate of over 2%. The population growth poses a serious threat for food and fiber production, sufficient shelter, and negative effects on the environment. These conflicting realities of declining rate of production growth and increasing populations are serious sources of concern for all of us.

Recently new productivity-enhancement technologies have been developed in the region that use natural resources more efficiently and minimize environmental degradation. Effective linkages, however, are required to accelerate the transmission of these innovations to the user community. A more participatory approach among an expanded set of stakeholders will be needed to accelerate adoption of more efficient technologies quickly; otherwise it will be extremely difficult to fulfill food demand of the population of this region. This chapter looks at some of the new resource-conserving innovations being promoted in Pakistan and the role of farmers and farm associations in this new alliance of collaborators.

II. WATER RESOURCE CONSTRAINTS

As populations grow, water becomes a major limitation for meeting food requirements. Extraordinary stress is placed on freshwater systems, particularly in arid and semi-arid regions. In addition, freshwater ecosystems are being intensively modified and degraded by human activity. So much water is consumed for domestic, industrial, and agricultural uses that the natural flow of major rivers such as the Colorado, Yellow, Indus, and Amu Darya no longer reach the sea during the dry season.

New estimates of water scarcity calculated by the World Resources Institute in collaboration with the University of New Hampshire show that 2.3 billion people live in river basins under water stress with annual per capita water availability below 1,700 m^3. Of these, 1.7 billion people reside in highly stressed river basins where water availability falls below 1,000 m^3 per capita annually, and chronic water shortages threaten food production and hinder economic development. Assuming that current consumption patterns will continue, at least 3.5 billion people or 48% of the world's projected population will live in water-stressed river basins by 2025.

Water is, therefore, certain to remain a major topic of discussion during coming years, and it seems likely that climate change will be a perennial topic at global gatherings of resource conservation and environmental policy-makers. As such, it is highly important to develop better understanding of water scarcity and its future trends. It is also necessary to consider possible strategies for increasing water productivity (crop per drop) leading to more efficient management of scarce water resources.

III. NEED TO CONSERVE AGRICULTURAL RESOURCES IN SOUTH ASIA

Proponents of the Green Revolution generally argue that developing countries should opt for an agro-industrial model that relies on standardized technologies and ever-increasing use of external inputs to provide additional food supplies for growing populations. A vast majority of farmers and researchers, however, propose that developing countries should favor an agro-ecological model, such as conservation agriculture, which emphasizes biodiversity; recycling of nutrients; synergy among crops, animals, soils, and other biological components; and regeneration, preservation, and increased efficiency in the use of natural resources. For an example, "The more you till, the more you harvest" will not remain a sustainable option. The priorities and options will, therefore, have to be rethought. Data are accumulating that yields in rice–wheat systems are stagnant and even declining (Hobbs and Morris, 1996). A strategy to bring more area under cultivation is expensive and in many parts of South Asia not feasible because of the unavailability of sufficient irrigation water. Therefore, increasing yield per unit area is imperative to meet future grain demands. Resource–conserving technologies like laser land leveling, no-tillage, and permanent-bed planting are available and are being promoted to achieve these objectives (Choudhary et al., 1994, 1997).

Conservation agriculture, a term coined by FAO, is a sustainable crop management approach that results in more efficient resource-conserving agricultural production. It seeks to conserve, improve, and make more efficient use of natural resources through integrated management of available soil, water, crop, and other biological resources in combination with selected external inputs. It utilizes some of the benefits of organic farming but does not prevent the use of chemical inputs if required. Conservation agriculture in its fullest sense includes use of an organic soil cover (residue mulch), minimal soil disturbance (uses biological tillage instead), and appropriate crop rotations to avoid weed, disease, and pest problems. In South Asia, farmers are just overcoming the mindset of minimal soil disturbance and still need to add the surface mulch to the system and utilize the resource-conserving practices in all the crops grown in a system. The latter is the next step as machinery is made available to allow farmers to plant into loose residues. Once all components of the system are used, the full benefit of conservation agriculture can be realized.

A. No-Tillage Technology

One of the main reasons for low yield of wheat in the rice–wheat system is late planting, a result of farmers using a late-maturing quality-grain basmati rice variety and excessive land preparation given by farmers to get a fine seedbed for wheat planting after rice harvest (Fujisaka et al., 1994). After rice harvest, sufficient residual moisture is generally available to establish wheat. Conventional multiple tillage operations reduces this soil moisture, leading to, in many cases, an extra irrigation (locally termed *Rauni* or a preplanting land soaking) to bring the field back to a suitable moisture for seed germination. This causes major delays in wheat sowing, which ultimately affects the final wheat yield. Decreases in wheat yield at the rate of 1% per day for every day's delay in sowing after mid-November is well documented (Ortiz-Monasterio, 1994; Randhawa et al., 1981). Moreover, the broadcast method of wheat planting results in seed placement at many different depths and into different soil moistures, leading to poor crop stand and lower yield (Hobbs and Gupta, 2003).

Reduced-tillage, minimum tillage, or no-tillage concepts are not new to many developed countries, and examples of adoption of this technology abound. Many benefits can be cited, including reduced production costs, erosion, and other land degradation. No-tillage technology establishes crops without tillage or seedbed preparation (Fig. 1). A no-tillage drill is used and is capable of seeding through the surface residue and provides good seed–soil contact. In Pakistan, a no-tillage drill was developed in the 1990s based on a prototype imported from New Zealand (Peter Aitchison Industries) and later indigenized with technical assistance from Massey University (New Zealand), the International Maize and Wheat Improvement Center (CIMMYT), and the Rice–Wheat Consortium for

(a) **(b)**

Figure 1   (a) Local no-till drill; (b) no-till wheat field.

Indo-Gangetic Plains (RWC) in collaboration with the Pakistan Agricultural Research Council (PARC), Islamabad, On-Farm Water Management (OFWM) of Punjab Department of Agriculture, Lahore, and local Pakistan manufacturers.

Initial trials in the mid-1980s were confined to progressive and large farmers (Aslam et al., 1993). Although yield results were impressive, uptake of the technology remained limited. This was partly because of the high cost of the first no-tillage seed drills and their limited availability, besides general apprehension of the farmers about the new technology. It could, therefore, not be popularized among the farming community in spite of a lot of benefits.

On-Farm Water Management (OFWM[*]) took responsibility for its introduction to farmers from 1996 to 1997. This time, small farmers were targeted, and simpler, less costly, and locally manufactured drills were made commercially available. Wheat was grown on 50 acres (20 ha) at 12 sites with 5 government-owned no-tillage drills during 1996 to 1997. The technology was rapidly accepted by the farmers since it reduced costs of production, conserved resources (water mainly), and improved yields. The area under no-tillage increased exponentially over the next six-year (1997–2003) period as shown in Fig. 2. In 2002 to 2003, adoption reached 189,983 ha planted with 2,200 farmer-owned drills.

The technique has also been found useful for raising wheat yields in rice-harvested fields where uncertain rains and excessive soil moisture do not permit timely sowing of wheat. No-tillage is playing a significant role in improving

---

* On-Farm Water Management (OFWM) is a Punjab province department with the mandate to extend improved water-productivity technologies to farmers.

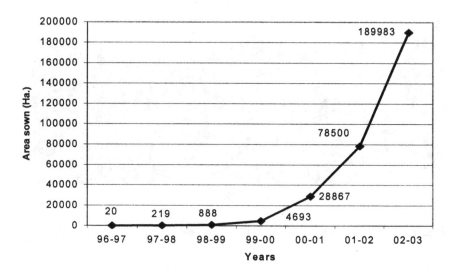

Figure 2 Rate of adoption of no-tillage wheat in Pakistan's Punjab province.

productivity and sustainability of the rice–wheat system by increasing farm income and improving livelihoods.

IV. IMPACT ASSESSMENT OF NO-TILLAGE TECHNOLOGY

The impact of no-tillage technology was studied by monitoring and gathering data from on-farm experiments with farmers with paired one-acre plots; one acre was no-tillage and the other conventional. The study was conducted at two sites (Muridke and Mona in Punjab Province) with five farms at each site. The soils were typical alluvial soils with silty loam to sandy loam textures. The data in Table 1 show the average data collected from all 10 paired plots from both sites over 2 years.

A. Cultivation Cost

On average, four ploughing with a tyned cultivator, two to three discings, and two plankings (wooden harrow) are used in conventional cultivation to prepare wheat seedbeds incurring an average cost of US$55 against a no-tillage planting cost $8 per hectare (Table 1). The data reveal that farmers can save cultivation costs of about $47 per hectare by adoption of no-tillage technology.

Table 1 Comparison of No-Tillage (ZT) and Conventional Tillage (CT)

Sr. no.	Item	ZT	CT	Difference
1	Cultivation Cost ($/ha)	8	55	$47
2	Energy saving (diesel consumption liters/ha)	14	68	54 liters
3	Irrigation water saving (ha-cm)	13.6	18.3	−4.7 ha-cm
4	Cultivation time (hours/ha)	2.7	16.7	− 12 hr
5	Weeds infestation (#/m^2)	9.8 a[a]	17.4 b	−7.6
6	Plant population (#/m^2)	119 a	115 a	+4.0
7	Plant tillers (#/m^2)	542 a	462 b	+80
8	Grain yield (t/ha)	4.90 a	4.43 b	+ 0.47 t

[a] Means within items followed by the same letter do not differ significantly at $P < 0.05$ using a paired T-test.

B. Energy Savings (Diesel Consumption Liters/Acre)

The traditional tillage system requires many operations to prepare a seedbed for planting wheat. This is partly due to the system used for growing rice, where the soils are puddled (plowed wet) to destroy soil structure and lower water percolation. This is done to pond water that is used to control weeds. The result is a hard, poorly structured soil after rice harvest. Seedbed preparation for wheat therefore uses large quantities of diesel for tractor operations. This resource is increasing in cost and makes cropping less profitable. No-tillage technology requires just one pass over the field, saving 80% of the fuel consumption (energy) (Table 1). This immediate cost reduction makes no-tillage attractive to farmers. But it also means less greenhouse gas emissions and so has environmental benefits. Reduction in tractor passes also reduces wear and tear on equipment and reduces compaction of the soil, leading to more favorable soil conditions for root development.

C. Irrigation Water Savings

Water is a major limiting factor for any crop and will become scarcer in the future. Soil moisture is important to germinate seeds, is a conduit for supplying nutrients to the roots, and plays a key role in photosynthesis and cooling of the plant. Water movement is determined by the distribution and size of soil pores and their degree of connectivity. No-tillage helps soil maintain a network of connected pores, leading to better water conductivity and drainage. Moreover, no-tillage technology can use soil moisture available after rice fields are harvested and thus save on this early irrigation. The improved soil moisture condition and use of a drill (conventional wheat planting is mostly done by the broadcasting method) mean more uniform seeding and germination and earlier emergence of the crop (Anonymous, 1999). The first irrigation also uses less water since the water moves across the field much faster than in plowed soil. This results in less waterlogging and yellowing of the plants; the yellowing results when too much water is applied in conventional systems. The data collected show a saving of 4–5 hectare-cm/ha of irrigation water because of savings in these irrigations (Table 1).

D. Cultivation Time

The impact of no-tillage was also noted in regard to saving cultivation time and the number of days needed to plant the crop after rice harvest. Fifteen more hours per hectare are needed for conventional plowing, but this can be extended over 10 to 20 extra days depending on the soil texture and moisture and availability of tractor power. No-tillage means less time pressure on the farmer during the busy rice-harvesting season, but even more important is the ability to plant the wheat crop close to its optimum planting date of mid-November. This timely planting factor is what gives no-till wheat a yield advantage over conventionally planted wheat.

E. Weed Infestation

The difference between conventional and no-tillage systems can also be seen in weed populations. It was observed that weed growth was 40% to 45% less in no-tillage plots as compared to fields sown with conventional technology (Table 1). Similar results have also been observed and reported by Verma and Srivastava (1994) and Singh (1995). It was further observed that the weeds in RCT plots were much weaker than those in conventional plots and were easy to control through herbicide application. The explanation for the lower weed counts is based on less soil disturbance and so less germination of the winter season weeds, especially *Phalaris minor*, a grassy weed associated with wheat planted after rice. The earlier planting also allows wheat to compete better with the weeds because *P. minor* germinates later in November when temperatures drop below a critical level.

F. Stem Borer Larvae

Many opponents of no-tillage wheat suggest that with the adoption of this technology on a large scale, more larvae of the yellow stem borer (YSB) and white stem borer (WSB) would survive in the base of undisturbed rice residue where it is known to hibernate and thus cause more damage to the subsequent rice crop. In fact, recommendations are formulated that insist that tillage is done to resolve this problem. Because of the seriousness of this claim, fields were monitored over the wheat season to explore the extent of the rice stem borer problem in the no-tillage-technology adopted areas. The rice stem borer was monitored by entomology staff of the national agricultural research center with samples collected in no-tillage and conventional tillage fields at random from various farmer fields. The data over several years have shown no significant difference in the population of stem borer larvae in the residues of rice stubbles in no-tillage and conventional tillage when measured in March although differences did exist just after rice harvest. Data did show that no-tillage does provide natural conditions favorable for the survival of beneficial insects (predators and parasites) through the anchored rice residues. These are assumed to feed on the rice stem borer, leading to control of stem borer and other pests.

G. Improvement in Soil Physical Condition and Fertility

Generally, all soils in the rice–wheat tract are deficient in organic matter (0.5%–1%). This deficiency leads to poor water- and fertilizer-use efficiency, stunted root growth, and poor crop stand, leading to lower crop yields. No-tillage allows rice crop residues to remain in the field and play a major role in improving organic matter content (up to 2%) of the surface soil. The lack of tillage also helps reduce oxidation of soil carbon and helps improve carbon content over time. It is hypothesized that if loose residue can be left on the surface as recommended in full conservation agriculture systems, this carbon sequestration would be even higher.

H. Plant Population and Tillers

Plant population was similar between no-till and conventional planted fields. But it was observed that broadcasting seed, common in the conventional system, led to reduced germination of seed. In fact, farmers use more seed in their conventional, broadcast system (125 kg seed/ha or more) than when using the no-tillage drill (calibrated to 100 kg seed/ha). A significant increase in tillering was observed in no-tillage plots compared to conventional plots due to more efficient use of fertilizer and less problems with waterlogging and yellowing of plants (Table 1).

I. Grain Yield

After performing farmer field trials for two consecutive years, it was observed that the wheat grain yield in no-tillage plots was statistically 10%–13% higher as compared to conventional plots, and there was an average yield increase of 470 kg/ha as compared to the conventional plots (Table 1).

Table 2 Resource Saving with No-Tillage Technology Calculated for Total Adopted
Area in Punjab of Pakistan in 2002–2003

Sr. no.	Description	Quantity/ amount
1	Technology adopted area (thousands of ha)	189.98
2	Water saving in adopted area @ 4 ha-cm/ha (million liters)	76
	Water saving value @ US$16.7/ha (million US$)	3.15
3	Diesel saving @ 54 liter per ha (million liters)	10.17
	Diesel saving @ US$0.35/liter (million US$)	3.60
4	Saving in cultivation cost @ US$47 per ha (million US$)	8.96
5	Wheat production increase @ 500 kg extra grain/ha (thousands of tons)	94.99
	Benefit extra grain in monetary terms @ US$125/ton (million US$)	11.87
Total saving (million US$)		**23.98**

J. Total Benefits

Table 2 calculates the impact of adopting no-tillage technology in terms of water, fossil fuel, cultivation costs, and increased yield for Pakistan. At current exchange rates the savings to Pakistan adds up to US$23.98 million for the 189,983 hectares sown to no-till or $126 per hectare. Other savings are not calculated in Table 2, including a value attached to reducing greenhouse gas (GHG) emissions, less herbicide and insecticide use, better fertilizer efficiency, less wear and tear on tractors and implements, and the value of improved soil properties and soil organic carbon that are shown to improve over time. The savings in diesel contribute significantly to less GHG emissions and to less import costs for the government of Pakistan. Note the $3.6 million savings in diesel used is included in the cultivation costs. Pakistan also expects that no-tillage will continue to expand in the coming years and reach one million hectares in 2 to 3 years. Application of no-tillage and other conservation agriculture practices across the system for wheat, rice, and other crops will also increase, leading to even more resource-efficient agriculture.

V. WHY THE NEED FOR NETWORKING CONSERVATION AGRICULTURE IN SOUTH ASIA?

Many countries have adopted resource conservation technologies. In the United States, Brazil, Australia, Canada, Argentina, Paraguay, and other countries, conservation agriculture and no-tillage are being adopted on an area of 66 million hectares (Derpsch, 2001). The reason for this success is the establishment of networks and farmer associations. Farmer associations such as Confederacion de las Americas de Asociaciones para la Agricultura Sostenible (CAAPAS) in Brazil and networks such as Rede Latinoamericana de Agricultura Conservacionista (RELACO) in Latin America, the South Africa Conservation Tillage Network (ACT), and the European Conservation Agriculture Federation (ECAF) are very effective bottom-up means of developing and disseminating conservation agriculture technology. Their formation made it possible to share the information on CA among various stakeholders. This exchange of experiences ultimately led to and is continuing to lead to rapid adoption of CA in Latin America, Africa, and Europe.

Although conservation agriculture is being promoted in South Asia, its pace of adoption is slower than needed because of the absence of similar farmer associations and networks listed above. Conservation agriculture practices being promoted in South Asia include no-tillage but also include permanent-bed planting and laser leveling. No-tillage is being practiced on about 500,000 ha in this region in 2002–2003, while laser land leveling has been done on only 40,000 ha. Bed-planting technology that results in 30%–50% savings in water is practiced on about 6,000 ha. These achievements have been made during the last six years. The current need is to start immediate steps for accelerating the adoption rate.

There are valuable experience and expertise on the practices and principles of conservation agriculture in South Asia. The effective collection, synthesis, and sharing of this knowledge will greatly stimulate adaptation and adoption of resource-conserving practices and systems throughout the region. Formation of networks and farmer groups in South Asia linked to this information would benefit the region and lead to faster adoption and better information on these new technological innovations.

The importance and need for a network for promotion of conservation agriculture practices in the region were emphasized in an international workshop, Conservation Tillage: A Viable Option for Sustainable Agriculture in Eurasia, held at Shortandy, Astana, Republic of Kazakhstan, September 19–24, 1999. Subsequently, its existence was endorsed by the International Workshop on Conservation Agriculture for Food Security and Environment Protection in Rice–Wheat Cropping Systems held at Lahore, Pakistan, February 6–9, 2001. One of its recommendations was that

> A Regional Network for Conservation Agriculture in the rice–wheat ecosystem in South Asia should be established similar to other networks around the world. Such a network would generate momentum for the adoption of CA in the region relevant to the diverse range of socio-economic and agroecological conditions. It would encourage exchange of expertise and experiences in conservation agriculture among stakeholders and member countries.

This network should promote an active integration of CA and IPM approaches in order to achieve a more holistic and system response to the sustainability and productivity required of farmers, in particular small farmers. The proposed network could be called the South Asia Conservation Agriculture Network (SACAN). It is proposed that such a network be coordinated by the On-Farm Water Management at Lahore, Pakistan.

Accordingly, a broad structure for SACAN outlining its mission, objectives, and management structure has been developed and proposed as a further step in this direction. A road map for follow-up actions was also suggested.

A. Mission Statement of SACAN

The mission statement of the South Asia Conservation Agriculture Network (SACAN) proposes helping to alleviate poverty and improve the environment through technology transfer and capacity building of different stakeholders by increasing the production and productivity as well as nutritional quality of food along with preserving and enhancing the natural resource base. SACAN will be committed to the advancement and promotion of conservation agriculture through research together with free exchange of experiences, knowledge, information, and data.

B. Prospective Partners

SACAN will be an international network of organizations and institutions. Its members will be equal partners interested in the promotion of conservation agriculture practices in the region designed to attain its goals. As a reinforced network, SACAN will promote interaction among its members by promoting cross-sectoral and multistakeholder dialogues at global, regional, and national levels. SACAN will pursue its mission in partnership with international, regional, and national agricultural researchers, scientists, experts, extension workers, policy-makers, government institutions/organizations, NGOs, machinery manufacturers, farmers, and other stakeholders.

C. Objectives of SACAN

The main objectives proposed for the network are as follows:

Act as a forum/platform/focal point for sharing/exchange of information and experiences among all stakeholders, interalia, including researchers, scientists, experts, policy-makers, extension workers, manufactures, and farmers, to increasingly apply methods of soil, water, and energy conser-

vation that increase productivity, ensure environmental protection, and
are economically viable.

Arrange workshops, symposia, discussion sessions, field days, and demonstration trials for adaptation and adoption of conservation agriculture.

Collect as well as prepare promotional material for various aspects of conservation agriculture and disseminate it among the farmers to facilitate
adoption of its practices.

Carry out studies, evaluations, and research on behalf of institutions and
donor agencies.

Support action at local, national, regional, and global levels that follows
the principles of sustainable agriculture resource management.

Identify gaps and stimulate stakeholders interaction to meet critical needs
within available land, water, and energy resources.

Encourage formation of national networks and associations aiming at promotion of conservation agriculture practices.

Interact with other regional and international networks for conservation
agriculture, such as the United Nation agencies, CGIAR institutes,
NARS, CIMMYT, IRRI, IWMI, ICARDA, RWC, universities, FAO,
and others.

D. Geographical Mandate and Management Structure

The South Asia Conservation Agriculture Network will promote its objectives
in Bangladesh, China, India, Iran, Kazakhstan, Myanmar, Nepal, Pakistan, Sri
Lanka, and Uzbekistan. There will be a steering committee (SC), a technical
committee (TC), various technical assistance committees (TACs), and a secretariat for operating the SACAN. The SC will comprise representatives of the partners
and will be responsible for making decisions on operational issues while the TC
will provide professional and scientific advice to the partners and will also safeguard the technical products of SACAN. The TACs will implement and promote
various activities, and the SACAN secretariat will provide all managerial and
coordinating services for smooth operation of the network activities.

E. Future Road Map

A road map has been prepared to help start up actions for launching the proposed
South Asia Conservation Agriculture Network. Some of its key elements are as
follows.

Establish a network secretariat.

Develop a network web site highlighting its mission, objectives, and functions and making available relevant, electronic-based information for
retrieval by stakeholders.

Arrange registration of the network as a nongovernment and nonprofit organization.

Establish linkages with prospective sources of information and similar networks in the world.

Draft network rules and bylaws.

It is proposed that a draft mission statement, proposed objectives and functions, and the suggested management structure of the SACAN be finalized. When consensus is reached, have the draft sent to

The international donor community to request technical and financial support for launching and future functioning of the network.

The host country governments to (1) provide international status to the network for providing traveling facilities (e.g., visa issuance) for experts visiting member countries in connection with its affairs as well as for providing other privileges extended to such organizations and (2) provide support for establishment of a network secretariat.

F. Conservation Agriculture Farmers Association of Pakistan (CAFAP)

The Conservation Agriculture Farmers Association of Pakistan (CAFAP) was established in 2001. It is a nongovernment organization (NGO) committed to promote resource conservation technologies in Pakistan. Presently, it has 5,000 members who interact with various stakeholders to exchange information for their mutual benefits. CAFAP organizes farmer gatherings, field days, seminars, and workshops to transfer knowledge of new technologies and get feedback for improvements and refinements.

VI. CONCLUSIONS

It can be concluded that adoption of resource conservation technologies offers effective and efficient use of available input resources, leading to improved land, water, and energy productivity. It provides opportunity for reducing the cost of crop production and improving produce quality to meet the challenges under the proposed WTO regime and results in sustainable agriculture and improving livelihoods of the farming community. No-tillage technology for wheat sowing in rice–wheat cropping systems is a viable solution for sustainable crop production and offers several benefits. The conventional technique of growing wheat requires excessive tillage (5 ploughings and 3 planking), whereas with no-tillage ploughing is not needed and saves considerable time and cost. The efficiency of fertilizer use was improved in no-tillage due to placement. Weed density was less in no-tillage fields compared to conventional plantings. No-tillage technology reduced fuel consumption and saved considerable foreign exchange used for fuel import and reduced greenhouse gas emissions significantly. Farm income improved with higher yield, less costs, and less wear and tear of machinery. Savings in irrigation water use with this technology is an important aspect since water is becoming a major resource limitation, especially following several drought years. Strategies are needed to adopt resource conservation technologies. Establishment of a regional network on conservation agriculture may be helpful in further accelerating the pace of adoption and promotion of resource conservation technologies in the South Asian region.

REFERENCES

Anonymous, 0 1999. Impact of resource conservation technology (no-tillage) on wheat production. On-Farm Water Management Annual Report 1998–99, Director General Agriculture (Water Management). Lahore, Pakistan.

Aslam, M., Majid, A., Hashmi, N.I., Hobbs, P.R. 1993. Improving wheat yield in the rice–wheat cropping system of the Punjab through no-tillage. *Pak. J. Agri. Res.* 14:8–11.

Choudhary, M.A., Baker, C.J. 1994. Overcoming constraints to conservation tillage in New Zealand. Conservation Tillage in Temperate Agroecosystems: Lewis Publishers. Boca Raton, FL, pp. 183–207.

Choudhary, M.A., Lal, R., Dick, W.A. 1997. Long-term tillage effects on runoff and soil erosion under simulated rainfall for a central Ohio soil. *Soil & Tillage Research* 42(3):175–184.

Derpsch, R. 2001. Conservation tillage, no-tillage and related technologies. *In* Conservation Agriculture: A Worldwide Challenge. García-Torres L., Benites J., Martínez-Vilela A., Eds: ECAF/FAO, Córdoba, Spain, pp. 161–170.

Fujisaka, S., Harrington, L.W., Hobbs, P.R. 1994. Rice–Wheat in South Asia: System and long-term priorities established through diagnostic research. *Agric. Systems* 46: 169–187.

Gill, M.A. 1999. Promotion of resource conservation tillage technologies in South Asia: An overview of Rice–Wheat Consortium activities in the Indo-Gangetic Plains. A report published by Directorate General Agriculture (Water Management) Punjab. Lahore, Pakistan.

Hobbs, P.R., Morris, M.L. 1996, Meeting South Asia's future food requirements from rice–wheat cropping systems: Priority issues facing researchers in the post-Green Revolution era. NRG Paper 96–01, CIMMYT, Mexico, D.F.

Hobbs, P.R., Gupta, R.K. 2003. Resource conserving technologies for wheat in rice–wheat systems. In *Improving the Productivity and Sustainability of Rice–Wheat Systems: Issues and Impact.* Ladha J.K. et al., Ed: ASA, Spec. Publ. 00. ASA Madison, WI (in press).

Oritz-Monasterio, J.I., Dhillon, S.S., Fischer, R.A. 1994. Date of sowing effects on grain yield and yield components of irrigated wheat cultivars and relationships with radiation and temperature in Ludhiana, India. *Field Crops Res.* 37:169–184.

Randhawa, A.S., Dhillon, S.S., Singh, D. 1981. Productivity of wheat varieties as influenced by the time of sowing. *J. Res., PAU, India* 18:227–233.

Singh, Y. 1995. Rice–Wheat Cropping System: NARP Basic Research Sub-Project. Progress Report (1994–95), G.B.P: University Agri. & Tech.. India.

Verma, U.N., Srivastava, V.C. 1994. Production technology of non-tilled wheat after puddled transplanted rice. *Indian J. Agri. Sci.* 64:277–284.

31

Reconciling Food Security with Environment Quality Through No-Till Farming

Rattan Lal and David O. Hansen
The Ohio State University
Columbus, Ohio, U.S.A.

Peter R. Hobbs and Norman Uphoff
Cornell University
Ithaca, New York, U.S.A.

No-till farming signifies a major shift in the conceptual basis of soil management for crop production.[*] It signifies much more than another method of seedbed preparation and the elimination of plowing. The concept challenges the scientific basis of "plowing" as an original, universal method of soil preparation. It questions the wisdom of repeatedly disturbing soil ecosystems. Nothing in nature matches the drastic perturbation of surface soils resulting from frequent plowing.

The concept of no-till farming embodies ecological principles of minimal or no soil disturbance, continuous maintenance of vegetal/mulch cover on the soil surface, strengthening natural processes that improve soil structure and recycle nutrient elements through the activity and species diversity of soil microflora and fauna, and maintaining/enhancing soil organic matter content. Crop residue mulch, minimal or no soil disturbance, and rotations are principal components of no-till farming. They require development of specific techniques of sowing in an undisturbed seedbed with loose residues, controlling weeds and other pests

---

[*] No-till farming is synonymous with the FAO term "conservation agriculture." It includes no-till or zero-till as means of minimizing soil disturbance and adds residue mulch and rotations within an agroecosystem perspective.

and pathogens, applying fertilizer, and using supplemental irrigation, etc. No-till, therefore, can only be used successfully if the supporting package of cultural practices is also available.

I. SOIL-SPECIFIC ADAPTATION OF NO-TILL FARMING

No-till farming can be adapted to a wide range of soils and environment for crop production. It is especially applicable on sloping lands that are prone to water erosion and on flat lands that are subject to wind erosion. It is an effective erosion control method provided that crop residue mulch is maintained on the soil surface.

Figure 1   A regular stand of no-till corn.

It is applicable to farming systems in which time saving for seedbed preparation is essential to obtain high yields and satisfactory grain quality. Its ability to reduce cost of fuel and other inputs in crop production is a key feature. Reductions in the use of soil water reserves, by reducing evaporation and improving water infiltration into the soil profile, is another important input saving in arid and semi-arid regions.

While no-till and other variants are being widely adopted in North America, South America, and Australia, its adoption has been minimal in Africa and in regions of rainfed agriculture in Asia. Progress is being made in its adaptation to sowing of wheat following rice in the rice–wheat system of South Asia. How-

Figure 2   A poor stand of no-till corn on a slot opened in a clayey soil that limited seed–soil contact and caused poor germination.

ever, the vast potential of this technology can only be realized through adaptive research on soil-site-specific problems. With appropriate modification to suit site-specific needs, no-till can be adapted to most soils, ecoregions, and cropping systems. Despite the versatility of its adaptation, no-till may not be the best alternative to obtain satisfactory crop yields in all soils and ecoregions.

Some tillage may be required to obtain a profitable crop yield on some soils and under certain climatic conditions. For example, no-till may not be the best method for preparing seedbeds in poorly drained, heavy-textured soils in cold regions where slow warming and anaerobiosis in spring can adversely affect crop stand and yield. Smearing of soil by a drill can leave a slot open that limits the seed–soil contact and restricts germination (Figs. 1 and 2). However, no-till combined with permanent-bed planting systems may reduce this problem. Similarly, no-till may not be the best option for growing root crops in shallow soils or in soils with low pH levels. High concentrations of Al^{+3} may require the deep incorporation of lime.

Despite these exceptions, no-till remains the best option for sowing wheat following rice in the alluvial soils of the Indo-Gangetic Plains in South Asia. Most of the problems with no-till in this region and system result from "fear of the unknown" rather than real constraints. Once tried, most obstacles can be overcome through simple innovations and site-specific adaptation. Nonetheless, it is important to define ecological limits such as soil type, climate, and farming/cropping systems to the use of no-till and its variants.

II. SPECIFIC ISSUES IDENTIFIED

The workshop identified several issues that need to be addressed.

A. Revisiting International Cooperation in the South Asian Region

During the 1960s, USAID along with the Rockefeller and Ford foundations, supported setting up agricultural universities under a land-grant system to build capacities to realize livelihood securities for a burgeoning population. Long-term training of scientists from these agricultural universities was promoted by a number of U.S. universities, with the most important input being training of local scientists. This effort was also supported by other forward-looking foundations and donor agencies. As a result, the Green Revolution was created in India. Today it has a surplus production of 60 million tons of food grain despite rapid demographic growth, with a new set of problems and challenges ahead. Most of the originally trained faculty has retired. Second-generation issues of globally competitive production, resource conservation, global warming, and improving access to food by poverty alleviation measures are the emerging concerns. Multilateral cooperation among different nations of the region will be required to achieve rural development and to realize peace through better understanding and mutual respecting of cultures. Human resource and institutional development are especially needed to rebuild the devastated agrarian economies of Afghanistan and other regions of Central Asia.

Decline in factor productivity in agricultural enterprises and improvements in input-use efficiency are the principal economic and environmental concerns of the region. Promoting efficient and resource conservation technologies is essential to minimize production costs, soil contamination, and water pollution and to prevent global warming. Rises in temperature due to greenhouse gas emissions are likely to redistribute water globally, change land-use patterns, create new plant diversity, and induce interregional human migration. The negative impacts of these trends are great for mankind and fully justify the restoration of multilateral cooperation, which was key to progress during the 1960s and should contribute to more effective solutions for these problems such as were achieved during the past.

B. Monitoring of Long-Term Consequences

There is need to establish ways to monitor how no-till farming impacts principal soils and ecoregions and soil and water quality. Understanding soil quality and soil resilience and the effect of management on them is necessary to achieve sustainable use of soil and water resources. The methodology must be carefully designed to establish meaningful cause–effect relationships among soil-quality indicators, on the one hand, and management systems, on the other. Specific issues with regard to no-till systems include the following:

> Root health, including soil pathogens and nematodes
> Rice establishment practices, e.g., alternative to puddling, that foster improvement over time in soil physical condition (soil structure) and soil organic matter dynamics
> Tillage systems and soil biology, including microbial biomass activity and species diversity of soil microflora and fauna, and the rate of turnover of soil organic matter.

A principal goal of monitoring is to strengthen the database that can improve the performance of agro-ecosystems under specific management practices such as no-till. Data must be collected to enhance the process-level understanding of alterations in soil and water quality, emission of greenhouse gases (CO_2, CH_4, N_2O), and efficiency of input use (fertilizer, irrigation) and to validate models. Long-term experiments can be helpful in coupling models with experiments, and modelers with experimentalists. By conducting the monitoring under actual farmer conditions, the results can be extrapolated with more assurance to other areas with similar circumstances for further impact.

C. Soil-Quality Assessment

Adaptation and adoption of no-till systems in RWS of South Asia to sow wheat on 500,000 ha in less than a decade have been revolutionary. However, no systematic assessment of changes in soil quality has been undertaken. Soil quality must be assessed with regards to the following parameters:

Soil structure, aggregation, mean weight diameter, etc.
Soil organic matter content, relative proportion of different soil fractions, and microbial processes
Soil water retention and transmission properties, infiltration capacity, water characteristic curves, and available water capacity
Salt balance in soil, especially with regard to secondary salinization
Groundwater balance, depletion and buildup, and linkage between groundwater and secondary salinization.

D. System of Rice Intensification (SRI)

The benefits of adopting the no-till system to wheat cannot be fully realized without the development of an intensive new production system. Many of the benefits of no-till in wheat are lost when the rice soil is puddled for the subsequent rice crop. No-till farming is needed for the entire system and should include rice and other nonrice crops to allow all the benefits to be obtained. This will create a challenge for the future, since weed management, development of aerobic rice varieties, water management, integrated pest management, and fertilizer issues will need attention in the rice phase, in particular.

An example of an innovative production system is the system of rice intensification (SRI) that involves a set of options for rice management, including the use of young seedlings, few seedlings per hill, wide hill spacing, and intermittent irrigation rather than continuous flooding. SRI may enhance diversity of soil micro-organisms and improve soil health. The labor requirements for SRI can present an initial barrier to adoption, but these can be reduced with time and experience, and the methodology has the potential to become labor-saving with appropriate mechanization. The benefits of SRI should be explored in combination with the RWS.

III. CHALLENGES AND OPPORTUNITIES IN NO-TILL FARMING

Principal challenges of no-till farming that need to be addressed through site-specific research include the following:

A. Systems Approach

No-till farming is more than just elimination of plowing. It is developing a complete package of agro-ecologically sound management practices, so that the no-till system fits into the overall scheme of the present and future trends of farming systems of specific regions. It is essential to develop a soil-specific package of cultural practices involving crop rotations and cover crops, weed control and pest management, seeding through crop residue, soil fertility management, irrigation, etc. The package must also consider socioeconomic and cultural factors such as labor requirements and gender issues. Impact assessment is needed and should go beyond just production and also include how changes impact the livelihoods of different social groups dependent on agriculture. A systems approach is also needed for effectively addressing questions related to soil health.

B. Seed Drill and Weed Control

Crop residue mulch must be maintained on the soil surface for best results with no-till farming. The presence of residue mulch conserves soil water, lowers the

Figure 3 Burning of rice straw in situ in Punjab. It is important to note that rice straw is not burned in most regions of South Asia. (Photo courtesy of Dr. Y. Singh, Dr. B.S. Ghuman, and other scientists of P.A.U. Ludhiana, India).

Figure 4   Wheat residue is stacked up at a threshing floor as shown in this picture near Ludhiana, India.

Figure 5   The wheat straw is carted away to cities and sold as fodder at as much as 40% of the price of wheat grains.

Figure 6   A heavy-duty no-till drill can cut through the residue and sow seed in a rough seedbed. However, such heavy machinery may not be suitable for the socioeconomic and cultural traditions of small-farm holders of South Asia.

soil temperature, and enhances soil organic matter content. Continuously burning (Fig. 3) or removing residue (Figs. 4 and 5) from the soil creates adverse impacts on soil and the environment. Therefore, developing a seed drill (Fig. 6) that can plant into the loose rice straw and establish the proper seed–soil contact is crucial for obtaining good stand and high yields and for improving soil quality. The benefits from having such a seed drill cannot be overemphasized, especially for minimizing the risks of poor crop stand (Figs. 7–9).

Weed control is another issue that needs to be addressed. Weeds are generally not a serious problem in fields of wheat sown after four months of flooding for rice cultivation. It is likely, however, that some persistent weeds may develop over time, especially in the rice phase. In addition to chemicals, innovative crop rotations may also be needed to alleviate the weed pressure. It is important to minimize dependence on herbicides. In addition to polluting waters, herbicides have also visible and hidden costs, including hidden carbon costs relating to their manufacture.

Because herbicides cannot be completely eliminated, the study of pathways, biodegradation, and transport processes in herbicides and pesticides remains im-

Figure 7 An uneven stand of no-till wheat (winter 2002–2003) through ice stubbles in Punjab, India.

portant. The use of herbicides may be needed in the initial transition stages to overcome serious weed problems, but it is expected that over time, herbicide use can be reduced and maybe even eliminated as the system stabilizes. Such research can be done by establishing long-term ecologically oriented monitoring of farmer fields. Indeed, development of alternative methods of weed control remains a high priority.

Figure 8 The stand of wheat through rice stubbles (winter 2002–2003) is better when stubbles are standing than on the ground.

Figure 9 A Pant Nagar no-till drill produces a good stand of no-till wheat, especially if the rice stubbles are standing, removed, or burned.

C. Crop Stand Establishment

Poor seedling establishment and sparse crop stand are the principal causes of low yields in no-till farming. The poor stand may be due to a wide range of factors including unavailability of the proper equipment or seed drill, poor seed–soil contact, low soil moisture, unfavorable soil temperature, pest infestation, and damage by birds and rodents. Numerous site-specific factors need to be addressed including time of sowing, seed rate, sowing depth, having a press-wheel to establish seed–soil contact, soil crusting, and interference by loose straw on the soil surface. Allelopathic effects may also need to be addressed.

D. Crop Diversification

Crop diversification through rotations is universally important for healthy and sustainable agriculture, and the rice–wheat system of South Asia is no exception. Diversification is a key to ecological stability. It is important to find viable/economic alternatives to rice so that the pressure on groundwater resources can be lessened. It is equally important to identify alternatives to wheat within the cropping system. In addition to agroforestry innovations, oil crops may play an important role in the local economy.

E. Soil Quality

Soil degradation is a major problem in need of attention. Specific soil quality issues in the rice–wheat region include depletion of soil organic matter and the mining of P, K, and micronutrients. The soil organic matter content of most soils of the rice–wheat system is low (0.2% to 0.3%) and has been exacerbated by continuous removal of crop residue. Enhancing soil organic matter content and achieving carbon sequestration are high priorities.

There is a strong interaction between soil organic matter content and soil structure. Coarse-textured soils with low soil organic matter content are prone to crusting and compaction, which must be alleviated by mechanical means. Structural degradation is exacerbated by puddling for rice cultivation (Figs. 10 and 11). Therefore, most soils in the rice–wheat region are structurally inert and characterized by weak aggregation of low stability. Preferential use of nitrogenous fertilizer in this cropping system has led to mining of soil P, K, and micronutrient reserves. No-till farming will not be successfully used in soils with degraded structure and nutrient imbalance and deficit, though both structure and nutrient availability can be improved over time with such cultivation methods.

Secondary salinization is another important process of soil degradation, especially in areas under canal irrigation. It is important to decrease the risks of salinity and sodicity. The potential impact of no-till farming on secondary

Figure 10 Puddling destroys soil structure and reduces seepage losses during rice cultivation. However, breakdown of aggregates may have an adverse impact on the following wheat. (Photo courtesy of Peter Hobbs.)

salinization needs to be studied, especially in terms of terrestrial carbon sequestration through soil reclamation.

F. Agricultural Sustainability

Sustainability as defined in terms of productivity and environment quality, including the quality of soil and water resources, remains a major challenge to agriculture. System performance indicators that can be used to assess sustainability need to be identified, especially with regard to the use of soil and water resources. Agronomic returns to no-till farming systems should be assessed in terms of (a) yield per unit area, time, or input, (b) soil organic matter dynamics, (c) soil quality, (d) groundwater resources, (e) quality of surface and groundwater, and (f) emission of carbon dioxide and other greenhouse gases, and other related indicators (IPCC, 2000; Lal, 2001, 2003). Appropriate system performance indicators need to be identified for evaluation. Crop yield *per se* is a simple indicator but not a good indicator of agricultural sustainability.

Figure 11 In addition to reducing seepage, puddling also facilitates transplanting of rice seedlings. (Photo courtesy of Peter Hobbs.)

G. Reducing Inputs

The rice–wheat system is based on a regime of high-energy inputs in terms of tillage, fertilizers and chemicals, irrigation, weeding, and other factors. Inputs should be minimized by reducing losses in the production process and by increasing efficiency. The goal is to sustain optimum economic returns while improving input-use efficiency. The excessive use of irrigation water has proven counterproductive and wasteful. Developing alternatives to flood irrigation and fertilizer broadcasting remains a high priority. The system of rice intensification discussed in Chapter 5 is particularly promising for reducing water and agrochemical inputs.

H. Quality of Life

Human suffering and drudgery accompany much of small-farm agriculture. While more commonly associated with subsistence agriculture, they are also present with smallholder agriculture that is tied to farming linked with commercial markets. A multidisciplinary approach is needed to address these concerns. Many of the activities associated with small-farm agriculture around the world are routine. Some of these can be replaced by alternative activities that are inherently more interesting and that may be less time-consuming, thereby providing more time to these farmers and their family members to pursue off-farm opportunities. Changing from plow-till to no-till has the potential to improve the quality of life of small farmers and their families in this way. It may enable a more varied set of agricultural production activities, and time savings associated with it may provide opportunities to pursue gainful employment in nonagricultural activities.

I. Social and Cultural Factors

Much smallholder agriculture is family farm agriculture that is permeated with complex cultural, ethnic, and gender assumptions and values, many of which are related to agronomic practices. The preparation of seedbeds for planting of crops has historically been deeply embedded in beliefs and customs. In some societies, the preparation of land is governed by norms and traditions that dictate who should prepare the land for planting, and how this is to be done. In some societies, this activity is viewed as something to be completed by men; in other societies, it is ascribed to women, and in others to the entire family. In certain societies, it is associated with a particular ethnic or religious group.

Changes from plow-till to no-till agriculture may have significant implications for the social organization of a particular society, and for this reason no-till agriculture may be rejected rather than because of technical or economic reasons. Adaptation of no-till principles and practices to specific geographic regions and farming systems should thus consider the social impacts of such changes.

In the present scenario of expanding populations and dwindling of land available for agriculture, it is not sufficient to just develop technology. It must also be provided to persons involved in agriculture in ways that are open to adaptation and improvement. In South Asia the large numbers of people who gain employment and livelihoods from farming include many members of resource-poor households. The new no-tillage techniques that can lower costs and raise net factor productivity need to be appropriate and available to a wide diversity of farmers. This will require new paradigms for extension and feedback that rely more on expanded alliances and closer links with communities through participatory and self-help approaches. More social science analysis will be needed to measure the impact of these new technologies on the livelihoods of various social groups so that zero-till farming can become a tool to help reduce poverty.

J. Soil Carbon Trading and Clean Development Mechanisms (CDM)

Use of crop residue mulch as part of no-till farming increases the potential for soil carbon sequestration in surface soil layers. Irrigated agriculture, as is the case in rice–wheat systems, offers potential to sequester inorganic carbon in the soil. Rates of sequestration may be reduced by hot and dry climatic conditions to a range of 50 to 100 kg/ha/y for organic carbon. With a total RWS cropped area of 13 Mha, the potential of soil C sequestration may be 0.6 to 1.2 MMT (million metric ton or Tg C/y). On a state or district level, carbon credits due to sequestration, may be traded under the International Emission Trading (IET, Article 17, and paragraphs 10 and 11 of Article 3) and the Clean Development Mechanism (CDM) of the Kyoto Protocol (Article 12 and paragraph 12 of Article 3) (Oberthür and Ott, 2000). Even a low payment of $2 to $4/ton could yield a revenue about $1 to $5 million/yr to provide an additional incentive to resource-poor farmers to improve soil quality through conversion to no-till farming. Treating carbon as a commodity, developing markets for it, and trading carbon credits could create added incentives for no-till farming. This creates conditions for reducing various greenhouse gas emissions: Carbon dioxide is reduced because less fuel is used for tractors and pumps; less methane is produced if rice paddies are converted from puddled anaerobic conditions to aerobic conditions; nitrous oxide emissions are lessened when fertilizer efficiency is improved.

K. Effective Partnerships

Effective partnerships that include national programs, CG centers, advanced institutions, NGOs, and the private sector are needed to efficiently accelerate the adoption of no-till farming in developing countries. The Rice–Wheat Consortium (RWC) is an ecoregional program of the CGIAR established in 1994 that has effectively facilitated collaboration between partners and promoted no-till farming in RWS of South Asia. Future programs could usefully join this alliance and benefit from the 10 years of institutional building and linkages.

IV. CONCLUSIONS

No-till farming is a management-intensive system. Plowing is replaced by better soil management with advanced planning for weed control, seeding and residue management, and control of pests and diseases. This also involves a different type of management, one more attuned to biological processes and dynamics. Therefore, educating farmers about the challenges as well as the foundations and merits of no-till farming is important.

No-till farming is an ecological approach to seedbed preparation, and differences in temperate and tropical ecoregions must be considered. Differences will exist in appropriate site-specific practices due to differing availability and cost of various inputs and other factors. Differences in climate, soil type and farm size, and cultural and ethnic factors must all be considered when developing new packages. All social groups should be able to benefit from adoption of no-till farming since more is produced at less cost. Even resource-poor farmers can benefit if the available technologies and extension system are tailored to their needs and if enabling factors are in place that allow them to experiment.

Major reasons for adopting no-till farming are to reduce the turnaround time between crops, increase input-use efficiency, enhance soil quality, and decrease risks of environmental pollution. However, farmers' decisions on adoption turn more on whether the innovation produces more at less cost and increases their profits. Institutions and professionals seeking to promote no-till farming need to help make it a risk-avoiding and a problem-solving strategy. It should also aim to alleviate specific constraints for farmers by reducing inputs, decreasing costs, and shortening the turnaround time between crops in a rotation.

Adoption of no-till farming is not a substitute for "good farming" practices. The latter includes crop rotations, cover crops, mixed farming, integrated nutrient management, integrated pest management, and judicious use of water to irrigate crops. A successful use of no-till farming implies minimal dependence on herbicides. Excessive dependence on herbicides can make no-till farming controversial as well as costly.

No-till is also a systemic and holistic approach to farming and resource management. The emphasis is on sustainable use of soil and water resources on a long-term basis. No-till is also soil- and site-specific, and it will not be universally applicable. Successful implementation of no-till depends on site-specific adaptation of the technology, and this process of adaptation needs to take account of social and economic factors if the technology is to be successfully adapted and adopted, leading to sustainable impacts on production, food security, and poverty.

REFERENCES

IPC. 2000. Land use, land use change and forestry. Intergovernment Panel on Climate Change, Special Report: Cambridge Univ. Press. Cambridge, U.K.

Lal, R. 2001. World cropland soils as source or sink for atmospheric carbon. *Adv. Agron.* 71:145–191.

Lal, R. 2003. Global potential of soil carbon sequestration to mitigate the greenhouse effect. *Crit. Rev. Plant Sci.* 22:151–184.

Oberthür, S., Ott, H.E. 2000. The Kyoto Protocol: International Climate Policy for the 21st Century: Springer-Verlag. Berlin.

Index

Printed in the United States
by Baker & Taylor Publisher Services